Going Wi-Fi

A PRACTICAL GUIDE TO PLANNING AND BUILDING AN 802.11 NETWORK

by Janice Reynolds

CMP Books
San Francisco and New York

Published by CMP Books
An imprint of CMP Media LLC
Main office: CMP Books, 600 Harrison St., San Francisco, CA 94107 USA
Phone: 415-947-6615; Fax: 415-947-6015
Sales office: 12 W. 21st St., New York, NY 10010 USA
www.cmpbooks.com
Email: books@cmp.com

ISBN: 1-57820-301-5

For individual orders, and for information on special discounts for quantity orders, please contact:
CMP Books Distribution Center, 6600 Silacci Way, Gilroy, CA 95020
Tel: 1-800-500-6875 or 408-848-3854; Fax: 408-848-5784
Email: cmp@rushorder.com; Web: www.cmpbooks.com

Distributed to the book trade in the U.S. by:
Publishers Group West, 1700 Fourth Street, Berkeley, California 94710

Distributed in Canada by:
Jaguar Book Group, 100 Armstrong Avenue, Georgetown, Ontario M6K 3E7 Canada

Cover design by Damien Castaneda
Text design by Brad Greene, Greene Design

Printed in the United States of America

03 04 05 06 07 5 4 3 2 1

Table of Contents

Introduction

Imagine anywhere, anytime Internet access.

Imagine a network that allows employees to enjoy a professional career yet work at home, without worrying about telephone jacks and cable outlets.

Imagine road warriors, armed with only a computing device (laptop or PDA), constantly on the move, quickly being able to sniff out wireless networks that provide high-speed Internet connections.

Imagine employees taking laptops to meetings, the cafeteria or even the warehouse—always in touch, always available.

Imagine Web surfing for a company's address and directions while sitting in traffic, or uploading a big presentation to the office while commuting by bus or train. Or perhaps browsing the Web while sitting on a park bench, or reviewing the latest news while you're waiting to pick up your kids from school. In other words, imagine being able to go anywhere with the means to communicate always at your fingertips.

Have I whetted your interest?

All of these exciting possibilities and many more become real with the help of "Wi-Fi" (short for "wireless fidelity") technology. The term Wi-Fi (which can be used interchangeably with IEEE 802.11a, 802.11b, 802.11g, Wireless Local Area Network, WirelessLAN, and WLAN) is the idiom commonly used to describe wireless networking technology.

The author is among those believing that Wi-Fi represents the proverbial "next big thing" in our culture. Creating a geographically extensive "Wireless Net" based on such technology just might help launch the next big business cycle.

At present, media discussions of Wi-Fi emphasize how forward-thinking individuals can set up and operate a network and gain high-speed Internet access without wires or a monthly line fee. With perhaps the spotlight being shared by domestic wireless Internet access with its tantalizing possibility: "Jetsonizing" one's apartment, mountain retreat, lakeside cabin or oceanview condo with such far-out applications as remotely controlling all domestic environmental settings (from lighting to climate control), telling the oven to pre-heat in anticipation of popping in the roast that's sitting in your refrigerator, or programming the TiVo to start recording the big football game, now, since you are stuck in traffic and won't be home in time for the kickoff.

But it's not just individuals who will benefit from Wi-Fi. The opportunities Wi-Fi offers to the business community are only limited by the stretch of the corporate imagination. Wi-Fi makes new business models possible—models that offer the potential of a quick profit for a small investment. Wi-Fi and its anytime, anywhere Internet capabilities make it a standout as the next most promising area where entrepreneurial spirit and technical creativity will burst wide open. Wi-Fi, like the Web, can flourish because its technology runs atop the freely accessible communications standards of the Internet.

Wi-Fi has the potential to let anyone with a computing device connect to the Internet at high speed without the need for pesky cables. Instead of moving data through a network using Ethernet cable, the most popular version of Wi-Fi (802.11b) transfers data using a band of radio frequencies around 2.4 gigahertz (GHz). This is the same range used by some cordless telephones. The data transfer rate of 801.11b is theoretically 11 Million bits per second (Mbps), but under normal conditions it's usually between 4 Mbps and 8 Mbps—far better than the 1.5 Mbps offered by a typical broadband connection.

Just as technical inventiveness and entrepreneurial daring opened the Web (and the Internet) to pioneering information providers and hundreds of millions of users, so too will Wi-Fi (at least, according to its proponents) make possible many cutting-edge innovations by making the Internet easier to connect to—at speeds comparable to today's fastest digital phone lines or cable modem hook-ups.

And guess what, it's already happening! Wi-Fi networks are open to the public in places like airport waiting lounges, cafés, coffee shops, bookstores, laundromats, brew pubs, maybe even your next-door neighbor's den. In fact, an estimated 15 million wireless Net surfers in the U.S now use Wi-Fi hubs. Although the U.S. currently has the largest number of Wi-Fi users, it is predicted that by 2006, 100 million users worldwide will be using some version of Wi-Fi.

As attractive as Wi-Fi is, however, IT managers must confront complex issues. For example, viability with next generation technology—deploying a network today that can be upgraded easily in the future. Then there's the perennial issue of security. A firewall costing thousands of dollars can be completely compromised by a single incorrectly configured access point (a transmitting device that connects a wired local area network to wireless computing devices, typically referred to as "clients"). WLANs could also fall victim to their own success as multiple network standards, including Bluetooth, crowd each other with interference. There are also Internet Protocol (IP) addressing issues, and locating access points across subnets can make it difficult to roam from one location to another without mobility middleware.

The good news is that solutions exist for most of these problems, and forthcoming standards will address many of today's limitations. The Wi-Fi technology available today is good and it will only get better as time goes by, but to be successful in using it requires careful navigation through an evolving landscape.

Section I: How Wi-Fi got to Where it is Today

Chapter 1: What is Wi-Fi

"Wi-Fi" is a play on the old audio term "Hi-Fi" (high fidelity). The term also has been trademarked by the Wi-Fi Alliance (formerly the Wireless Ethernet Compatibility Alliance). Today, Wi-Fi is most commonly used to describe a wireless local area network based on the IEEE 802.11 series of standards, which is a set of wireless technical specifications issued by the Institute of Electrical and Electronic Engineers (IEEE). The IEEE is an international professional organization for electrical and electronics engineers, with formal links with the International Organization for Standardization (more commonly known as the "ISO").

The IEEE 802.11 standards specify an "over-the-air" interface consisting of radio frequency (RF) technology to transmit and receive data between a wireless client and a base station or access point (an "infrastructure" configuration), as well as among two or more wireless clients that happen to be within communications range of each other (an "ad hoc" configuration).

The IEEE 802.11 standards resolve compatibility issues between manufacturers of wireless networking equipment operating in specific frequency bands within the unlicensed spectra of 2.4 GHz and 5 GHz. (Unlicensed spectrum refers to airwaves that haven't been allocated to an exclusive user.) This wireless, flexible data communications system can be implemented either as an extension to, or as an alternative for, a wired local area network (LAN). As such, it is a wireless networking standard that's accepted worldwide and, consequently, has rapidly gained acceptance as an alternative to conventional wireline technologies.

The term "Wi-Fi" began life a few years ago as the IEEE 802.11 High Rate (HR) Standard, which later became known as the IEEE 802.11b standard. Originally, "Wi-Fi" was used only in place of the 802.11b standard (operating at 2.4 GHz and 11 Mbps), in the same way that the term "Ethernet" is used in place of the wired LAN standard, IEEE 802.3. However, in October 2002, the Wi-Fi Alliance extended the "Wi-Fi" trademark to include both 802.11b and its higher-bandwidth brother, 802.11a (54 Mbps).

In case you're wondering, 802.11g will also fall under the Wi-Fi banner. Furthermore, since in many ways, 802.11g is a higher-bandwidth version of the 802.11b specification, the Wi-Fi Alliance also has stated that when it does certify 802.11g products as interoperable, the organization won't label them "802.11g." Instead, it plans on calling such products "54 Mbps 802.11b," because focus groups of potential customers have shown that users are confused by the alphabet soup nature of the IEEE's designations for the various 802.11 standards.

***Note:** Specific groups, known as "Task Groups" inside the 802.11 standards body oversee enhancements ranging from bandwidth to specific portions of the 802.11 standard. The appearance of each new "flavor" of the technology results in a new suffix to 802.11, i.e. 802.11a, 802.11b, up to 802.11n. While there is a method for the IEEE standards naming scheme, the alphanumeric nature of these standards can be confusing, which is why the moniker "Wi-Fi" caught on. A complete technical discussion of each individual specification or standard can be found in Chapter 6.*

Ultimately, the term "Wi-Fi" will probably end up as a generic description of all 802.11 networks, in much the same way that "Ethernet" is used to describe all IEEE 802.3 networks. Coincidentally, since 802.11 networks are considered wireless extensions of wired Ethernet, they can handle conventional networking protocols such as TCP/IP, AppleTalk and PC file sharing standards.

The opportunity for wireless data communication in the United States and elsewhere is huge. According to the research firm Gartner, Inc., as revealed in its Dataquest Market Analysis Perspective entitled, "Wireless Data in the United States: Pieces of the Puzzle are Missing, but a Picture is Taking Shape," more than 25 million of the U.S.'s 112.1 million workforce have a mobile job requirement and that number is steadily increasing.

HOTSPOTS ARE ALL THE RAGE

Imagine a new business model that offers the potential of a quick profit for a small investment. That's what HotSpots provide. North America, Singapore, Korea and Japan have led this Wi-Fi trend, offering traveling laptop users temporary wireless Internet access in various types of public gathering spots including, for example, hotels and cafés. According to Gartner Inc., the number of HotSpots will grow to more than 71,000 in 2003 from about 1200 in 2001.

Imagine a relatively inexpensive way of providing customers with a "value-add" service and while doing so keeping them in your place of business for longer, spending more. That's what a HotSpot offers.

The wireless LAN is extending its domain beyond the home and enterprise and is rapidly growing in popularity due to its public HotSpot application. A HotSpot is a specific geographic location in which an access point provides public wireless broadband network services to mobile visitors through a wireless LAN. HotSpots typically require that the end-user pay a fee before they can access the network and are usually situated in places that are heavily populated with mobile computer users, e.g. airports.

Wi-Fi HotSpots are targeted toward both business users and consumers, as is reflected in the wide range of different access packages offered by this new crop of service providers. Some HotSpot providers are targeting the business market by offering monthly and annual subscriptions. Others allow users to purchase as little as 30 minutes worth of wireless access for the equivalent of a few dollars. Still other HotSpots are in their trial phase, meaning they are initially providing free Internet access.

Unfortunately, a subscription to *one* HotSpot provider doesn't entitle the user to log on to HotSpots operated by a rival operator (although this is changing) and it must be realized that these individual networks typically have a short range (300 or so feet), although this range limit may soon be expanded. However, when you look at the Wi-Fi HotSpot market, think of the Internet and the potpourri of networks and Internet Service Providers (ISPs) who saw life in the early days. Over time, many of these providers, especially the smaller ones, consolidated and a seamless network emerged. It's expected that the same thing will happen in the Wi-Fi HotSpot marketplace.

According to Sky Dayton, CEO of wireless aggregator Boingo Wireless, "HotSpots are appearing worldwide. Grassroots organic growth and media attention have helped make the term 'Wi-Fi' a household word. And justifiably, because Wi-Fi is the first revolution in Internet access since the commercial Internet launched."

The numbers, while inconsistent, still have a tale to tell.

TeleAnalytics, a research firm, says that as of December 2002, there were 24,000 Wi-Fi HotSpots worldwide. Moreover, the hi-tech research firm In-Stat/MDR expects the Wi-Fi marketplace to expand to 42,000 sites worldwide by 2006, although Gartner offers a more optimistic figure of 152,000. Furthermore, many experts predict that the current mass of Wi-Fi installations in the U.S. and elsewhere could quickly expand to millions of access points representing even more millions of potential users. For instance, Canadian wireless industry observers predict that wireless access points will more than double every two years. In fact, Brantz Myers, of Cisco Systems Canada (a global firm that specializes in Internet networking services), predicts, "wireless networking will overtake copper and fiber lines in the near future."

In the U.S., the rapid emergence of the 802.11 standard has been such a remarkable, unplanned phenomenon that it has moved forward largely without the backing of major corporate service providers. Two good examples of this are Joltage and Boingo. These two companies (and others) are selling services that allow a computer user to sign up once and then access the Internet from a wide range of "HotSpots" via wireless access points located at various places around the country.

However, well-established corporations and telecommunication providers are starting to enter the market too. The cellular provider formerly known as VoiceStream, operating under its new moniker, T-Mobile, is busy dotting the country's urban centers with HotSpots, from Starbucks to hotels to airports.

In December 2002, Toronto start-up Spotnik Mobile secured an investment of $6 million from Telus Corp., enabling Spotnik to provide Telus customers access to wireless networks. Bell Canada also has big plans for Wi-Fi. Both providers see smart devices such as cell phones, PDAs, laptops and tablets working in each other's wireless worlds and, of course, they plan to charge plenty for it.

Unlike the U.S., where start-up companies often assume the role of the typical wireless public access provider, in the Pacific Rim it's the large established corporations, telecom carriers and ISPs that are leading the way. For example, McDonald's Corp. has announced that it plans to outfit 4000 of its fast-food restaurants in Japan with Wi-Fi networks. In Singapore—where interest in Wi-Fi is huge—two major carriers, SingTel and StarHub, set the pace. Singtel has invested more than $560,000 in wireless zones, while StarHub claims to have the largest public HotSpot in Suntec City. And SingNet Broadband, which has more than a 50 percent share of the domestic broadband market, has launched "Home Wireless Surf" to provide households with wireless broadband Internet connectivity anywhere within the home.

Although Wi-Fi has yet to gather steam in most of Europe (it took two years of working with national governments and Europe's standards bodies to open Europe's doors), again it's the established telcos who are in the forefront of this nascent market. For instance, Sweden's Telia AB, Finland's Sonera Oyj, and Norway's Telenor Mobile AS are in the vanguard of installing public access points in Scandinavia; among them, they have managed to install a few hundred HotSpots across the region.

Other European telecom operators, such as BT Group PLC in Britain, also provide HotSpots. BT already has announced that it plans to establish approximately 400 HotSpots in Britain by June 2003 with plans to expand to 4000 by 2005.

A number of European airports feature HotSpots, including Amsterdam's Schiphol Airport and the Copenhagen airport. And industry analysts state that the HotSpot market is steadily gaining momentum; by 2003 telecom operators and start-up companies in the Wireless Internet Service Provider (WISP) industry are expected to seize the low-cost opportunity that Wi-Fi offers to link up today's on-the-go populace.

What About the Future?

It is expected that the wireless public access business will be fragmented for the near future, with services being offered piecemeal by entrepreneurial start-up companies, ISPs, mobile operators, infrastructure operators, city networks, established corporations and organizations, property owners, and even underground organizations (offering free access in apartment buildings and neighborhoods, for example).

Currently, only a small number of people are making money off of Wi-Fi installations, but that will change in the near future. Don't believe me? Well, consider this: three industry powerhouses have bet that there's going to be big money in Wi-Fi. AT&T, IBM, and Intel, along with two venture capital firms, announced in December 2002 that they had formed a company called Cometa Networks that will provide wholesale broadband access services nationwide.

So, although the industry is in its early stages and still faces many challenges and questions about its development, Wi-Fi has changed the paradigm in which the communications industry and its technology operates. And, once Wi-Fi can provide roaming virtually anywhere, and can overcome Quality of Service (QoS) and security issues, it will achieve its true potential. When Wi-Fi does this (and it will), it will quickly hit critical mass—making it ubiquitous—a part of our culture.

Wi-Fi's success is enhanced by the plummeting cost of a Wi-Fi installation, making such a network a no-risk financial decision, especially for small businesses and companies in hard-to-wire locations. But another prime factor that's driving the Wi-Fi boom are a new collection of chips from Intel that are designed to transform laptops and tablet PCs into portable offices. Products bearing these chips carry the Centrino label and have built-in support for both the 802.11b and 802.11a standards. After those chips and their products proliferate, it will be almost impossible to find a device that's not 802.11 compliant. Then all of those wireless Internet access providers, suffering because there weren't enough 802.11 devices in the market, will be in a commanding position.

Craig Mathias, a principal with the Farpoint Group, an advisory firm specializing in wireless communications and mobile computing, predicts, "In the next five years, we'll all be using [Wireless Internet Access]. It will be as big as cellular in terms of number of subscribers. Whether you use a wireless LAN at home, in the office, or in public, the login and security is all the same."

A WIRELESS NET

While public and private wireless LANs are a major market for Wi-Fi technology, the real promise of Wi-Fi lies in its ability to string together many WLANs in order to implement a larger wireless wide area network (WWAN), just like the Internet (it even interfaces with the Internet), but with one major difference—this network needs no wires, giving end-users complete freedom and mobility. Since it uses unlicensed or open spectrum, the costs involved in building such a wireless WAN are very low. So perhaps the most promising and exciting aspect of Wi-Fi is that this technology can enable a grassroots revolution in wireless communications, especially for data. Wi-Fi might even usurp the expensive 2.5G, 2.75G and 3G networks. More on that later!

In some places, Wi-Fi networks are already spreading into every venue, providing very inexpensive "last mile" access to the Internet to anyone with Wi-Fi capability. While currently traditional communications operators (e.g. Verizon, Qwest, BT) still provide the landline connections to the Internet itself, even the connections could begin to go wireless as vendors start offering Wi-Fi hubs that are designed to communicate with each other, and to pass traffic one to another. To give the reader a sampling of how the Wireless Net is already taking off, consider the following:

In a business model reminiscent of the early 90s, wireless aggregator Boingo Wireless has announced that it is looking for a few good HotSpot operators. Boingo states it has a way to let ordinary people turn their broadband access points into a moneymaking opportunity, using its aggregation software and a truly grassroots effort. And because of Wi-Fi's inherent low barrier to entry, a typical system could cost as little as $500, not including Internet access. This $500 buys what Boingo calls its "Hot Spot in a Box," which is basically a wireless access controller that incorporates access point functionality and which is pre-configured with Boingo's system communication tools for billing and authentication, authorization and accounting. Boingo said its second-generation version of this hardware is due out sometime in 2003, and will supposedly sport a $300 price tag.

Another start-up is Deep Blue Wireless. It offers access to the Internet via Wi-Fi at over 50 HotSpots in the San Francisco area and provides inexpensive Wi-Fi roaming at more than 600 locations across the U.S.

Through a partnership with T-Mobile (the relaunched version of VoiceStream), Starbucks coffee shops have launched a network of HotSpots for their customers' use. These HotSpots are available in 1200 locations throughout the U.S., with a further 800 planned by mid-2003. Pilot projects are also underway in London and Berlin, with more in the offering. "This service is a natural extension of the Starbucks coffeehouse experience, which has always been about making connections with the people and information that are important to us over a cup of coffee," said Howard Schultz, Starbucks chairman and chief global strategist. "Mobile professionals across the globe have been waiting for just such an offering: high-speed wireless Internet access in a familiar and widely available location that keeps them connected while on the road, or between the home and office. It's the right service offered in the right environment." In early 2003, a spokesperson for Starbucks, which considers its fee-based HotSpots "a new line of business," said that the HotSpots were really paying off for the company, with Wi-Fi users staying about nine times longer than about 70% of its typical customers—45 minutes compared to five minutes or less. Users are typically male professionals and although some are frequent national travelers, a surprising number are "local mobile professionals" such as real estate agents, who meet their customers at a Starbucks where they can conduct their business via web-based information and listings.

In late June 2003, Schultz announced that the installation of wireless Internet service at Starbuck stores was one of the primary reasons for what he called the "stunning" sales results his company reported for its second quarter. He credited the unexpectedly strong 10 percent comparable-store sales increase seen in the five weeks ending June 29 (with overall sales soaring 25 percent) in part to "popular programs such as the Starbucks card and Wi-Fi network" (and also to "market-defining beverage innovations").

Even after companies such as Boingo build out their networks, to be successful, they must expand their customer base, the more the better. For many in the HotSpot provider space, corporate customers are their gold standard. Established businesses such as bookstores, coffee shops and auto repair shops, along with residential and office complexes, are sought out to serve as initial HotSpot venue operators.

Next, these disparate networks must be connected together to build out a virtual wireless network that individuals can connect to via a single login-password infrastructure, so as to provide a comprehensive, encompassing wireless network giving individual customers the roaming capabilities that they need and demand. Cellular networks, at one time, were in the same fix. However, it wasn't long before regional cellular network operators entered into mutual "roaming agreements" to enable their customers to obtain service outside their home region. The same is expected to occur in the WISP market space.

3G AND WI-FI?

One has distance, the other has speed; together, they might become the "dynamic duo."

Although true 3G cellular technology is still in its infancy, it is predicted to give cellular customers users access to the Internet at a speed of up to 2 Mbps (although 144 Mbps will more likely be the data rate experienced by most end-users—a rate that is far slower than Wi-Fi's *minimum* speed of one Mbps). Furthermore, 3G speeds can't even come close to Wi-Fi's *maximum throughput* speeds of 7 Mbps to 32 Mbps.

Wi-Fi has the cellular industry worried. A recent report by the investment research firm ARCchart found that Wi-Fi could pose a risk to the success of next-generation (e.g. 3G) cellular operators. The research firm reported that Wi-Fi could eat up as much as 64 percent of 3G revenues in the next four years. It's easy to see why 2.5 and potential 3G operators are beginning to fret about competition from Wi-Fi.

However, 2.5 and 3G networks have one asset Wi-Fi networks lack—the ability to communicate over long distances. If these two mobile communication sectors could work as a team, though, there might be the tantalizing makings of a partnership. Given the relative strengths of both technologies, it's easy to find analysts, Wi-Fi operators, and even big wireless carriers who argue that it makes sense for the two technologies to come together.

Many cellular providers are quietly "test driving" the technology before promoting it as part of their service package, but not all—cellular provider T-Mobile has entered the market in a big way. Another cellular provider, Sprint PCS, has indicated that it definitely views Wi-Fi as complementary to its Sprint PCS Vision service. However, at the time of this writing Sprint had no specific plans in place for a 3G-to-Wi-Fi product. But that's not to say that the idea isn't under active consideration. According to its spokesperson, Jennifer Walsh, "Any service that creates the larger demand for wireless coverage, we see as a positive thing."

In South Korea, home of some of the most advanced and popular wireless data services, Korea Telecom, the nation's largest telecommunications company, is currently selling Wi-Fi access in addition to its regular cellular phone service. The company has installed 1000 access points in major cities and plans to deploy thousands more. The telco is forecasting 3.6 million South Koreans will use the networks by 2005.

Numerous problems arise from a cellular / Wi-Fi pairing. Perhaps the trickiest is that Wi-Fi vendors and Wi-Fi network operators face a catch-22: the general public isn't going to invest in dual-mode Wi-Fi/cellular devices until Wi-Fi service is widespread and Wi-Fi isn't likely to become pervasive until there is a decent base of such devices so as to enable a groundswell of subscribers to use Wi-Fi networks. But also, any networks that try to offer both cellular and Wi-Fi currently must use cobbled together gear, making it difficult for both the providers and the customers.

There is much more on the feasibility of teaming up these two technologies in Chapter 13.

A BIT OF HISTORY

Now it's time to take a journey into wireless data communications' past. While some would say that wireless data communications began with such primitive devices as waving lanterns by night or sending smoke signals, others would perhaps move it up a bit and begin the journey on May 24, 1844, the date Samuel Morse sent the message, "What hath God wrought?" from the old Supreme Court chamber in the United States Capitol to his partner in Baltimore.

***Note:** An interesting historical tidbit is that Morse allowed Annie G. Ellsworth, the young daughter of a friend, to choose the words of the message, and she selected a verse from Numbers XXIII, 23: "What hath God wrought?" which was recorded via raised dots and dashes onto paper tape and then translated later by an operator.*

Still other historians might use 1896 as wireless data communications' starting point. That was the year Guglielmo Marconi had the idea that invisible electric waves could be used for telegraphic signaling and telegraphed Morse code two miles using "radiator" electric waves.

A more relevant history would probably begin with a group at the University of Hawaii who brought together network technologies and radio communications for the first time in 1971 with a wireless network known as the ALOHANET. The network connected seven computing sites across four islands. That timeline would then progress on to the 1980s, when another wireless networking pioneer group, consisting of early amateur radio hobbyists, designed and built Terminal Node Controllers (TNCs) that act much like today's telephone modems. The hobbyists then interfaced their computers to radio equipment, thus converting digital computer signals to ones that a radio could broadcast over the airwaves.

But it was in 1985 that perhaps the most important event in wireless networking history occurred. That is the year that the United States Federal Communications Commission (FCC) made commercial development of radio-based LAN components possible by authorizing the public use of the Industrial, Scientific and Medical (ISM) band, a set of radio frequencies centered around 2.4 GHz which are universally acknowledged to be available for use by wireless technologies.

***Note:** Interestingly, the 2.4 GHz part of the ISM band had previously been called the "Junk Band" because, years before, 2.43 GHz was allocated for use by microwave ovens and no one ever expected that any application would "co-occupy" a band filled with microwave oven emissions.*

The ISM band is very attractive to wireless network vendors because it provides a part of the spectrum upon which to base their products, and end-users do not have to obtain FCC licenses to operate the products. This FCC action provided the impetus for the development of wireless LAN components, which led to today's Wi-Fi phenomenon.

The 1980s also provided one other incentive for wireless networking—the laptop computer and other small, portable computing devices, such as handheld computers and personal digital assistants (PDAs). Today, these are the typical computing devices used to communicate with Wi-Fi networks.

In 1990, NCR and Motorola began shipping wireless network interface cards (also known as wireless NICs, wireless adapters, or wireless PC cards) and soon other vendors followed suit. Without a standard, however, vendors began developing proprietary radios and access points. It wasn't long before the need for a wireless standard became evident. The IEEE 802 Working Group, the same Working Group responsible for wired LAN standards such as Ethernet (802.3) and Token Ring (802.5), stepped up to the plate and in July 1997 the 802.11 specification was completed.

In 1997, another Wi-Fi milestone occurred. That's the year the FCC allocated three 100 MHz unlicensed spectral bands between 5.15 GHz and 5.8 GHz (the "U-NII bands"). As with the ISM bands, users can operate within the U-NII bands without a license, and to encourage innovation and free market forces the FCC has imposed virtually no restrictions on how this band is used, except for out of band emissions and transmission power levels. These are the same bands that 802.11a uses.

In 1999, the Wi-Fi Alliance, a nonprofit organization originally formed as the Wireless Ethernet Compatibility Alliance (WECA), took on the task of certifying equipment manufactured by its membership as conforming to 802.11b standard (it later added 802.11a, and now had added 802.11g, to its certification program). To that end, the Wi-

Fi Alliance has instituted a test suite that defines how member products are tested to certify that they are interoperable with other Wi-Fi Certified products. Products awarded the Wi-Fi Certified logo have undergone strict, rigorous, and independent testing at one of four labs, located in Tokyo, Japan, San Jose, CA, Winnersh, U.K. or Taipei, Taiwan. Because of its rigorous testing standards, the Wi-Fi interoperability certification program has become the international standard for providing high-quality interoperability testing for 802.11-based products.

Note: *In case you are wondering, IEEE merely promulgates standards; it doesn't actually test for equipment interoperability.*

When the Wi-Fi Alliance's "Wi-Fi Certified" trademark is stamped on the packaging of wireless products (see Fig. 1.1), whether laptops and networking cards, hubs, access points, or whatever, it indicates compatibility with the 802.11 family of standards. A user with a "Wi-Fi Certified" product can use any brand of access point with any other brand of client hardware that is also certified. Typically, however, any Wi-Fi product using the same radio frequency (for example, 2.5 GHz for 802.11b, 5 GHz for 802.11a), will usually work with any other, even if not actually "Wi-Fi Certified."

The Wi-Fi Alliance is diligently keeping apace with the technological advances in the wireless arena. In late 2002, the Wi-Fi Alliance began to certify dual band 802.11a/b products in an effort to ensure that dual-band wireless LAN clients can efficiently find and hop onto the fastest network available. And, in early 2003, noting Wi-Fi's rapid advancement in the wireless service provider sector, the Wi-Fi Alliance launched a new program, which it calls the "Wi-Fi ZONE program," to create a global brand for easier

UNLICENSED SPECTRUM

Unlicensed spectrum has two main advantages:

1. Since there is no licensing procedure, deployment can be fast and inexpensive. This makes it practical to mass market inexpensive wireless systems (otherwise the cost of a single license would be a significant part of a system's overall deployment costs).
2. It's shared, which is essential for wireless systems that support devices that are moved from place to place, like laptops, PDAs, and phones.

The broad and uncoordinated use of unlicensed spectrum, however, raises management issues, such as:

- It is in limited supply so overuse is bound to occur.
- Some technologies conflict in their implementation and use, e.g. 802.11b networks and 2.4 GHz wireless phones.
- Technical and end-user competition can be momentarily problematic.
- Incompatible technologies are built on specific frequencies.
- Special interests want to claim and license it as their own resource. So far, cooler heads have prevailed.

Maintaining or expanding this resource and allowing innovative engineers to solve many of the problems that arise due to the growing popularity of Wi-Fi can achieve greater public good.

Figure 1.1 The Wi-Fi Alliance's Wi-Fi Certified Trademark. Products bearing this trademark have been certified by the Wi-Fi Alliance as meeting its interoperability standards

Figure 1.2 The Wi-Fi Alliance's Wi-Fi Zone brand.

recognition of public access HotSpots. The program includes setting a minimum standard of quality for HotSpots before they can label themselves a Wi-Fi ZONE or display the ZONE logo (see Fig. 1.2).

The Wi-Fi ZONE brand could be used in conjunction with whatever brands are also applicable to the venue provider, e.g. Starbucks Coffee shops could show the Wi-Fi ZONE brand along with the T-Mobile logo to indicate they have public wireless access.

That is just the first phase of the Wi-Fi Alliance's ZONE program. The second phase is slated to include use of what it calls "the Wi-Fi ZONE Finder tool," a searchable database of qualified HotSpots (similar to what's available at www.80211hotspots .com). For more information on the Wi-Fi ZONE program visit www.wi-fizone.org.

CONCLUSION

Projections for the Wi-Fi market are among the rosiest in the information technology community. At the December 2002, 802.11 Planet Conference & Expo, industry experts agreed that the $2 billion Wi-Fi industry should expand at a compounded growth rate of 30 percent to nearly a $6 billion industry, this puts it in line with the kind of growth found in the mobile PC industry.

The business community (and even the residential market) likes Wi-Fi's ease of use, low costs and quick time-to-use. But that's just the tip of the iceberg. Unlike 3G technology, which is a "top-down," carrier-driven system that takes time to roll-out and very expensive to provision, Wi-Fi is a "bottom-up" technology that is easy to roll-out and very affordable—Wi-Fi's technical pieces are inexpensive and networks benefit from the use of open spectrum while providing high-speed connectivity and offering rapid deployment.

Finally, entrepreneurs find Wi-Fi to be an ideal technology framework—they can go in and set up wireless hubs in neighborhoods to provide data communication services for a low initial investment. Services run by wireless Internet access providers can resemble traditional Internet service, designed for a range of commercial clients who want to get their public facility online. In fact, the wordsmiths have already crafted a new term for this industry—Wireless Internet Service Providers or WISPs. There are many business models for the WISP industry. Some are start-up regional (and national) service providers; others are venue owners (hotels, airports, cafés, convention centers) who have outfitted their own facilities to accommodate their visitors/customers.

Consumers are constantly faced with choices, whether it's what café they frequent, what hotel they stay in, or what business center they use. The choices they make are driven by many factors, but if they know a particular establishment offers a public Wi-Fi network for their use, it may provide one more reason to choose that establishment over another. Ultimately, Wi-Fi will become an "amenity" like any other—customers will simply expect it to be there, like a free glass of water, public restrooms, and electrical outlets.

Chapter 2: The Promise of Wi-Fi

As Wi-Fi technology matures, it's easy to see its advantages: speed (which can be as high as 54 Mbps), reliability, mobility, and easy integration into existing wired networks. Other benefits include cross-vendor interoperability, practical interference-free communication over reasonable distances, and even security that can be augmented by several emerging technologies.

As Wi-Fi becomes ubiquitous, the personal opportunities for jumping onto the information highway will be simple. Downloading a map and directions while you are on the road will be easy. Pull into a "HotSpot" at a service station and receive the map on your PDA (personal digital assistant), or perhaps your Wi-Fi-enabled automobile can receive the information. You will be able to check your email and your voicemail while "on-the-go," play games with a cyberpartner in Ohio or Australia while waiting at an airport, download music as you're listening to it on your MP3 player, or remotely control your appliances. The choices will be endless.

Corporate employees can be mobile; hospital professionals can be more efficient and effective; road warriors can keep their information, documents and literature up-to-date; field service personnel can obtain real-time information on customers' orders, repair requests and existing maintenance contracts; and home users can enjoy Wi-Fi's easy installation and the wire-free environment it provides.

Once you go wireless, you will never want to use a cables again. All that's needed is a Wi-Fi-enabled computing device and, perhaps, access to an access point (whether it's in the home or a corporate wireless network, HotSpot or FreeSpot).

FREEDOM & MOBILITY

In the case of wireless local area networks (WLANs), freedom derives from mobility. Mobility comes in two forms. One is itinerancy, which refers to the ability to use computers wherever you happen to stop, and the other is roaming, which is the ability to use computers on the move (or more to the point perhaps, not having to log in each time you move but being able to continue from where you left off). Both can be provided by a WLAN, though, as the reader shall learn, roaming is more complex to provide.

Wi-Fi, with its promise of mobility, opens up an entirely new dimension of user freedom. The cellular phone industry whetted consumers' appetite for mobility, and now this newly-nomadic crowd craves similar mobility (and the ensuing freedom) in all of their communications needs, whether it's voice or data. Wi-Fi is thus the next logical step in the communications evolution, as it gives users freedom to achieve the following:

- Physical mobility while maintaining connectivity in their place of work, home, neighborhood and eventually, the world.
- To grow their network without the necessity of installing new cables and wires.
- To move an office or business without incurring the huge costs normally associated with local area network (LAN) installations.

Once Wi-Fi access points are installed strategically throughout a building, campus, neighborhood, region, or even nationally, users simply insert an wireless network interface card (NIC) into their computer, load software, and voila!—freedom to move about from one location to another with their Wi-Fi-enabled computing device in tow. As long as such suitably equipped users can receive a signal, they can access the Internet. But that's just the beginning of what one can do with Wi-Fi.

In an educational environment, a Wi-Fi network affords students and faculty access not only to the Internet, but also to library resources, class schedules, assignments, and so forth, whether in their dorm room, study room, classroom, or sitting under a tree.

In a corporate WLAN, the user can access the Internet *and* access all appropriate corporate servers, whether the user is in the office, their boss's office, a conference room or the office cafeteria. In other words, wireless LAN systems can provide end-users access to real-time information anywhere within an organization. This ability to work any place, any time, fosters teamwork. Exchange of data is no longer confined to cabled areas. WLANs promote higher productivity and can help a company provide service opportunities not possible with wired networks.

Consider this: wireless LANs allow users to stay connected to their network for approximately one and three-quarter more hours per day than with a wired LAN. Users equipped with a laptop and a wireless connection can roam their office building without losing their connection, or having to log-in again on a new machine in a different location. This translates to a very real increase in productivity, as much as 22% for the average end-user—at least according to NOP World Technology, a leading technology research firm. That same research firm also found that:

- 63% of respondents report that wireless LAN technology improves the accuracy of everyday tasks—with 51% of healthcare organizations finding significant improvements in accuracy (50% increased accuracy on average). That is crucial, given the life and death implications of improved accuracy at the point of care.
- 87% of respondents believe that a wireless LAN improves their quality of life, taking into account attributes such as increased flexibility, productivity and timesavings—with 43% overall believing that this improvement is significant.
- HotSpot usage is gaining momentum; 60% of respondents are familiar with the concept, with 16% currently accessing the Internet via HotSpots. Among those aware of HotSpots (but not currently using it) a further 54% are interested in taking advantage of the capability.
- On average, 16% of employees within organizations with wireless LANs in place have

access to the technology. 56% of the respondents who use the technology use it either constantly or (at a minimum) daily.

INSTALLATION MAGIC

There's hardly a network architect or IT staffer who hasn't thought, at some point that "networking would be great—if it wasn't for the wires." Thus, perhaps the most obvious advantage of Wi-Fi is that it doesn't require extensive cabling and patching.

Are you moving your organization to a new building? Do you need to set up a new office and add it to your existing network? Are you adding a new department in the same building or simply inserting a few machines into your network? Wi-Fi solutions can substantially simplify all of those tasks.

Even when you add a wireless LAN as an extension to an existing wired LAN, the access point for the WLAN can be placed where the cable for the LAN is located; there is no need for additional cabling. You only need to increase the number of access points to accommodate any growth spurts the company might experience.

Installation of a WLAN does not involve complicated undertakings such as constructing raised floors, cable channels, or coverings on the floor or walls designed to conceal cables.

Ease of installation of wireless LANs also reduces costs, the only major requirements are to connect the access point to a wired network connection and then install the WLAN software on the mobile computing devices, if needed. Time and labor can be saved because adding a new user into a WLAN takes just a few minutes—and after the initial installation, IP addresses need not be changed if users move to new locations.

Wireless LANs should be the preferred method for network installation in challenging locations, such as the following:

* **Difficult locations.** Wireless technology allows businesses to use locations where it is difficult or impossible to lay cabling. For instance, when trying to lay cable in an area that is impacted by a freeway, some locales have right-of-way restrictions that must be dug around. Another example is two buildings on opposite sides of a street—with Wi-Fi there can be LAN access within minutes, avoiding the time and expense of laying cable under the road.
* **Architecturally challenging buildings.** How do you retrofit an older building with today's networking infrastructure? If the building is built of stone and bricks with thick interior walls, or is otherwise architecturally unique, your options to run network cabling may be limited. In addition, many old buildings don't have a "false" ceiling, meaning there is no easy access to spaces above the visible ceilings to install cabling. In such architecturally "difficult" buildings, you may need thousands of feet of cable and many hours of painstaking (and expensive) labor to connect a LAN. Above all, retrofitting older buildings with networking capabilities using Wi-Fi technologies lowers the overall cost of ownership and facilitates faster, simpler and more flexible installation. With Wi-Fi, except for connecting access points, physical wires no longer need to be run through walls or between floors.
* **Hazardous Materials.** Where there are hazardous materials (such as asbestos particles) that might disturb the environment when cable is installed.

Consider wireless LANs when upgrading campus networks. Access point devices have advanced technologically allowing long distance, scalable connections with the wired campus infrastructure, so that end-users (e.g. students and faculty) can roam the campus at will and yet maintain their network connection.

Furthermore, Wi-Fi offers installation flexibility in both the networking of dedicated computer labs and mobile computer labs where the primary component is a computer "drop" allowing network attachment via a wireless access point mounted on a mobile cart. In an educational environment, such carts also provide a secure refuge for notebook computers when they're not in use, keeping them charged and ready to roll out for the next class. Moreover, some wireless carts even feature additional ports for peripheral connections, such as printers, projectors and scanners, making it a true computer lab on wheels. Such carts include the "Simply Mobile" carts from Toshiba, MobileSchool from DataVision-Prologix, and the Wireless Information Networks (WIN) MobileLAN ONE wireless system, a motorized, wireless notebook cart with built-in power strips, a wireless access point and antenna system, an internal cooling fan, and an industrial-grade locking system.

IMPRESSIVE TOC AND ROI

The total cost of ownership (TCO) for a wireless LAN is lower than a wired LAN. For example, to install a wired LAN you might have to build cable channels, lay cables to different floors to access everyone's desk, and install switches for the cables. All of this adds substantially to a LAN's costs.

According to a recent industry study that NOP World Technology conducted on behalf of Cisco, Inc, wherein it polled more than 300 U.S.-based organizations, wireless networking has a measurable impact on return on investment (ROI). Organizations with more than 100 employees reported that they saved, on average, $164,000 annually on cabling costs and labor. Those savings do not include the financial benefits of increased productivity, which can increase an organization's return on their wireless LAN investment by even more substantial amounts.

The simple and flexible architecture of WLANs greatly reduces network management costs related to moves, adds and changes.

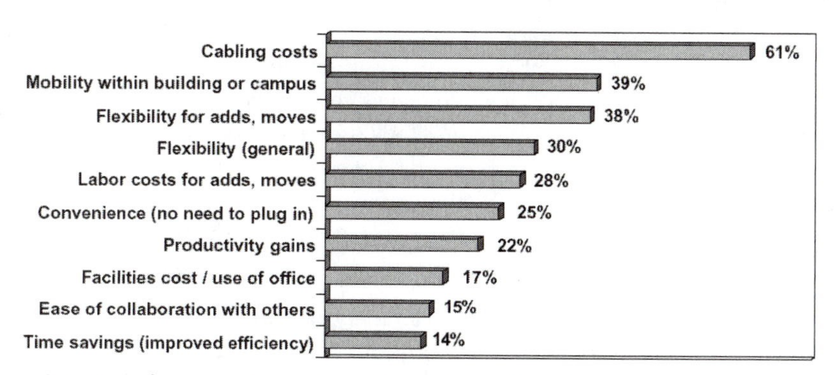

Figure 2.1 Per industry study conducted by NOP World Technology on behalf of Cisco, Inc. *Graphic courtesy of NOP World Technology.*

Wi-Fi is particularly important for fast growing organizations that may, in the near future, need a larger physical plant to accommodate their growth. If these businesses initially installed a wired LAN, a move to a larger facility would necessitate a costly new, or an additional, wired network installation. With Wi-Fi, the only additional costs are the expenses incurred for more access points to serve a larger area encompassing more users.

In short, install a wireless network and you never have to leave your networking investment behind.

Consider this: vendors such as Cisco and Agere are selling Wi-Fi technology as a way for corporations to cut costs and boost productivity. According to Cisco, the world's largest maker of wireless equipment for the enterprise, the cost of wireless gear, installation, and support comes to about $500 per year for each user, or $1 to $2 a day. Some other facts Cisco touts are as follows: assume the average corporate employee costs $100,000 to $300,000 a year in salary and benefits, a company can recoup its $500 investment by squeezing just a few extra minutes of work a day out of each one. And Larry Birenbaum, general manager of Cisco's Ethernet Access Group, has been quoted as saying, "If you believe you can increase an employee's productivity by 1 to 2 minutes a day, you've paid back the cost of wireless."

THE BUSINESS COMMUNITY BENEFITS

Today's business community is demanding more rapid, flexible and cost efficient ways of accessing information and network infrastructure. Wireless LANs meet those demands. Wi-Fi solutions enable the business community to realize flexibility and real-time access to information for any employee who needs to be constantly connected. The ease and speed of connecting and disconnecting wireless devices gives organizations a reliable, scalable and easy-to-integrate tool that can increase productivity and save money.

In today's ever-changing workplace, a wireless LAN combines the power of freedom and the ability for employees to access corporate digital resources (calendars, schedules, literature, documents, and other critical operating data) *and* the Internet *and* their email—wherever and whenever they need it.

In many organizations, employees spend, on average, close to 50 percent of their time away from their desk. For those entities, the mobility of wireless is a necessity—rather than simply a "nice-to-have."

Industry studies have sliced and diced all the small ways Wi-Fi is helping to save time and to gain efficiencies. For instance, the ability to go through email while waiting for an appointment, checking a database for real-time data during a meeting with a client, or present networked data to colleagues during a conference.

- Sage Research found that on average, a WLAN user could save up to 8 hours per week.
- Microsoft has presented subjective user data, which indicates that users saved 0.5 to 1.5 hours per day, or an average of 4 hours per week.
- The Gartner Group found that professional wireless users reported increased productivity of up to 25 hours per week.

If an organization is already supporting a variety of mobile devices, it can offer users higher-performing connectivity and access at lower cost through wireless networking. (See Fig. 2.3.)

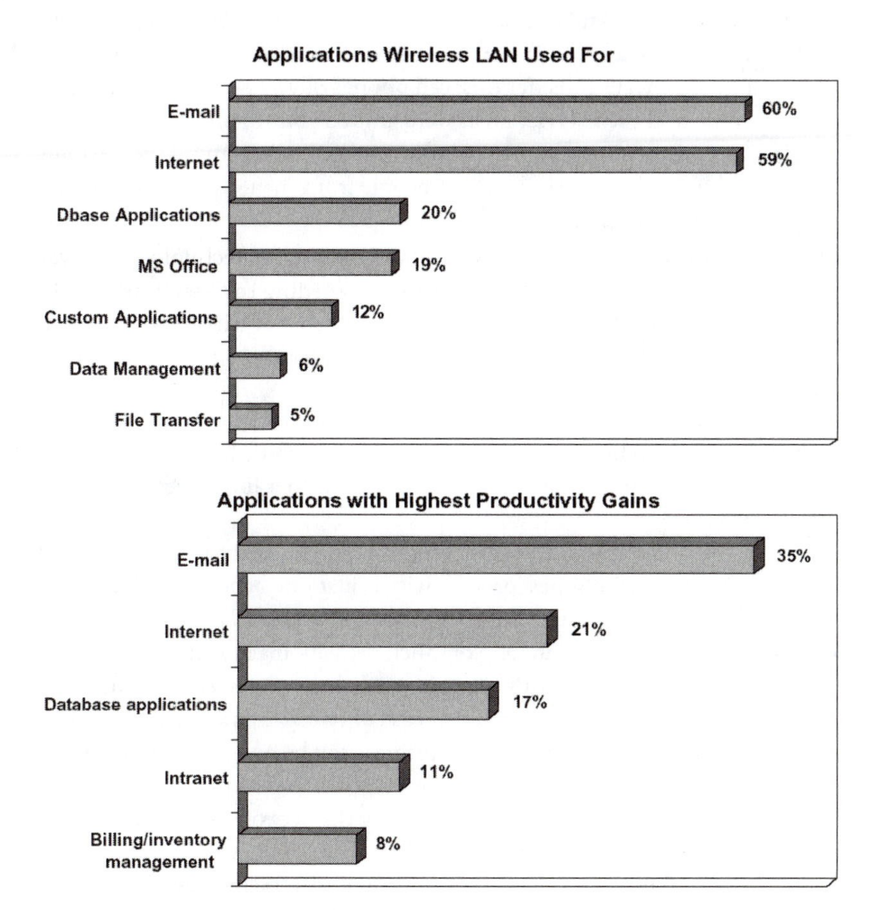

Figure 2.2 Per industry study conducted by NOP World Technology on behalf of Cisco, Inc. *Graphic courtesy of NOP World Technology.*

Mobile Device	Monthly Charges Per User
Cell Phone	$100
Wireless Handheld E-mail Device	$85
Telephone	$70
Wireless	$48 ($28 support, $20 infrastructure)
Notebook vs. PC	$38
Audio Bridging	$20

Prepared by Intel Corporation (2002)

Figure 2.3 It costs less to support a productive worker on a WLAN than on other popular mobile devices.

Even after all that has been said and written about it, some readers will still question whether Wi-Fi is *really* "ready for prime time," since it is so different from conventional network technology. Well, consider this: cell phones of years gone by were bulky, their reception was poor and their service areas were limited—but they still were a "must have" device. Wi-Fi stands in an analogous position today: it's becoming increasingly popular in many workplaces because it unchains people from their desks (sound familiar?). With Wi-Fi, users can take their laptop/PDA/tablet to the cafeteria, a café, the park, or even start their evening commute, and still be in touch. For Internet junkies and workaholics, Wi-Fi is a dream come true. According to *Business Week*, employees with access to Wi-Fi stay online an average of 105 minutes more each day and one company told the publication that it had seen a 20 percent productivity gain due to installation of Wi-Fi technology within its networks.

One way companies can squeeze even more minutes of productivity from their employees is by keeping them busy while they are traveling. Some 600 U.S. hotels offer wireless networks today, a number that the research firm, Gartner Inc., expects will grow to 5800 by the year 2005.

Similar productivity gains are what spearheaded the universal business adoption of cellular phones. But, unlike cell phones, productivity is just one benefit of Wi-Fi.

For system administrators, Wi-Fi technology improves operational flexibility: WLANs do not require extensive cabling and patching or much network management. Wi-Fi even saves technical support time because (1) it's a simple technology, (2) you can pre-configure clients, and (3) it can be used as backup media in case of failure of a wired component. Another feature that many find appealing is that printers can be hooked up wirelessly to the network and placed on carts so they can be wheeled to different locations.

Finally, for business owners and managers, extending the network to temporary contractors and guests is easy. Increasingly, mobile users bring their own Wi-Fi-enabled computers with them, so they can easily access their email and the Internet from a suitably-equipped location via a Wi-Fi network.

Small Businesses: For the manager of a small business or office, wireless means you might not need a full-time IT person, or additional cables, or any need to expand the router. Wi-Fi is easy to administer, and to add or move a workstation all that's needed is to power up the computer, and the employee is back on the network.

Seasonal Businesses: For owners of a seasonal business, Wi-Fi provides a network that offers quick and easy set up, whenever and wherever it's needed. Moves to a different or larger location are fast and simple—just unplug the network and computers and plug them in at the new location. For example, with the help of Wi-Fi solutions, it's possible to temporarily set up a store on a barge or a retired cruise ship in order to serve a seasonal tourist crowd.

Service Businesses: For business owners who depend on customers walking through their doors (rather than a competitor's door) and whose business model dictates that the customer stays on the premises for a while (e.g. coffee shops, auto repair shops, hotels), making their property Wi-Fi friendly might be an astute business move. As more and more devices ship with built-in Wi-Fi capability, providing wireless Internet access has the potential to attract customers, which can ultimately result in higher sales and increased customer satisfaction and loyalty.

Hotels find that having a wireless network in-house is attractive to business travelers, while commercial property owners find Wi-Fi an excellent sales tool, and airports are discovering that Wi-Fi has become a necessity for air travelers. There are many people who want to be work-enabled no matter where they are and connecting via Wi-Fi is fast becoming the method of choice for Net nomads everywhere.

Wi-Fi access providers are teaming up with service business owners. To give the reader an idea of the growth in this value-add area, consider this: the Gartner Group expects close to 116,000 "coffee-shop" networks by the end of 2004, many more than the less than a 1200 or so that existed at the beginning of 2002. Once a Wi-Fi network (e.g. HotSpot) has been installed, a café owner might find that the service attracts telecommuters, who stay longer reading their email (and perhaps even surfing the Web, if alone) while dining. Customers who once spent $5 for a quick lunch are now spending more, e.g. perhaps ordering dessert while answering an email, or ordering an after dinner drink to relax as they check out the latest movie trailers.

So, if your business has the potential to boost its revenues by enticing customers to stay around a bit longer, a Wi-Fi installation may be in your future.

Libraries: Patrons could bring their own laptops into the library to do their research. Library staff could take advantage of the mobility granted by a wireless network with projects such as inventory control, or remote circulation. Bibliographic instruction could "go mobile"—classes could be taught in any area within the range of an AP.

Educational Institutions: Increasingly students bring their own computers to school—laptops, PDAs, and even the tablet PC are classroom bound. They want to be able to use them on campus, anytime, anywhere. But, providing sufficient network points in places where they want to use them—libraries, study areas, classrooms and cafés—can be expensive and inflexible if a traditional wired LAN set-up is used. However, with a wireless LAN on campus, the user can go anywhere and stay connected, including places that might not have been wired such as corridors, cafés and even outside. Many educational institutions that use WLANs have included coverage in café and (in locales with reasonable climates) outdoors. Experience shows that this is very popular with students and they do make effective use of such spaces for online learning.

It's also worth noting that when students have wireless LAN cards in their mobile computing device, they can share data between the their computers even in the absence of a network by using "peer-to-peer" networking (the "ad hoc" mode), i.e. the computers talks directly with each other without servers and routers, similar to the model that made "Napster" a household word.

Healthcare: Organizations in the healthcare field, in particular, have found that a Wi-Fi network improves the collection and maintenance of accurate and up-to-date data. Instant access to real time records and resources can impact the way in which a physician delivers treatment to a patient and can make a difference in a patient's health outcome.

Wireless networking offers healthcare organizations the ability to achieve significantly greater accuracy in everyday tasks and can help to eliminate the paper overload. The anytime, anywhere aspect of wireless communications allows increased access to accurate information when it is needed most. In the healthcare industry this could mean the difference between life and death.

Companies with Field Staff: Sales personnel can make rapid (and better) decisions where it counts—in the customer's office. They can access information (custom pricing information, for example) and expert help (e.g. will a specific product be suitable for a specific environment) from the home office, outside vendors, and beyond. Response times to customer inquires can be cut dramatically, improving overall sales. If the field staff includes repair crews, Wi-Fi can allow them not only to stay in touch with the dispatcher, home office and sales staff (repair crews probably see more customers than even the average salesperson), but also to access necessary schematics and other information, therefore "upping the odds" that the repair will be completed in one service call. Repair crews also could fix some equipment before it malfunctions—the equipment itself calls in the repair. Billing too could be initiated directly from the customer's location.

Warehouses and Distribution Centers: Wi-Fi has been used in these locations extensively since the latter-half of the 1990s. Managers quickly saw the benefit of connected mobility. With a Wi-Fi system in place, goods received can be tracked from the moment they enter until they are shipped out—receiving, put-away, inventory, and quality control. Wireless networks enable employees to quickly and easily check product costs and availability, as well as to log products via their laptops or a handheld computer without having to walk back and forth to a central PC. A typical Wi-Fi network might consist of a combination of forklift-mounted data terminals and handheld data terminals with integrated barcode scanners. This allows for communication between the terminals and the host computer, which enables scenarios such as forklift operators at various locations being dynamically directed to complete specific tasks. There is much that can be accomplished with a Wi-Fi-enabled operation: greater productivity, better inventory control, fewer errors in orders and shipments, overall savings in operational costs, and more.

Manufacturers: Wi-Fi has already become a necessity in the manufacturing industry. In many cases, factories are just too large to be served by conventional wiring. Also, wired systems are vulnerable. As an example, Boeing opted for Wi-Fi after it found that errant trucks and forklifts crushed cables, disrupted communications. Wi-Fi can improve productivity, enabling such things as access to corporate records from any location or viewing electronic blueprints *in situ*.

Disaster Management: A disaster could potentially paralyze a business. The key requirements for a disaster recovery communications network are mobility, reliability, security, and ease-of-use for both the IT staff and the end-users. During a disaster-related crisis, the staff will mostly likely be mobile—dealing with matters outside their normal job description in order to get things back to some semblance of normalcy. In such situations, Wi-Fi is the optimal choice. Wi-Fi-enabled devices such as, laptops, handhelds, and PDAs (along with interactive pagers and cell phones) can be the lifeline needed for governmental agencies, disaster relief organizations, and businesses to stay afloat during a crisis.

When an organization is suddenly forced to temporarily relocate because of a disaster, it usually doesn't know ahead of time what the new office space will look like—it might be a school gym, a church, an empty warehouse or even a barge floating on a river. But the organization in question probably won't be able to alter the temporary space, nor in all likelihood even want to incur the expense of doing so. During such crises there generally isn't time to pull a lot of telephone and LAN cables to bring about even the vague appearance of a traditional corporate IT communications infrastructure. Thus, an organization's communications network needs to be easy to set up and capable of working in a wide variety of building spaces. Just what Wi-Fi does best.

Temporary Locations: Wireless LANs are great for businesses that regulary occupy temporary lease space; for training sites that constantly require remodeling to suit specific training situations; and for temporary registration booths at conventions, trade shows and other like events.

Wi-Fi installations are gaining traction in the business arena and the number of wirelessly enabled devices continues to grow and near-term investment in basic vertical-specific solutions (e.g. patient care in the healthcare industry and warehouse management in distribution) provides a solid foundation for future growth. As the number of productivity-enhancing applications increase, the business community will continue to invest more time and money in Wi-Fi solutions, bringing them one step closer to the truly wireless enterprise.

SPAWNING START-UPS

Wireless networking start-ups are hot; venture capitalists are throwing what little money they have left at wireless entrepreneurs in the hope that anytime anywhere computing is the telecommunications industry's best shot at bootstrapping itself out of the economic doldrums of recent years. There is no shortage of entrepreneurs to back. Start-ups are testing a variety of different business models, such as:

- Delivering Wi-Fi Internet access through public access points (i.e. HotSpots) in airports, hotels and coffee shops
- A fresh crop of wireless aggregators whose aim is to corral the multiple individual HotSpots that are appearing around the country, allowing users to gain access at any of the HotSpots under that provider's umbrella. Good examples of this business model are Boingo and HereUare.
- Other entrepreneurs are working to make Wi-Fi signal transmission cheaper and smarter.
- There are also firms that have begun to provide back-end services to Wi-Fi service providers such as roaming capabilities, authentication and billing.
- The market also clearly needs WLAN integrators.

Just about anyone who knows anything about the technology has probably daydreamed ideas for new ventures that exploit it. It doesn't even take that much imagination. Fast connection speed and mobility will engender innovations not yet thought of, and free bandwidth and an expanding user base will fuel innovation–perfect for startups and entrepreneurs.

"Wi-Fi Internet access may not be anywhere, anytime, but in a few years it is going to be in most of the places you are likely to need it," predicts one wireless consultant. And, according to David Farber, professor of telecommunications at the University of Pennsylvania and one of the industry's top networking experts, Wi-Fi is about to go mainstream, becoming an integral part of western culture.

Wi-Fi technology has the entrepreneurial crowd going wild. It's amazing how many business plans these days are built around this exciting technology. Wi-Fi allows for a lot of creativity: you can create a network, design applications, and build a business without having to hire Washington lawyers.

Entrepreneurs find Wi-Fi's most admirable attributes to include the following:

* Fast (to both set up and use).
* Inexpensive (under $200 for a small installation).
* Easy to install and use.
* Operates on unlicensed spectrum, so no extra monthly costs on top of the charge for a broadband connection are incurred.
* Most high-end laptops now come enabled for Wi-Fi.

All of the above add up to a near no-risk financial decision. It's not surprising that so many organizations are hoping to capitalize on this technology that allows people to tap into the Net from anywhere to retrieve email and surf the Web at lightning speed.

A PANACEA FOR DEVELOPING COUNTRIES

This era has been called the Information Age. But most developing and emerging countries even now are battling to fully enter the Industrial Age. Access to relevant up-to-date information is essential to these countries in terms of advancing their struggling business sector and helping their scientific community (especially in the medical arena) to form closer bonds with their peers and to keep abreast of the latest techniques and technology. These governments know that without a robust communications infrastructure there can be no information systems or Internet access—they will be isolated in a world that is more and more dependent on high-speed data exchange.

Developing countries such as Estonia, India, Tanzania, Brazil, and Costa Rica have already found that regions with technology-supportive infrastructures, policies and an educated work force are better equipped to reap the benefits of economic development than those that don't have such an infrastructure.

Telecompetition, Inc. (a research firm that specializes in market size data for sub-national areas) released a study in late 2002 predicting that by 2010 the use of wireless devices for data will grow twice as quickly in the world's developing and emerging nations than in the so-called "developed nations." In fact, the report predicts that by 2010 there will be 729 million mobile data subscribers in both the developing and emerging economies, which is about the same as the number of similar subscribers in developed nations. That's because mobile data access in developing and emerging countries is driven by a fundamental need for basic communication services—a need that doesn't exist in developed countries.

Eileen Healy, president of Telecompetition says, "Developing and emerging economies are highly motivated to build an infrastructure as quickly as possible. Adequate communications infrastructure is widely recognized as a key success factor for emerging economies. Mobile networks can often be built more quickly than a traditional wireline infrastructure, and the new IP-based mobile technologies will provide a more cost-effective way for developing countries to expand both voice and data communications."

Wi-Fi technology does not require the expensive initial investment in cable networks that have traditionally carried the brunt of Internet data. Therefore, many developing countries look upon the technology as a way to jump-start their move into the Information Age without undue financial strain on their small budgets. In these countries, businesses and governments have limited resources to support a widespread voice communications infrastructure, much less advanced computer networks. New technologies such as Wi-Fi hold the promise of providing these countries and their citizens with economical access to global data networks.

Some of those countries are already starting to explore the many possibilities afforded by wireless technology. The impetus: An expensive network of copper wires isn't necessary, Wi-Fi allows governments to wirelessly beam information to remote villages, either via a wireless mesh network, or a wireless-enabled bus, or even by satellites that normally fly virtually unused over these countries.

A Wireless Mesh Network

There are many areas around the world where Wi-Fi Nets can be found enabling people to roam, free from wires and still able to access the Internet at high speed. An article written by John Markoff for the March 3, 2002 edition of *The New York Times*, entitled "The Corner Internet Network vs. the Cellular Giants," (which I paraphrase) perhaps says it best:

Currently, most Wi-Fi networks serve as individual beacons that provide wireless Internet connections to portable computing devices situated within 300 feet/152 meters or so of a transmitter, writes Markoff. But what happens when these access points reach a mass that can be woven together to provide anytime anywhere Internet coverage–a high-speed wireless data network built from the bottom up, rather than from the top down (as cellular networks are built). Many industry experts believe that it is possible to take the tens of thousands of local wireless access points and lash them together into a single anarchic wireless network with connections passed from one Wi-Fi node to another, similar to the way cellular phone signals pass from cell to cell.

This ubiquitous wireless network would be modeled closely on the original nature of the Internet, which grew by chaining together separate computer networks. The technology that enables this ubiquitous wireless net is called "wireless mesh routing," i.e. a network architecture built around a "viral telecommunications network" comprised of numerous WLANs that work in unison to provide a pervasive envelope of connectivity. A mesh network connects computers and other devices with each other in a way that enables and extends the reach of all users throughout the network. These networks contain clients (e.g. laptops, PDAs, phones) and access points that act as relays and routers to pass traffic on to similar devices. Traffic bounces along a series of these devices until it gets to its destination, much like packets on the Internet.

Mesh networks also have special design characteristics that make them desirable for certain areas or applications. For instance, a line of sight between connecting devices is not required and devices are free to move around arbitrarily. Mesh networking is viewed as a low-cost method for quickly building a national (or even global) network infrastructure, since, with directional antennae, Wi-Fi can be extended to work for 40 kilometers (25.8 miles)–a distance that would allow rural areas to be linked via wireless mesh routing techniques. Once in place, such a wireless net would offer the promise of a vastly more powerful collaboration–driven by the same technical and social forces that originally built the Internet.

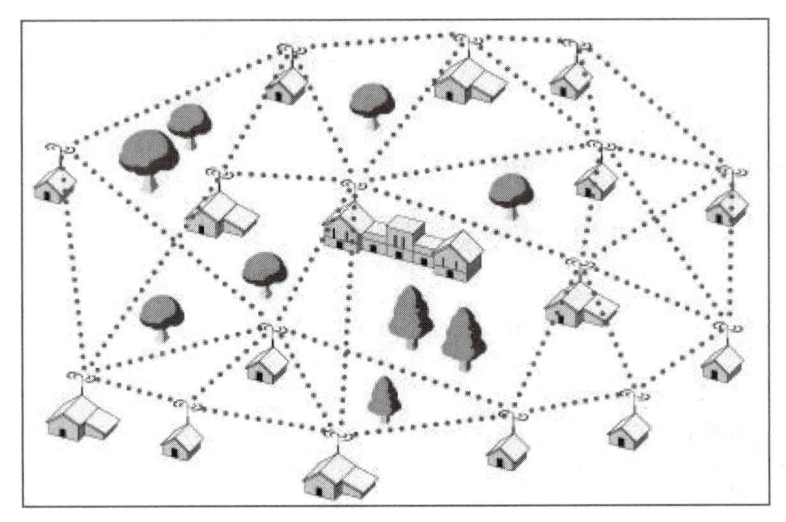

Figure 2.4 In a mesh network, each wireless node serves as both an access point and wireless router, creating multiple pathways for the wireless signal. Mesh networks have no single point of failure and can be designed to route around line-of-sight obstacles that can interfere with other wireless network topologies. (Graphic courtesy of Proxicast.)

Such networks have the critical advantage of economy of scale. In contrast to cellular data networks, in which every customer is an added cost and a drain on system resources, with wireless mesh networks, the more users who join the network the better the network's performance, so long as they contribute to the infrastructure (by relaying data traffic).

Like the Internet, wireless mesh routing is potentially a proverbial "disruptive technology" in that it is likely to upset the existing order by using the same powerful economies of cost and scale that initially drove the growth of the Internet's subset, the Web. It has the power to potentially transform communications, especially in the many emerging markets of the world—Indonesia is a good example. Thus, Wi-Fi, a technology that the world's developed markets consider to be an "alternate" technology, could become a "mainstream" technology in developing countries, enabling them to jump-start their communications infrastructure into the modern era via high-speed Wi-Fi-enabled communications.

All this has big implications for telecommunications in the developing world. Think for a moment of a "communication utopia" consisting of (1) very low cost, mass-manufactured Wi-Fi PC cards and hubs (we're already there), (2) hubs spread around urban clusters in the developing world—including low-income settlements (it's already beginning to happen), (3) Voice-over-IP, so that voice calls could be made; these calls, of course would be connected to landline and cellular networks through gateways (already possible).

In many ways, Wi-Fi enables a perfect business model for a regional communications company—(a) infrastructure that eliminates most, if not all, of the labor costs of a conventional communications service provider and (b) aggregate capital investment that is mostly borne by the individual customer since they pay for their own set-up costs.

Bridging the Digital Divide

The United Nations Secretary General, Kofi Annan, believes that information and communication technologies (like Wi-Fi) can improve the lives of people in developing countries. According to Annan, information and communication technologies (ICT) can help the poor work their way out of poverty, while at the same time benefiting the world community as a whole. Bridging the digital divide, in developing nations, is a formidable task that requires not only leadership, but also a major commitment of resources. Innovations such as Wi-Fi, and other low-cost technologies and business models are now being explored as a means to provide cheap, fast, and eventually free access to the Internet.

In September 2002, Annan delivered his remarks at the opening of the third meeting of the United Nations Information and Communication Technologies Task Force. In those remarks, he stated, in part:

"There is a vast potential for investment growth in the developing countries. Information and communication technologies can help us turn this potential into concrete opportunities that will help the poor work their way out of poverty, while, at the same time, benefiting the world community as a whole....

"With innovations such as wireless fidelity—commonly known as Wi-Fi—and other low-cost technologies and business models that are now being explored, we should aim to provide cheap, fast and, eventually, free access to the Internet. But investments will still be necessary, not only to ensure that people have the technical skills and the literacy level needed to use information technology facilities and service them, but also to create content that reflects the interests of that part of the world....

"As regards access, developing countries often do not have enough phone lines, which in any case, provide little bandwidth, and, by developing country standards, are expensive to run. Radio [Wi-Fi] is a possibility, indeed a vital one."

And, in November 2002 in his "Challenge to the Silicon Valley," Annan said, among other things, that:

"We need to think of ways to bring wireless fidelity [Wi-Fi] applications to the developing world, so as to make use of unlicensed radio spectrum to deliver cheap and fast Internet access....

"Information technology is extremely cost-effective compared with other forms of capital. Modest yet key investments in basic education and access can achieve remarkable results. Estonia and Costa Rica are well-known examples of how successful IT strategies can help accelerate growth and raise income levels. But even some of the least-developed countries, such as Mali and Bangladesh, have shown how determined leadership and innovative approaches can, with international support, connect remote and rural areas to the Internet and mobile telephony."

Despite the rapid growth in information and communications technology, the Internet still remains a distant dream for much of the world's population. The main reasons are lack of infrastructure, especially in the developing countries, and the exorbitant cost of connectivity. Wireless technologies can alleviate this situation.

Another proponent for wireless is Dr. Onno Purbo, the Jakarta-based expert on information and communication technologies. According to Dr. Purbo, connecting to the Internet using a simple aluminum antenna and a wireless network card could be the best way to narrow Indonesia's digital divide and bolster economic development. In late 2001, he told an audience at the Ottawa headquarters of the International Development Research Centre (IDRC), "The goal is to see a knowledge-based society in Indonesia. We need to transform Indonesia's people into knowledge producers rather than knowledge consumers. And the fastest way to do that is through the Internet." That's why Purbo tirelessly promotes the use of Wi-Fi-enabled Internet connections and why he has held numerous workshops across his country to teach thousands of people how to build their own Wi-Fi systems.

Purbo's own high-speed Internet connection is through a wireless local area network and an aluminum antenna. The antenna, which he fastened to the back of his house, uses radio waves to reach access points up to eight kilometers/5 miles away. Not including his computer, the Wi-Fi system cost under $300, a price tag that is shrinking by the day.

"The bandwidth is free and can be resold in Internet cafés. People only pay for the connection, and they can share [that cost] by setting up a neighborhood network," says Purbo. He estimates there are about 2500 Indonesians throughout the country's archipelago who currently have Wi-Fi connections. Purbo laments, "There are an estimated four million [Internet] users among the country's 231 million people. Even within the university community, only 200 of the country's 1300 institutions are on the Internet." Widespread access remains a challenge.

Purbo has set himself a personal goal: to boost those numbers to at least six million in the next few years. To reach his goal, he plans on spreading the message of Wi-Fi connectivity through a series of workshops. After Purbo completes his Eisenhower Fellowship in Washington, his plans are to start his third annual road show of seminars and workshops where he will discuss how the use of voice transmitters, or telephony, can be incorporated into a Wi-Fi system. His plan is to set up Internet chat rooms where Indonesian students can regularly reach students from other countries. He hopes the dialogue between these students will help transform Indonesia into a knowledge-based economy.

Estonia

Let's take a closer look Estonia, a sort of "poster child" for jumpstarting Internet usage via Wi-Fi in developing and emerging nations. People in Estonia can access the Internet from about 500 Public Internet Access Points (i.e. HotSpots). Each HotSpot has a special traffic sign showing its location with the @ symbol.

Most HotSpots are located in libraries. One can easily determine where the nearest HotSpot (also referred to as a Public Internet Access Point or PIAP) is located by going to www.regio.delfi.ee/ipunktid. About 60 public locations (city squares, hotels, pubs, airports etc.) are currently covered with high-speed wireless Internet access.

Consider the Estonian government's very impressive figures from November 1, 2002:

- 41 per cent of the population are regular Internet users.
- 30 per cent of the population have a computer at home, 59 per cent of home computers are connected to the Internet.
- All Estonian schools are connected to the Internet.
- There are about 500 Public Internet Access Points in Estonia, 36 per 100,000 people (one of the highest numbers in Europe).
- Incomes can be declared to the Tax Board via Internet.

- Expenditures made in the state budget can be followed on the Internet in real-time.
- The Government has changed cabinet meetings to paperless sessions using a web-based document system.
- 40 per cent of Estonian residents conduct their everyday banking via the Internet.
- 61 per cent of the population are mobile phone subscribers.
- All of Estonia is covered by digital mobile phone networks.

According to the RIPE Network Coordination Centre (a Regional Internet Registry that provides allocation and registration services which supports the operation of the Internet globally), Estonia maintains the highest Internet connected hosts / population ratio in Central and Eastern Europe, and it is also ahead of most of the European Union countries.

The Potential Is There

Professor Nicholas Negroponte, director of Media Lab at the Massachusetts Institute of Technology (MIT), an organization that promulgates information and communications technology for developing and underdeveloped countries, predicts that Wi-Fi technology will revolutionize the telecom industry. Another Wi-Fi supporter, Intel's Craig Barrett, says that the technology will be lapped up by developing countries to spiral broadband growth. This ardent supporter has also been quoted as saying, "developing countries will lead the broadband revolution and in India this revolution will start with technologies like Asymmetric Digital Subscriber Line (ADSL) and 802.11b."

Information and communication technologies have enormous potential to meet a nation's development challenges, if government, business and the nonprofit sectors work together in strategic partnership. And while technology, including Wi-Fi, doesn't offer a panacea for all of the developing world's problems, detailed analysis does reveal ample evidence that when used in the right way and for the right purposes, information and communications technologies can have a dramatic impact on achieving specific national development goals and strategies.

Of course, as Bill Gates points out, what deprived people need is not computers but fundamentals such as medicine. And, Kevin Watkins of Oxfam reminds us that in much of sub-Saharan Africa more than half of primary-age children are denied the opportunity of even a rudimentary primary education and less than one third make it to secondary school. What both of these men (and others) say is true, and such depressing facts aren't irrelevant. But, Internet access could enable easy dissemination of vital medical information and provide for remote diagnosis. It also offers the potential for improving education—that is if appropriate and inexpensive methods of information delivery (such as Wi-Fi) are utilized.

CONCLUSION

In the early days of this technology, a Wi-Fi card cost $300 to $500 and it took incredible perseverance to get the software working, and a valiant struggle thereafter to persuade the operating system to behave correctly. That, thankfully, is now all in the past. Today, Wi-Fi is *truly* plug-and-play. Anyone can use this technology—your grandmother, your artistic aunt, your technophobic father.

The advantages and benefits of Wi-Fi are such that it's about to "go mainstream," becoming a part of everyday experience. Soon Wi-Fi will be built into all computing devices; most portable devices are already Wi-Fi-enabled (printers just joined the Wi-Fi club). It will be as prevalent as dial-up networking—something you don't think much about because it's always available.

For Wi-Fi to perform all of these miracles the vendor community must develop and deploy low-cost, mobile communication devices—this is vital to the success of Wi-Fi in developing and emerging nations. Such devices will, in turn, become the "computers of tomorrow." Optimally, the end-user's cost should be no more than $100 per device. This is plausible, assuming that emerging markets should create a demand for 10 million such units a year, allowing economy of scale to lower prices, thus making Wi-Fi-enabled, low-cost, portable computing devices the standard device used by the next billion or so people to join the projected Wireless Net.

Then, of course, content developers and enterprises will need to put together applications that can leverage all of this wireless technology. They must think of innovative ways in which millions of users can interact via portable, high-speed devices.

Chapter 3: Fueling the Wi-Fi Fire

Wi-Fi brings to fruition a long dreamed about ideal—a populace interconnected via intelligent devices. With the push of a button, you're communicating: attending an IM (instant message) meeting, videoconferencing, sending a presentation to a colleague (which you put the final touches on while traveling between Paris and New York), sending your secretary your mark-up of a legal brief as you wait in a train station—the list of possibilities is endless.

"Wi-Fi use is experiencing explosive growth. . . .It will fundamentally change the way people use technology and enable high-speed Internet access anytime, anywhere for business and consumer use." These positive remarks are by Intel executive vice president and Intel Capital president, Les Vadasz. Industry icon Bill Gates, who is also a Wi-Fi fan, predicts that wireless networking will become commonplace in the next ten years.

The entire communications industry has taken a keen interest in spurring acceptance of this wireless networking technology. Dell Computer equips practically all of its new laptops with Wi-Fi connections; Sony Corp. puts Wi-Fi into a host of electronic gadgets; and Intel is not only investing heavily in Wi-Fi startups, but also recently released new chips with Wi-Fi transceivers. A sampling of industry interest in Wi-Fi, includes:

- Buffalo Technology Inc, D-link Systems Inc., Linksys Group Inc., Proxim Inc. and others sell Wi-Fi gear directly to the consumer.
- Boingo Wireless Inc., FatPort Corp., HereUAre Communications Inc. and Surf Sip Inc. as well as others in the Wireless Internet Service Provider (WISP) industry enable Internet access via HotSpots situated in various strategic locations.
- Agere Systems, Cisco Systems, Dell Computers, Enterasys Systems, Hewlett-Packard Co., IBM, Intel, Microsoft, Symbol Technologies, and others sell software, networking devices and other Wi-Fi gear to a mixed market.
- Texas Instruments, Intersil, Broadcom and others manufacture the chipsets.
- The "specialty" players provide applications such as security (e.g. products offers by companies like Agere, Radius and Odyssey), Quality of Service (e.g. Cirrus Logic and Juniper Networks), voice (e.g. Telesym and Symbol), etc.

Today, major manufacturers, vendors and even telecommunications providers are hatching Wi-Fi plans. This is good; large, established companies within the communications business community must provide support, aid, and encouragement to generate the synergy necessary to propel Wi-Fi to a position of universal acceptance and availability. Just look at what some of these heavyweights are doing toward that end.

Intel: This industry icon is very upbeat about Wi-Fi. Pat Gelsinger, Intel vice president and chief technology officer, says, "Wi-Fi is one of the ripest areas for innovation in the industry. Intel will continue to play a key role in its development through investments, research, industry programs and products."

Intel backs all of these optimistic statements with massive expenditures. As of the beginning of 2003, Intel's strategic investment program, Intel Capital, had invested approximately $25 million in more than ten companies working in the Wi-Fi space. Intel also established, in 1999, the Intel Communications Fund (managed by Intel Capital) to focus on accelerating Intel voice and data communications initiatives. The $500 million fund will devote $150 million specifically for investment in Wi-Fi companies worldwide. Attention will be focused on companies developing hardware and software products and services that enable user-friendly and secure wireless network connections, simpler billing procedures, creating a robust infrastructure and new ways to connect while on the road. (For more information about the fund, visit www.intel.com/capital/portfolio/funds/icf.htm.)

Hewlett-Packard: This is another company adding fuel to the Wi-Fi fire. It has launched a global initiative designed to deploy HotSpot infrastructure in such public sites as cafés, airports, hotels and restaurants. HP's HotSpot implementation strategy is to provide access points, antennae, security and network integration, as well as subscription services and applications through worldwide partnerships.

Michael Flanagan, HP's worldwide wireless LAN solutions program manager, said, during a comment on the company's push into Wi-Fi, "Adoption of the Wi-Fi standard is increasing more rapidly even than projections made earlier this year [2002]." He went on to say, "In the enterprise area, especially in hotels, airports and convention centers, companies are discovering they can use the same infrastructure for their employees and guests."

Cisco Systems: This industry giant has taken a slightly different track. It is busy forming partnerships, alliances and making acquisitions to forward its mission of providing Wi-Fi gear and infrastructure to the enterprise marketplace. For instance, Cisco acquired Radiata and in the process got its hands on some very spiffy 802.11a technology. The company also recently acquired the home networking vendor Linksys. Moreover, as the reader shall learn, the company has entered into some judicious partnerships and alliances, which too should serve to further Wi-Fi's popularity.

IBM: This famous technology company ("Big Blue") is making a major push into the enterprise Wi-Fi market, especially through deals with other industry giants like Nokia and Cisco. Adel Al-Saleh, general manager of wireless e-business at IBM, does a good job of explaining Big Blue's position on Wi-Fi. His comment on the Cisco alliance is very telling: "Cisco has a significant share in the LAN market in America, and we are jointly going after that space." And while commenting on the Nokia partnership, Al-Saleh said, "We have spent a lot of time taking wireless LAN to the enterprise, and we need their technology to make more data services available in that market."

***Note:** IBM has installed Wi-Fi networks and provided 6400 new IBM desktop PCs equipped with wireless NICs for more than 600 Boys & Girls Clubs across the U.S. This outpouring of community spirit is providing Internet access for more than 200,000 youngsters. "These technology centers are helping us level the playing field for our Boys & Girls Club members," said Bill Regehr, senior vice president of Boys & Girls Clubs of America. "For some of them, it's the first time they've ever had the opportunity to use a computer or access the Web after school."*

Ericsson: This well-known, worldwide supplier of end-to-end solutions for mobile and broadband Internet access has an aggressive wireless vision: an "all communicating" world via voice, data, images, and video. This fits well with the continuing proliferation of Wi-Fi installations. Supporting this vision is Ericsson's "Always Best Connected" concept, which refers to an environment where an end-user is always able to connect to a network using the best technology available at any specific time and location.

Ericsson believes that integrating Wi-Fi-enabled WLANs (wireless local area networks) and public HotSpots with the cellular industry's 2G and 3G networks is an attractive business concept. This developmental thrust has prompted Ericsson to enter into some interesting partnerships: Agere Systems (formerly the microelectronics division of Lucent Technologies) has teamed with Ericsson to offer a package of products and services intended to help Internet Service Providers (ISPs) cultivate the Wi-Fi crowd. These two companies, in turn, have entered into a partnership agreement with the hardware vendor, Proxim, to help telecom operators integrate Wi-Fi HotSpots with cellular networks.

In 2003, its development efforts led to the introduction of Ericsson's Mobile Operator WLAN product package, which enables operators to integrate WLAN with their existing 2G and 3G mobile businesses. It works with existing infrastructure, subscriber management, billing and authentication systems. The product allows network operators to offer customers enticing added service packages, combining features from both WLAN and mobile phone technologies, and creating new revenue streams from locally adapted content.

In such environments, the standard GSM mobile phone Subscriber Identification Module, popularly known as a "SIM" (a removable chip the size of a postage stamp), will be an important component of the system, since it can hold user information and be used in tracking usage. SIM technology can also provide security, enable unified billing for operators and support roaming capabilities, making access to a public WLAN as easy as accessing a cellular network.

Microsoft: Of course, any discussion of industry titans wouldn't be complete without mentioning Microsoft. According to Microsoft chairman, Bill Gates, the company's vision for the near future revolves around wireless networks. Gates considers the technology one of the most important innova-

Figure 3.1 Subscriber Identification Modules (SIMs) could play an important role in expanding Wi-Fi's acceptance.

tions of the past five years, ergo his statement: "If any one technology has emerged in the past few years that will be explosive in its impact, it's 802.11."

This software king is working overtime to gain a lead in the Wi-Fi marketplace. Perhaps the most significant is the latest Microsoft operating system, Windows XP, which was specifically designed to make wireless networking simple. For example, much of the software needed to run Wi-Fi is already available in Microsoft's Windows XP. This helps to make using Wi-Fi as easy to use as a CD-ROM drive.

Microsoft's strategy for capturing a large share of this budding market is three-pronged: to support device hardware manufacturers with its Pocket PC architecture; to partner with important wireless infrastructure providers, such as AT&T; and to encourage application development through ISV (Independent Software Vendor) partnerships and support for SQL Server CE (a database that runs anywhere the Windows CE operating system will run).

Wi-Fi Alliance: Another powerful influence in the Wi-Fi world is the Wi-Fi Alliance, an industry group charged with monitoring standards for wireless networking. It was founded in 1999 as the Wireless Ethernet Compatibility Alliance (WECA) through the efforts of Cisco, Intersil, Agere, Nokia, and Symbol in a bid to present an united front in their efforts to bring about one globally accepted standard for high-speed, wireless, local area networking–at the time it was the IEEE 802.11b standard. Membership in the Alliance is continuing to grow. Today, the Wi-Fi Alliance's Board Member Companies consist of such industry bigwigs as Agere, Cisco, Dell, Intermec Technologies, Intel, Intersil, Microsoft, Nokia, Philips, Sony, Symbol Technologies, and Texas Instruments.

The Wi-Fi Alliance offers a test suite that defines how member products are tested to certify that they are interoperable with other "Wi-Fi Certified" products. Once a device has successfully passed through one of Wi-Fi's testing facilities, the compliant hardware is authorized to display the organization's "Wi-Fi Certified" mark. (See Fig. 1.2.) Membership in the Wi-Fi Alliance is open to all companies that support the 802.11 family of standards.

Since the certification program began in March 2000, its more than 200 members have received Wi-Fi certification for over 550 products. The assurance of compatibility that comes with this certification program has gone a long way in making home and business users comfortable buying into the technology.

Wi-Fi Smart Card Consortium: This is a new organization with a stated mission of defining specifications for worldwide access to Wi-Fi networks with smart card security and related capabilities. Its launch was announced in a December 3, 2002 press release issued by its founding members: Cisco, INRIA, SchlumbergerSema and Ucopia. The reader knows Cisco but may not be as informed about the other three founding members. INRIA is a French research organization that has lent its prestige to the initiative. SchlumbergerSema is a division of Schlumberger Limited, a global technology services company. Ucopia is a start-up company providing enterprise mobility management software and is the driving force in the creation of the WLAN Smart Card Consortium. The first plenary meeting of the consortium was held February 4-5, 2003 in Paris.

Ubiquitous Network Forum: Wi-Fi received a tremendous boost in 2002 when Japan (one of the world's most developed wireless markets) announced several proposed Wi-Fi initiatives. The announcement included the Ubiquitous Network Forum, which has

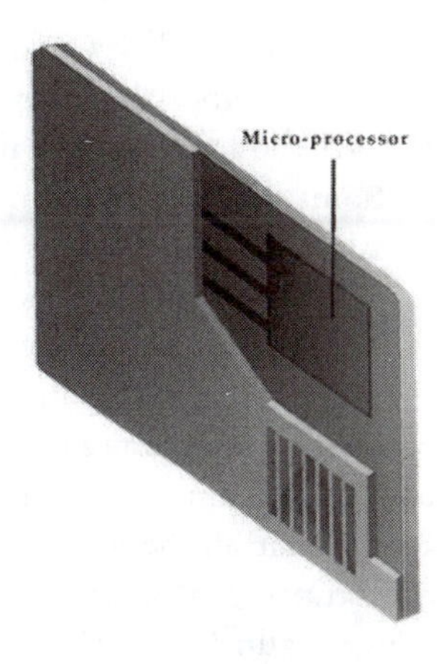

Figure 3.2 A smart card consists of a piece of plastic similar in size to a credit card. Embedded in it is a tiny computer microprocessor and memory chip(s), such as a SIM.

the stated mission of developing new high-speed wireless infrastructures and generating ¥80 trillion in new markets by 2010.

The roll call could go on and on. What's important is the cooperation and synergy that all of this activity creates. It fuels innovation, which in turn feeds energy into the Wi-Fi stampede.

SYNERGY HELPS TO HIT A HOME RUN

The heavy hitters are involved, the rookies have embraced Wi-Fi big time, and the entrepreneurs are finding new and exciting ways to exploit the technology. But for the synergy to come full circle, the financial community must also join the Wi-Fi team. Grassroots organizations and start-up companies have limited budgets; that makes it difficult for such entities to get funding and to find other companies interested in them, their projects, and/or their products. Happily, the venture capital market finds Wi-Fi compelling and some within that industry are taking on the role of "team scout." Thus, contrary to the slump occurring in other investment areas, investors can be found to fund Wi-Fi projects and start-ups.

But, just as with the great American pastime of baseball, it takes a combination of "all-stars" and "rookies" to make up a winning team. A team of nothing but homerun hitters always has problems making it to the World Series—there's talent to spare and the hunger to win may be there, but something is lacking. Many times the missing ingredient is the infectious excitement generated by the rookies. That same kind of synergy—the establishment teaming up with the start-ups—is what will send Wi-Fi through the roof.

In 2001, a rookie company, Wayport, Inc., a provider of Wi-Fi wireless and wired high-speed Internet service to business travelers through its nationwide network of over 400 hotels and a number of airports, grew its revenue by pioneering the next evolution of the industry: wireless broadband roaming. To promote continued adoption of Wi-Fi and to make it easier for travelers to roam between service providers' networks, Wayport is pursuing roaming agreements with several providers, including cellular carriers. Wayport is also working with strategic partners, including homerun hitters, Microsoft, Sony, IBM and Dell, to drive wireless adoption.

Another good example of synergy is the top of the lineup collaboration between Intel and Intersil. Intersil Corporation (a market leader in the 802.11b chip space) and industry titan Intel Corporation joined together in an effort to accelerate deployment of current and future Wi-Fi technology and products. To that end, Intel incorporated Intersil's PRISM 2.5 chipset in its PRO/Wireless 2011B LAN products—a suite of products that can wirelessly transmit voice, data and video content at speeds up to 11 Mbps.

Today, Wi-Fi equipment makers, from newcomers like Pronto Wireless to veterans such as Cisco Systems, are focusing on improving the speed and quality of their equipment's functionality. Cisco, for instance, is developing a Wi-Fi network interface card (NIC) for laptops that won't drain a battery by constantly looking for an access point. Instead, the access point will "wake" the card into action when it's near a network.

Some within the communications industry are also working to provide a smooth transition between cellular and Wi-Fi networks. These include the aforementioned Ericsson, the communications industry's icon Lucent Technology, and software developer NetMotion Wireless.

What about Bluetooth, that alliance between mobile communications and mobile computing companies that led to a communications standard allowing wireless data communications at a 10 meters (33 feet) range? Well, Bluetooth technology hasn't been forgotten in all the rush to embrace Wi-Fi. For example, chipset developers Cambridge Silicon Radio, Mobilian and Silicon Wave (in partnership with chipset leader Intersil) have all launched dual-mode Bluetooth and Wi-Fi (802.11b) solutions. And Intersil / Cambridge Silicon Radio / Smart Modular Technologies have formed an alliance in the hopes of bringing a mini-PCI board to market, sometime in 2003, that combines 802.11b wireless LAN and Bluetooth technology. The importance of these announcements lies in the blending of two technologies once considered mutually exclusive—both technologies operate in the same part of the radio frequency spectrum (2.4 GHz), resulting in some interference between the two (although the level of interference and its impact on the reliability of both technologies is currently open to debate).

Developers are excited because the resulting technology offers Wi-Fi (and Bluetooth) developers an easier integration path. Co-location solutions will also reduce the overall cost of implementing both technologies as components can be shared, although Frost & Sullivan and other analysts believe that such dual-mode solutions need further development before they can be said to completely solve the Wi-Fi/Bluetooth conundrum. Still, they do admit that such solutions represent a major step in the right direction.

Bringing the focus back to Wi-Fi—synergy is at work everywhere. Note Cisco's announcement in late 2002, wherein it informed the public that it is partnering with Intersil to provide an IEEE 802.11g client reference design. That announcement sparked life in this almost lifeless specification. 802.11g endured innumerable delays as it made its way through the standardization process, and these delays caused a total lack of inter-

Figure 3.3 A silicon chip is a tiny electronic circuit on a piece of silicon crystal. It contains hundreds of thousands of micro-miniature electronic circuit components, which are packed and interconnected in multiple layers within a single chip. Then the surface of the chip is overlaid with a grid of metallic contacts used to wire the chip to other electronic devices. And all of this is done in an area less than 2.5 millimeters square. (There are 2.54 centimeters to an inch, and each centimeter is 10 millimeters long.) These components can perform control, logic, and memory functions. Silicon chips are found in the printed circuits of, for example, personal computers, televisions, automobiles, appliances, etc.

est in the new specification within the industry, mainly due to the anticipated release of 802.11a products. In fact, there was even some doubt as to whether 802.11g end-products would ever be brought to market. Never mind that chip giant, Intersil, has worked very hard to advance Wi-Fi and 802.11g in the marketplace; indeed, the manufacturer has been promoting its 802.11g OFDM solution since early 2001. Now, Intersil has a powerful ally—Cisco. With Cisco's backing of the technology, 802.11g will certainly gain traction in the WLAN market, since Cisco's reach and influence in the business networking market is extensive. As illustration, witness the early 2003 upswing in announcements of chipsets containing 802.11g technology (albeit originally built upon a *draft*, not the ratified version, of the 802.11g specification) and the resulting products that have begun to hit store shelves.

Once you have the synergy going full steam, the inventors and engineers suddenly have the upper hand, devising new gear, products and applications for fledgling markets such as Wi-Fi.

EVOLVING TECHNOLOGY

The potential of Wi-Fi is limited only by our imaginations. But one of the biggest roadblock for WLAN services *had been* the relatively small number of devices that were Wi-Fi-compliant. That impediment has been overcome—the marketplace is overflowing with Wi-Fi-enabled devices. Manufacturers are rushing to put wireless-enabled devices on store shelves so users can surf, email, send and receive large data files, view quality video, and listen to ear popping audio—all while on the go—using new and better smart phones, Pocket PC phones, PDAs, notebooks/laptops, tablet PCs, MP3 players, and other devices—devices that are now only a glimmer in the designer's mind.

Companies are also racing to come up with all kinds of innovative applications and products that can take advantage of Wi-Fi's capabilities: antennae that extend Wi-Fi's range, chip designs and signal processing algorithms that can mitigate the loss of speed over long distances, methods to carry voice as well as data traffic on Wi-Fi networks, devices, and applications that can do just about, well, anything.

The Wi-Fi market has grown quickly from a very small piece of the overall communications sector with only a few vertically integrated manufacturers and vendors, to today's quickly evolving marketplace. It seems that the entire communications industry is itching to get in on the action and integrate Wi-Fi into their product lines.

Chip Manufacturers

Intersil Corp., Broadcom Corp. and Agere Systems Inc. were the manufacturers in the forefront of the Wi-Fi phenomenon; at least as far as specialty chipsets were concerned. These manufacturers now face growing competition in the Wi-Fi market, not only from their counterparts in Asia but also from Intel, and others. In fact, Intel could quickly take the lead due to the introduction of its new collection of chips (under the name of "Centrino") that are designed to transform laptops and tablet PCs into portable offices.

Along with the trend toward Wi-Fi specific chipsets, there are sub-specialties that are helping Wi-Fi to gain worldwide acceptance. Within these sub-categories there are:

- Chips that help reduce power needs.
- Dual-band and multi-standard chips (802.11a/b, 802.11b/g and 802.11 a/b/g).

* Chips that can provide lower cost solutions.
* Chips that provide added security enhancements.

With chip manufacturers jumping headfirst into the red hot embedded Wi-Fi market, seamless wireless functionality is being upgraded and enhanced in everything from PDAs, cell phones and laptops to antenna technology, network cards, and more.

The reader might ask, why are chip manufacturers so important to Wi-Fi? The primary reason is power—Wi-Fi is a power hog. The transmission and receipt of voice or data is the single most power-hungry mode of any wireless device, and Wi-Fi is no exception. The processing power necessary for end-users to use a Wi-Fi-enabled device to send and receive data drains the small batteries that are commonplace in most mobile computing devices. In the past, this, and the size of the average chipset, made Wi-Fi impractical for small devices such as PDAs and cell phones.

Chip manufacturers, however, have made it to first base—they have found ways to reduce the size of their chips (although more needs to be done). Some are expected to round second, in the near future, with newly designed chipsets that should help to put a lid on Wi-Fi's power gobbling ways.

The chipset industry, while not hitting a home run, is at least on base. Many of the latest-generation Wi-Fi chipset designs do cut power needs by introducing performance enhancement techniques into the design of the chipset, for example:

* Texas Instruments's revised version of the ACX100 WLAN chip consumes only a tenth as much power in standby mode as its predecessor.
* Advanced Micro Devices (AMD)'s Alchemy Solutions Am1772 wireless LAN chipset and mini-PCI card for notebooks is designed to be ultra-efficient and ultra-compact.
* Intel's Calexico, which is similar to the Alchemy chipset except it is a hybrid 802.11a/b combo, has been deployed as part of the company's Centrino line.

With the proliferation of Wi-Fi specifications (i.e. 802.11a, b and g), device manufacturers are also clamoring for multi-standard chipsets. Again, the chip manufacturers have stepped up to the plate, as in the following examples:

* Broadcom has launched a simultaneous dual-band 802.11a/b chipset using an all-CMOS design. It is a three-chip design that comprises the BCM4309 baseband/MAC chip, which supports the 802.11b and 802.11a protocols as well as IEEE-802.1X security, the BCM2050 2.4 GHz radio interface chip, and the BCM2060 5 GHz radio chip. The BCM4309 dynamically selects the best performance available at either 2.4 or 5 GHz. This enables end-users to make a network connection and an ad hoc user connection simultaneously, among other things.
* Intersil's PRISM Duette, a dual-band 802.11a and 802.11g wireless LAN solution that is capable of transmitting high-speed video, voice, and data. It is fully backward compatible with the installed base of over 15 million 802.11b systems worldwide.
* AMD is expected to introduce its next-generation of Wi-Fi chips with a multi-standard design, i.e. a hybrid 802.11b/a/g chip.

There's lots more that could be said about how the chip manufacturers have embraced Wi-Fi but. . . suffice it to say that these guys have exceeded expectations and through their efforts, Wi-Fi has taken a giant step forward.

Shipments of chipsets for Wi-Fi grew to 23 - 25 million for the year 2002 versus less than 8 million in 2001, at least according to a year-end 2002 report from the market research group Allied Business Intelligence Inc. (ABI). The ABI report, entitled, "Wi-Fi Integrated Circuits: Industry Dynamics, Market Segmentation and Vendor Analysis for 802.11a/b/g," indicates the following:

- By 2004 revenue from dual-band chip sets will exceed that of 802.11a, b or g.
- Chipsets will be targeted for embedded implementations particularly in battery-constrained devices such as PDAs and mobile handsets.
- 802.11g chips will have a significant impact in the SoHo/retail market and will comprise 18 percent of chipsets shipped in 2003.

The emergence of multiple 802.11 variants (a, b and g) and constant innovations in chip design are creating an extremely dynamic Wi-Fi integrated circuit market.

Extending the Range

This section shows the reader the latest innovations in extending the range of a Wi-Fi network, in providing dual-mode networking, and in transmission security. It doesn't cover typical range extender antennae such as those offered by HP/Compaq, Avaya and others. There is a general discussion of antennae in Chapter 19.

What's New?

Wi-Fi's limited range of transmission causes some concern in the marketplace. Again, many individuals and companies are working to resolve the problem. Let's look at some of the more interesting and promising antenna solutions. (It's interesting to note that most of the companies working on this issue are start-ups.)

EtherLinx: Reminiscent of the early PC days, EtherLinx Communications, a small start-up working out of a garage in Campbell, California, has come up with a way to extend Wi-Fi's range from a mere few hundred feet to up to 50 miles (80.4 kilometers) and still maintain incredibly high data transmission speeds. The data transfer rate over a long distance has been successfully tested at 10 Mbps, but when it reaches the market EtherLinx will offer only 2 Mbps (which is still faster than a T-1 line or most DSL services) to ensure that they provide their customers with consistent quality of service.

The key to EtherLinx's success is a single, small antenna that transmits data faster than either Cable Modem or DSL service to a tiny $150 receiver mounted to the side of the end-user's premises. Then that receiver is hardwired to a PC using standard CAT 5 wiring. Individual repeater devices can retransmit signals, so non-line-of-site signal paths can be handled by deploying repeaters in a mesh or other topology away from the base site.

The company says its technology even addresses two of the other most notable challenges for Wi-Fi networks—security and privacy. That is because transmitted data can be received only by the company's equipment, and each receiver has a unique address. This makes it relatively easy to detect, and then track, anyone who's trying to gain unauthorized access.

During 2002, EtherLinx operated a small, for-pay trial in Oakland, CA and is currently slated to begin covering the small city of Campbell, CA (an area in which DSL service is not available) in the near future. Interestingly, Congressman Mike Honda, D-

CA, met with the company's founders, after which he scheduled a demo and presentation in Washington, D.C. Honda says he wants to see how quickly this technology can be deployed elsewhere in the nation.

Other companies—Nokia, Iospan Wireless, and Navini Networks—are trying to duplicate EtherLinx's achievements; but to date none has met with EtherLinx's success.

Vivato: This startup company has introduced two FCC-approved switches that can increase the range and capacity of Wi-Fi antenna systems. Instead of providing service to only several dozen people situated within a few hundred feet of the transmitter, the Vivato switches enable an antenna to be accessible to legions of users from long distances. In an open environment, the outdoor switch can transmit to a client four miles (6.4 kilometers) away, while the indoor switch has a range of approximately 2000 feet (609.5 meters).

The switches, which are about 3.5 feet by 1.5 feet by 2 inches (1.06 meters x 45.72 centimeters x 5.08 centimeters), create directional beams rather than radiating energy "isotropically" (in all directions). This makes it possible to greatly extend the range of the system at low power and reach more users. At the same time, the switches can maintain an average transmission speed of between 6 and 11 Mbps and provide security using state-of-the-art encryption, the latest authentication methods, and virtual private networking. The company also thinks that the sensitivity of their technology makes the system ideal for detecting potential intruders.

The technology Vivato uses actually stems from 1950s research on a phased-array antenna for the military. That technology makes it possible to electronically "steer" numerous radio beams from a single point. The multi-element antenna array is assembled using ordinary antenna elements. Special circuitry and algorithms perform the digital signal processing required to implement the digital beam steering operations, which consist of applying the array spatial signal processing algorithms to form the digital antenna array pattern from the multiple antenna inputs.

When beams are focused properly, the signal strength is increased, and when you use a large number of them, you can greatly increase an antenna's traffic capacity. In practice, the switches will determine each authorized Wi-Fi-enabled device's location so the system can send a high-powered signal straight to that specific device. This is contrary to the conventional Wi-Fi transmission technology where a single, low-powered signal is transmitted in all directions and only some of it is gathered by the receiver's antenna.

A Vivato spokesperson says that the company expects their technology to be especially suited to office buildings and campus environments (e.g. universities and hospitals) because it enables so many people to use a single Wi-Fi Internet connection simultaneously. Ken Beba, the company's chairman and CEO, is very enthusiastic about the new switch. He predicts that the product "will change the way people think about the physics of Wi-Fi."

Vivato's indoor switch is designed for placement in, say, the corner of a large office where it can provide wireless service throughout a building. This centralized approach of transmitting a series of beams is in sharp contrast to other companies that try to extend the range of Wi-Fi by creating meshes of overlapping access points.

***Note:** Even though France has yet to certify Vivato's technology, the 2003 Cannes Film Festival had wireless broadband access, thanks to Vivato, Intel Corporation and France's KAST Telecom. This group created one of the world's largest Wi-Fi HotSpots using the Vivato Wi-Fi switch, enabling anyone with a Wi-Fi-compatible computing device to gain wireless high-speed Internet access as they strolled along the Croisette and the Bay of Cannes.*

Moreover, although Vivato's initial audience is corporate buyers, extending the geographic range of wireless Internet networks could help open the market to WISPs and even traditional ISPs looking for alternatives to high-speed Cable Modem and DSL connections.

Mesh technology: Vivato is only one of several start-ups working to extend the range and performance of Wi-Fi networks. As many as two dozen other companies have appeared with alternative approaches, many of which involve mesh networking technology, which uses multiple antennae to pass data around like a bucket brigade.

Typical of these innovators is MeshNetworks, Inc., a company founded in 2000 to commercialize technology originally developed for military application. Soon after receiving the FCC's regulatory approval of its Mesh Enabled Architecture (MEA) and MeshLAN Multi-Hopping 802.11 network products, the company began shipping its products to customers.

Wi-Fi networks use access points to create wireless zones with a radius of about 300 feet (91.4 meters) around each access point. MeshNetworks uses a technique called "hopping" to extend the range of a Wi-Fi network. Users of the company's MEA and MeshLAN products send and receive networking signals that "hop" from one mobile computing device to another until they find one hooked up to an access point. According to the company, this technique can extend a network's range to about 1500 feet (457.2 meters), which is enough to cover an office building using only a single MeshNetworks access point.

MeshNetworks' target market is the very large enterprise, especially companies converting a campus or office building from a wired to a wireless setup. Gemma Paolo, a wireless analyst with In-Stat/MDR, extrapolates that that market is ripe for MeshLAN technology. But, at the same time, Paolo suggests that currently such a network would have to be a really big deployment to make economic sense. Paolo opines that "with the cost of access points coming down so low," the MeshLAN method might not be worth it. Of course, MeshNetworks disagrees. Rick Rotondo, its vice president of technical marketing, says that the company's product improves overall performance and cuts costs, too. He points out that MeshLAN modems don't just find an access point to use; they find the one currently subject to the least traffic.

Other companies are also working on mesh network technology. For instance, another start-up company, SkyPilot Network, is making equipment and providing mesh networking service for the residential and small business market. Nokia, an industry giant, is also scouting the mesh networking market.

Smart antenna technology: California Amplifier's adaptive digital beamforming technology, which it calls "RASTER," has great potential for driving broader implementation of Wi-Fi products in enterprise, HotSpot, and campus environments. RASTER uses multiple antennae and powerful digital signal processing to dynamically form an antenna pattern optimized to a desired client (e.g. computing device) at any given instant, based on interference and other wireless channel conditions. RASTER technology is integrated with 802.11's Physical Layer and Media Access Control (MAC) processing, allowing an access point to form, transmit and receive beams; cancel interference; and better utilize high multipath, non-line-of-sight channels, thus serving to significantly enhance a network's coverage and data throughput, particularly in areas where poor signal quality is an issue.

***Note:** California Amplifier is better known for its broad line of integrated microwave solutions used primarily in conjunction with satellite television and terrestrial broadband wireless applications, than for smart antenna technology that can enhance Wi-Fi's performance.*

In a typical Wi-Fi system, an access point acts as a hub—coordinating communications between many client stations. RASTER resides at the access point; thus, unlike other approaches to increase throughput and coverage, products based on RASTER technology will not require any changes to the 802.11 standards, and is expected to work with commercially available NIC cards.

However, development of RASTER-enhanced products that are cost-competitive with standard access points requires integration into a next generation 802.11 chip. To that end, California Amplifier is currently discussing licensing agreement terms with 802.11 integrated circuit developers. Keep your eyes peeled for announcements concerning RASTER chipsets.

Multi-mode Antennae

Wi-Fi and cellular seem to have the makings of a perfect partnership. However, this partnership requires multi-mode antenna technology. In other words, an antenna that can receive signals from both a cellular network and a Wi-Fi network. At least one company has come up with a solution. In a breakthrough that indicates the shape of things to come, SkyCross Inc., a trailblazer of next-generation antenna technology, has announced that it has developed a single embedded antenna that supports cellular, GPS, Wi-Fi and Bluetooth applications. The company claims that its antenna performs as well or better than antennae that support only single functions. Not only does the SkyCross antenna allow multi-mode mobile devices, but its small footprint encourages device manufacturers to create smaller, more compact equipment.

Many Wi-Fi networks operate in dual-mode, 802.11b and 802.11a (and 802.11g is expected to be included in the near future). Antennae that can send signals over both sets of frequencies (2.4 GHz and 5 GHz) are needed. Orinoco, an established vendor offering various Wi-Fi gear, has an antenna with the necessary flexibility. Its AP-2000 5 GHz Kit consists of a radio card and a special multi-mode antenna that mounts directly on the access point and supports all three flavors of 802.11—a, b and g.

Boosting Wi-Fi's Security Quotient via Antenna Technology

An "optical antenna" is like a standard electronic antenna in that both collect electromagnetic energy (light or radio waves) from a volume of space and channel it through to a receiving element, or, conversely, transmit energy originating from a small source over a large area. Thanks to a carefully designed, geometrically-shaped lens component, optical antennae can be compact and yet can be used to direct radio frequency (RF) energy along beam-like paths, thus enhancing a Wi-Fi network's security quotient. This is the type of optical antenna developed by researchers in the U.K. at the University of Warwick's engineering department. It consists of a combination of precise curvatures on the lens and a multi-layered filter. Although optical antennae are already on the market, the University of Warwick design brings such precision to the technology that it can detect a signal on a single electromagnetic wavelength.

***Note:** This research effort has also led to the development of a new compact RF antenna for 2G GSM and 3G UMTS cellular telephones.*

Ultimately, such advanced optical antennae can provide particular strategic advantages in areas, such as Wi-Fi networks, where large amounts of information need to be sent quickly and securely. Because of the way Wi-fi networks are set up and the way that ordinary half-wave and quarter-wave dipole antennae radiate electromagnetic waves in all directions, it is possible for just about anyone to tap into a network without the knowledge or permission of the network administrators. (There are even terms to describe such activity: "warwalking" and "wardriving") However, in the case of an optical antenna that transmits and receives infrared signals (i.e. the invisible portion of the electromagnetic spectrum that lies between visible light and radio waves), the beams can be more tightly controlled.

Unlike typical radio frequencies, which can pass right through walls, infrared energy makes a network more secure because it is contained within a room and doesn't easily leak out through the walls and windows (it is even possible to coat windows so they fully reflect infrared energy). It's also possible to create a precise beam between one point and another, which doesn't diverge (spread out) much in comparison to a conventional radio frequency beam. Concentrating the signal energy in a beam this way makes it possible to transfer Wi-Fi network data over distances of up to three miles. Moreover, since optical technology provides greater bandwidth over greater distance than lower frequency radio technology, more information can be transferred during a measured time interval.

The University of Warwick's optical antenna has been licensed to Optical Antenna Solutions, a company based in England. One of the first ideas under development is for credit card payment systems, e.g., equip credit cards with infrared links for use at gas pumps and supermarkets.

APPLICATIONS

We've seen how the manufacturing sector is doing its part for the Wi-Fi cause. Now all that's needed is a compelling application that will make Wi-Fi the next "must have" tech product.

Wi-Fi must move beyond its current "one trick pony" slot (mobility/speed) if it is to continue its growth pattern. Currently, everyone views Wi-Fi as just a high-speed data transfer network that provides the end-user with mobility. But even the cell phone industry realizes that mobility alone isn't enough—more is needed to keep the buying public interested.

Industry experts say Wi-Fi's predicament could soon parallel some of the problems facing the traditional telcos after the break-up of the Bell monopoly in the 1980s. For years, the only telephone application was voice calls; this left the telcos with only one source of income. Competition (made possible by the Bell divestiture) forced the incumbent telcos (i.e. the Baby Bells) to look for new, compelling services that their customers would immediately latch onto. And that, dear readers, is how we got such value-added, "enhanced" services such as call waiting, call forwarding, "one number follow me," three-way calling, etc.

At the moment, Wi-Fi's only application is high-speed data transfer. It needs a true "killer app"—a specific application that practically compels the general populace to embrace Wi-Fi.

In their book, *Unleashing the Killer App* (Harvard Business School Press, 1998), Larry Downes and Chunka Mui define a killer app as "a new application so powerful that it transforms industries, redefines markets, and annihilates the competition. The compass, the steam engine, the cotton gin, and the Model T were all killer apps."

A killer app has important first-order effects (it does something better than anything that has come before it); but it also engenders even more important (and unpredictable) second-order effects—it can change the very workings of our daily lives. For example, do you remember life before email? Do you recall the huge, yawning gap in communications functionality that existed between the direct, confrontation phone call and the incredibly slow, impersonal letter sent by parcel post?

The term "killer app" has become so commonplace that whenever a new personal computer technology emerges, someone almost immediately jumps up and asks, "What's the killer app?" Some of the better-known data-related killer apps to appear in the last two decades or so have been: the web browser for the Internet, WordStar, WordPerfect and Lotus 1-2-3 for the new PC market, and Visi-calc for early Apple Computers.

The lucky among us with keen observational powers have perhaps some small chance of seeing a killer app coming and profiting from it, says Downes and Mui, because "A killer app is a new good or service that establishes an entirely new category and, by being first, dominates it, returning several hundred percent on the initial investment."

Problems arise, however, when a killer app is hypothesized before its underlying technology is sufficiently advanced (and inexpensive) for it to spawn. Back in the 1990s, when the telecommunications guru Harry Newton was publishing the magazine *Computer Telephony* (now called *Communications Convergence*), killer apps in the "computer telephony" arena were in abundance. Each month, Newton and his staff would treat the magazine's readers to a new, wondrous killer app idea made possible by the then nascent field of computer telephony, also known as computer telephone integration (CTI), which is basically the application of computer intelligence to telephony. But many times it took a bit of time before the killer app "exploded," particularly those for call centers (e.g. predictive dialers, automatic call distribution, "screen pops" of caller information instantly extracted from a database and presented to a call center agent).

Other computer telephony-based killer apps, such as unified messaging, while immediately identified as such, took several years for the technology to be perfected and the general public has only now begun to adopt it in earnest. Unified messaging is the general convergence in the world of communications. This advanced message management solution is applicable to all media types—PSTN, cellular, wireless—since it provides access to any message, anytime, anywhere, from any device.

Each killer app helps its related technology to skyrocket in popularity. But the killer app might not be a single application. The idea among the Wi-Fi industry is that to obtain more market penetration, all that may be needed is to tie together and productize what's already in place (e.g. gaming / entertainment / mobility and high-speed). AOL did this when it threw the doors to the Internet wide open for the general public—it neatly bundled the Internet, "content," and "ease of use" into a product you could acquire for a monthly fee.

Brainstorming about the next killer app is so compelling because of the many interesting ways the concepts of "location," "24x7 access," and "wireless technology" can be combined and molded into interesting new services. After all, with the increasing adoption of broadband services—including Wi-Fi—the proverbial information superhighway is now open for business, ready to provide device-to-device communications, with little or no human intervention required.

Visual Media

Could digital photography, video and related gear be Wi-Fi's ticket to fame and fortune? There is a lot of action in this market space.

Cameras: Sanyo Japan demonstrated a prototype of a Wi-Fi-enabled camera at the Networld + Interop 2002 Tokyo show. The camera, which is based on the 1.5 megapixel DSC-SX560, is designed for use with an IEEE 802.11b wireless CompactFlash card for connection to both public Wi-Fi networks as well as private indoor wireless LANs. The camera can be configured to upload the image immediately to a server on a Wi-Fi network.

Projectors: It's becoming more and more common to find networked projectors that can communicate via Wi-Fi. Some feature onboard computers for PC-less presentations and can transfer and store files. Still other projector makers see networking as an option that can be added to units through devices that can be connected to new and existing models. For instance:

- Sony Electronics Inc.'s VPL-FE110 networked projector features a built-in Windows CE operating system, wired or wireless networking capabilities and 64 MB of onboard file storage. The unit can directly connect to a LAN via a CAT-5 interface or wirelessly via 802.11b by inserting a Wi-Fi card into its PCMCIA slot (now also called a "PC slot").
- Sharp Electronics Corp. created a user-friendly wireless unit in its NotevisionM25X (PG-M25X), which sends wireless images to a projector via an 802.11b network.
- NEC Solutions America Inc.'s LT series (models LT220, 240 and 260) wireless projectors feature ImageXpress technology for connection to a wireless system via 802.11b or a wired LAN using 10Base-T cabling. The projectors offer remote diagnostic and control capabilities, and allow users to both print what is onscreen and transmit data to the units.
- Sanyo Fisher Company introduced its new Wi-Fi-enabled products at the 2002 Infocomm in Las Vegas. One of Sanyo's newest proprietary technologies, the optional Wireless Imager, supports three new ultraportable projector models that are designed for use in the conference rooms or on the road, each unit weighs less than 10 pounds and is housed in a sleek, durable magnesium alloy cabinet. The imager enables the

three new models—PLC-XU37, PLC-XU32 and PLC-SU32—to use IEEE 802.11b technology to transfer data from a computer to the projector without VGA or control cables for the PC.

Is Location Really Everything?

Perhaps what will send Wi-Fi adoption rates through the stratosphere will be "location aware" technology—providing conveniences like directions to the nearest restaurant or movie theater. Think about the convenience that location aware technology can provide. For instance, what if your automobile had an always-on digital service powered by Wi-Fi technology? It could constantly monitor diagnostics for the car, letting you know when the air in a tire is low, a tailpipe clogged, or it could provide continuous location information, so you would never become lost in your travels. You could even keep track of your teenager, that is, if he or she were driving a Wi-Fi-enabled car with location aware technology onboard. Driving around and hungry? You could check out information on the local area restaurants, even the menus and prices, and then make a reservation when you see something interesting. Of course, all of this marvelous technology requires interactive web access.

***Note:** The downside to location aware technology is that it could be used for obnoxious "spam." The Wi-Fi aficionados may find themselves assaulted with unwanted ads as they surf the Web while snuggled in their bed waiting for the Sand Man to visit peaceful slumber upon them. But never fear, I guarantee you that somewhere, some innovative person is working on a counter to this type of intrusive data feed.*

Another tack that location-aware technology could take is the one propounded by the positioning technology firm, Ekahau. Its Ekahau Positioning Engine (EPE) 2.0 can locate Wi-Fi enabled devices—including Voice-over-IP (VoIP) telephones—to within about a meter (3.28 feet), at least according to Ekahau. This technology, which was developed by a team at the University of Helsinki, has been on the market since 2001.

The Ekahau product can be used in places such as a supermarket where shoppers could set network-connected carts to notify them of aisles with special offers. It also could be used to indicate the nearest helpful shop assistant. Additionally, the software could help a company improve productivity. For example, in a warehouse, staff could use the software to locate the nearest employee possessing the ability to carry out a specific task.

Given the low cost of rolling out a private Wi-Fi network with three access points (for triangulation purposes), Ekahau reckons there is a market for networks used primarily for location-based purposes as opposed to carrying other data.

Voice over Wi-Fi

Is the marriage of voice with Wi-Fi the magic bullet for which the public is searching? While transmitting voice over a wireless LAN isn't revolutionary (companies such as Symbol Technology and SpectraLink have been doing it for a few years), what is new is delivering voice outside the restricted area of a corporate WLAN.

TeleSym, a voice-over-IP (VoIP) technology company, is offering voice-over-Wi-Fi technology that has resolved the kind of quality issues that dog most VoIP applications. TeleSym has minimized the impact of latency by using Edge QoS real-time latency management in the client software. Other features include Caller ID and telephone dialing using a Microsoft Outlook contact database. The TeleSym voice-over-Wi-Fi offering can be used in two different deployments:

1. In a closed Wi-Fi-enabled workgroup environment such as a hospital or warehouse, workers can use their handhelds or notebooks for regular one-on-one phone conversations. A push-to-talk capability allows users to set up conference calls to all members of a workgroup by pressing a button that is located on the side of all Pocket PC 2002 devices.
2. With the addition of the SymPhone Connector on the server, users can also connect over the Internet back to their corporate PBX system, and from there they can place regular outbound calls. This type of set-up means that employees no longer have to carry multiple pieces of hardware, such as a cell phone for voice and a PDA to look up information.

Symbol Technology, Inc. offers a handheld that can transmit both voice and data over 802.11b networks and H.323 telephony infrastructures. The NetVision DataPhone is typically marketed to vertical industries such as retail, manufacturing and healthcare. According to the company, a classic application for the wireless phone is retail, where it can be used to scan products, access inventory records and enable personnel to wander throughout the store. The phone, of course, can also access the Web to provide access to data and support telephony.

BroadSoft is partnering with SJlabs, a provider of soft clients (software telephones) for mobile devices and industrial terminals, to develop a VoIP application. SJlabs' new Sjphone is the first Session Initiation Protocol (SIP)-based soft client for PDAs (i.e. software that extends the functionality of a PDA so that it can also act as a telephone). The combination of technologies from these two innovative developers should enable Wi-Fi users to get, for the first time, a complete suite of business telephony services over any PDA, as long as it's based on Microsoft's CE operating system running within an 802.11 network.

"We believe the addition of Wi-Fi voice to PDAs is another example of how using BroadWorks can help service providers to further differentiate themselves," says Michael Tessler, president and CEO of BroadSoft. He adds that the combination of 802.11's broadband speed capabilities and BroadWorks' wide array of business services will enable PDAs to provide much more than traditional phone services.

Global IP Sound's VON 2002 is a telephony program that runs on a standard PDA, such as HP's iPaq, via an 802.11b connection to the Internet. The application provides terrific sound since the software resolves 8 kHz of audio versus 3 kHz for conventional telephony applications and provides what the company calls "Edge QoS"—a combination of techniques that serve to reduce loss of quality due to packet jitter, data errors and packet loss. During the product demo the software ran on several different types of computing devices (e.g. laptops and desktops). Since the version that does 8 kHz audio at 80 Kbps has relatively low complexity, it could even run in telephone-like appliances with inexpensive processors.

Industry analysts predict that by 2007 close to 60 percent of the total PDA and other handheld devices sold will have dual-mode cellular / Wi-Fi capabilities. Consequently, it's not much of a leap to forecast that voice-over-Wi-Fi applications will become a popular Wi-Fi application.

If VoIP is the killer app that "closes the deal" for massive Wi-Fi acceptance, then within five years the solution to sub-par cellular service will be for subscribers *not* to use it. By that time, there will surely be enough Wi-Fi access points deployed so that VoIP phones could reign supreme.

Security and QoS

In spite of Wi-Fi's many stellar qualities, security and quality of service (QoS) represent a bit of a black hole for this technology. Yet, security is not as bad as all of the pundits would have you believe. Keep in mind that:

* Wired and wireless LANs share some of the same security issues, e.g. vulnerability to hackers and eavesdroppers.
* In order for a WLAN to be compromised, the modulation techniques, *and* radio domains, *and* channels and subchannels, *and* security ID, *and* passwords must be known.
* All certified Wi-Fi products include Wireless Equivalent Privacy (WEP) that can encrypt all data that passes between an access point and a wireless card. While, not perfect, if WEP is properly enabled, it usually will block casual snoops.
* IT managers can lock out users down to the wireless station level, thus users can be included or locked out at any time.

Admittedly more needs to be done. That's why there is a hotbed of activity going on throughout the industry as innovative engineers try to up the security quotient of the average Wi-Fi network.

To give the reader an idea of what is happening in the security and QoS area:

* Symbol Technologies, Inc., a global leader in wireless mobile computing, has come out with its Symbol Mobius Wireless System, which is designed to enhance QoS and deliver application-specific security to Wi-Fi networks, among other things.
* Cisco Systems, the leading global provider of networking gear, offers its Wi-Fi compliant Aironet technology, which is specifically designed to provide advanced security features to protect a Wi-Fi network against warwalkers and malicious hackers.
* The IEEE 802.11 Working Group has Task Groups working diligently on specifications, to address both of these issues: 802.11e for QoS and 802.11i for Security.

***Note:** Most of the QoS improvements are being implemented through software or firmware enhancements of the MAC so the equipment you buy today should be easy to update later with the latest and greatest in QoS and security.*

Since these two issues are so critical to Wi-Fi's continued growth, and since there is so much that can be said about both, Chapters 16 and 17 are devoted to the ins and outs of QoS and security, respectively.

DEVICES

Wi-Fi capability has been provided, for the most part, via add-ons to existing computing devices, e.g. PC card (for a laptop), PCI card (for a PC), smart card (for a PDA or cell phone). This current situation is quickly changing, however. Devices made specifically for Wi-Fi are becoming easier and easier to find. In fact, it won't be long before Wi-Fi technology shows up in just about anything that uses information and has a battery or fuel cell: cell phones, GameBoys, MP3 players, Walkmans, watches, automobiles, and on and on. Such universal availability should aid Wi-Fi in its march toward universal acceptance.

Wi-Fi has even gone the wearable route! Vocera Communications is merging telephone networks with Wi-Fi's ability to deliver data over short ranges. The company has teamed with chipmaker Intersil to create a two ounce rectangular "badge" that can be worn on the user's clothing, preferably close to the face (e.g. collar or sleeve). Since the badge enables instant two-way voice conversation without the need to remember a phone number or manipulate a handset, it can replace the bulky handsets used by most within the target market. This new breed of phone network uses Wi-Fi to provide hands-free, voice-activated communications throughout any 802.11b networked building or campus and, in doing so, provides instant two-way voice communications within that environment.

The Vocera Communications system is made up of two elements: the Vocera Server Software and the Communications Badge.

The Vocera Server Software runs on a standard Windows 2000 server and houses the centralized system intelligence—the Call Manager, User Manager, and Connection Manager programs as well as speech recognition software and various kinds of database software.

The Vocera Communications Badge is controlled using natural spoken language. To initiate a conversation with two people, say, Steve and Madelyn, the user would simply say, "Get me Steve and Madelyn." On those occasions when a live conversation is not necessary, text messages and alerts can be sent to the LCD screen on the Communications Badge.

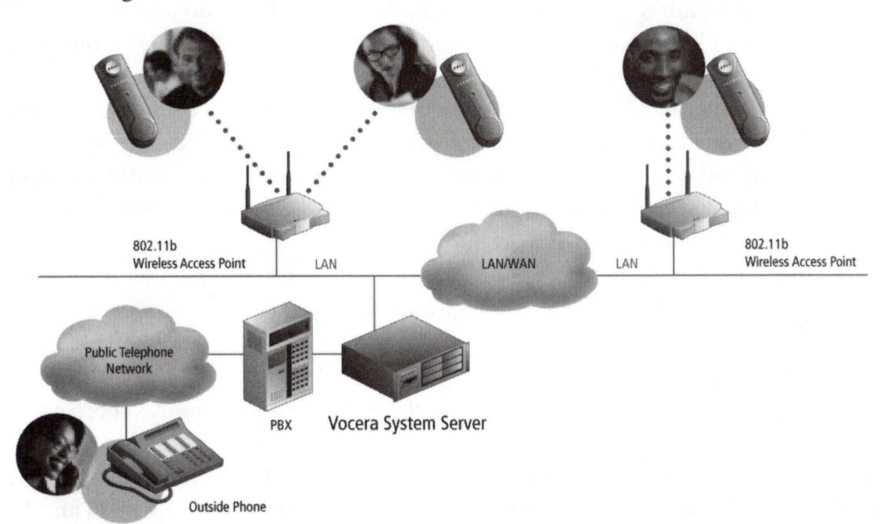

Figure 3.4 Vocera Communications Network Diagram. *Graphic courtesy of Vocera Communications.*

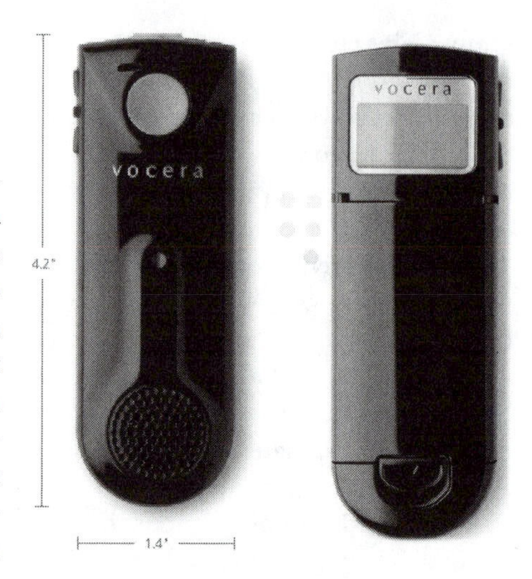

Figure 3.5 The badge on the right shows the speaker and the badge on the left displays the LCD screen.

The famous American chain store, Target, is outfitting its retail outlets with a Vocera system and Disney World has the technology under consideration as a way to allow parents keep track of their children when visiting its theme parks. Also, don't be surprised if you find a Vocera system in use in healthcare facilities, where the transmissions of traditional cell phones might interfere with the operation of certain medical equipment. In fact, the Vocera system has been successfully trialed in a San Francisco healthcare facility.

Computing Devices

Apple, operating well ahead of the curve on Wi-Fi, has been integrating wireless connectivity into its computers for years. In fact, Apple should be given credit for helping to establish 802.11 as the *de facto* standard for wireless networking.

Today, however, there really is no clear market leader. Most major notebook/laptop computer makers have followed Apple's lead, shipping products with embedded 802.11 NICs (Network Interface Cards). This means that it's incredibly inexpensive for the end-user to enjoy Wi-Fi services. For example, Fujitsu's Lifebook and its 3COM wireless NIC (also known as a wireless adapter) are making a strong showing in an ever more crowded market of Wi-Fi-enabled notebook computers. And, by the time you read this book, Dell will have reached its goal of Wi-Fi-enabling all of its laptops. IBM, HP-Compaq and Sony are also jostling for a leadership position.

There are essentially two methods by which notebook/laptop computers are Wi-Fi-enabled: embedded and attached.

Embedded: Two prime examples of embedded Wi-Fi solutions are IBM's ThinkPad A31 and Toshiba's Satellite Pro 4600. But soon all computing devices will have embedded Wi-Fi capabilities, at least if Intel has its way.

Attached: In the past, the most popular method to give a portable computing device Wi-Fi ability is to add an adapter card. Inexpensive PC Card-based adapters from Dell, Compaq and others account for most of the legacy base of notebooks/laptops with 802.11 functionality.

Notebooks and laptop computers aren't the only computing devices sporting Wi-Fi capabilities:

- At the industry's 2002 Comdex show, HP introduced its iPaq (a small PDA-like device). This little unit offers a color screen along with its built-in Wi-Fi and Bluetooth connectivity. It even features a universal remote for home and office electronics!
- Toshiba's e740 PDA supports Wi-Fi right out of the box.

* Microsoft's new Tablet PC comes with built-in Wi-Fi.
* HP also offers a Wi-Fi-enabled printer.
* A wireless netcam is available from D-Link.

That's just the tip of the iceberg. The increase in Wi-Fi-enabled devices (computing and otherwise), along with manufacturers providing smaller, faster and cheaper devices, will drive demand for Wi-Fi access, in a variety of arenas, worldwide.

CONCLUSION

No one doubts that Wi-Fi will experience dramatic growth. However, as the reader should now understand, the devil is in the details. To usher this technology and its pervasive applications into homes and businesses requires the resources of many companies, working in concert. After all, an entire ecosystem must be created.

Will the industry work as a team and make it to the playoffs? It's possible, even probable. Still, the way the author sees it—at this writing—the compelling reason why most people *must* have a Wi-Fi network, at work or in their home is still lacking. So, although the industry as a whole has taken a giant leap forward in its effort to fuel the Wi-Fi flame, additional "fanning" is still needed.

Perhaps we're looking in the wrong direction. Maybe it's the cellular industry that will fan consumers' desire for all things Wi-Fi. With the advent of a cellular / Wi-Fi partnership, Wi-Fi will find itself catapulted into the "must have" category. (Keep this in mind when you read Chapter 13, which is devoted to this very subject.)

Finally, for Wi-Fi to continue its upward-spiraling growth pattern, the regulatory agencies must do their part. They must continue to govern Wi-Fi with a light touch—if they feel the need to impose a heavy regulatory hand, they can easily stop innovation.

Chapter 4: The Regulators' Role

The successful, long-term development of the Wi-Fi industry is contingent upon the availability of spectrum for its expansion. But for decades spectrum has been considered a scarce resource—a resource that governments must carefully control and manage. To that end, every country has established some sort of regulatory body to oversee its strict regulation.

Some regulatory bodies will aid Wi-Fi in its rise to prominence in an effort to establish their respective countries as infotech hubs, others will not. Most regulatory agencies, however, will take a balanced stance, acting to prevent spectrum overcrowding while at the same time promoting Wi-Fi interoperability and QoS standards.

Over the last decade or so, high-speed wireless local area networks (WLANs) were made possible because various regulatory agencies decided to set aside a small swath of unlicensed radio frequencies and to allow anyone who followed a simple set of rules to use that spectrum with no administrative process or fee for access. Much of Wi-Fi's popularity is due to the fact that it can operate in this unlicensed spectrum and this popularity is creating a new generation of people who use Wi-Fi and exploit its unique technology through creative applications.

There is pressure being put to bear on regulatory agencies, however, to inhibit this mainly grassroots innovation. If history can be our guide, government officials are at this moment being lobbied to impose changes in the regulations—changes that could be detrimental to Wi-Fi's growth. Large telcos, cellular providers, broadcasters and the military constantly challenge regulatory agencies' spectrum allocations.

The worry is that as the result of those challenges, the regulators will step in and end all of the creativity Wi-Fi has wrought. Thus, it's these governmental regulatory agencies that are the key to the continuance (or the dampening) of the Wi-Fi boom.

Before we look at how the various regulatory agencies are dealing with the Wi-Fi phenomenon, we first need to discuss what's at stake—spectrum. Do you understand what we mean by "spectrum"? If not, here is a very brief tutorial.

✪ WHAT IS SPECTRUM?

Spectrum is a conceptual tool used to organize and map a set of physical properties to delineate electromagnetic waves, which are produced by electric and magnetic fields,

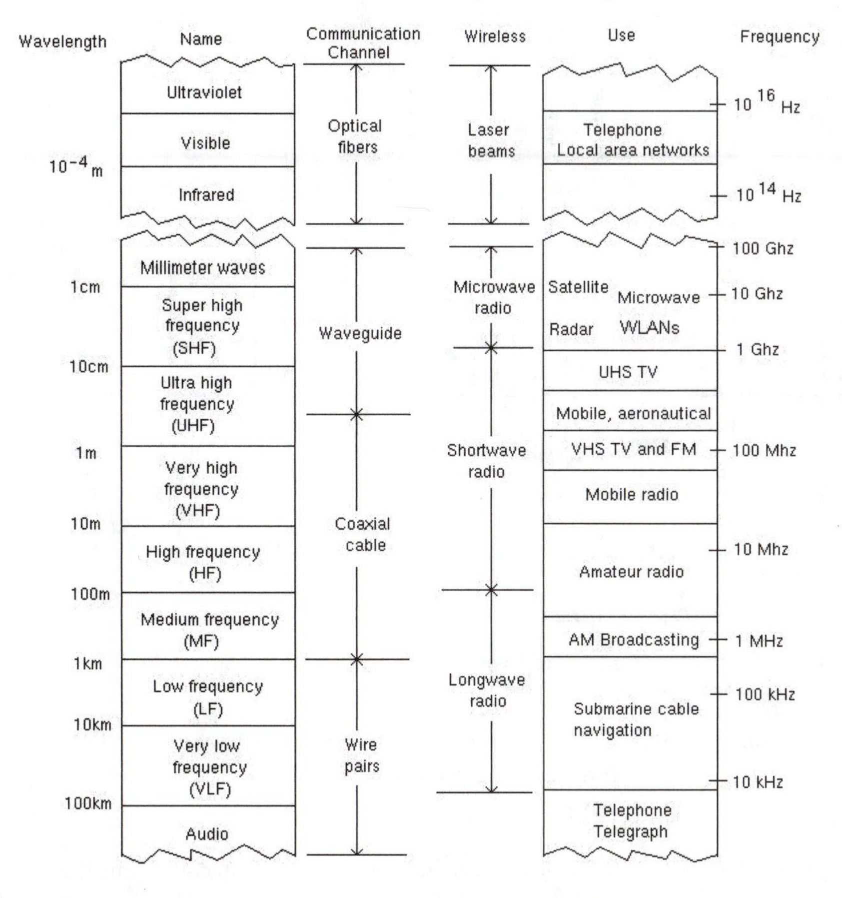

Figure 4.1 The electromagnetic spectrum. While this book deals with wireless data transmission, and this section discusses spectrum within that connotation, the reader should understand that "spectrum" is a general term that is used to encompass both the spatial and temporal properties of any medium, including your telco's copper wiring, fiber optic cable, coaxial cable, and ambient air.

and which move through space at different frequencies. These frequencies are measured in Hertz (Hz), which is equivalent to the number of waves or cycles per second. (Megahertz (MHz) refers to 1000 kilohertz and gigahertz GHz refers to one million kilohertz.)

The set of all possible frequencies is called the electromagnetic spectrum, which spans a wide range of physical energies such as radio, light and x-rays.

The subset of frequencies between 3 kilohertz (kHz) and 30 gigahertz (GHz) is known as the "radio spectrum." This spectrum is used by radio waves for communication, or to accomplish work such as a microwave oven heating food or radar detecting a storm. "Bands" refer to the ranges of frequencies within the radio spectrum.

Radio services (i.e. categories of radio use) utilize the "radio frequencies" within the electromagnetic spectrum. Examples of such radio services include the Cellular Radiotelephone Service, the Television Broadcast Service, and the Aviation Radio Service. But there are dozens of radio services, e.g. "governmental" (such as defense and space explo-

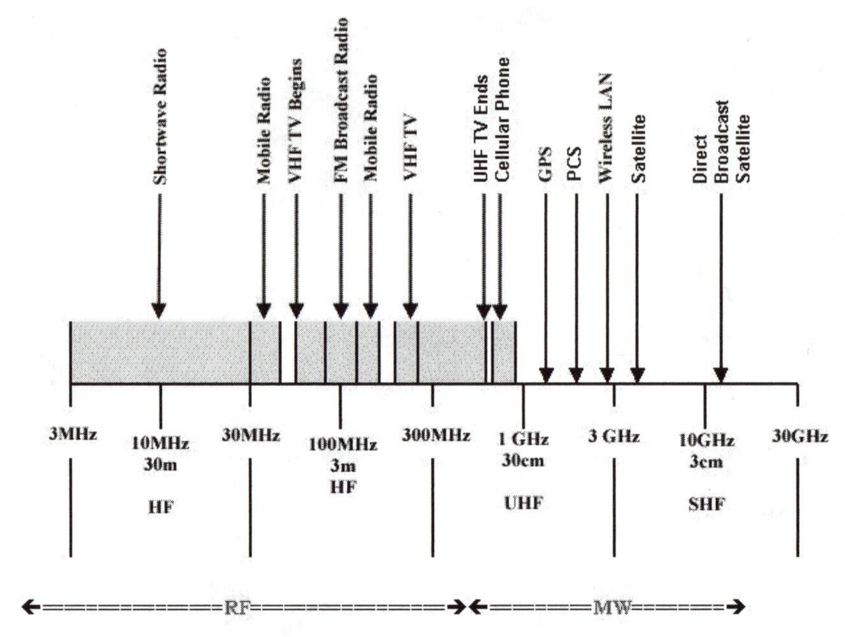

Figure 4.2 The "Radio Frequencies" within the Electromagnetic Spectrum.

ration), "non-governmental" (police radio or home satellite TV), and even flexible radio services that are generic and without a prescribed purpose.

A HISTORY LESSON

From the earliest use of radio for ship-to-shore and ship-to-ship communications it was apparent that in time there would be a need to coordinate the use of the airwaves that these transmissions traversed. Without such coordination, radio communications would become chaotic. Mutual interference would make radio reception so unreliable as to be virtually useless.

The number of services that occupy the radio spectrum has grown considerably since the first wireless Morse Code transmissions of the late 19th century. Most readers, however, probably take spectrum for granted and perhaps don't fully understand just how important spectrum is to society. With a little thought, however, it's easy to understand that it is what makes television, communications for emergency services, marine transportation, and the space program possible.

But this spectrum is strictly regulated. To understand the full importance of the influence that governmental regulatory bodies have over services that traverse the airwaves, it's good to have a little knowledge of what happened in the past with regards to spectrum regulation.

The notion that electrical communications devices need regulation has its origins in the mid-1800s long before commercial wireless use of spectrum began (indeed long before electromagnetic waves were recognized as such). But the most influential regulations, inasmuch as Wi-Fi is concerned, came about when it was realized that radio equipment was easily thrown off when signals of the same frequency from more than one source overlapped. The regulators referred to this phenomenon as "interference." But, in this instance, they got the term wrong. The waves sent out by different transmitters don't actually interfere with each other; instead, they pass right through each other unchanged. The interference occurs at the receiver when its antenna picks up multiple signals of the same frequency and has trouble telling them apart. With suitable, unique encoding schemes, however, more than one signal can share the same set of frequencies. In other words, interference is a function of the intelligence (or lack thereof) designed into the transmitter and receiver, not purely a result of what happens in the airwaves.

***Note:** Today's receivers are a lot more intelligent than they were 90 years ago. Technical advances enable modern radio signals to be coded digitally so that they can be easily separated from each other. Thus, it's argued that regulators no longer need to continue their outmoded policies of chopping the airwaves into distinct regions of frequency and geography.*

For the purposes of this section (and book), the pivotal historical date is 1865, since that was when representatives from 20 European States decided to meet in Paris, France and work out a framework agreement for the international operation of the telegraph. Now that telegraph lines were beginning to span oceans and connect different countries, international tariffs and standard operating instructions had to be agreed upon by all the nations involved.

The members of this first International Telegraph Convention took it upon themselves to draw up a set of common rules that standardized equipment, which would guarantee generalized interconnection. On the 17th of May 1865 the delegates signed an agreement introducing new charging zones, new telegram categories, etc. The convention also adopted uniform operating instructions, which had hitherto been different from one country to another, and it laid down common international tariff and accounting rules. This gathering also established the International Telegraph Union to facilitate subsequent amendments to the initial framework agreement and they adopted the first spectrum regulations, commonly referred to as the "Telegraph Regulations."

The next date of importance within the context of this book is 1896, the year wireless telegraphy, the first type of radiocommunication, was invented. Wireless telegraphy quickly became invaluable to the maritime industry and it wasn't long before an international set of regulations governing wireless telegraphy was implemented. The first "International Radiotelegraph Conference" (attended by 29 member countries) was convened in Berlin in 1906. These rules, which have since been amended and revised by numerous radioconferences held throughout the years, are now known as the "Radio Regulations." They specify the basic regulatory principles still in common use today: the grouping of services into dedicated frequency bands; national allocations reported to a central registration agency, the Berne Bureau; and the partitioning of spectrum frequencies so that those below 188 kHz are to be used for long distance communications and frequencies between 188 kHz and 500 kHz are reserved for military use.

It wasn't long before governments worldwide began declaring that spectrum was a public resource over which they had the authority to license. For instance, in the U.S., the Radio Act of 1912 was adopted to address the issue of radio wave interference by limiting the wavelengths over which stations could transmit. That same year, the International Radiotelegraph Conference (IRC) published the first attempt at an International Table of Frequency Allocations. Member states' adherence to the Table was nonmandatory.

The next important step in international spectrum regulation and management occurred in 1920 at the Preliminary World Conference on Electrical Communications held in Washington, D.C. Several important proposals were introduced at that meeting. They include the merger of the Telegraph and Radiotelegraph conventions into a single Universal Electrical Communications Union (which was a blueprint for what would become today's International Telecommunications Union) and a spectrum allocation process that would require international approval before use.

When the International Radiotelegraph Union held its 1927 meeting in Washington, D.C., 80 countries participated. At that meeting a number of important issues were decided. The organization's responsibility was broadened beyond maritime transmitters to all radio transmitters. A scheme was established to allocate frequency bands to various radio services that were in existence at the time (fixed, maritime and aeronautical mobile, broadcasting, amateur, and experimental) so as to ensure greater efficiency of operation in view of the increase in the number of radiocommunication services and the technical peculiarities of each service. Also, the first comprehensive International Table of Allocations was introduced and the International Radio Technical Consulting Committee (CCIR) was established to study technical issues.

Then in 1932, at a joint Telegraph and Radiotelegraph Conference in Madrid, the two conventions were merged into a single organization entitled the International Telecommunications Union or ITU. The new name, which didn't become effective until January 1, 1934, was chosen to properly reflect the full scope of the ITU's responsibilities, which by that time covered all forms of communication, wire and wireless, radio, optical systems and other electromagnetic systems. In 1947, the ITU became a specialized agency of the United Nations.

Note:* *Today the ITU is an intergovernmental organization, within which the public and private sectors cooperate for the development of telecommunications. The ITU adopts the international regulations and treaties governing all terrestrial and space usage of the frequency spectrum, as well as the use of the geostationary-satellite orbit, by which countries adopt their national legislation. It also develops standards to facilitate the interconnection of telecommunication systems on a worldwide scale regardless of the type of technology used.

In the U.S., the Radio Act of 1927 attempted to deal with the increasingly complicated issue of spectrum regulation, but it was the Communications Act of 1934 that gave birth to the Federal Communications Commission and gave the agency authority to regulate *every aspect* of any spectrum not used by the U.S. government.

Another year that holds particular importance to the spectrum regulatory system is 1947. That was when the 1947 ITU Atlantic City conference approved the ITU's United Nations status; established the International Frequency Registration Board (IFRB) to manage the frequency spectrum (which was becoming increasingly complicated); and mandated that adherence to the International Table of Frequency Allocations (introduced in 1912) would be compulsory. That Table assigns, to each service using radio waves, specific frequency bands with a view to avoiding interference between stations—

in communications between aircraft and control towers, car telephones, ships at sea and coast stations, radio stations or spacecraft, earth-based stations, and so forth. More important, full membership status was limited to sovereign nations, with each nation allotted one vote, regardless of its size. "Sector" membership status was open to telecommunications operators, broadcasters, and organizations. (Prior to the 1947 Conference, the latter group had full membership status.) This membership reorganization substantially increased the influence of small or developing countries in a manner totally unrelated to industrial wealth or telecommunications usage.

***Note:** The International Table of Frequency Allocations is reproduced in columns 1-3 of the FCC's Table of Frequency Allocations, which can be found on the FCC's website at (www.fcc.gov/oet/spectrum/table). Also, the ITU Radio Regulations, Articles 1, 8—Frequency Allocations, have been faithfully reproduced at www.kloth.net/informations/freq-itu.htm.*

In 1956, the International Telephone Consultative Committee (founded in 1924) and the International Telegraph Consultative Committee (founded in 1925) were merged to form a very influential European agency called the Comité Consultatif International Télégraphique et Téléphonique (CCITT), which translates to the "International Telephone & Telegraph Consultative Committee." The CCITT was charged with the responsibility of coordinating the technical studies, tests and measurements being carried out in the various fields of telecommunications, as well as drawing up international standards.

In 1959, the initial Radio Regulations were entirely re-written, for the first time, by the Geneva Administrative Radio Conference. This same year also saw the establishment of the European Conference of Postal and Telecommunications Administrations, which is commonly known as "CEPT." The original members were the incumbent monopoly-holding postal and telecommunications administrations, but today CEPT's activities include cooperation on commercial, operational, regulatory and technical standardization issues.

The Asia Pacific Telecommunity (APT) was created in 1979 in Bangkok. The APT has four major activities: technical assistance to the developing countries of the region, a regional information and communications technology (ICT) forum, a telecommunications standards forum, and a radio regulatory forum, which is known as the APG.

In the early 1990s, many of the regulatory agencies around the world began to auction spectrum, which generated hundreds of billions of dollars in revenue for their governments. With "good" spectrum seeming increasingly scarce, communications providers were forced to pledge huge sums of money to acquire available spectrum, with little hope that they would ever recoup their costs.

With the beginning of the new millennium, CEPT felt some reorganization was needed within its ranks if it was to meet the upcoming challenges of convergence in the radio and telecommunications market. Two of its offices, ERO (European Radiocommunications Office) and ETO (European Telecommunications Office), were brought under the umbrella of a new entity, the Electronic Communications Committee (ECC).

On January 31, 2002, the ERO launched a new European Frequency Information System (EFIS), which is available to the public on the Internet either via the ERO website (www.ero.dk) or the EFIS website (www.efis.dk). With this tool, the ERO hopes to provide a valuable service to all parties with an interest in spectrum utilization. The EFIS will also contribute to the CEPT policy objectives of harmonization and transparency as well the European Union's policy objectives laid down in the Decision of the Council and European Parliament on Radio Spectrum Policy. (You can find a copy of this document at www.etsi.org/public-interest/Documents/Legislation/2002_676.pdf.)

With EFIS you can search for and compare spectrum utilization across Europe as well as related information such as CEPT activities, radio interface specifications, and other national or international regulations. Another of ERO's objectives is to develop proposals for a "European Table of Frequency Allocations and Utilizations" for the frequencies 29.7 MHz to 105 GHz. This table is slated for implementation by June 2008. A significant proportion of the input to develop such a table was provided by a process of Detailed Spectrum Investigations (DSI), where portions of the radio spectrum were studied in depth to identify current use and future requirements. The results of the DSI study were published in ERC Report 25, which can be found at www.ero.dk/doc98/official/pdf/rep025.pdf.

In fact, much of the information the reader might need on European spectrum allotment, allocation, assignment, and management can be found at either www.ero.dk or www.eto.dk. Both websites contain a treasure trove of up-to-date information.

SPECTRUM ALLOCATION

The world's governments have historically made spectrum scarce and expensive by splitting it up and restricting its usage. "Spectrum allocation" refers to the governmental function of apportioning bands to services such as radio, TV and WLANs. Although various governments have declared their control over all spectrum and set up laws and rules to govern the use of such spectrum, legally speaking, no one can really "own" spectrum.

International spectrum allocation is accomplished by making an entry in the aforementioned official International Table of Frequency Allocations that is maintained by the United Nations' International Telecommunication Union (ITU). As the principal international institution for achieving agreement and cooperation among nations on the use of telecommunications, the ITU frequently changes this table through conferences and negotiations.

Proposals to modify the International Table of Frequency Allocations are based largely on the requirements and desires of individual countries to operate particular radio services in particular bands. Needs differ among countries. Decisions are driven in many cases by such arguments as economic and national importance of one service vis-à-vis another. However, allocation decisions are also dictated by technical considerations, in particular, those allocations involving the sharing of frequencies between two or more services. Many times the arguments are based on complex technical factors such as acceptable interference levels and noise, power flux densities, antenna patterns, and so forth. Nonetheless, countries are sovereign with regard to the use of the radio spectrum (and regulation thereof) within national borders and have no obligation to adopt or follow the International Table of Frequency Allocations within their borders.

The mechanisms of the ITU are designed to achieve the maximum utilization of the electromagnetic spectrum by the widest range of users, and to avoid a situation where one user is accommodated at the expense of another. Adherence to ITU agreements is voluntary and cannot be enforced by higher authority. There are also no sanctions in place to compel an ITU member to abide by ITU rules, although membership in ITU entails a treaty obligation to conform to the collective decision of its members.

Domestic spectrum allocation is accomplished by making an entry into an official national Table of Allocations, which is maintained by each individual country's designated regulatory agency. Domestic allocations generally conform to, or at least do not conflict with, the International Table of Frequency Allocations. In fact, the ITU's allocations can strongly impact domestic spectrum decisions. In some instances, however, multinational manufacturing interests can virtually dictate domestic allocations.

***Note:** A license or assignment issued by a governmental regulatory body is a renewable contract between a specific government and a licensee. A license serves to authorize the licensed party (licensee) to use frequencies and bands within a specific radio service, usually in a particular geographic location, although some licenses convey nationwide privileges.*

SPECTRUM MANAGEMENT

Wi-Fi operates under a spectrum management system that is 90 years old, albeit with the help of some late 20th century contributions by the regulators. There have been, historically, four core assumptions underlying all governmental spectrum management policies, none of which hold relevance in today's society:

1. Unregulated radio interference will lead to chaos.
2. Spectrum is a scarce resource.
3. Government command and control of the scarce spectrum resource is the only way chaos can be avoided.
4. The public interest centers on government choosing the "highest and best use" for the spectrum.

These archaic assumptions have guided regulatory decision-making for far too long. Let's look at some of the erroneous decisions, as they pertain to these four assumptions. We'll use the FCC Commissioner, Michael Powell's October 12, 2002 speech entitled, "Broadband Migration III: New Directions in Wireless Policy," given at the Silicon Flatirons Telecommunications Program at the University of Colorado at Boulder, as our guide as we investigate the details of these four assumptions. In that speech, Commissioner Powell gave voice to his belief that today's environment has strained the previously outlined assumptions to the breaking point. Powell also said, "Modern technology has fundamentally changed the nature and extent of spectrum use. So the real question is, how do we fundamentally alter our spectrum policy to adapt to this reality? The good news is that while the proliferation of technology strains the old paradigm, it is also technology that will ultimately free spectrum from its former shackles."

Interference. There are many different ways to deal with radio interference outside of the strict spectrum allocation model. Technologies such as Dynamic Frequency Selection (DFS), Transmitter Power Control (TPC), and spread spectrum can permit far more intensive use of a frequency band since they can keep the level of interference down. Imposing limits on certain technical parameters such as output power can also minimize interference.

Powell states, "Due to the complexity of interference issues and the RF environment, interference protection solutions may be largely technology driven." He further said, "Interference is not solely 'caused' by transmitters, which many seem to assume–and on which our regulations are almost exclusively based. Instead, interference is often more a product of receivers; that is receivers are too dumb or too sensitive or too cheap to filter out unwanted signals. Yet, our decades old rules have generally ignored receivers."

He recommends, "The time has come to consider an entirely new paradigm for interference protection. A more forward looking approach requires that there be a clear quantitative application of what is acceptable interference for both license holders and the devices that can cause interference. Transmitters would be required to ensure that the interference level–or 'interference temperature'–is not exceeded. Receivers would be required to tolerate an interference level."

Scarcity. Spectrum appears scarce only because current regulations put draconian limitations on its use. As a former FCC chief economist once quipped, "The only way you can waste spectrum is not to use it. And the FCC has wasted enormous amounts of spectrum. It is a classic vicious cycle: spectrum is scarce, so FCC regulation is necessary, and FCC regulation ensures that spectrum stays scarce."

According to Powell, "Much of the Commission's spectrum policy was driven by the assumption of acute spectral scarcity–the assumption that there is never enough for those who want it. Under this view, spectrum is so scarce that government rather than market forces must determine who gets to use the spectrum and for what. The spectrum scarcity argument shaped the Supreme Court's Red Lion decision [Red Lion v. FCC, 395 U.S. 367 (1969)], which gave the Commission broad discretion to regulate broadcast media on the premise that spectrum is a unique and scarce resource."

He went on to say that "the presumptions of Red Lion and similar broadcasting regulation based on scarcity have been called into doubt by the proliferation of media sources," and that "we question the continued utility of the pervasive scarcity assumption for spectrum based services." He added that recent studies by the FCC's Enforcement Bureau have shown that "most of spectrum is not in use most of the time."

The Commissioner even acknowledges that "innovative technologies like software defined radio and adaptive transmitters can bring additional spectrum into the pool of spectrum available for use." Although he does say that "scarcity will not be replaced by abundance; there will still be places and times when services are spectrum constrained. However, scarcity need no longer be the lodestar by which we guide the spectrum ship of state."

Powell's recommendation: There is a substantial amount of "white space" that's out there, which is not being used by anybody. In particular, certain timeslots can lay dormant for as much as second at a time, and sometimes much more in the case of point-to-point communications, such as a large trucking company's mobile communication network. "One way the Commission can take advantage of this white space is by facilitating access in the time dimension."

Methods such as overlay techniques (where more than one service shares a licensed band) and interference minimizing technology can serve to move spectrum management from "command and control" to a more democratic policy, including open spectrum, i.e. establishing a spectrum "commons" or "park" to accommodate unlicensed services. Even a policy of re-allocation of occupied bands (where sharing isn't possible) could be put in place. Such relocation has occurred in the past. A perfect example is that of the FCC enabling the Personal Communications Service (PCS) to be built on spectrum formerly occupied by "fixed microwave" licensees, who used their spectrum for statewide industrial functions. PCS auction winners paid those licensees to move to alternative bands that were, fortunately, available. Is it any wonder this almost century old system is no longer adequate to serve the world we live in today?

Government Command and Control. The "highest and best use" approach has always been problematic. Creating broad service classifications and applying block allocations can serve to ease administrative duties. Block allocation also simplifies control. Spectrum can be dedicated to desirable ends, such as fostering a centralized vision of new technology (e.g., interactive television initially flopped in a massive display of the limits of central planning). Also, consider the nationwide block allocation of UHF television channels 14 through 69—even in the largest metropolitan areas, less than a handful of UHF stations actually broadcast, resulting in a vast amount of prime spectrum nationwide lying idle. The same situation can be found elsewhere in various service classifications. But block allocation can also create overcrowding: administrative miscalculation of enormous and intensive demand for spectrum in an allocated band can result not only in overcrowding, but also interference and delays. A good example is the vast range of frequencies from 500 kHz to 30 GHz, which are most desirable because of the physical characteristics of the waveforms that invite innovative modulation techniques to enable them to be used in a number of unique ways. But perhaps the reason the block system is still in place, despite all of its inequities, is because it protects incumbents and is a barrier to entry for competitors.

Commissioner Powell suggests that "government spectrum policy continues to be constrained by allocation and licensing systems from a bygone era." He also says, however, that "In the last twenty years, two alternative models to command and control have developed, and both have flexibility at their core. First, the 'exclusive use' or quasi property rights model, which provides exclusive, licensed rights to flexible use frequencies, subject only to limitations on harmful interference. These rights are freely transferable. Second, the 'commons' or 'open access' model, which allows users to share frequencies on an unlicensed basis, with usage rights that are governed by technical standards but with no right to protection from interference." Powell feels that "the Commission has employed both models with significant success" and concludes that "we will undoubtedly use both models as we move forward."

He even recommends that going forward, "license holders should be granted the maximum flexibility to use—or allow others to use—the spectrum, within technical constraints, to provide any services demanded by the public. With this flexibility, service providers can be expected to move spectrum quickly to its highest and best use." Powell however, does add the caveat, "Such flexibility should not come at the cost of clearly defined rules."

Public Interest Standard. Powell also addresses the meaning of the phrase "public interest, convenience or necessity." He states, "The public interest must reflect the realities of the marketplace and current spectrum use. Today, I would suggest that full and complete consumer choice of wireless devices and services is the very meaning of the public interest. Certainly government telling consumers what types of services and devices they should have or own is not my view of the public's interest." He also says that public interest goals include "national defense, public safety, and critical infrastructure."

TECHNOLOGY TO THE RESCUE

Most industry experts are of the opinion that new technologies make the current system of spectrum allotment, allocation, assignment and management obsolete. Early receivers and transmission schemes were such that interference between competing services within the same frequency band was a common problem. However, the development and implementation of new technologies (e.g. spread spectrum and digital radio technology) allows for the use of sophisticated receiver and transmission schemes, enabling multiple signals to be transmitted at the same frequency, interference-free.

In a world of exclusive spectrum, the development of digital radio technology could be stunted. As long as a license is required to transmit, there is no incentive for innovators to design creative ways for radio transmitters to coexist.

But look at what happened when governmental agencies set aside small spectrum bands (2.4 GHz and 5 GHz) for unlicensed uses. Though initially employed for consumer electronics and industrial applications, these bands have proven to be a fertile ground for new services. Despite the limited bandwidth and other restrictions that hamstring services using these unlicensed bands, new industries (e.g. cordless phones and WLANs) have flourished in this shared unlicensed spectrum. In fact, it's the success of Wi-Fi-enabled WLANs that has spawned the growing interest in unlicensed spectrum.

Still, the current swirl of excitement surrounding Wi-Fi is nothing compared to what might occur if significant swaths of spectrum were opened up for use as a commons. "We could have the greatest wave of innovation since the Internet if we could unlock the spectrum to explore the new possibilities," says David Reed, one of the early architects of the Internet. Other high-profile proponents of this concept include Paul Baran, the inventor of packet switching; cyber-law expert Lawrence Lessig; and futurist George Gilder.

The technological innovations that have occurred in both hardware and software (e.g. smart receivers, picocell designs, new antenna technology, and even quantum communications) should make it practicable to loosen the regulators' steel grip on spectrum, allowing bandwidth to become cheaper and ubiquitous. By using these technological marvels, instead of expensive networks, numerous wireless services could coexist in the same frequency bands. Such methods can serve to accommodate vast numbers of users—forcing a rethinking of the traditional view of the radio spectrum as a scarce resource.

OPEN SPECTRUM, ANYONE?

We can glimpse the possibilities of open spectrum when we view the success of existing unlicensed bands, such as the ISM (a set of radio frequencies centered around 2.4 GHz) and the U-NII bands (three 100 MHz unlicensed spectral bands between 5.15 GHz and 5.8 GHz), which are open for anyone to transmit within certain technical parameters such as power limits.

As the reader should now understand, technologies developed in recent years make it relatively easy to allow more than one user to occupy the same range of frequencies at the same time. Those innovative techniques serve to obviate the need for exclusive licensing. Instead, they could support a true open spectrum environment, which hopefully would allow the same degree of openness, flexibility and scalability for communication that the Internet promotes for applications and content. Thus, like the Internet, open spectrum is an idea with tremendous commercial potential. Open spectrum would also help to ensure that Wi-Fi technologies continue to thrive.

There are two ways to implement an open spectrum environment. The first is to designate specific bands for unlicensed devices, with general rules to foster co-existence among users. This is the method that allows Wi-Fi to flourish in the 2.4 GHz and 5 GHz bands. The second mechanism is to "overlay" unlicensed technologies in existing bands, with the proviso that the "sitting tenant's" services will not be disturbed. This approach effectively manufactures new capacity by increasing spectrum efficiency. Overlays can be achieved either by using an extremely weak signal or by employing agile radios that are able to identify and move around competing transmissions.

Both unlicensed spectrum and overlays have their place. Eventually overlay approaches will be more significant because they can work across the entire spectrum rather than requiring the creation of designated "commons" or "parks." But for this to come about requires the removal of limitations in existing rules, creation of additional unlicensed bands, establishment of rules to facilitate additional forms of overlay, and funding for research into next-generation technologies to support this paradigm.

***Note:** A spectrum "commons" or "park" refers to an area of spectrum where users multilaterally coordinate their communications using communication protocols embedded in end-user equipment, without the need for a licensing process.*

However, the notion of using smart technology (and smart devices) instead of intelligent networks runs counter to the received wisdom within the communications industry. Moreover, companies that spent huge amounts of money acquiring spectrum licenses and then built out expensive networks based on that spectrum will naturally oppose any change to the *status quo*.

Another roadblock to open spectrum is that current spectrum rules don't provide enough unlicensed spectrum or the right equipment parameters for a true spectrum commons. For instance, current regulations prevent spread spectrum devices from overlaying on licensed bands at low power.

***Note:** The FCC proposed the use of overlays during the 1980s and voiced its support for the idea that spread spectrum radios could transmit effectively at such low power that they would be unnoticeable to high-power licensed services. Unfortunately, that proposal drew so much opposition and concern about interference from major industry players that the agency decided not to go forward with it. A more recent FCC proceeding on ultrawideband technology (a variant of spread spectrum that operates on very wide channels with extremely low power) generated similarly intense responses. It's not only carriers who are concerned about changes to the status quo. Public-safety organizations, the military, the airline industry and global positioning system (GPS) vendors are also opponents.*

The limited amount of available open spectrum is keeping the Wi-Fi industry from realizing its full potential—from growing as explosively as, say, the computer industry, which saw tremendous growth once open source software became the norm. Of course, the other technological revolution that continues to grow in an open environment is the Internet—the open access communications scheme of the Internet has spurred innovation that few (if any) developments in history can match.

It is hoped that regulatory agencies will see the light and provision more open spectrum so that Wi-Fi and other innovative wireless technologies may follow a similar path. Admittedly, this will not happen overnight. So far, few regulatory bodies have shown a willingness to take such a monumental step and most, for the near future, will continue to issue limited, exclusive licenses affordable only by large corporations.

On a more positive note, in late 2002, the FCC's Spectrum Policy Task Force recommended modernization of the spectrum management rules. This would allow the system to evolve from the traditional government "command-and-control" model to a more flexible, consumer-oriented approach, according to the FCC's November 7, 2002 statement that accompanied the task force's report. (For more information on the task force and to obtain copies of the reports, go to www.fcc.gov/sptf.)

And, in Europe, the CEPT states that it endorses the principle of adopting a harmonized European Table of Frequency Allocations and Utilizations. The European Common Frequency Allocation Table, which covers major usage of the frequency bands, including Wi-Fi-enabled services, is slated to be adopted by CEPT administrations by 2008. (The Table can be found at the ERO website—www.ero.dk.)

Regulatory agencies worldwide have begun to reevaluate the current scheme of spectrum allocation. Yet any changes must be made with a clear understanding of the implications that new technology brings to the table. By grasping the benefits that these new technologies provide, it will be easier for these regulatory bodies to take affirmative steps to free up more spectrum by way of the commons and overlay models.

At the moment, however, it appears that rendering the allocation system obsolete is unlikely to occur in the near to medium term (e.g. CEPT's 2008 timeline). Interference concerns and existing treaty obligations will continue to require specifying ranges of radio frequencies for the foreseeable future.

But, in the long term, an enlightened model of spectrum management *could* serve to reduce spectrum scarcity—due to the efficiency-enhancing technologies available today. This is similar to how Digital Subscriber Line (DSL) technologies have served to reduce copper exhaust (i.e. scarcity of copper lines) in many urban areas.

CONCLUSION

The connected populace is demanding more broadband access, wireless service providers are scrambling for more capacity, and wired service providers are seeking new revenue streams. Many are considering the solution that is literally all around them—the airwaves.

While regulators do need to loosen their grip on spectrum, it's an undeniable fact that spectrum is a valuable resource that needs to be *managed* in order to ensure that radio services are able to operate on a non-interference basis. But, when you look at spectrum and Wi-Fi, there is another undeniable fact—for Wi-Fi to continue its amazing growth pattern, it needs more spectrum than what has (currently) been made available.

Exclusive licensing may have been the only approach to managing spectrum in the 20th century. But with today's technical community offering solutions that are smart enough to distinguish between signals, allowing users to share the airwaves without the need for restrictive licensing, a change in policy is needed. By making more efficient use of the available spectrum, the capacity constraints that could curtail Wi-Fi's race to prominence can effectively be removed. This requires that instead of treating spectrum as a scarce physical resource, regulators must allow for methods to be put in place to "stretch" its capacity. This will, in turn, pave the way for open spectrum. Then at least a reasonable amount of unlicensed spectrum can be made available to all as a commons, or unlicensed park, and non-intrusive overlay techniques can be used wherein new radio services can co-exist with traditional licensed services in the same frequency band.

Such changes would almost certainly result in the development of new applications and services and at the same time reduce prices, foster competition, and create business opportunities.

If the Wi-Fi industry can reclaim even just a small amount of spectrum from its legacy owners, there will be more room for innovators to build more wireless networks. Achieving a completely open spectrum regime will take years, or even decades, but that is all the more reason to begin the transition now.

Section II: The Technology

Chapter 5: Modulation Techniques

Equipment that adheres to the specifications as set out in the 802.11 series of standards can use the unlicensed ISM and U-NII frequency bands *only* if certain rules are observed regarding transmitted power and signal modulation. For instance the original 802.11 standard was crafted for use in the ISM 2.4 GHz to 2.485 GHz band, because there was (and is) no licensing requirement so long as the transmitted power of the 802.11 equipment is no more than one Watt and the transmitter uses a "Spread Spectrum" transmission technique. The one Watt power restriction serves to limit the range where one radio may interfere with another. The spread spectrum requirement is intended to make the WLAN signal appear as background noise to a narrowband or narrow spectrum receiver.

The term "signal modulation" refers to techniques whereby data is superimposed on or "encoded" into a carrier signal wave by means of a process, which is referred to as "modulation." This modulation process changes the signal wave so that it varies in its signal pattern. All "signal modulation" is accomplished in either of two main ways: analog and digital.

All 802.11 technologies use digital modulation. In fact, there are a number of digital modulation techniques used by the 802.11 series. The most important though are spread spectrum, especially Direct Sequence Spread Spectrum (DSSS), Complementary Code Keying (CCK), and Orthogonal Frequency Division Multiplexing (OFDM). We will now take a closer look at each of these three signal modulation techniques.

AN INTRODUCTION TO SPREAD SPECTRUM

The premier technology that overthrows the fundamental assumption that spectrum must be licensed to avoid interference is "Spread Spectrum," a technology that spreads a radio signal out over an entire frequency range, preventing concentration of the signal in any one place, allowing large numbers of users to share the same bandwidth. The military developed it for use in reliable, secure, mission-critical communications systems, based on ideas worked out in the 1940s to overcome intentional interference by hostile jamming and eavesdropping.

In 1924 Alfred N.Goldsmith filed the earliest U.S. patent that can be construed as being spread spectrum. Later, in 1940, Hedy Lamarr, the actress, created the concept of spread spectrum and two years later received a U.S. patent for a "secret communication system." The patent was issued to her and George Antheil, a film-score and avant-garde composer, to whom Lamarr had turned for help in perfecting her idea. The Lamarr-Antheil system uses a (primitive and probably unworkable) mechanical switching device similar to a player piano roll to shift frequencies faster than an enemy could follow them.

In Robert Scholt's highly acclaimed historical paper entitled "The Origins of Spread Spectrum Communications," Scholt states that one of the earliest applications of spread spectrum was the communications link between Roosevelt and Churchill during World War II, although at the time the technique was not known as spread spectrum. The full text of the Scholt paper can be found in the *IEEE Transactions on Communications, Vol. Com-30, No. 5.*, May 1982, 822-854. (Robert Scholt received the 1983 Leonard G. Abraham Prize Paper Award and the 1984 IEEE Donald G. Fink Prize award for his work.)

After the invention of the transistor and microelectronics, the U.S. Army began using spread spectrum in the 1950's for Electronic Counter Countermeasures and in electronic warfare because of its characteristics of Low Probability of Interference (LPI), Low Probability of Detection (LPD), and anti-jam capability.

More recently, spread spectrum has been combined with digital technology for spy-proof and noise-resistant battlefield communications. In 1962, Sylvania installed it on ships sent to blockade Cuba. (Indeed, it was Sylvania that came up with the name "spread spectrum.") But spread spectrum also has been utilized in a few, special non-military applications. The NASA space shuttle S-Band communications links implemented spread spectrum for interference avoidance. The spread spectrum-based Tracking and Data Relay Satellite System (TDRSS) program was used for multiple access purposes in which several users simultaneously share the same satellite repeater power and bandwidth.

Wireless equipment engineers also realized that spread spectrum technology offered many solutions to questions that had previously stalled the development of credible wireless networking. These innovative engineers realized that spread spectrum techniques could allow wireless systems to *reuse* frequencies within the Institutional Scientific Medical (ISM) band without interference. That was important, because the discovery meant that low-cost, FCC license-free wireless networking products could become a reality. By the mid-1990s the FCC had a number of wireless networking standards presented for its review.

The term "spread spectrum" arose from the characteristic broad spectral shape of the transmitted signal. Although spread spectrum techniques are widely used today in mainstream wireless sys-

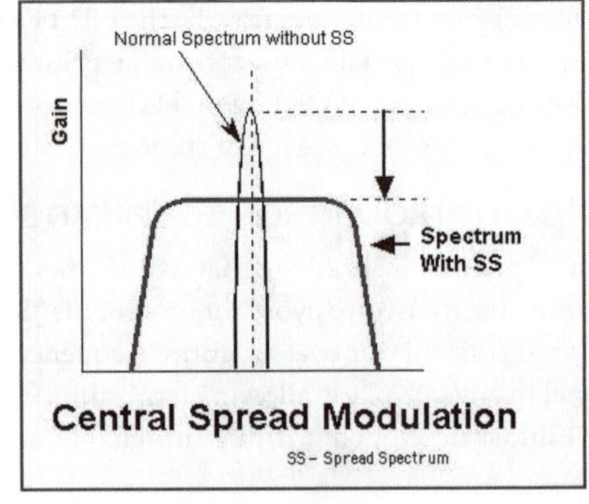

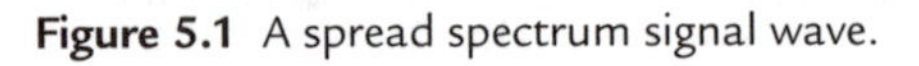
Figure 5.1 A spread spectrum signal wave.

tems, its full potential wasn't realized until there was enough computing power available to exploit its full capabilities.

In a spread spectrum system, signals are chopped up, distributed across many frequency bands and then reassembled on the other end. Spread spectrum allows two or more users to share the same frequency band, sometimes at the same time, sometimes at different times, depending on the system used. It's reminiscent of the way Internet routers allow millions of people to share the same data networks. However, the Internet processes data packets in a decentralized way, and spread spectrum enables receivers to distinguish overlapping communications based on codes that identify data packets belonging to specific users. The smarter the devices are–at both ends of the transmission–the more users a system can accommodate. This means that a Wi-Fi transmission won't necessarily interfere with a cellular voice communications service using the same spectrum segment. Thus, there isn't any need to give any one service exclusive use of a particular band of spectrum; all that's needed are mechanisms to ensure that equipment used to transmit and receive does so cooperatively, rather than in a contentious manner.

Spread spectrum is designed to trade off bandwidth efficiency (i.e. the signal is spread over a broad range of frequencies) for reliability, integrity, and security. This means that more bandwidth is consumed than in the case of narrowband transmission, but this tradeoff produces a signal that is, in effect, louder and thus easier to detect, provided that the receiver knows the parameters of the spread spectrum signal being broadcast. A receiver that is tuned to the right frequency and possessing the proper decoding procedure can extract the message, thus allowing a greater number of users to share the bandwidth. (For a receiver that does not have these capabilities, a spread spectrum signal just looks like background noise.)

***Note:** In the past, systems that used the radio spectrum relied primarily on narrowband modulation techniques whereby all of the power in a transmitted signal is confined to a very narrow portion of the frequency bandwidth. The problem with these techniques is that an interfering frequency at or near the transmitting frequency can cause interference, rendering the signal unrecoverable. Amplitude Modulation (AM) is an example of a narrowband technique in which the amplitude (volume) of the carrier signal is made stronger or weaker based on the information in the signal to be transmitted. The large amounts of power associated with AM allow the signal to travel long distances before it attenuates to an undetectable level. That's why AM radio stations' broadcast signals can be received over long distances*

In order to qualify as a spread spectrum signal, the following criteria must be met:

1. The transmitted signal bandwidth is greater than the minimal information bandwidth needed to successfully transmit the signal.
2. Some function other than the information itself is being employed to determine the resultant transmitted bandwidth.

There are many benefits to using spread spectrum. For instance, since the signal is spread over a large bandwidth, its duration on any particular frequency in the allocated frequency segment is just a fraction of a second, and its "power density" is very low. This means that the average detectable power on any given channel is extremely low, and other devices using the same channel won't notice the spread spectrum transmission.

And since the transmitted spread spectrum signal is uniformly spread over a wide range of frequencies, the signal steers clear of interference and noise from other signals.

Interestingly, the bandwidth at these frequencies can be "reused," because when spread spectrum coexists with narrowband signals, the only result is a slight increase in the "noise floor" in any given slice of spectrum used by the conventional narrowband transceivers. In other words, narrowband transceivers can continue to use their respective slices of the spectrum, and any overlapping spread spectrum signals will only be detected as a slight background noise. Moreover, as we've seen, spread spectrum transceivers can also share the same frequency band with other spread spectrum transceivers.

Spread spectrum communications are normally secure, because spread spectrum produces a pseudo-random signal, so the transmitted signal appears as noise to other receivers—only receivers possessing the proper duplicate pseudo-random noise code sequence are able to recover the signal. Hence, multiple stations can simultaneously transmit spread spectrum and narrowband broadcasts.

The reader might wonder, if spread spectrum technology is supposedly secure and Wi-Fi uses it, why is there such concern over Wi-Fi security? Many Wi-Fi vendors initially did claim that it was difficult or impossible to "despread" or demodulate the signals, but this is not actually the case; spread spectrum techniques are not *absolutely* secure. Wi-Fi's spread spectrum technology modulates the RF signal and spreads the transmission over the entire frequency band allocated for Wi-Fi communication. The spread spectrum technique Wi-Fi uses adds a redundant pattern to each bit transmitted and only the sender and receiver hardware are supposed to know the particular code used to generate these patterns. In actuality, intruders can crack this code relatively easily, since spread spectrum was incorporated into Wi-Fi not for reasons of security, but because of the improved signal-to-noise ratio and resistance to interference.

***Note:** Don't worry, the industry is working diligently to address Wi-Fi's security dilemma and have come up with some very innovative solutions as discussed in Chapters 6 and 17.*

Getting back now to our perusal of spread spectrum, there are a number of incarnations of spread spectrum modulations, but we will only discuss, very briefly, two—frequency hopping and direct sequence—both have advantages and disadvantages associated with them. It's these two spread spectrum techniques that play a large part in the acceptance of the IEEE 802.11 standard for the implementation of wireless LANs.

DSSS and FHSS

Direct Sequence Spread Spectrum (DSSS) works by taking a modulated narrowband carrier rapidly modulating its phase so that it expands, occupying a bandwidth much greater than the information bandwidth of the signal. This modulation/expansion is performed by multiplying the information signal by a code called a "spreading code" or pseudo-noise (PN) spreading sequence, a deterministic, computed, binary sequence of bits (1s and 0s) that nevertheless appears to be random. This code is generated by a pseudorandom generator that produces one out of millions of possible binary number sequences of a fixed length. After a given number of bits are produced, this code repeats itself exactly. This "almost random" code has a bit rate significantly greater than the information bit rate, and when it is used to modulate the carrier phase it spreads the signal energy over a wider portion of the frequency spectrum than would the narrowband signal alone. The amount that the signal is spread is determined by the ratio of the bit rate of the spreading sequence divided by the data rate of the information signal (this ratio is also known as the "processing gain") The bandwidth of the spread signal is the product of the bandwidth of the unspread signal and the processing gain. Each bit of the pseudorandom number sequence is often referred to as a "chip," and the bit rate of the sequence is called the "chip rate." The receiver then uses the same PN sequence to "despread" and "decrypt" the original signal by multiplying the signal with a locally generated reproduction of the PN sequence.

Since the transmission is spread across a wide frequency band (a result of the spreading process), transmission power is lower than that of narrowband transmissions. Thus, to other radio services operating in the same band, the signal appears to be low power background noise. But since the signal is low power and spread across a wide frequency range, the spread spectrum signal itself is susceptible to noise. However, in cases of signal corruption the redundant data helps to recover the original signal; the number of chips is directly proportional to the immunity from interference.

Unlike DSSS, which chops the data into small pieces and spreads them across the frequency domain, in Frequency Hopping Spread Spectrum (FHSS) systems, the carrier frequency of the transmitter abruptly changes (or hops) as dictated by a generated pseudorandom code sequence. A short burst of data is transmitted on a narrowband

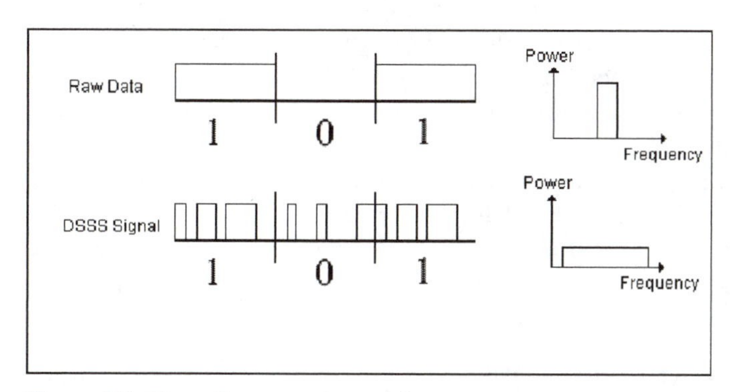

Figure 5.2 Direct Sequence Spread Spectrum.

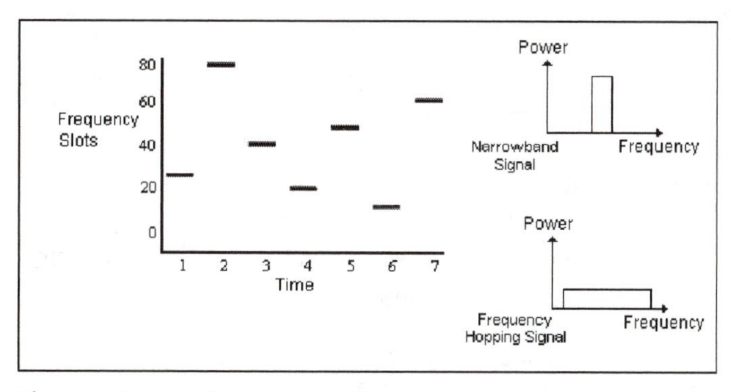

Figure 5.3 Spread Spectrum using the frequency hopping technique.

and then the transmitter quickly returns to another frequency and transmits again. The sequence of hops the transmitter makes is pseudo-random, but the receiver knows the pattern, enabling it to receive each short burst of data at the proper frequency and time and thus reproduce the original constant signal. Since the transmitter and receiver are synchronized, the stream of data appears to be constant. (To other radio services, the FHSS signal appears as a short burst of noise.)

There are certain rules governing how a FHSS device must behave to ensure that a device doesn't use too much bandwidth or linger too long on a single frequency. For instance, in North America, the ISM band is separated into 75 hopping channels and the power transmitted on each channel must not exceed one Watt.

Frequency hopping devices have different characteristics when compared to DSSS devices. First, DSSS provides a higher throughput and is more immune to interference than frequency hopping. However, there is a tradeoff: DSSS uses two to three times more power and tends to be more costly than FHSS. But, since FHSS's immunity to interference tends to be lower than DSSS signals, if a burst of data is corrupted on one hop then the entire data packet must be sent again. Still, frequency hopping does have one major advantage over direct sequence–several access points can coexist in the same area. Consequently, if an access point is struggling to cope with a large number of users, another access point can be added to take some of the load. This cannot be done with DSSS access points as they would block each other from transmitting.

Such techniques, however, do trade off power against bandwidth and range. A narrowly focused, high-power signal overwhelms other signals. To understand this, consider one of the raucous Sunday morning talk shows, where everyone is speaking at once. As each pundit tries to get their thought across, each person begins to speak ever louder, causing a domino effect as everyone at the table begins to yell louder and louder in an effort to be heard. In the ensuing carnage, no one can be heard, so no one gets his or her message across. On the other hand, if a low-power signal is spread over a wide range of frequencies or is sent a shorter distance, many communications can coexist. The same occurs in the case of a "pundits' corner"–if everyone speaks in a calm, controlled manner, everyone's point of view can be heard and considered.

802.11'S MODULATION TECHNIQUES

The original 802.11 standard specified two different spread spectrum transmission techniques: DSSS and FHSS. All radio equipment use the 2.4 GHz ISM band, and systems based on the *original* 802.11 standard provide data rates up to 2 Mbps. This is possible because DSSS utilizes an 11-bit chipping code called the Barker Sequence for signal spreading with modulation being achieved using either binary phase shift keying (BPSK) or quadrature phase shift keying (QPSK) techniques. (For FHSS, a modulation technique called Gaussian frequency shift keying or GFSK is employed.) Furthermore, in the U.S. DSSS deployments provide 11 independent channels by using different predefined chipping codes. (FHSS based implementations provide for 78 different logical channels through different hopping patterns, although in reality fewer channels would be actually usable due to frequency separation requirements.)

FHSS was dropped from the 802.11b specification because it was felt that "direct spread" could handle the tradeoff between wireless devices coexisting with other users, while extracting the greatest capacity from systems that are both power and band limited. Later this aspect of 802.11b underwent modification after the FCC indicated in a Notice of Proposed Rule Making from the FCC published in the year 2001: ET 99-231; FNPRM & ORDER 05/11/01 (adopted 05/10/01); FCC 01-158 *Amendment of Part 15 of the Commission's Rules Regarding Spread Spectrum Devices, Wi-LAN, Inc. et al.* that it would consider relaxing the spread spectrum requirement on the ISM band in order to abandon the peaceful "coexistence of equipment" requirement (interference rejection) in favor of support for greater wireless network capacity (higher bit-rate transmissions). Therefore, for high bit rates above 2 Mbps (5.5 Mbps to 11 Mbps and higher) 802.11b's purely spread spectrum techniques have been supplanted by CCK modulation so as to provide 4 or 8 bits per transmission symbol. The combination of QPSK and CCK is what enables 802.11b's maximum data rate of 11 Mbps. Lower data rates are accommodated through a dynamic rate shifting scheme. Also, the reader should note that 802.11g supports CCK modulation so as to provide backwards compatibility with 802.11b. (As an option for faster link rates, 802.11g also allows packet binary convolutional coding (PBCC) modulation.)

CCK

CCK allows for multi-channel operation in the 2.4 GHz band by virtue of using the existing 802.11 1 and 2 Mbps DSSS channelization scheme. The spreading employs the same chipping rate and spectrum shape as the 802.11 Barker Sequence, allowing for three non-interfering channels in the 2.4 to 2.483 GHz band. Thus CCK modulation provides for a spectrum similar to that of the original 802.11 systems, at least at the low bandwidths. This allows interoperability with the original 802.11's DSSS modulation technique and it also allows for 802.11b multi-channel operation in the 2.4 GHz band using the existing 802.11 DSSS channel structure scheme.

CCK modulation consists of a set of 64 eight-bit code words. As a set, these code words have unique mathematical properties that allow them to be accurately distinguished from one another by a receiver even in the presence of substantial noise and multipath interference (e.g., interference caused by receiving multiple radio reflections within a building). The 5.5 Mbps rate uses CCK to encode 4 bits per symbol, while the 11 Mbps rate encodes 8 bits per symbol. Actually, to attain 11 Mbps CCK modulation,

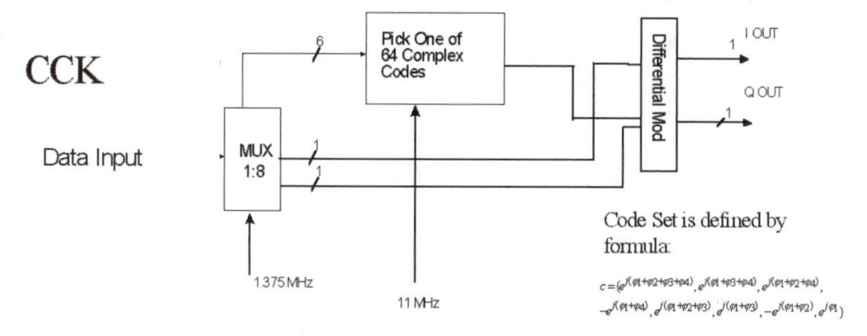

Figure 5.4 This graphic shows how the CCK modulation is formed.
Graphic courtesy of Intersil.

6 bits of the 8 are used to select one of 64 symbols of 8 chip length for the symbol and the other 2 bits are used by QPSK to modulate the entire symbol. This results in modulating 8 bits onto each symbol. The chipping rate is maintained at 11 million chip bits per second for all modes. Both speeds use QPSK as the modulation technique and signal at 1.375 million symbols per second.

The FCC regulations for the ISM band require at least 10 decibel (dB) of processing gain (11 dB for 802.11), which is normally achieved with spread spectrum techniques. CCK can achieve this gain too without having to be a conventional spread spectrum signal. Rather than using one or two 11-bit Barker sequences, CCK uses a series of codes called "complementary sequences." Because there are 64 unique code word sets that can be used to encode the signal, up to 6 bits can be represented by any one particular code word (instead of the single bit represented by a Barker symbol).

The wireless radio transmitter device generates a 2.4 GHz carrier wave (2.4 to 2.483 GHz) and modulates that wave using a various techniques, depending on the circumstances. For a 1 Mbps transmission, BPSK is used (one phase shift for each bit). To accomplish 2 Mbps or greater transmission, more sophisticated QPSK is used. QPSK can encode two bits of information in the same space as BPSK encodes one. The tradeoff is the need for increased power or else one must decrease the range to maintain signal quality.

Unfortunately, the FCC regulates the output power of portable radios to just one Watt; therefore, as the 802.11 transceiver moves away from the radio, the radio must adapt to the situation by using a less complex (and slower) encoding mechanism to send data. Ironically, the CCK code word is modulated with the same QPSK technology that was used in 2 Mbps wireless direct spread radios. This enables an additional 2 bits of information to be encoded in each symbol. Eight binary "chip" numbers are sent for each 6 bits, but each symbol encodes 8 bits thanks to the QPSK modulation. So, for a 1 Mbps transmission, 11 million chip bits per second times 2 MHz equals 22 MHz of spectrum. Likewise, for a 2 Mbps transmission, 2 bits per symbol are modulated with QPSK, 11 million chips per second, and thus you need 22 MHz of spectrum. In short, to transmit at a bit rate of 11 Mbps, you need 22 MHz of frequency spectrum.

OFDM

The other modulation technique that plays a large part in the IEEE 802.11 series of standards for wireless LANs is orthogonal frequency division multiplexing (OFDM). OFDM is a digital modulation method in which a signal is split into several narrowband channels at different frequencies. The technology was first conceived in the 1960s and 1970s during research into minimizing interference among channels near each other in frequency. Both 802.11a and 802.11g use OFDM.

With OFDM, priority is given to minimizing the interference, or crosstalk, among the channels and symbols comprising the data stream. Less importance is placed on perfecting individual channels. In some respects, OFDM is similar to conventional frequency division multiplexing, although the way the signals are modulated and demodulated differ.

The demand for high-speed wireless networking exposed the limits of spread spectrum technologies. Because of their relatively inefficient use of bandwidth, the original spread spectrum systems cannot satisfy the higher data rates that technology operating in the 5 GHz U-NII bands can accommodate. That is why the more efficient and robust OFDM was added to the Wi-Fi modulation mix. OFDM, sometimes referred to as multi-carrier or discrete multi-tone modulation, utilizes multiple sub-carriers to transport information in from one particular user to another. An OFDM-based system divides a high-speed serial information signal into multiple lower-speed sub-signals that the system transmits simultaneously at different frequencies in parallel.

The benefits of OFDM are high spectral efficiency, resiliency to RF interference, and lower multi-path distortion. The orthogonal nature of OFDM allows sub-channels to overlap, having a positive effect on spectral efficiency (see Fig. 5.5). Each one of the sub-carriers transporting information is just far enough apart from each other to theoretically avoid interference.

This parallel-form of transmission over multiple sub-carriers enables OFDM-based WLANs to operate at higher aggregate data rates, such as up to 54 Mbps with 802.11a/11g-compliant implementations. In addition, interfering RF signals will only destroy the portion of the OFDM transmitted signal related to the frequency of the interfering signal.

***Note:** The better the quality of the signal at 5 MHz, the more complex the modulation technique that can be supported within each sub-carrier channel and the more bits, therefore, can be impressed on the RF signal. The coding techniques and data rates that are specified to work in tandem with OFDM in 802.11a include BPSK at 125 Kbps per channel for a total of 6 Mbps across all 48 data channels, QPSK at for 250 Kbps per channel for a total of 12 Mbps, 16QAM (16-level quadrature amplitude modulation) at 500 Kbps per channel for a total of 24 Mbps, and 64QAM (64-level QAM) at 1.125 Mbps per channel for a total of 54 Mbps.*

OFDM exhibits lower multipath distortion (delay spread), since the high-speed composite's sub-signals are sent at lower data rates. Because of the lower data rate trans-

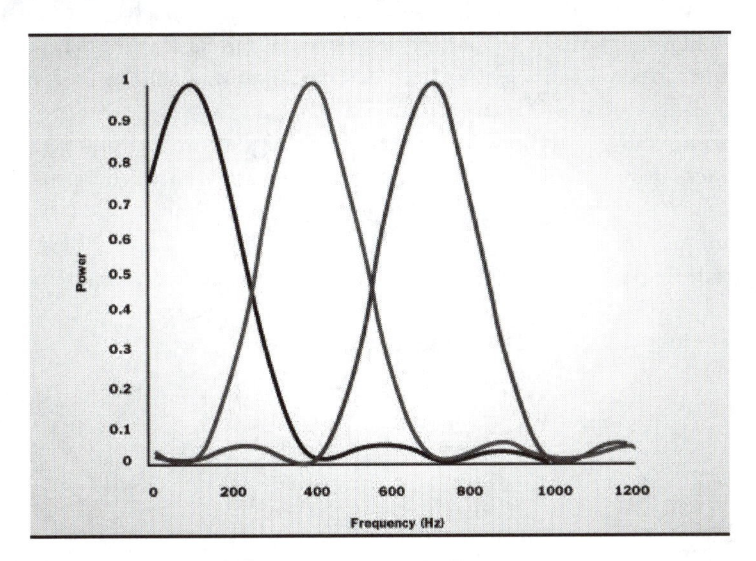

Figure 5.5 The orthogonal nature of OFDM's efficient use of bandwidth.

missions, multipath-based delays are not nearly as significant as they would be with a single-channel high-rate system. For example, a narrowband signal sent at a high rate over a single channel will likely experience greater negative effects from delay spread because the transmitted symbols are closer together.

In fact, the information content of a narrowband signal can be completely lost at the receiver if the multipath distortion causes the frequency response to experience a null at the transmission frequency. The use of the multi-carrier OFDM significantly reduces this problem.

Multipath distortion can also cause inter-symbol interference, which occurs when one signal overlaps with an adjacent signal. OFDM signals typically have a time guard of 800 nanoseconds (ns), however, which provides good performance on channels having delay spreads up to 250 ns. This is good enough for all but the harshest environments. Delay spread due to multi-path propagation is generally less than 50 ns in homes, 100 ns in offices, and 300 ns in industrial environments.

Many wired and wireless standards bodies have adopted OFDM for a variety of applications. For example, OFDM is used in European digital audio broadcast services. The technology also lends itself to digital television, and OFDM is the basis for the global Asymmetric Digital Subscriber Line (ADSL) standard. ADSL uses a variant of OFDM called discrete multi-tone (DMT), where the system sends several sub-carriers in parallel using the Inverse Fast Fourier Transform (IFFT) technique and receives the sub-carriers using the Fast Fourier Transform (FFT) techniques. In wireless LANs, OFDM is at the heart of the IEEE 802.11a and g specifications and European HiperLAN/2 standard. All implement OFDM in a similar way with the main difference being how the standards perform convolutional encoding.

CONCLUSION

The author had a somewhat "chicken before the egg" dilemma. Should the reader learn a bit about Wi-Fi's modulation techniques before moving on to a technical discussion of the individual standards or should the discourse on standards come before the modulation tutorial. As you can see, it was decided modulation would be first. Hopefully, the reader will agree that this is the correct order.

Many of the traditional modulation techniques hinder throughput capabilities of wireless devices by restricting the number of bits that can be encoded by analog signal values. But the desire for increased throughput over wireless links has led to new, innovative techniques for signal modulation at the OSI's Physical Layer. These new techniques provide for parallel data transmission while making the transmission more robust and resilient against channel errors.

Look for more innovations in this area in the future.

Chapter 6: The Wi-Fi Standards Spelled Out

A wireless local area network (WLAN) is a shared-medium communications network that broadcasts information over wireless links to be received by all stations (e.g. computing devices). Most of today's WLANs are built upon the IEEE 802.11 family of standards. These standards define technical specifications or guidelines advocated by specific organizations in order to establish consistency in hardware and/or software development.

Because of the alphabet soup nature of the 802.11 series, all 802.11 standards / specifications / technologies are commonly referred to as "Wi-Fi" (the original 802.11 specification is the exception). It is noted, however, that "Wi-Fi" is also a trademark of the Wi-Fi Alliance, a nonprofit organization originally formed as the Wireless Ethernet Compatibility Alliance (WECA).

Wi-Fi stands apart from other wireless technologies because it operates on unlicensed spectrum, which means that large telecommunications companies don't control it. That freedom allows for a "cauldron of innovation" that is leading to a brave new world for Wi-Fi technology in homes and businesses.

Now mainstream, Wi-Fi technology for broadband wireless networking is cheap, easy to deploy and is found throughout the world. To understand how and why all of this is happening, it's important to understand a bit about Wi-Fi's technology, its standards, history, and even its future.

THE WLAN FAMILY TREE

Before 802.11, wireless LAN technologies were low-speed proprietary offerings. Many used the unlicensed 902-928 MHz ISM (Industrial, Scientific and Medical) band and utilized a spread spectrum modulation technique (see previous chapter) to minimize interference.

Although slow and proprietary (most operated around 500 kilobits per second), these products offered the freedom and flexibility that only wireless networking can provide. So those early WLAN products found a home in industries where mobile workers could use handheld devices for inventory management, data collection, and to deliver patient information right to the bedside. By 1991, Aironet, Symbol Technologies and others were marketing their proprietary 900 MHz wireless LAN technologies in earnest.

It wasn't long before wireless vendors began developing proprietary 2.4 GHz WLAN products, opening up additional markets, especially in the warehousing and educational sector. Educational institutions were just beginning to embrace the Computer Age by wiring classrooms. However, many institutions quickly learned the difficulty of wiring rooms in older, established universities occupying 18th and 19th century buildings. For those institutions, wireless networking effectively overcame the physical limitations encountered on many campuses.

The IEEE

Next, the reader should be aware of the importance of the Institute of Electrical and Electronic Engineers (IEEE), an international professional organization for electrical and electronics engineers, with formal links with the International Organization for Standardization (more commonly known as the "ISO"). Within the IEEE 802 Committee are a number of Working Groups that strive to define different aspects of LANs and MANs.

IEEE became aware of the first non-standard wireless LAN protocols, which operated in the 900 MHz band, in the late 1980s. Thereafter the IEEE 802 LAN/MAN Standards Committee set on a course toward the development of a set of standards for wireless LANs.

In 1990, the IEEE 802 Executive Committee established the 802.11 Working Group to create a wireless local area network standard. Under the chairmanship of Vic Hayes, an engineer from NCR (one of the leading global technology companies), the 802.11 Working Group began to develop a WLAN specification. During the course of the project, that Working Group found it necessary to form individual Task Groups to work on different aspects of the 802.11 standard. The standards that these Task Groups define are suffixed by the Task Group's letter (e.g. 802.11a Task Group). That is how the story of the now-standard designations of "802.11a," "802.11b," "802.11g," etc., came to exist.

Now let's look back to when the progenitor of today's Wi-Fi technology saw its first breath of life.

The Original 802.11 Specification

On June 26, 1997, seven years after the formation of the 802.11 Working Group, the IEEE Standards Board approved the first 802.11 standard. It was thereafter published on November 18, 1997.

The original 802.11 specification, which provides data transfer rates of 1 and 2 Mbps, specifies an over-the-air interface between a wireless client and a base station (i.e. access point) or between two wireless clients. It specifies an operating frequency in the unlicensed 2.40-2.483 GHz ISM (Industrial, Scientific, and Medical) band. But in 1997, for a wireless networking product to operate in the ISM bands, it must use some form of "spread spectrum" modulation, which, as it name implies, "spreads" a signal's power over a wide band of frequencies. This original 802.11 specification provided one of two incompatible forms of spread spectrum modulation, either frequency hopping or direct sequence. Since both techniques spread the signal over more than one frequency in order to boost signal-to-noise performance (a technique called "process gain"), it wastes bandwidth. Moreover, the specification required WLANs that operate at the 1 Mbps level to

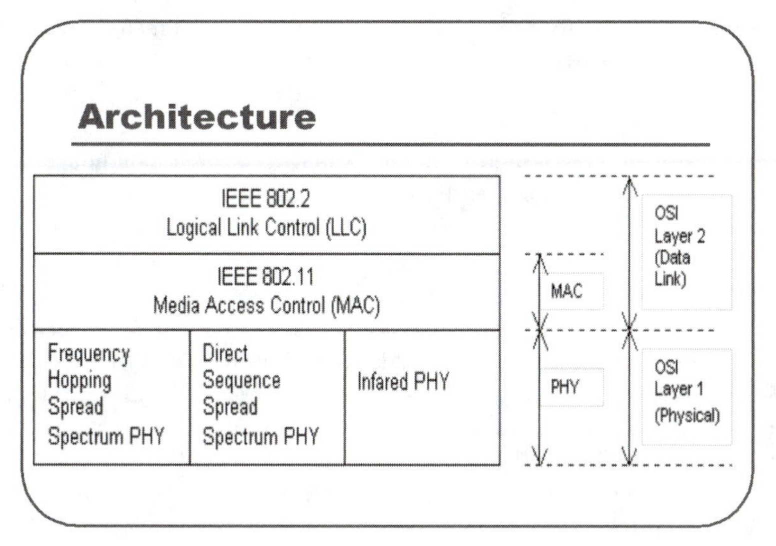

Figure 6.1 The OSI model in relation to the IEEE's 802.11 specifications.

employ a binary phase shift keying (BPSK) modulation scheme, while systems that are designed to operate at the 2 Mbps level must operate off a quadrature phase-shift keying (QPSK) modulation scheme.

The initial 802.11 specification also describes the OSI's (Open System Interconnection) Data Link Layer's MAC (Medium Access Control) sublayer, and the PHY (Physical) Layer definitions. To date, all of the ensuing amendments and supplemental 802.11 standards are either enhancements to the original MAC for QoS (Quality of Service) and security, or an extension to the original PHY for high-speed data transmission.

What's most important about 802.11 though, is that this standard laid the groundwork for future technologies. Today's wireless networking boom owes its emergence to the innovative path starting from that original IEEE WLAN specification.

The Ethernet Connection

In the wired world, 802.3 Ethernet is the predominant LAN technology. Its evolution is analogous to Wi-Fi. To fully appreciate Wi-Fi's role in today's networking environments, you must understand the Wi-Fi / Ethernet relationship.

The IEEE's 802 Executive Committee mandate is to design specifications to standardize the physical path over which computers communicate. Toward that end, the Committee formed the 802.3 Working Group to define the physical media and the working characteristics for physical communication in a LAN environment. The result was an Ethernet-like standard published in 1985. This first publication was named "IEEE 802.3 Carrier Sense Multiple Access with Collision Detection Access Method and Physical Layer Specifications." As such, it defines the Physical Layer and the Data Link Layer's MAC sublayer, including specifying cabling options, data transmission method, and means for controlling access to the cable.

Originally providing for ten megabit per second (Mbps) transfer rates, the 802.3 Working Group continues to keep pace with the data rate and throughput requirements of contemporary LANs. That's why the 802.3 standard, albeit with many extensions, still governs how computers communicate today.

This set of specifications is more commonly known as "Ethernet," although the IEEE's document doesn't refer to 802.3 as "Ethernet," per se. "Ethernet" is a specific product trademarked by Xerox, whereas 802.3 is a set of standards that can be used by anyone.

802.3 Ethernet provides an evolving, high-speed, widely available and interoperable networking standard. Furthermore, the IEEE 802.3 standard is open. An open standard is defined as a technical standard that has been accepted by a *bona fide* standards organization such as the American National Standards Institute (ANSI) or the European Technical Standards Institute (ETSI). The open standard status of 802.3 serves to decrease barriers to market entry, and results in a wide range of suppliers, products, and price points from which end-users can choose to build their LANs. And, most importantly, conformance to the Ethernet standards allows for interoperability, enabling users to select individual products from a multiplicity of vendors, secure in the knowledge that the products will work together.

Where the 802.3 Ethernet standard allows for data transmission over twisted-pair and coaxial cable, the 802.11 WLAN standard allows for transmission over different media, e.g. radio frequency and infrared light. But since 802.11 networks are considered wireless extensions of wired Ethernet, the 802.11 specifications were designed to work within a mixed Ethernet/Wi-Fi environment. Thus, 802.11 networks can handle conventional wired networking protocols such as TCP/IP, AppleTalk, and PC file sharing standards.

Wi-Fi also seems to be heading down the same "generic term" path of Ethernet, i.e. it is being used to describe all 802.11 networks, much the same way that "Ethernet" describes all IEEE 802.3 networks.

The Birth of Wi-Fi

The Wi-Fi story begins at the Institute of Electrical and Electronic Engineers (IEEE). The IEEE's 802 LAN/MAN Standards Committee defines standards for Local Area Networks (LANs) and Metropolitan Area Networks (MANs), including Ethernet, Token Ring and Wi-Fi.

Since the initial IEEE 802.11 standard was adopted officially in 1997, we have seen the data rate rise from 1 or 2 Mbps to 11 Mbps, and then to 54 Mbps.

Even as it ratified the 802.11 standard, the IEEE 802 Executive Committee knew that as the world became more bandwidth-hungry, a more robust and faster wireless networking technology would be needed. Therefore, the Committee continued its work. Within 24 months, the Working Group approved not one, but two, Project Authorization Requests for higher rate PHY Layer extensions to 802.11; both were designed to work with the existing 802.11 MAC Layer. One was IEEE 802.11a operating in the U-NII bands at 5 GHz, and the other was IEEE 802.11b operating in the ISM band at 2.4 GHz. In late 1999, the IEEE published these two supplements.

That's about the time the term "Wi-Fi" entered the picture. Originally, the term "Wi-Fi," as promulgated by WECA (now known as the Wi-Fi Alliance), was to be used only in place of the 2.4 GHz 802.11b standard. However, in October 2002, the Wi-Fi Alliance extended the "Wi-Fi" trademark to include the non-compatible 802.11a standard, which supports bandwidths up to 54 Mbps. Furthermore, the Alliance has indicated that since 802.11g is finalized, it is taking the necessary steps to also extend the "Wi-Fi" trademark to products built upon that standard's specifications.

WHY "b" BEFORE "a"?

There's some confusion about why 802.11b products came before 802.11a. So let's figure out what happened—you will need some history and a scorecard to keep the versions straight.

First, the reader must understand that while there are strong similarities between 802.11a and 802.11b, there are also important differences. Next, you need to appreciate that the IEEE had to find a means (1) for attaining both the high data rates needed to make WLANs a viable option, which in turn requires spectrum (an allocation of frequencies that the specification can use), and (2) for optimizing certain power requirements, which must be low enough to prevent interference with other users of the same frequencies, but at the same time strong enough to allow for consistent data rate speeds. The IEEE succeeded. Both 802.11a and 802.11b products address those needs, and, in some instances, 802.11a does a superior job. But that alone wasn't enough to propel 802.11a into a dominant role.

Hindering 802.11a's acceptance in the marketplace is its overall cost of installation, which is significantly higher than the cost of an 802.11b network. The distance that signals travel is related to their frequency—the lower the frequency, the further they will travel (for a given amount of power). This means that considerably more access points (APs) are needed for wireless LANs that use the higher frequency U-NII band than for those that use the lower frequency ISM band.

Another reason that 802.11b gained dominance in the marketplace was that after the IEEE's approval of 802.11a, its U-NII spectra, which is a huge amount of spectrum (more than many conventional broadcast television channel bandwidths combined), were put under scrutiny, not only by regulatory agencies, but also by the manufacturing sector. This gave 802.11b time to develop a strong installation base.

The 802.11a's OFDM waveform also initially caused some consternation in the manufacturing sector, because it is more difficult to implement than the forms of modulation used by 802.11b. (It's interesting to note that when 802.11a was ratified in 1999, very few designers had been exposed to OFDM.)

Finally, and perhaps the most important reason why 802.11b became a international de facto standard for WLANs and why its gear arrived first, was because there was already near worldwide acceptance of the 802.11b's 2.4 GHz range for low power, and usually license-free, operation. Cordless phones, microwave ovens, and baby monitors already use this spectrum (as do HomeRF networking and Bluetooth devices). Conversely, in many parts of the world, 802.11a's spectra weren't (and in some places still aren't) readily available for the public's use (the U.S. being one of the notable exceptions).

AN ALPHABET SOUP

This alphabet soup of standards can lead to confusion in the marketplace. Specific Tasking Groups inside the 802.11 Working Group oversee enhancements ranging from bandwidth to specific technical aspects of the 802.11 standard. The appearance of each new "flavor" of the technology results in a new suffix to 802.11, i.e. 802.11a, 802.11b, up to 802.11n (at the time of this writing).

These amendments and supplements help to assure that the 802.11 series can resolve compatibility issues between manufacturers of wireless local area network equipment operating in specific frequency bands. The series also provides the ability for a wireless data communications system to be implemented either as an extension to, or as an alternative for, a wired local area network.

Now let's take an in depth look at each member in the 802.11 family.

802.11a

When the IEEE ratified the 802.11a and 802.11b wireless networking communications standards in 1999, its goal was to create a standards-based technology that could span multiple physical encoding types, frequencies and applications in the same way the 802.3 Ethernet standard has been successfully applied to 10 Mbps, 100 Mbps and 1000 Gbps technology, over fiber and various kinds of copper. However, unlike its more popular sibling, 802.11b, it took a while to commercialize 802.11a. (For the reasons why, see the "Why 'b' Before 'a'" text box.) But that is old news; 802.11a has overcome its slow start and is now a viable player in the wireless networking marketplace.

The 802.11a standard, which supports data rates of up to 54 Mbps, is the Fast Ethernet analog to the 11 Mbps 802.11b. Like Ethernet and Fast Ethernet, 802.11b and 802.11a use an identical MAC (Media Access Control). However, while Fast Ethernet uses the same Physical Layer encoding scheme as Ethernet (only faster), 802.11a uses an entirely different encoding scheme than 802.11b. That encoding scheme is OFDM.

This specification also delivers other major changes. 802.11a specifies operation in the 5 GHz Unlicensed National Information Infrastructure (U-NII) bands—5.15-5.25 GHz, 5.25-5.35 GHz, and 5.725-5.825 GHz. It offers data rates that range from 6 to 54 Mbps. (Devices utilizing 802.11a are required to support speeds of 6, 12, and 24 Mbps; most now include 48, 36, 18 and 9 Mbps support as well). This standard also offers higher capacity and less radio frequency interference with other types of devices than its slower sibling. Within this spectrum, there are twelve 20 MHz channels (802.11a uses eight of these channels) and each band offers different output power limits.

***Note:** The U-NII bands have several advantages over ISM, one of the most notable being their ability to support 10 Mbps and higher data transfer rates. The U-NII bands also are dedicated solely to high data rate communications.*

Because of its operation in the larger and currently less crowded U-NII bands, 802.11a may offer less potential for RF interference than the other PHY standards (i.e. 802.11b and 802.11g) that utilize the smaller and more congested ISM band. Also, since it offers high data rate capabilities, 802.11a easily can support voice and multimedia applications and densely populated user environments.

Transmit Frequencies

The IEEE 802.11a Task Group elected to use a version of OFDM proposed by NTT and Lucent, which is known as "Coded Orthogonal Frequency Division Multiplexing," although this book and others simply refer to the modulation technique as OFDM. This OFDM version sends a stream of data symbols in a massively parallel fashion over multiple sub-carriers, which essentially are small slices of RF spectrum within a designated channel within a designated carrier frequency band.

The IEEE 802.11a standard specifies an OFDM Physical Layer that splits an information signal across 52 separate sub-carriers to provide transmission of data at a rate of 6, 9, 12, 18, 24, 36, 48, or 54 Mbps, with the 6, 12, and 24 Mbps data rate capability being mandatory. Four of the 52 sub-carriers are pilot sub-carriers that the system uses as a reference to disregard frequency or phase shifts of the signal during transmission. A pseudo binary sequence is sent through the pilot sub-channels to prevent the generation of spectral lines. The remaining 48 sub-carriers provide separate wireless pathways for sending the information in a parallel fashion. The resulting sub-carrier frequency spacing is 0.3125 MHz (for a 20 MHz-wide channel with 64 possible sub-carrier frequency slots).

In the U.S., the FCC allocated 300 MHz of spectrum for unlicensed operation in the 5 GHz block, 200 MHz of which is at 5.15 MHz to 5.35 MHz, with the other 100 MHz at 5.725 MHz to 5.825 MHz. The spectrum is split into three working "domains." The first 100 MHz in the lower section is restricted to a maximum power output of 50 mW (milliwatts). The second 100 MHz has a more generous 250 mW power budget, while the top 100 MHz is delegated for outdoor applications, with a maximum of one Watt power output. (See Fig. 6.2.)

***Note:** Unlike 802.11a, 802.11b cards can radiate as much as one Watt of power, outdoor or indoor (at least in the U.S.), although most modern radio cards radiate only a fraction—30 mW—of the maximum available power, due to the need for battery conservation and heat dissipation.*

As previously stated, the 802.11a standard operates in the 5 GHz frequency range, with a total bandwidth available for IEEE 802.11a applications of 300 MHz. That is almost four times that of the ISM band, which offers only 83 MHz of spectrum in the 2.4 GHz range for 802.11b devices.

Band	Channel numbers	Frequency (MHz)	Maximum output power (Up to 6 dBi antenna gain)
U-NII lower band 5.15 to 5.25 MHz	36	5180	40mW (2.5mW/MHz)
	40	5200	
	44	5220	
	48	5240	
U-NII middle band 5.25 to 5.35 MHz	52	5260	200mW (12.5mW/MHz)
	56	5280	
	60	5300	
	64	5320	
U-NII upper band 5.725 to 5.825 MHz	149	5745	800mW (50mW/MHz)
	153	5765	
	157	5785	
	161	5805	

Figure 6.2 802.11a's OFDM operating bands, channels, transmit frequencies and maximum output power. The low and middle bands are intended for in-building applications, and the high band for outdoor use (e.g., building-to-building).

As shown in Fig. 6.2, there are twelve 20 MHz channels, and each band has different output power limits. In the U.S., the Code of Federal Regulations, Title 47, Section 15.407, regulates these frequencies.

***Note:** The 802.11a standard requires receivers to have a minimum sensitivity ranging from -82 to -65 dBm (decibels referenced to 1 milliwatt), depending on the chosen data rate. Because of the relatively low power limits in the lower frequency bands, WLAN designers should carefully consider range requirements of the application before choosing a particular band. For a fuller understanding of RF power values, see Appendix II.*

The 802.11a standard gains some of its performance from the higher frequencies at which it operates. The laws of information theory tie frequency, radiated power, and distance together in an inverse relationship. Thus, moving up from 2.4 GHz to the 5 GHz spectrum allows for shorter practical distances, given the same radiated power and encoding scheme. In addition, the encoding mechanism used to convert data into analog radio waves can encode one or more bits per radio cycle (more commonly referred to as "Hertz"). By rotating and manipulating the radio signal, vendors can encode more information in the same time slice.

To ensure that the remote host can decode these more complex radio signals, you must use more power at the source to compensate for signal distortion and fade. The 802.11a technology overcomes some of the distance loss by increasing the equivalent (or effective) isotropic (or isotropically) radiated power (EIRP) to the maximum allowable value of 50 mW.

***Note:** According to Newton's Telecom Dictionary, the EIRP is the product of the power supplied to the transmitting antenna and the antenna gain in a given direction relative to an isotropic antenna radiator. EIRP may be expressed in Watts or dB (above one Watt).*

Physical Signals

The 802.11a designers hoped that OFDM could give 802.11a systems 802.11b-like transmission distances. And at lower speeds it can. Furthermore, the variation of OFDM that 802.11a uses was developed specifically for indoor wireless use and thus offers performance much superior to that of spread spectrum solutions.

802.11a's OFDM techniques work by breaking one high-speed data carrier into several lower-speed sub-carriers, which are then transmitted in parallel to each other. Each 20 MHz wide high-speed carrier is broken up into 52 sub-channels, each approximately 300 KHz wide. Coded OFDM uses 48 of these sub-channels for data, while the remaining four are used for error correction. 802.11a's encoding scheme and error correction technique also serve to aid 802.11a in its efforts to deliver higher data rates and a high degree of multipath reflection recovery.

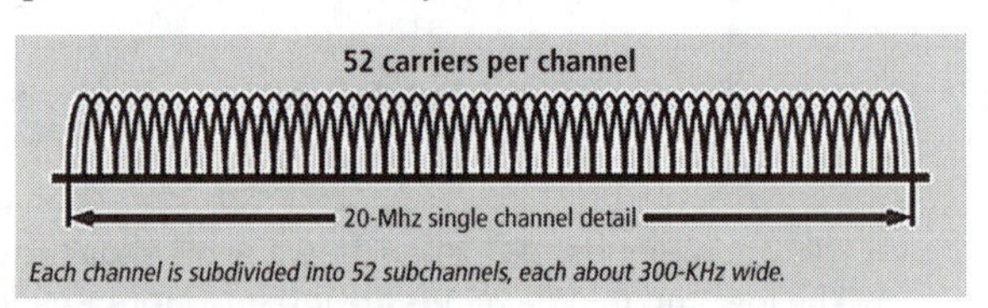

Figure 6.3 802.11a sub-channels.

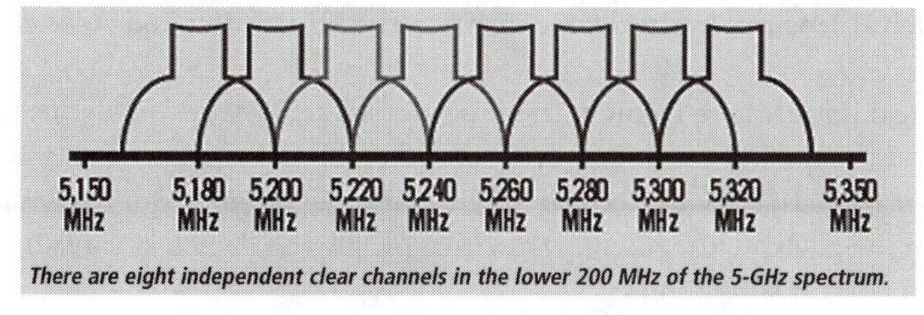

There are eight independent clear channels in the lower 200 MHz of the 5-GHz spectrum.

Figure 6.4 802.11a independent clear channels.

At the low end of the speed gradient, binary phase shift keying (BPSK) is used to encode 125 Kbps of data per channel, resulting in a 6000 Kbps, or 6 Mbps, data rate. But if you use quadrature phase shift keying (QPSK), it is possible to double the amount of data encoded to 250 Kbps per channel, yielding a 12 Mbps data rate. And by using 16-level quadrature amplitude modulation (QAM) encoding 4 bits per hertz, you can achieve a data rate of 24 Mbps. (See Fig. 6.5.) The 802.11a standard specifies that all 802.11a-compliant products must support those three basic data rates. But the standard also allows vendors to extend the modulation scheme beyond 24 Mbps. However, remember that the more bits per cycle (hertz) that are encoded, the more susceptible the signal will be to interference and fading, and ultimately, the shorter the range, unless power output is increased.

802.11a's top data rate of 54 Mbps is achieved by using 64 QAM, which yields 8 bits per cycle or 10 bits per cycle, for a total of up to 1.125 Mbps per 300 KHz channel. (With 48 channels, that results in a 54 Mbps data rate.)

As you might expect, the higher speeds work only under the best of circumstances and only over very short distances, with the exact distance being fairly unpredictable because so many factors can influence it.

The symbol rate refers to the rate of transmission of a symbol, or set of bits. The delay spread is the variation in timing between receipt of the signals associated with a given symbol. This delay spread is caused by multipath fading. It follows that the symbol rate must be slowed down enough that each symbol transmission is longer than the delay

Bit Rate	Bits per Symbol	Coding Rate	Coding Gain (G_c)	Effective SNR	Modulation
6 Mb/s	1	1/2	NA	SNR*G_c	BPSK
9 Mb/s	1	3/4	NA	SNR*G_c	BPSK
12 Mb/s	2	1/2	5.5 dB	SNR/2*G_c	QPSK
18 Mb/s	2	3/4	7.0 dB	SNR/2*G_c	QPSK
24 Mb/s	4	1/2	8.5 dB	SNR/4*G_c	16QAM
36 Mb/s	4	3/4	4.5 dB	SNR/4*G_c	16QAM
48 Mb/s	6	2/3	5.5 dB	SNR/6*G_c	64QAM
54 Mb/s	6	3/4	6.5 dB	SNR/6*G_c	64QAM

Figure 6.5 Parameters for 802.11a transmission rates.

spread, which is sensitive to the degree to which the transmitter and receiver have clear line-of-sight.

Since the data rates are so high, the modulation techniques so sophisticated and sensitive, and the quality of the airwaves so uncertain and tenuous, 802.11a specifications include forward error correction (FEC). Through the embedding of some redundant data in the payload, the receiving device typically can detect, isolate, diagnose and correct errors in transmission.

While this approach inherently adds some overhead to the transmission and further reduces throughput, it largely obviates the need for bandwidth-intensive retransmissions. Although it also necessitates the embedding of some additional intelligence in the receiving device, the costs of doing so are fairly modest in contemporary terms. On balance, this approach is much more efficient and cost-effective than its alternatives.

At the MAC sublayer, 802.11a makes use of the Carrier Sense Multiple Access with Collision Avoidance (CSMA/CA) protocol, which requires that each attached device broadcast a request to send (RTS) frame before transmitting. If the RTS frame gets through, the destination device responds with a clear to send (CTS) frame. All other devices on the network honor this reservation, and the transmission ensues. While RTS serves to make CSMA/CA more reliable, it does so at a cost. The RTS/CTS frames consume bandwidth, which means that this approach is somewhat overhead intensive and, therefore, affects total throughput. Also, the additional programmed logic makes CSMA/CA somewhat more expensive.

CSMA/CA is used in 802.11a systems because the allotted RF spectrum is so limited between the shared access point and the client workstations that an unacceptable level of collisions would otherwise result. So while there are some drawbacks to using CSMA/CA, on balance, it is more cost-effective than its alternatives.

PHY and MAC

The primary purpose of the OFDM PHY is to transmit media access control (MAC) protocol data units (MPDUs) as directed by the 802.11 MAC sublayer. The OFDM PHY is divided into two sublayers: the physical layer convergence protocol (PLCP) and the physical medium dependent (PMD).

The MAC sublayer communicates with the PLCP via specific primitives through a PHY service access point. When the MAC sublayer instructs, the PLCP prepares MPDUs for transmission. The PLCP also delivers incoming frames from the wireless medium to the MAC sublayer. The PLCP sublayer minimizes the dependence of the MAC sublayer on the PMD sublayer by mapping MPDUs into a frame format suitable for transmission by the PMD.

Under the direction of the PLCP, the PMD provides actual transmission and reception of PHY entities between two stations through the wireless medium. To provide this service, the PMD interfaces directly with the air medium and provides modulation and demodulation of the frame transmissions. The PLCP and PMD communicate using service primitives to govern the transmission and reception functions.

Frame formats and protocol. The PLCP preamble field (see Fig. 6.6) is present for the receiver to acquire an incoming OFDM signal and synchronize the demodulator. The

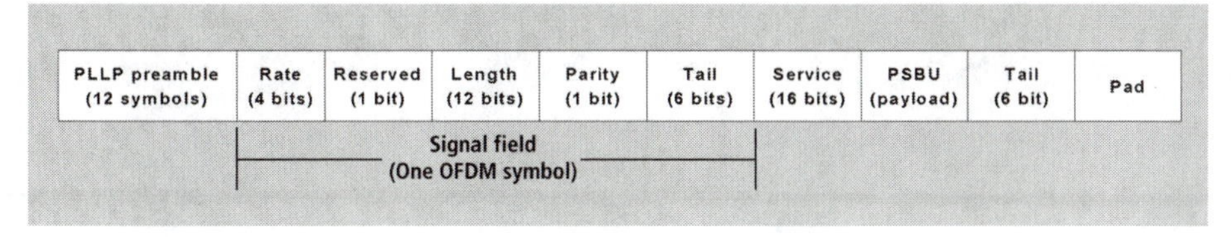

Figure 6.6 This graphic illustrates the frame format for an 802.11a frame.

preamble consists of 12 symbols. Ten of the symbols are short for establishing automatic gain control (AGC) and the coarse frequency estimate of the carrier signal. The receiver uses the long symbols for fine-tuning. With this preamble, it takes 16 microseconds to train the receiver after first receiving the frame.

The signal field consists of 24 bits, defining data rate and frame length. The 802.11a version of OFDM uses a combination of BPSK, QPSK, and QAM, depending on the chosen data rate (see Fig. 6.5). The length field identifies the number of octets in the frame. The PLCP preamble and signal field is convolutionally encoded and sent at 6 Mbps using BPSK, no matter what data rate the signal field indicates.

The convolutional encoding rate depends on the chosen data rate (see Fig 6.7). The parity field is one bit based on positive (even) parity, and the tail field consists of six bits (all zeros) appended to the symbol to bring the convolutional encoder to zero state.

Note. *Convolutional encoding refers to the convolutional code that is generated by inputting a bit of data, giving the commutator a complete revolution, and then repeating the process for successive input bits to produce a convolutionally encoded output. The convolutional code is produced by a convolutional coder, which is a coder with memory. The convolutional coder accepts k binary symbols at its input and produces n binary symbols at its output, where the n output are affected by v+k input symbols. Memory is incorporated because v>0. Code rate R=k/n. Typical values: k, n: 1 - 8; v: 2—60; R: 0.25—0.75.*

The service field consists of 16 bits, with the first six bits as zeros to synchronize the descrambler in the receiver, and the remaining nine bits (all 0s) are reserved for future use. The PLCP service data unit (PSDU) is the payload from the MAC sublayer being sent. The pad field contains at least six bits, but it is actually the number of bits that make

Data rate (Mbps)	Modulation	Coding Rate	Coded bits per subcarrier	Code bits per OFDM symbol	Data bits per OFDM symbol
6	BPSK	1/2	1	48	24
9	BPSK	3/4	1	48	36
12	QPSK	1/2	2	6	48
18	QPSK	3/4	2	96	72
24	16-QAM	1/2	4	192	96
36	16-QAM	1/2	4	192	144
48	16-QAM	3/4	4	288	192
54	64-QAM	2/3	6	288	216

Figure 6.7 Modulation techniques of OFDM as used in 802.11a.

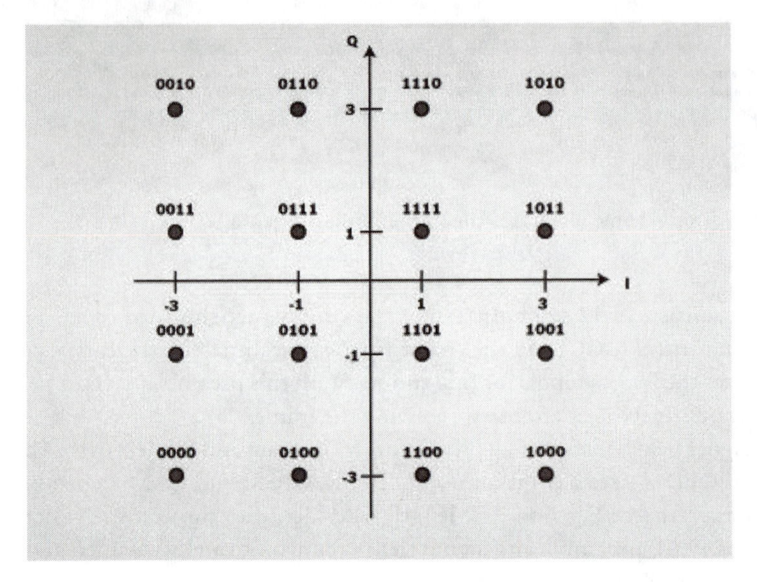

Figure 6.8 Constellation map for the 16-QAM 802.11a modulation.

the data field a multiple of the number of coded bits in an OFDM symbol (48, 96, 192, or 288). A data scrambler using a 127-bit sequence generator scrambles all bits in the data field to randomize the bit patterns in order to avoid long streams of 1s and 0s.

With 802.11a OFDM modulation, the binary serial signal is divided into groups (symbols) of one, two, four, or six bits, depending on the data rate chosen, and converted into complex numbers representing applicable constellation points. If a data rate of 24 Mbps is chosen, for example, then the PLCP maps the data bits to a 16-QAM constellation (see Fig. 6.8). The constellation is gray-coded.

After mapping, the PLCP normalizes the complex numbers to achieve the same average power for all mappings. The PLCP assigns each symbol, having a duration of four microseconds, to a particular sub-carrier. An inverse Fast Fourier Transform (FFT) combines the sub-carriers before transmission.

As with other 802.11-based PHYs, the PLCP implements a clear channel assessment protocol by reporting a medium busy or clear signal to the MAC sublayer via a primitive through the service access point. The MAC sublayer uses this information to determine whether to issue instructions to actually transmit an MDSU.

802.11a's version of OFDM with its high degree of spectral efficiency and resiliency to interference and multipath distortion, and existing inclusions in the leading higher rate WLAN standards, provides a strong base for the development of newer broadband wireless networks.

802.11a in Europe

For a time after its introduction, 802.11a was either limited or forbidden by most European nations, since many reserved the 5 GHz band for military or security use. When

WLAN vendors like Siemens, Philips Electronics, and Ericsson complained to the ETSI about the lack of spectrum for their 802.11a equipment, the North Atlantic Treaty Organization (NATO) interceded, pointing out to the ETSI that it uses parts of the 5 GHZ band for radar and satellites. A long battle ensued. The outcome was HiperLAN/2, which the ETSI created as a standard method for wireless LANs to use the 5 GHz spectrum in Europe.

However, HiperLAN/2 didn't catch on in the marketplace. Nonetheless, it did serve one purpose—to clear the path for 802.11a to gain a foothold Europe. Since HiperLAN/2 devices have not found a market, at the behest of some 802.11a supporters, ETSI has since modified its position. The ensuing modifications allows equipment that's not based on the HiperLAN/2 standard to be used in the 5 GHz range (with some exceptions) as long as the products:

1. Follow the IEEE 802.11a specifications.
2. Have the ability to "sense" when radar or other types of broadcasts enter the spectrum and have the intelligence to avoid them. This technique is known as Dynamic Frequency Selection (DFS).
3. Use Transmit Power Control (TPC) to reduce a radio signal's power depending on how close a device with an 802.11a card is to the access point.

Chipsets used in 802.11a products destined for Europe now include both DFS and TPC. In the U.S., negotiations are underway between the FCC and the Department of Defense to accommodate similar measures.

While 802.11a products that meet the ETSI requirements can be sold in nine European countries—Austria, Belgium, Denmark, Finland, France, Greece, Ireland, Norway and Portugal—these products still face a few more local regulatory hurdles. After meeting the ETSI certification demands, 802.11a equipment makers must incorporate the specific requirements of regulatory agencies within each European country in which they want to sell their 802.11a equipment. These countries often have requirements other than those imposed by the ETSI.

At this writing, most of the European countries have officially allowed 802.11a broadcasts in the 5.15 GHz to 5.35 GHz range, and the 5.47 GHz to 5.725 GHz range. A few countries, however, have carved out different areas within those two ranges in which Wi-Fi can operate, and made some channels within the spectrum off-limits.

Due to local regulatory requirements, as of mid-2002, 802.11a products were only available in France, the UK, the Netherlands, Belgium, Denmark, and Sweden. And, to give you an idea of how muddled this subject is in Europe, in mid-2002, Proxim (a prominent Wi-Fi vendor) traveled to Spain to demonstrate its new 802.11a equipment, and the government refused its request to use the appropriate spectrum.

The Future

What about HiperLan/2? Well, it is still clinging to life, but just barely. As you shall learn, other 802.11 standards are addressing the 5 GHz, 802.11a / HiperLAN/2 issues, e.g. 802.11d, h and j. As for 802.11a, it is expected that it will find worldwide acceptance, although some industry pundits opine that 802.11g will quash 802.11a's growth in Europe and elsewhere.

OTHER WLAN TECHNOLOGIES

ETSI's HiperLAN: The European Telecommunications Standards Institute (ETSI) is one of the world's most respected standards bodies, publishing many telecommunications standards. One of those standards is HiperLAN, which grew out of efforts to develop a wireless version of the wired ATM networking technology. The original HiperLAN standard was approved by ETSI in February 2000. That standard operates at rates up to 20 Mbps.

Like Wi-Fi, HiperLAN is a family of standards: HiperLAN, HiperLAN/2, HiperAccess (also known as HiperLAN/3), and HiperLINK (or HiperLAN/4). Note, however, that the latter two standards aren't applicable to local area networks.

Also like Wi-Fi, HiperLAN and HiperLAN/2 specifications define the Data Link Layer and the Physical Layer, along with the Data Link Layer's two sublayers: the Logical Link Control (LLC) and the Medium Access Control (MAC).

While the original HiperLAN standard and its successor, HiperLAN/2, are still on the books; they seem to be archaic. Furthermore, although there are supporters who market this technology for local area networking, it looks like HiperLAN/2 may merge with the IEEE's 802.11a.

	PHY	MAC	Raw rate (Mbps)
IEEE 802.11	Frequency Hopping, Direct Sequence	Carrier Sense Multiple Access Collision Avoidance (CSMA/CA)	1 or 2
IEEE 802.11b	Complementary Code Keying Direct Sequence	CSMA/CA	11
IEEE 802.11g	Orthogonal Frequency Division Multiplexing	CSMA/CA	54
IEEE 802.11a	OFDM	CSMA/CA	54
HiperLAN1	GMSK	Three phase priority driven	23.5
HiperLAN2	OFDM	Time Division Multiple Access	54

Figure 6.9 Although HiperLan/2 has lain dormant since its ratification, its predecessor has had some success outside the U.S.

Japan's MMAC: In Japan, the Multimedia Mobile Access Communication (MMAC) Systems Promotion Council group is developing specifications for advanced types of wireless systems. However the IEEE is developing additional standards, under the guidance of Task Group 802.11j, to meet Japanese regulatory guidelines, and therefore MMAC, like the HiperLAN standards, is unlikely to meet the market momentum of Wi-Fi. (See "802.11j" in this chapter.)

802.11b

The most commonly deployed WLANs conform to the IEEE 802.11b specification. Not only are they increasingly deployed in private enterprise applications, but also in public applications such as airports and coffee shops. This standard includes three transmission options, one of which is infrared-based, and two of which are RF-based. (This book covers only the RF options.)

This PHY specification provides High Rate Direct Sequencing Spread Spectrum (DSSS) modulation in the 2.4 GHz band (the same ISM band that the original 802.11 uses), and offers the potential of three simultaneous channels. The IEEE 802.11b Task Group also enhanced the original 802.11's DSSS PHY to include 5.5 Mbps and 11 Mbps data rates, in addition to the one Mbps and two Mbps data rates of the original 802.11 standard. The Task Group did this by employing DSSS modulation using the Barker code chipping sequence. Each bit is encoded into an 11-bit Barker code (e.g., 10110111000), with each resulting data object forming a chip. The chip is put on a carrier frequency (i.e., a small frequency range that carries the signal) in the 2.4 GHz range, and the waveform is modulated using one of several techniques. 802.11b systems running at one Mbps use Barker code and binary phase shift keying (BPSK) modulation, and those running at two Mbps use Barker code and quadrature phase-shift keying (QPSK) modulation.

Then in 2001, the FCC indicated in a Notice of Proposed Rule Making from the FCC (ET 99-231; FNPRM & ORDER 05/11/01; FCC 01-158 *Amendment of Part 15 of the Commission's Rules Regarding Spread Spectrum Devices, Wi-LAN, Inc. et al.*), that it would consider relaxing the spread spectrum requirement on the ISM band. This Amendment abandons the peaceful "coexistence of equipment" requirement (interference rejection) in favor of support for greater wireless network capacity (higher bit-rate transmissions). The 802.11b Task Group responded to this action by modifying the specification. For high bit rates above 2 Mbps (i.e. 5.5 Mbps to 11 Mbps), 802.11b's purely spread spectrum techniques was supplanted by complementary code keying (CCK) modulation so as to provide 4 or 8 bits per transmission symbol, which when combined with QPSK, allows 802.11b to easily reach its maximum data rate of 11 Mbps.

Thus current 802.11b networks use CCK to provide the higher data rates. To support very noisy environments as well as extended range, 802.11b wireless LANs also are capable of "dynamic rate shifting," which allows data rates to be automatically adjusted to compensate for the varying nature of the radio channel. Initially, the equipment tries to connect at the full 11 Mbps rate. If the devices move beyond the optimal range for 11 Mbps operation, or if considerable interference is encountered, then the 802.11b devices will "fall back" and transmit at lower speeds; first to 5.5, then to 2, and finally to 1 Mbps. Likewise, if the device moves closer or if the interference disappears, then the connection will automatically increase to 11 Mbps. Rate shifting is a Physical Layer mechanism that's transparent to the user and the upper layers of the OSI protocol stack.

Furthermore, systems running at 5.5 Mbps and 11 Mbps use a combination of CCK and QPSK. CCK involves 64 unique code sequences, each of which supports 6 bits per code word. The CCK code word is then modulated onto the RF carrier using QPSK, which allows another two bits to be encoded for each 6-bit symbol. Therefore, each 6-bit symbol contains 8 bits (i.e. 1 byte).

Because of the complex encoding, it is much more difficult to discern which of the 64 code words is coming across the airwaves. Furthermore, the radio receiver design is significantly more difficult. In fact, while a 1 Mbps or 2 Mbps radio has one correlator (the device responsible for lining up the various signals bouncing around and turning them into a bitstream), 11 Mbps radio must have 64 such devices.

PHY and MAC

Like 802.11a, the 802.11b PHY Layer is split into two sublayers: PLCP (Physical Layer Convergence Protocol) and the PMD (Physical Medium Dependent). The PMD takes care of the encoding. The PLCP presents a common interface for higher-level drivers to write to, and provides carrier sense and CCA (Clear Channel Assessment), which is the signal that the MAC sublayer needs to determine whether the medium is currently in use. (See Fig. 6.10.)

The PLCP consists of a 144-bit preamble that is used for synchronization to determine radio gain and to establish CCA. The preamble comprises 128 bits of synchronization (scrambled 1 bits), followed by a 16-bit field consisting of the pattern 1111001110100000. This sequence, which is called the SFD or start frame delimiter, marks the start of every frame.

The next 48 bits are collectively known as the PLCP header. The header contains four fields: signal, service, length, and header error check (HEC). The signal field indicates how fast the payload will be transmitted (1, 2, 5.5 or 11 Mbps). The service field is reserved for future use. The length field indicates the length of the ensuing payload, and the HEC is a 16-bit CRC of the 48-bit header.

The PLCP is always transmitted at 1 Mbps, which serves not only to complicate things, but also degrades performance, because 24 bytes of each packet are sent at 1 Mbps. This means that the PLCP introduces 24 bytes of overhead into each wireless packet before the packet's destination is even considered. In comparison, Ethernet introduces only 8 bytes of data. Because the 192-bit header payload is transmitted at 1 Mbps, 802.11b is at best only 85 percent efficient at the Physical Layer.

The MAC sublayer's most basic ability is to sense a quiet time on the network before transmitting. Once the host has determined that the medium has been idle for a minimum time period, it may transmit a packet. This minimum time period is known as "distributed coordination function inter-frame spacing" or "DIFS." If the medium is busy, the node must wait for a time equal to DIFS, plus a random number of slot times. The time between the end of the DIFS period and the beginning of the next frame is known as the contention window.

Each station listens to the network, and the first station to finish its allocated number of slot times begins transmitting. If any other station hears the first station talk, it stops counting down its back-off timer. When the network is idle again, it resumes the countdown. In addition to the basic back-off algorithm, 802.11b adds a back-off timer

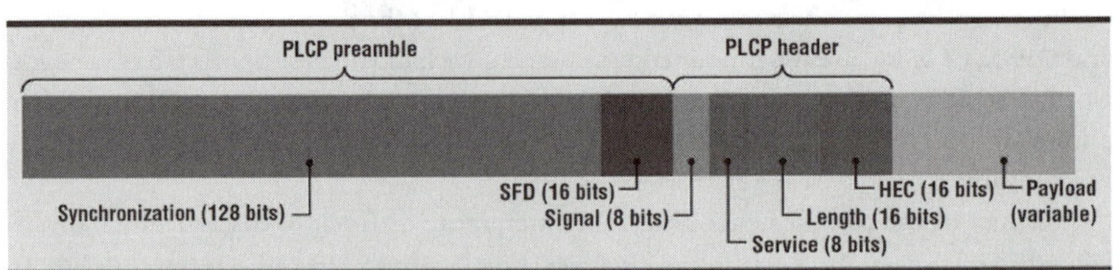

Figure 6.10 The IEEE 802.11 PHY frame using DSSS.

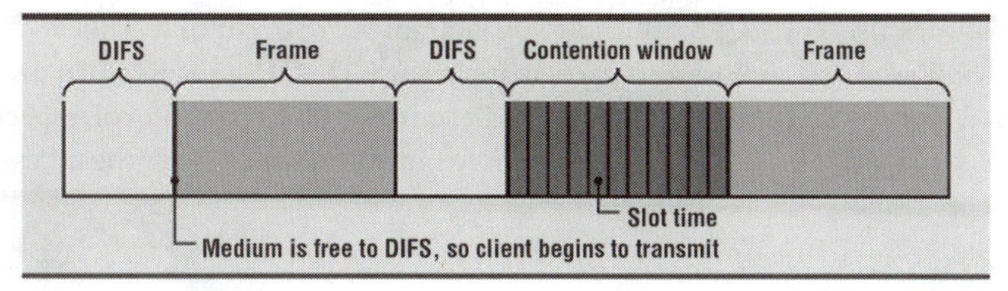

Figure 6.11 802.11b contention window.

that ensures fairness. Each node starts a random back-off timer when waiting for the contention window. This timer ticks down to zero while waiting in the contention window. Each node gets a new random timer when it wants to transmit. This timer isn't reset until the node has transmitted (see Fig. 6.11).

The Hidden-Node Problem

The hidden-node problem can occur where walls and other structures create obscure radio coverage areas. To handle this situation, an RTS/CTS (request to send/clear to send) is specified as an optional feature of the IEEE 802.11b standard. RTS/CTS solves the hidden-node problem in the following fashion:

Referring to Fig. 6.12, when node A wants to transmit data to node B, it first sends an RTS packet. The RTS packet includes the address of the receiver of the data transmission ensuing and the duration of the whole transmission, including the ACK related to it. Node B hears this request (as do nodes D and E). Node A must use the standard transmission method to obtain access to send the RTS packet. Once the receiving host receives the packet, that host replies with a CTS message that includes the same duration of the session about to happen. When node B replies with this CTS message, node C (and F and G) hears this response and is made aware of the potential collision, and will hold its data for the appropriate amount of time, preventing a collision. If every node

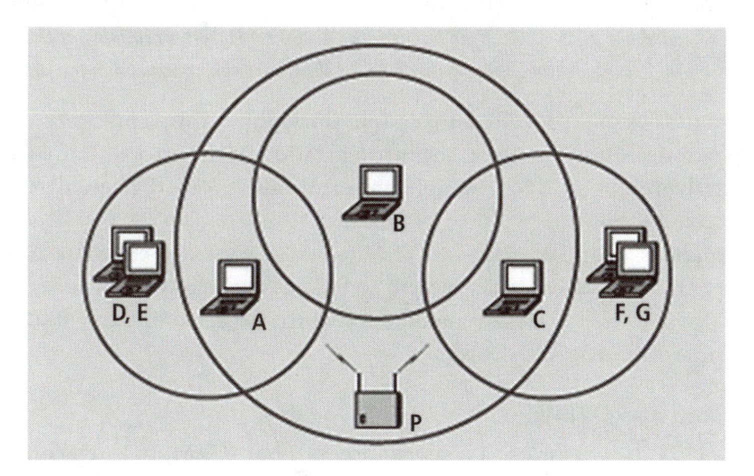

Figure 6.12 The hidden node problem. Workstations A, B and C can all see wireless access point P. Workstations A and B can see one another, and B and C can see one another, but A can't see C.

on the network uses RTS/CTS, collisions are guaranteed to occur only while in the contention window. Access points also participate in the RTS/CTS process when necessary.

RTS/CTS, however, adds significant overhead to the 802.11b protocol, especially at small packet sizes. If used, RTS/CTS thresholds must be set on both the access point and the client side.

Power Level Influences

The FCC limits the power output of the 802.11b system to one watt EIRP equivalent (or effective) isotropic (or isotropically) radiated power. At this low power level the physical distance between the transmitting devices becomes an issue due to signal attenuation, with error performance suffering as the distance increases. (Note: 100 meters / 328 feet is a pretty good rule of thumb for an 802.11b WLAN with clear line-of-sight.)

Any dense physical obstructions between transmitter and receiver add considerably to the problem. Therefore the devices adapt to longer distances, physical obstructions and other factors that impact signal strength by using a less complex encoding technique, resulting in lower signaling speed, which translates into a lower data rate.

For example, a system running at 11 Mbps using CCK and QPSK might throttle back to 5.5 Mbps by halving the signaling rate as the distances increase, as doors and walls get in the way, and as bit errors increase.

The situation may get even worse when you move your laptop out to poolside on a sunny summer afternoon. The distance from an access point and interference encountered as the signal travels that distance affects a signal's quality. Thus the system might throttle back to two Mbps using only QPSK, or even one Mbps using BPSK, to keep the connection from degrading. This process is much the same as that used by conventional fallback modems that might initiate a call at 56 Kbps (actually 53.3 Kbps), and "fall back" to rates of perhaps 28.8 Kbps or 14.4 Kbps as the quality of the dial-up PSTN connection degrades.

***Note:** The actual throughput of an 802.11b system is much less than the raw bandwidth. Physical Layer overhead consumes 30%-50% of the bandwidth. An 802.11b system running at the full rate of 11 Mbps therefore, provides throughput of only 6 to 7 Mbps, assuming overhead in the range of 40%.*

If there are a lot of errors in transmission, throughput drops precipitously, as the receiving station must advise the transmitting station of the errored frames and then wait for retransmissions. If, for example, the error rate is 50%, the actual throughput drops to about 2 Mbps.

This scenario is a blend of a best-case 11 Mbps and a worst-case error rate. In actuality, such an error rate would cause the system to fall back to a lower transmission rate of perhaps 2 Mbps, at which rate the combination of the Barker code and QPSK would be used and the error rate would drop.

Power Savings and DTIMs

By default, 802.11b systems use a constant access mode (CAM) to listen constantly to the network and get the data they need. When power utilization is an issue, however, the workstations and access points can be configured for polled access mode (PAM). With this, the client devices on the network wake up at a regular interval and listen for a special packet called a traffic information map (TIM) from the access point. In between TIMs, the client radio shuts off and thus conserves power. All the devices on the network share the same wake-up period, as they must all wake up at exactly the same time to hear the TIM from the access point.

The TIM informs certain clients that data is waiting at the access point. A client card stays awake when the TIM indicates it has messages buffered at the access point until those messages are transferred, and then the card goes to sleep again. The access point buffers the data for each card until it receives a poll request from the destination station. Once the data is exchanged, the station goes back into power-saving mode until the next TIM is transmitted. Tests run by *Network Computing* magazine discovered that PAM mode could save power by as much as 1000 percent, depending on the volume of traffic on the network.

The access point indicates the presence of broadcast traffic with a delivery traffic information map (DTIM) packet. The DTIM timer is always a multiple of the TIM timer and is often adjustable at the access point. Setting that value high cuts down on the amount of time the station must stay awake checking for broadcast traffic. However, a higher DTIM timer means that the radio will stay on longer to receive DTIM traffic when it does come up in the time cycle.

In a typical enterprise environment, two or more access points will provide signals to a single client. The client is responsible for choosing the most appropriate access point based on the signal strength, network utilization and other factors. When a station determines the existing signal is poor, it begins scanning for another access point. This can be done by listening passively or actively probing each channel and waiting for a response.

Once information has been received, the station selects the most appropriate signal and sends an association request to the new access point. If the new access point sends an association response, the client has successfully transferred to a new access point. This is called "make, then break" behavior.

The Channels

The 802.11b specification divides the assigned RF spectrum into 14 channels. The FCC allows the use of 11 channels (1 through 11). Since the U.S. 2.4 GHz band is only 83 MHz wide, and the 802.11b channels are 25 MHz wide, only three channels can be used simultaneously. So not only is the amount of spectrum highly limited, but not all of it is used.

Furthermore, there also is overlap between adjacent channels (e.g., channels two and three), which affects performance. This overlap, therefore, requires that any given system maintain maximum channel separation from other systems in proximity.

When designing international wireless LANs, you must choose channels with the least common denominator because different local regulatory bodies allow the use of a varying number of channels. For example, Japan allows the use of only one channel, but the U.K. allows the use of channels 1 through 13. However, none allow the use of all 14.

The transmitter's modulator translates the spread signal into an analog form with a center frequency corresponding to the radio channel chosen by the user. The following table identifies the center frequency of each channel:

802.11B TRANSMIT FREQUENCIES

Channel	Frequency (GHz)
1	2.412
2	2.417
3	2.422
4	2.427
5	2.432
6	2.437
7	2.442
8	2.447
9	2.452
10	2.457
11	2.462
12	2.467
13	2.472
14	2.484

The Future

It is expected that over the next few years, 802.11b networks will be slowly phased out. There is still much debate over whether this popular WLAN standard's successor will be 802.11a or 802.11g.

802.11g

The IEEE approved 802.11g on June 12, 2003 after many long and often-heated debates. The final 802.11g specification seems technically sound and is as much a tribute to the IEEE's democratic principles as to the organization's persistence. The proposal offers a solid migration path from 802.11b to higher data rate wireless networking, with a data transfer rate that equals that of 802.11a, while providing a theoretically greater reach than 802.11a systems.

The charter of the 802.11g Task Group was to develop a higher speed extension (up to 54 Mbps) to the 802.11b PHY, while operating in the 2.4 GHz band and implementing all mandatory elements of the IEEE 802.11b PHY standard. Nevertheless, in early 2002, the 802.11g Task Group opted not to use 802.11b's DSSS modulation technique and instead decided to use OFDM as the basis for providing the higher data rate extensions. However, to provide backwards compatibility with 802.11b, the specification supports Complementary Code Keying (CCK) modulation (which 802.11b also uses) and, as an option for faster link rates, it also allows packet binary convolutional coding (PBCC) modulation. See Fig. 6.13, which sets out 802.11g's modulation schemes and their corresponding data rates.

IEEE 802.11g mode	Throughput with 1,500B packets using DCF (Mbits/s)
1 Barker	0.9
2 Barker	1.8
5.5 CCK or PBCC	4.5
11 CCK or PBCC	7.9
22 PBCC	12.5
24 OFDM	15.3
36 OFDM	19.8
54 OFDM	24.3

Figure 6.13 802.11g modulation schemes and corresponding data rates.

Both mandatory and optional aspects are included in the 802.11g standard. The mandatory aspects include the use of OFDM to support higher data rates and support for CCK to ensure backward compatibility with existing 802.11b radios. The optional elements are CCK/OFDM and packet binary convolutional coding (PBCC). Developers may elect to include either optional element or omit both options entirely.

The Mandatory Elements

IEEE 802.11a and 802.11g now share a common high-rate waveform (coded OFDM) and offer complementary advantages to end-users. 802.11a systems enjoy more spectrum at 5 GHz, thus allowing for more channels and, by extension, more users. On the other hand, 802.11g systems provide backward compatibility with existing Wi-Fi devices and offer a range advantage relative to systems operating at 5 GHz.

Every packet of transmitted data can be thought of as consisting of two main parts: a preamble/header and a payload. The preamble/header alerts all radios sharing a common channel that data transmission is beginning. The preamble is a known sequence of 1's and 0's and allows radios to get ready to receive data. The header immediately follows the preamble and conveys several important pieces of information, including the length (in microseconds) of the payload. Other radios will not begin transmission during this

A DUAL-BAND FUTURE

The emergence of IEEE 802.11g is extremely beneficial for the WLAN market. In the longer term, 802.11g represents an important step toward the realization of dual-band (2.4 GHz and 5 GHz) radios.

The use of OFDM in the 2.4 GHz band will facilitate the development of dual-band radios. The reason is quite simple: developers of dual-band radios will need OFDM capability for 5-GHz operations and CCK capability to support Wi-Fi at 2.4 GHz. By using OFDM at 2.4 GHz, implementing 802.11g in a dual-band device will require no additional complex hardware.

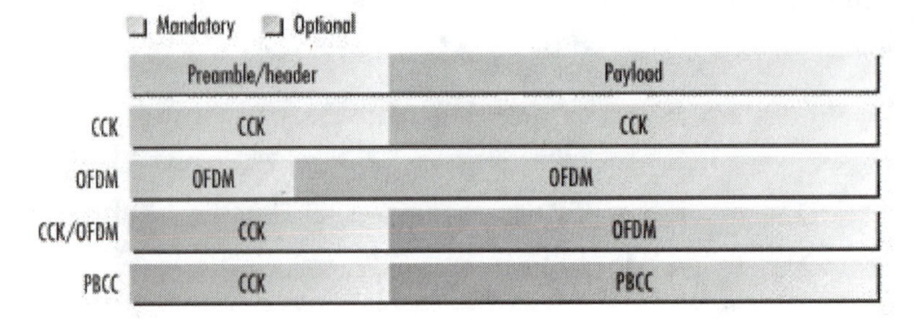

Figure 6.14 802.11g is a PHY extension to the 802.11b standard, although there are many differences between the two. For instance, 802.11g differs in packet format. While the only mandatory modes are CCK for backward compatibility with existing 11b radios and OFDM for higher data rates, developers can choose two optional elements, CCK/OFDM and packet binary convolutional coding (PBCC).

period, thus preventing a network collision. The preamble/header and the payload are normally sent using the same modulation format (CCK for example).

Since CCK is the modulation format for current IEEE 802.11b systems, the preamble/header and the payload can both be transmitted using CCK modulation. CCK, however, is a single carrier waveform, whereby data is transmitted by modulating a single radio frequency or carrier. On the other hand, OFDM, as a multi-carrier access scheme, splits up the data among several closely spaced sub-carriers and modulates each carrier using the same binary phase shift keying (BPSK) as used in 802.11b. This multi-carrier feature helps OFDM provide very reliable operation, even in the presence of severe signal distortion resulting from multipath. In addition, OFDM systems can support higher data rates than single carrier systems, without incurring a huge penalty in terms of complexity. So, for data rates up to 11 Mbps, CCK is a good option; however, as data rates go higher, OFDM becomes the clear choice.

OFDM employs a much shorter preamble length than CCK–it is just 16 microseconds in length, as compared to 72 microseconds for CCK. This shorter preamble reduces network overhead.

The Optional Elements

The optional elements included in the 802.11g specification are as follows:

- CCK/OFDM is a hybrid of CCK and OFDM designed to facilitate use of the OFDM waveform while supporting backward compatibility with existing CCK radios. CCK is used to transmit the packet preamble/header and OFDM is used to transmit the payload. CCK/OFDM supports data rates up to 54 Mbps. The distinctions between OFDM and CCK/OFDM are explained in more detail below.
- PBCC is a "single carrier" solution. This waveform can also be described as a hybrid. It uses the CCK to transmit the header/preamble portion of each packet and PBCC to transmit the payload. PBCC supports data rates up to 33 Mbps.

With CCK/OFDM, the CCK header alerts all legacy Wi-Fi devices that a transmission is beginning, and informs those devices of the duration (in microseconds) of that

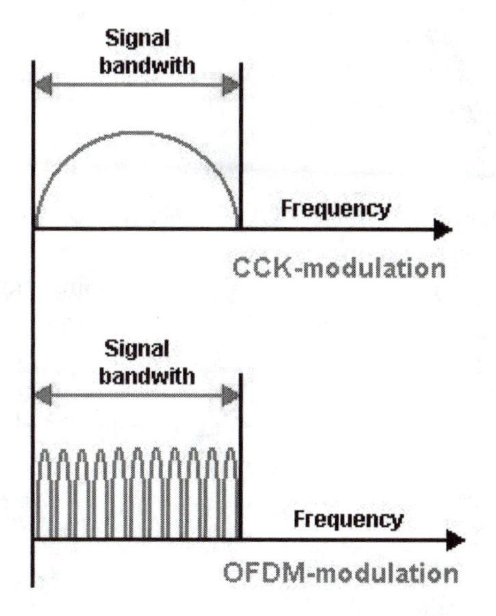

Figure 6.15 This graphic shows the difference in CCK and OFDM modulated waveforms.

transmission. The payload can then be transmitted at a much higher rate using OFDM. Even though existing Wi-Fi devices will not receive the payload, collisions are prevented because the preamble/header is transmitted using CCK.

The mandatory OFDM waveform can also coexist and operate with non 802.11g devices. However, a different method referred to as "request to send/clear to send" or "RTS/CTS" is required. This method is described in greater detail below.

Like CCK, PBCC is a single-carrier system, but that's where the similarity ends. It is a more complex signal constellation (8-PSK for PBCC vs. QPSK for CCK) and it employs a different code structure. PBCC can also be thought of as a hybrid waveform because it uses a CCK preamble/header with a PBCC payload. Note though that the maximum data rate for the PBCC option is 33 Mbps. It is very likely that most IEEE 802.11g radios will implement only the mandatory modes. Thus the remainder of this section describes how radios using OFDM modulation (OFDM preamble/header and OFDM payload) can interoperate with existing Wi-Fi radios (CCK preamble/header and CCK payload).

Normally, all of the radios on a given channel share access to the airwaves by means of a "listen-before-talk" mechanism, referred to as "carrier sense multiple access/collision avoidance" or "CSMA/CA." In simple terms, the radios listen to determine if another device is transmitting. Each radio on a channel waits until there is no other transmission in progress before beginning to transmit. There are additional provisions to reduce the probability that more than one radio will attempt to transmit at the same moment.

Thus the salient points of the 802.11g standard are:

- Support for CCK modulation is mandatory in order to ensure backward compatibility with existing 802.11b systems.
- Support for OFDM modulation is mandatory for data rates >20 Mbps.
- Both CCK/OFDM and PBCC are optional. Vendors can choose to implement PBCC, CCK/OFDM, or not to use either option.

Given that the mandatory OFDM waveform is capable of data rates up to 54 Mbps, it is very likely that many IEEE 802.11g radios will implement the mandatory modes only, and not include either of the optional elements.

Interoperability

Now we will look at how radios using OFDM modulation (OFDM Preamble/Header and OFDM Payload) can interoperate with existing 802.11b radios (CCK Preamble/Header and CCK Payload). Not an easy task, since to do so, they must address the problem that existing 802.11b devices can only receive CCK transmissions, including answering the question of if existing radios cannot receive OFDM transmissions, how they will avoid colliding with those same transmissions. Furthermore, the CSMA/CA mechanism is not suitable when CCK radios and OFDM radios operate on the same channel.

Fortunately, a mechanism already exists in the 802.11 protocol that addresses these problems very efficiently. That mechanism is *request to send/clear to send* (RTS/CTS). Let's examine how RTS/CTS works with 802.11g devices.

Normally, all radios sharing a given channel (including the access point) can "hear" one another. However, this is not always the case. There are instances when all radios can hear and be heard by the access point (AP), but they cannot hear each other. Under those conditions, the listen-before-talk mechanism would break down because radios might detect a clear channel, and begin transmitting to the AP, while the AP is already in the process of receiving another transmission from a "hidden" radio. This hidden node requires the use of the RTS/CTS feature (as discussed in the 802.11b section). Under the RTS/CTS mechanism, each node must send an RTS message to the AP and receive a CTS reply before transmission can begin.

The situation of CCK and OFDM radios operating on the same channel is analogous to the 802.11b hidden node problem—CCK radios cannot detect OFDM transmissions. And the 802.11g specification uses the same mechanism, RTS/CTS, to address the issue. But, with RTS/CTS at the helm, 802.11g OFDM radios can operate on the same channel as existing 802.11b radios, without collision.

While the RTS/CTS mechanism results in additional network overhead, the penalty is fairly modest. The benefit is a migration path to higher data rates for radios operating in the 2.4 GHz band. In the future, when it is expected that networks may make exclusive use of OFDM in the 2.4 GHz band, there will no longer be a need to use RTS/CTS.

Range

Since 802.11g uses OFDM, if OFDM's benefits are extrapolated to the lower frequency 2.4 GHz band, its range should be around 50 percent greater than that of either 802.11a or 802.11b. Also, since the coverage area depends on the range squared, 802.11g could cover the same area as the other systems with fewer than half as many access points. Thus, in the long term, 802.11g's range could be its greatest selling point.

Of course, increased range isn't always a benefit. Because every user shares the available bandwidth, a larger range just spreads it out more thinly. This means that 802.11g is a good choice in environments containing few users, or where users don't need a high-speed connection, e.g. facilities such as warehouses, but probably not offices or homes.

Crowded areas such as conference centers and airports need the highest density of coverage they can get, and will eventually move to 802.11a. However, with the large 802.11b installed base, this group will likely to stick with 802.11b for a little while longer. Since IEEE 802.11g is compatible with this installed base, and since products built upon the final 802.11g specification are available, expect to see a dog fight between dual-mode 802.11a/b systems and 802.11g. After all, it took a long time for 802.11g to become ratified, so the dual-mode systems have had a chance to gain a foothold in the marketplace.

***Note:** You should also note that 802.11g handles signal reflection much better than 802.11b. Radio signals bounce off different material at different angles and speeds. A receiver must reconcile all the different reflections of the signal, which because of reflection, arrive at slightly different times. 802.11g (and 802.11a) slices up the spectrum in such a way that receivers can handle these reflections in a more effective way than 802.11b.*

The longer range also causes another problem—the signal is more likely to "leak." If you haven't set up a secure system, intruders can crack into your network from further away. It also means that you're probably jamming somebody else's airwaves. Both are issues in skyscraper office buildings that house several companies. However, this problem can be overcome by using access points with directional antennae, which focus their transmission and reception on a specific area. The most common directional antennae radiate in an arc rather than a full sphere. They can attach to a wall and only provide coverage on one side of it. More complex antennae are available that can adjust to cover different shaped regions, but these usually require trained radio engineers to set up.

Directional antennae are frequency-specific, which could lead some users to choose 802.11g over 802.11a. The former is based on the same frequency as 802.11b, and hence could re-use the same antenna; the latter would need a new one. A dual-mode 802.11a/b access point requires two separate antennae. This applies to regular (omni-directional) antennae too, but these are cheap to mass-produce. There's one built into every interface card, and vendors don't see any problem in miniaturizing them enough to produce dual-mode cards.

Caveat

Although the IEEE standards board has ratified 802.11g, the FCC will need to approve the use of OFDM in the 2.4 GHz band (a necessary action when one messes with the PHY). Also, it is expected that 802.11g will run into quite a few roadblocks on its way to global acceptance, due to the local regulatory hurdles that this new standard must overcome.

Furthermore, since the lack of a finalized standard didn't stop leading vendors from introducing new 802.11g-compliant gear, some of the 802.11g products on the shelves may be built around a *draft version* of the standard. Thus those products are proprietary products. That means that if you use an 802.11g product that doesn't sport a Wi-Fi certification label, there might be serious interoperability problems in the future, when Wi-Fi certified products 802.11g products hit the market. Or perhaps not—those products may work flawlessly with the tested and certified 802.11g gear that should be available by late 2003. But, as an early 2003 Garner report warns, organizations that jump the gun on 802.11g are taking a significant risk—a risk they should factor into the true cost of their 802.11g investments.

A Standards War?

While both 802.11a and 802.11g offer theoretical bandwidths of up to 54 Mbps in the labs, 802.11a delivers its load over the 5 GHz spectrum, while 802.11g uses the 2.4 GHz spectrum and is therefore reverse compatible with the dominant 802.11b standard. However, while early tests show 802.11g achieving slightly better range than 802.11a, unlike 802.11a, 11g fails to maintain its throughput performance at the outer extremes of its range.

That's just a taste of why there are such heated debates concerning the use of 2.4 GHz "b" and "g" versus 5 GHz "a," not only within the industry, but also among its experts and the media. As illustration, Apple has shunned 802.11a in favor of 802.11g. But others in the industry—most notably Intel—have been pushing 802.11a as a faster follow-on to the popular 802.11b.

Tom Mitchell, CEO of RadioLAN Marketing Group, a company that makes bridging equipment, heartily disagrees with the 802.11g cheerleaders. According to Mitchell, in bridging, "compatibility is a bad thing." Companies don't want 802.11b or g gear to sniff out their wireless data. He goes on to state that 5 GHz reduces interference and is more controlled than the more prevalent 2.4 GHz band.

However, Cho Yong-cheon, CEO of wireless LAN equipment maker Acrowave, sees both sides of the coin, "802.11g will likely get ahead of 802.11a with its interoperability, but 802.11a will eventually win over 802.11g because it is more frequency efficient." He goes on to say, "We plan to develop both types to better cope with the market, however."

It's not hard to find leading chip manufacturers rushing to cash in on 802.11g: Intersil and Broadcom already have 802.11g chips in products and some of those products should already be on store shelves. Rich Redelfs, president and CEO of Atheros, says that although his company shipped over one million 802.11a chipsets in 2002, the company is gradually shifting its focus toward combo 802.11a/g boards. Redelfs says over time, the market for 802.11a or 802.11b products will shrink as 802.11a/g combination chipsets dominate. But Atheros also states that both 802.11a and 802.11b will find new lives in embedded devices and as part of home multimedia networks.

Among the pundits, there's Will Strauss, an analyst for research firm Forward Concepts, who is of the opinion that "802.11a has reached a dead end." At the same time, you can find analysts such as those at research and consulting firm Frost & Sullivan who state just the opposite—they believe that there will be a migration towards higher speed WLAN technologies operating in the 5 GHz band because the spectrum generally contains less potential interference. Also disagreeing with those saying that 802.11a-only products are destined to be just a footnote in the history of wireless networking is Allen Nogee, analyst for In-Stat/MDR. According to Nogee, 802.11a's large data pipe and multiple available channels make it ideal for video streaming, conferencing or education.

Even the U.S. Department of Defense is joining the fray. The DOD is calling for limits on the use of the middle of the 5 GHz band in the United States. The Pentagon argues that such gadgets interfere with military radar.

The Future

As the reader should now understand, there is a lot of room for debate on these issues and the author has much more to say about the pros and cons of all three standards in Chapter 10. For now it's enough to say that, in most cases, a 2.4 GHz installation is the way to go for common office applications, since 2.4 GHz products are inexpensive and capable of supporting most application requirements. But there will always be situations that can strongly benefit from the use of 5 GHz, e.g. densely populated environments and networks that support multimedia applications.

802.11c

This missing amendment was to have provided the information needed to ensure proper bridge operations, but it was never published. Instead, a bridging layer was incorporated into the 820.11d standard.

802.11d

This is referred to as the "global harmonization standard." It was ratified in 2001. 802.11d defines Physical Layer requirements that conform to international regulatory requirements. As such it promotes worldwide use of Wi-Fi.

When the original 802.11 standard first became available, only a handful of regulatory domains, i.e. U.S., Europe, and Japan, had rules in place for the operation of wireless LANs based on the 802.11 series of standards. To support a widespread adoption of Wi-Fi, the 802.11d Task Group's crafted a definition of PHY requirements that would satisfy the unique regulatory requirements for channelization demanded by various local regulatory bodies throughout the world.

Making new use of existing spectrum always requires testing and assurances that new users won't interfere with legacy users and new applications won't affect adjacent spectrum bands. In some regions, the spectrum used by Wi-Fi equipment is adjacent to commercial radar systems. In other areas, Wi-Fi devices are thought to be prone to interference from cordless phones or microwaves.

The 802.11d Task Group has an ongoing charter to define PHY requirements that satisfy various regulatory needs as they develop. This is especially important for operation in the 5 GHz bands because the use of those frequencies differ widely from one country to another.

802.11e

This specification promises to bring Quality of Service (QoS) to Wi-Fi networks. The 802.11 Working Group realized that the original 802.11 standard and its amendments, a, b, and g, don't provide for an effective mechanism to prioritize traffic. Without such a mechanism, there can't be any strong quality of service, which means that Wi-Fi can't optimize the transmission of audio and video.

The 802.11e Task Group defines QoS by providing classes of service with managed levels of QoS for data, voice, and video applications. It does this by addressing the issue of QoS in the Data Link Layer's MAC sublayer. In particular, the Task Group is working on a prioritized scheme that can be used to ensure that high priority users get more bandwidth allocation than low priority users. To do this, it replaces the Ethernet-like

MAC sublayer with a coordinated Time Division Multiple Access (TDMA) scheme, and adds extra error-correction to prioritized traffic. This is accomplished by the use of what is being termed a "hybrid coordination function" (HFC) at the MAC. The HCF uses an enhanced, contention-based channel access method that operates at the computing device concurrently with a polled channel access mechanism that operates at the access point.

Because 802.11e falls within the MAC sublayer, it will be common to all 802.11 PHYs standards (e.g. 802.11a, b, and g) and be backward compatible with all existing wireless LANs based on the 802.11 series of standards. As a result, the lack of a finalized 802.11e specification shouldn't impact a decision on which Wi-Fi flavor to use when deploying a new WLAN. It should be relatively easy to upgrade any existing access points to comply with 802.11e, once it is ratified, through relatively simple firmware upgrades.

Still, 802.11e is critical for service providers who want to offer audio and/or video on demand and/or voice-over-IP service. "You're never going to be able to stream high-quality video from your PC to your TV in the home without Quality of Service. If you want wireless networking to grow beyond resource sharing in the home, you have to ensure quality of service in content delivery," says Barry Davis, Intel's director of platform architecture in the Mobile Communications Division.

Quality of Service also ranks high for enterprises, particularly as voice-over-IP (VoIP) becomes more important in wireless communications.

***Note:** Many WLAN manufacturers have targeted QoS as a feature to differentiate their products, but be advised that the QoS features being offered with these products are proprietary. So, if you are using such a solution to deploy a WLAN, you must standardize on that vendor's products.*

There have been innumerable delays, thanks to arguments over how many classes of service should be provided and exactly how they should be implemented. However, it appears as if most of the issues have been resolved and that 802.11e might even be ratified by the time you read this.

802.11f

The IEEE approved 802.11f on June 12, 2003. 802.11f is not a specification, per se. Instead, it's a "recommended practice" document, meaning that vendor compliance is completely voluntary. The document was drafted with the goal of improving the handover mechanism in Wi-Fi networks, so that end-users can maintain a connection while roaming between two different switched segments (radio channels), or between access points attached to two different networks. This is vital if Wi-Fi networks are to offer the same mobility that cell phone users take for granted.

To reach the stated goal, the 802.11f Task Group found a method that will provide access point (AP) interoperability within a multivendor WLAN network. Thus the 802.11f document defines the registration of access points within a network and the interchange of information between APs when a user is handed over from one access point to another.

The original 802.11 Working Group specifically left out the definition for this element in order to provide flexibility in working with different distribution systems (i.e. wired backbones that interconnect access points). However, this omission has caused unforeseen problems. For instance, users roaming between APs may experience a loss of some data packets, especially when moving between APs manufactured by different vendors because many times they don't easily interoperate to support roaming.

802.11f serves to reduce vendor lock-in and allow multi-vendor infrastructures, and still provide the necessary information that access points need to support 802.11 distribution system functions, such as roaming.

In the absence of 802.11f, when installing a WLAN, you should employ the same vendor for access points to ensure interoperability for roaming users (although a mix of access point vendors might work correctly, especially if the APs are Wi-Fi-certified).

***Note:** The inclusion of 802.11f in access point design will open up WLAN design options and add some interoperability assurance when selecting access point vendors.*

It remains to be seen, however, whether manufacturers will adopt 802.11e features. "Why do you need it," scoffed one Cisco wireless LAN marketing executive at a recent industry trade show.

802.11h

This specification addresses the requirements of the European regulatory bodies in that it attempts to add better control over transmission power and radio channel selection to 802.11a. As such, 802.11h is actually "spectrum managed 802.11a."

In Europe (and elsewhere), there's worry about 802.11a transmissions interfering with satellite communications, which have "primary use" designation—most countries authorize WLANs for "secondary use" only.

European radio regulations for the 5 GHz band require products to have transmission power control (TPC) and dynamic frequency selection (DFS). TPC limits the transmitted power to the minimum levels needed to reach the farthest user. DFS selects the radio channel at the access point to minimize interference with other systems, particularly radar. Thus 802.11h is being written to avoid interference through the use of DCS and TPC, which makes it similar to HiperLAN/2, the dormant European-based competitor to 802.11a. To implement DCS and TPC, the 802.11h Task Group has developed associated practices that affect both the MAC and the PHY.

Ratification of 802.11h will provide better acceptability within Europe for IEEE-compliant 5 GHz WLAN products. (OEMs and systems firms including 802.11 interfaces in their products hopefully can expect pan-European approval of 802.11h sometime during the second half of 2003.)

Although European countries, such as the Netherlands and the U.K., currently allow the use of 802.11a under the condition that TPC and DFS must also be present, pan-European approval of the 802.11h standard (along with 802.11e) could be just the ticket to making 802.11a acceptable to many, if not all, local regulatory bodies.

802.11i

Security is wireless networking's major weakness. Of course, vendors have not improved matters by shipping products with default security features set as "none." The 802.11i Task Group is hard at work on a specification that will dramatically enhance the security features provided in a WLAN system. It has been working on this specification for more than two years, but now the final specification seems to be at hand. After two years of rousing debate, the body responsible for the Wi-Fi standard is finally putting the finishing touches on its new security standard, IEEE 802.11i.

802.11i will address many of Wi-Fi's security issues. To accomplish this task, it is expected that since most 802.11 security mechanisms are seated in the Data Link Layer's MAC sublayer, the 802.11i Task Group will update the MAC, and that those updates will then apply to all 802.11 PHY standards—802.11a, 802.11b, and 802.11g.

802.11i focuses on several security enhancements to temporarily support Wired Equivalent Privacy (WEP). These enhancements are collectively known as the Temporal Key Integrity Protocol (TKIP). The security enhancements include an authentication protocol, key-hashing function, combined with a real message integrity check (to avert forgery), and dynamic key management (i.e. rekeying). The 802.11i Task Group also is taking the necessary steps to assure that TKIP is backward compatible with WEP.

This section will examine the key technical elements that have been defined by the 802.11i Task Group to date. While these elements might change, the information set forth herein should provide the reader with some insight into the security features that 802.11i promises.

***Note:** I want to thank CMP Media LLP and Dennis Eaton, the author of "Diving into the 802.11i Spec: A Tutorial" (dated November 26, 2002), which was published on the CMP Media LLP's CommsDesign website, inasmuch as the author used the information gleaned from that article to write the following text.*

For this discussion, we will divide the proposed 802.11i specification into three main sections, organized into two layers. On the lower level are improved encryption algorithms in the form of the temporal key integrity protocol (TKIP) and the counter mode with CBC-MAC protocol (CCMP). Both of these encryption protocols provide enhanced data integrity over WEP, with TKIP targeted at legacy equipment and CCMP targeted at future WLAN equipment.

Above TKIP and CCMP sits what is now referred to as "802.1X," a standard for port-based access control developed by a different body within the IEEE 802 organization. As used in 802.11i, 802.1X provides a framework for robust user authentication and encryption key distribution, both features originally missing from the original 802.11 standard.

These three security meastures work together to form an overall security system. However, to understand how all three of these methods fit together, the reader must first understand how they operate individually. So, let's take a look at each one in more detail. We will start with 802.1X.

802.1X. IEEE 802.1X is a standard for port-based network access control. It can be applied to both wired and wireless networks. Further, it provides a framework for the centralized authentication of users or stations, for encryption key distribution, and it can be used to restrict access to a network until the end-user is authenticated. 802.1X also is used in conjunction with one of a number of upper layer authentication protocols (discussed later), to perform verification of credentials and generation of encryption keys. Thus, the standard's flexibility allows for multiple authentication algorithms, and since it is an open standard, multiple vendors can offer innovative enhancements.

There are three primary roles played by enterprise equipment in an 802.1X system. The *authenticator* (in a wireless network, this is typically the access point) is the port that enforces the authentication process and routes the traffic to the appropriate entities on the network. The *supplicant*, which in a wireless network is usually the computing device, is the port requesting access to the network. The *authentication server* (AS) is a third entity that performs the actual authentication of the credentials supplied by the supplicant. The AS is typically a separate entity on the wired side of the network, but could also reside directly in the authenticator. The most common type of authentication server in use today to authorize remote users is RADIUS, although other authentication services could be used since a particular authentication server to be used is not specified in the 802.1X standard.

The Controlled/Uncontrolled Port Concept. An 802.1X operation can be understood using the concept of a controlled port and uncontrolled port. (See Fig. 6.16.)

The controlled and uncontrolled ports are logical entities and are the same physical connection to the network. Whether a frame traveling through the access point (AP) is routed through the controlled or uncontrolled port is determined by the authentication state of the client device.

Prior to authentication by the AS, the AP will only allow the client to communicate with the AS. After successful authentication by the AS, the AP will also allow the client to access other services available on the network.

The actual authentication data exchanged is a function of the upper layer authentication protocol used (discussed below); the message protocol and routing of the messages is controlled by 802.1X.

It's important to note that a mutual authentication process is used, and both the network and the client are authenticated to each other. As part of the authentication process, the MAC level encryption keys used by the chosen encryption protocol will be generated. 802.1X is then used to plumb the encryption keys down to the MAC on both the AP and the client device.

In 802.1X-enabled WLAN systems, two sets of keys are generated, *session keys* (also referred to as pairwise keys) and *group keys* (also referred to as groupwise keys). Group keys are shared amongst all the clients connected to the same AP and are used for multicast traffic. Session keys are unique to each association between an individual client and the AP, and create a private virtual port between a client and the AP.

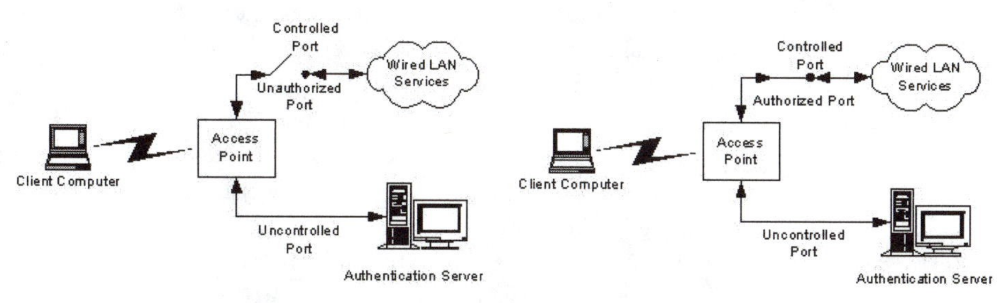

Figure 6.16 802.1X state before (left) and after (right) successful mutual authentication.

802.1X enhances the enterprise security model by providing the following improvements over standard WEP:

* It provides support for a centralized security management model.
* The primary encryption keys are unique to each station so the traffic on any single key is significantly reduced.
* When used with an AS, the encryption keys are generated dynamically and don't require a network administrator for configuration or intervention by the user (this is analogous to the use of dynamic IP addresses versus static IP addresses on the network).
* It provides support for strong upper layer authentication.

802.1X in a SOHO Network. In the home and small business environment most users are not expected to have a RADIUS server available for authentication. In this case, the 802.11i standard uses 802.1X in a pre-shared key configuration, however most of the previous concepts and operation remain the same.

When operating with AS support, a master key, called the pairwise master key (PMK), is generated via the exchange between the client and the AS. The PMK is used as source material for generation of the lower level keys used by the MAC layer encryption. When an AS is not present, the PMK is manually entered into each device in the WLAN and serves as a pre-shared key for authentication and source material of the lower level encryption keys. The user model is more analogous to standard WEP in this case, since it requires manual distribution and configuration of a shared secret; however, this should be adequate for most small deployments.

When used in pre-shared key mode, session keys are still provided and the improved encryption methods discussed below are fully supported. It's important to note that upper-layer authentication is not supported, and the security of the network is broken, if the shared key is ever compromised. In many small deployment scenarios, these tradeoffs are likely acceptable in exchange for ease of deployment and configuration of the Wi-Fi equipment.

Encryption. The 802.11i standard provides two improved encryption algorithms to replace WEP. TKIP and CCMP are set out in the standard. Furthermore, the standard is written in such a way that it is able to support the addition of new encryption protocols, should they be required in the future. Thus a WLAN is able to support the simultaneous use of more than one encryption protocol with the client and AP using the highest level of security that both can mutually support.

However, a true 802.11i system uses either the TKIP or CCMP protocol for all equipment—not both. A WLAN that supports the simultaneous use of WEP along with the CCMP or TKIP encryption protocols is called a "transitional network," and is assumed to be a temporary configuration for the purposes of converting all clients to a TKIP- or CCMP-based security solution.

Let's take a closer look at these two encryption algorithms. We'll start with TKIP.

TKIP: This encryption method was designed to address all the known attacks and deficiencies in the WEP algorithm, while still maintaining backward compatibility with legacy hardware. It was designed to be made available as a firmware or software upgrade to existing hardware, so that users would be able to upgrade their level of security without replacing existing equipment or purchasing new hardware. TKIP provides an upgrade path by offering an additional protocol or a wrapper around WEP.

TKIP is composed of the following elements:

* A message integrity code (MIC) provides a keyed cryptographic checksum, using the source and destination MAC addresses and the plaintext data of the 802.11 frame (or MAC service data unit (MSDU) in IEEE nomenclature). This protects against forgery attacks.
* Countermeasures to bound the probability of successful forgery and the amount of information that an attacker can learn about a particular key.
* A 48-bit IV and an IV sequence counter to address replay attacks. Fragmented packets (MAC protocol data units (MPDUs) in IEEE nomenclature) received out of order are dropped by the receiver.
* Per packet key mixing of the IV is used to break up the correlation used by weak key attacks.

The structure of a TKIP-encrypted MPDU is shown in Fig. 6.17. As mentioned previously, TKIP uses an extended 48-bit IV called the TKIP sequence counter (TSC). The use of a 48-bit TSC extends the life of the temporal key (discussed below) and eliminates the need to re-key the temporal key during a single association. Since the TSC is updated with each packet, 248 packets can be exchanged using a single temporal key before key reuse would occur. Under steady, heavy traffic conditions, it would take approximately 100 years for key reuse to occur.

The TSC is constructed from the first and second bytes from the original WEP IV and the 4 bytes provided in the extended IV. TKIP extends the length of a WEP encrypted MPDU by 12 bytes—4 bytes for the extended IV information and 8 bytes for the MIC.

The TKIP *encapsulation process* is shown in Fig. 6.18. Temporal and MIC keys are used, which are derived from the PMK generated as part of the 802.1X exchange discussed previously.

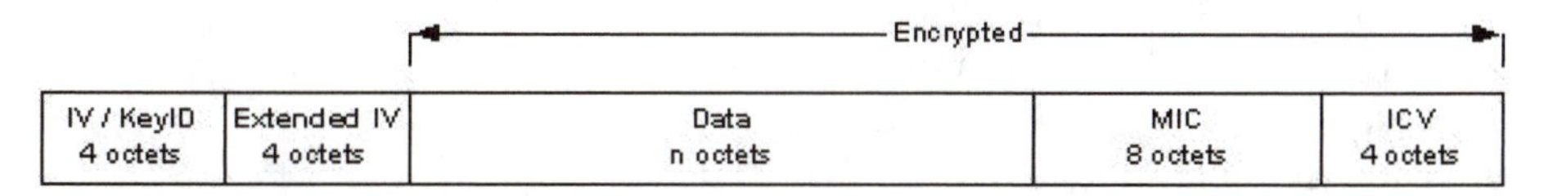

Figure 6.17 MPDU format after TKIP encryption.

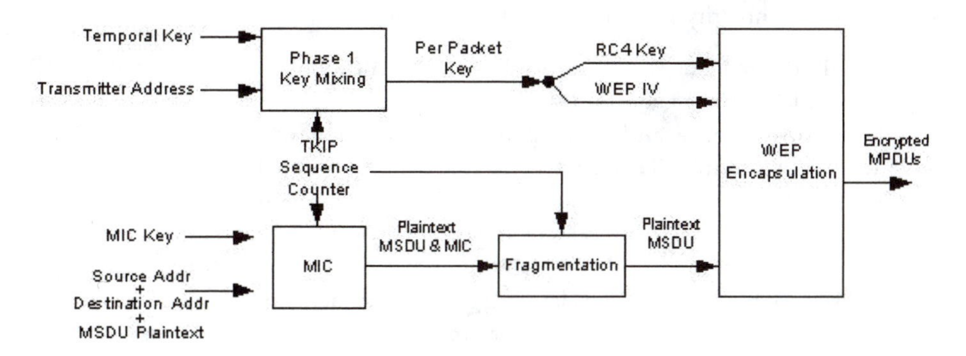

Figure 6.18 This graphic depicts the TKIP encapsulation process.

The temporal key, transmitter address, and TSC combine in a two-phase key mixing function to generate a per packet key to be used to seed the WEP engine for encryption. The per packet key is 128 bits long, and is split into a 104-bit RC4 key and a 24-bit IV for presentation to the WEP engine.

The MIC is calculated over the source and destination MAC addresses and the MSDU plaintext, after being seeded by the MIC key and the TSC. By computing the MIC over the source and destination addresses, the packet data is keyed to the sender and receiver preventing attacks based on packet forgery.

The MIC function, nicknamed "Michael," is a one-way cryptographic hash function, not a simple CRC-32 as is used in computing the WEP integrity check vector (ICV). This makes it much more difficult for an attacker to successfully intercept and alter packets in a denial of service attack. If necessary, the MSDU is fragmented into MPDUs, incrementing the TSC for each fragment, before encryption by the WEP engine.

The *decapsulation process* is essentially the same as the process illustrated in Fig. 6.18 with the following exceptions. After recovery of the TSC from the received packet, the TSC is examined to ensure that the packet just received has a TSC value greater than the previously received packet. If it does not, the packet is discarded in order to prevent potential replay attacks.

Also, after the MIC value has been calculated based on the received and decrypted MSDU, the calculated MIC value is compared to the received MIC value. If the MIC values do not match, the MSDU is discarded and countermeasures are then invoked. These countermeasures consist primarily of rekeying the temporal key, while controlling the rate at which this happens and sending alerts to network administration for follow-up.

To summarize TKIP:

- It was designed as a wrapper around WEP, so as to mask WEP's weaknesses by preventing data forgery, replay attacks, encryption misuse, and key reuse.
- It can be implemented in software.
- It reuses existing WEP hardware.
- It runs WEP as a sub-component.
- It doesn't unduly degrade a system's performance.
- It uses the 128 bit encryption key, with the AP and client device using the same key (TKIP's per-packet key construction makes this kosher), along with two 64-bit data integrity keys–one for the AP and the other for the client device, so each can use different data integrity keys for transmit.

Thus TKIP can be used with legacy equipment. Although not as robust as CCMP, TKIP does allow network administrators to avoid incurring additional expense for new equipment, while significantly upping the security quotient of their existing WLAN.

CCMP: In addition to TKIP encryption, the 802.11i draft defines a new encryption method based on the advanced encryption standard (AES). AES-based encryption can be used in a number of different modes or algorithms. The mode that has been chosen for 802.11 is the counter mode with CBC-MAC (CCM). The counter mode delivers data privacy while the CBC-MAC delivers data integrity and authentication.

AES is a symmetric iterated block cipher, meaning that the same key is used for both encryption and decryption, multiple passes are made over the data for encryption, and the clear text is encrypted in discrete fixed length blocks. The AES standard uses 128-

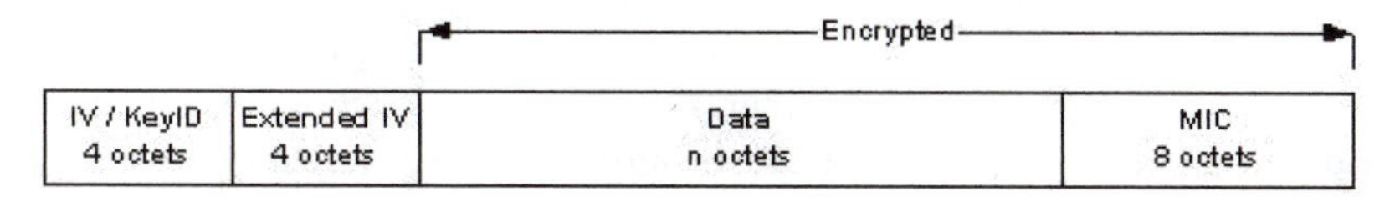

Figure 6.19 Format of a CCMP encrypted MPDU. The packet is expanded by 16 bytes over an unencrypted frame, and is identical to a TKIP frame, with the exception of the legacy WEP ICV included in a TKIP frame.

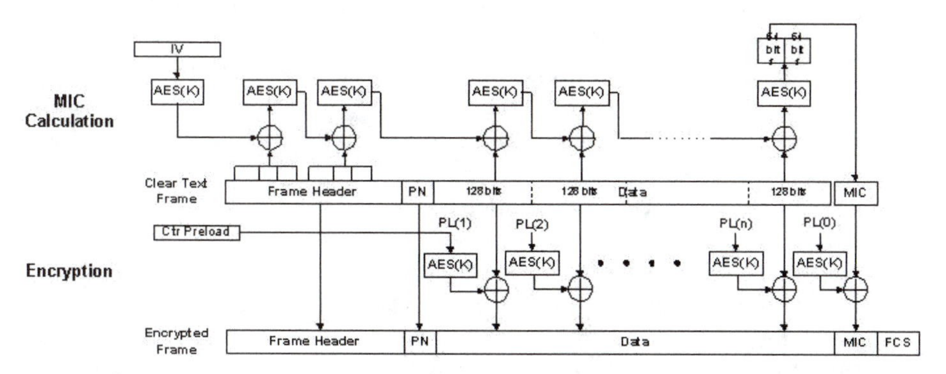

Figure 6.20 Diagram of the CCMP encapsulation process.

bit blocks for encryption. For 802.11, the encryption key length is also fixed at 128 bits. Unlike TKIP, CCMP is mandatory for anyone implementing 802.11i.

Like TKIP, CCMP also uses a 48-bit IV called a packet number (PN). The packet number is used along with other information to initialize the AES cipher for both the MIC calculation and the frame encryption. (Fig. 6.20 shows the CCMP encapsulation process.)

The AES encryption blocks in both the MIC calculation and the packet encryption use the same temporal encryption key (K in Fig. 6.20). As with TKIP, the temporal key is derived from the master key that was derived as part of the 802.1X exchange discussed previously.

The MIC calculation and encryption proceed along parallel paths as shown in Fig.6.20. The MIC calculation is seeded with an IV formed by a flag value, the PN, and other data pulled from the header of the frame. This IV is fed into an AES block and its output is XORed (to hide the plaintext process) with select elements from the frame header, which is then fed into the next AES block. This process continues over the remainder of the frame header and down the length of the packet data to compute a final 128-bit CBC-MAC value. The upper 64 bits of this MAC are extracted and used in the final MIC appended to the encrypted frame.

***Note:** XOR is the common expression used for the "eXclusive OR" binary operation. (It is pronounced "Ex-Ore.") XOR is a fundamental binary operation frequently used in cryptographic algorithms. It operates on binary digits, bits, which take the value 0 or 1. A XOR B is equal to 0 if and only if A = B, i.e. 0 XOR 0 = 0, 0 XOR 1 = 1, 1 XOR 0 = 1, 1 XOR 1 = 0. It's a very fast operation in computers, and it has the useful property that if A XOR B = C, then B XOR C = A as well as A XOR C = B. When viewed in terms of wireless LAN security, the usefulness of XORing becomes more apparent if you look at it as: plaintext XOR keystream = ciphertext. It is then obvious that ciphertext XOR keystream = plaintext, which thus implements a simple and robust reversible transformation of plaintext into ciphertext using a keystream—the fundamentals used in stream ciphers.*

The encryption process is seeded by a counter preload also formed from the PN, a flag value, data from the frame header, and a counter value which is initialized to 1. This preload value is fed to the AES block and its output is XORed with 128 bits of clear text from the unencrypted frame. The counter value is incremented by one and this process is repeated for the next block of 128 bits of clear text. This process continues down the length of the frame until the entire frame has been encrypted. The final counter value is set to 0 and input to an AES block, whose output is XORed with the MIC value computed previously before appending to the end of the encrypted frame for transmission.

The CCMP decapsulation process is not shown but is essentially the reverse of the encapsulation process of Fig. 6.20. A final step is added to compare the value of the computed MIC to that received before the decrypted frame is passed on by the MAC.

To summarize CCMP:

- It provides a long-term security solution for Wi-Fi networks.
- It is based on AES in counter mode encryption with CBC-MAC data origin authenticity, otherwise known as CCM, which is authenticated encryption combining counter mode and CBC-MAC, using a single key (a 128 bit block cipher that was designed for IEEE 802.11i). This allows CCM to provide authenticity and privacy via a CBC-MAC of the plaintext, as appended to the plaintext, to form an encoded plaintext (the encoded plaintext is encrypted to CTR mode), although it can leave any number of initial blocks of the plaintext unencrypted.
- It needs only one fresh 128-bit key, and the same 128-bit Temporal key is used by both the AP and the client device (CBC-MAC IV, CTR constructions make this kosher); the key is configured by 802.1X. Furthermore CCM encrypts packet data payload and protects packet selected header fields from modification.
- It requires new hardware because of AES, e.g. new AP hardware and perhaps new client device hardware, especially for hand-held devices.
- It is a brand new protocol and thus offers few concessions to WEP.
- It protects MPDUs = fragments of 802.2 frames.

DATA TRANSFER SUMMARY

	WEP	TKIP	CCMP
Cipher	RC4	RC4	AES
Key Size	40 or 104 bits	128 bits encryption 64 bit authentication	128 bits
Key Life	24-bit IV, wrap	48-bit IV	48-bit IV
Packet Key Integrity	Concat	Mixing Function	Not Needed
Data	CRC-32	Michael	CCM
Header	None	Michael	CCM
Replay	None	Use IV	Use IV
Key Management	None	EAP-based	EAP-based

* It is intended only for packet environment.
* It does not attempt to accommodate streams.

Thus although for the most part CCMP requires the purchase of new equipment before it can be used, it does provide CCM, which offers wireless networks a provably secure mode of operation.

Authentication Protocols. Upper layer authentication (ULA) protocols are not specified in the 802.11i standard, but will be an integral part of the security system in the majority of deployments. The reason ULA protocols are not included in the 802.11i standard is because, as the name implies, they operate at higher layers of the OSI model and are therefore outside the scope of the 802.11 standards, which operate only at the OSI lower layers—the Physical Layer and the Data Link Layer's MAC sublayer.

There are a number of popular ULA protocols in use today, primarily in the enterprise environment where the network infrastructure is in place to support their use. The ULA protocols are used to provide a mutual authentication exchange between the client and an authentication server residing somewhere on the network, and to generate session keys to be used between the client and the AP over the wireless link.

The ULAs work in conjunction with 802.1X, where 802.1X is used to enforce their use and route the messages properly, and the ULA protocols define the actual authentication exchange that takes place. In most cases, a RADIUS server will be used for authentication since many companies already use RADIUS for their dial-up users.

Some of the more popular authentication protocols include: the extensible authentication protocol with transport layer security (EAP-TLS), the protected extensible authentication protocol (PEAP), the extensible authentication protocol with tunneled transport layer security (EAP-TTLS), and the lightweight extensible authentication protocol (LEAP).

EAP-TLS is a certificate-based authentication protocol and is supported natively in Windows XP. It requires initial configuration by a network administrator to establish the certificate(s) on the user's machine and the authentication server, but no user intervention is required thereafter. The certificates are digital signatures that are used in conjunction with public key encryption techniques to verify the identity of the client.

During an EAP-TLS exchange, the client and authentication server exchange credentials and random data in order to simultaneously synthesize the encryption keys at both ends of the link. Once this has been completed, the server sends the encryption keys to the AP through a secure RADIUS channel, and the AP exchanges messages with the client to plumb the encryption keys down to the MAC encryption layer.

PEAP is an IETF draft standard and can be used to provide a secure password based authentication mechanism. Although it has not been implemented in any products to date, this is likely to change in the near future.

In a PEAP exchange, only the authentication server is required to have a certificate. After the initial communication with the authentication server, the public key from the AS certificate is sent to the client computer. The client computer then generates a master encryption key, encrypting this key using the AS's public key and sending the encrypted key to the AS.

Now that the master key is on both ends of the channel, this key can be used as source material to establish a secure tunnel between the AS and the client, over which any subsequent authentication method can be used to authenticate the client computer to the AS. In many cases is it expected that this will be some form of a password-based authentication protocol.

EAP-TTLS is also an IETF draft standard and can be used to provide password-based authentication of the client computer. EAP-TTLS is very similar in operation to PEAP, and has been implemented in some RADIUS server and supplicant software designed for use in 802.11 WLAN networks.

LEAP is a proprietary standard developed by Cisco Systems, and was designed to be portable across a variety of wireless platforms. It has gained popularity due to the fact that it was the first, and for a long time the only, password-based authentication scheme. It also provides this support across several different client operating system platforms.

LEAP is based on a straightforward challenge-password hash exchange, where the authentication server issues a challenge to the client and then the client returns the password to the authentication server, after first hashing it with the challenge text sent by the AS.

Since 802.11i updates MAC, once it's ratified, the installed base should be able to upgrade existing access points with firmware upgrades. But note that the implementation of new encryption methods like the Advanced Encryption Standard (AES) might require new hardware.

There is no estimated timeline for the ratification of this specification. While it may be finalized sometime toward the end of 2003, don't count on it. For now, owners of WLANs can provide stronger forms of security that go well beyond WEP, by implementing proprietary security mechanisms available from access points vendors. The problem with this idea is that the network providers will probably need to deploy network cards and access points from the same vendor. Another approach is to set up an IPSec virtual private network (VPN) to run over the wireless LAN.

***Note:** One of the mistakes that WLAN manufacturers might make when evaluating the individual security elements of 802.11i is to consider them as individual security silos. They mustn't. All of the 802.11i "pieces" work together to form an overall security system. Taken individually and out of the context of the overall system, any single "piece" could be shown to have security weaknesses.*

802.11j

This is a newly proposed standard. As it now stands, the 802.11j Task Group is mandated to draft a specification that will meet international regulatory requirements, specifically 4.9-5 GHz operation in Japan, which will requirement enhancement to 802.11a PHY and 802.11 MAC. Basically, 802.11j is the equivalent of 802.11h, but it is designed for the Japanese regulatory environment.

802.11k

WLAN QoS stands to benefit from another standard proposal, tentatively labeled 802.11k. The proposed new standard would allow the gathering of detailed information about the communications link between stations and clients. It would standardize the way all 802.11 networks report radio and network performance conditions to other parts of the network stack, to applications, and to administrators and operators for the purpose of network management, fault finding, and other diagnostics. For example, if a network administrator had all the qualitative information about a station, including its performance capabilities, he or she could then know how to provision it downstream.

The gist of 802.11k is to strengthen 802.11e (QoS) by overlaying 802.11k technology. The 802.11k Task Group only came into existence in early 2003, so its work has just begun. The vision of the 802.11k Task Group is to let higher applications see information about wireless access points and clients, even if they're on different subnets. This is an important step in making an enterprise wireless LAN a unified, consistent system, instead of a loose collection of individual subnets. The goal is to make low-level measurements from the PHY and MAC layers of the wireless LAN available to higher-level applications, which can then make decisions and take actions based on this data.

To quote an "unapproved draft" document IEEE document entitled "Radio Resource Measurement Vision and Architecture," dated January 2003:

"The 802.11k vision is to make PHY and MAC layer measurements available to upper layers. This means that it is expected that the upper layers can and will make decisions about the radio environment and what can be accomplished in that environment.

"The most important information is that about the Access Points and the PCMCIA cards (STAs in 802.11 jargon). This information includes all the APs and STA radios that can be seen even if they are not on the same subnets."

The Radio Resource Measurement (RRM) standard will define the requirements for measuring the radio environment, such as:

- Data, Voice, and Video. The Internet has changed and continues to evolve to include voice and video in the data environment it has grown from. The bursty nature of the Internet has been somewhat tamed and garnered to include the added benefits of voice and video in making realistic representations of the real world through networking and computing. Therefore the primary requirement is to enable that voice and video to extend to the radio and wireless LAN environment.
- Rogue Access Points. Companies that are delving into wireless LANs on a big scale must have some means of controlling and managing their environments. This includes the unlicensed radio environment that has become prevalent with the use of wireless LANs. One of the primary requirements is to be able to identify who is in your radio environment and what they are transmitting.
- Quantify WLAN radio topology for AFS and TPC. Also inherent in needing to understand and manage your radio environment are the radio measurements that precipitate a change either in frequency or in the power with which you are transmitting.
- Measure BSS overlap to feed mitigation (802.11e) and help balance coverage, capacity, and QoS. Another requirement is to balance the radio environment to its maximum efficiency. The measurements needed to accomplish this purpose are the loads on the access points and the stations themselves. By measuring or asking for the measurements of the other access points and stations in an area, the WLAN acts as a synchronized network and enables maximum use of that network.
- Quantify each station's local performance to assist admissions control (802.11e) and to facilitate roaming and load balancing. By understanding and measuring the performance of the individual pieces of the network, the whole can facilitate roaming and load balancing of the entire mechanism.

* Detect non-802.11 interference and quantify noise to facilitate adjustments in WLAN configuration (radar). Enabling measurements and adjustments to the radio environment is a requirement if the system is to be capable of understanding its environment.

In practice, the protocol elements that will be specified in 802.11k will be MAC and PHY extensions—most likely MLME and PLME services. The standard will also probably deal with protocol, not decision-making or algorithms. For example, a set of measurements may be defined, but there will be no specific rule as to when these measurements should be made, or how the results should be used.

At this moment, the standard is envisioned more as a means for providing a toolkit for RRM. For instance, specific RRM requirements include:

* Capabilities, measurements, and statistics, i.e. to identify an RRM-capable access point or station, and for a management entity to request that an access point make appropriate measurements to determine the radio environment. Stations would have the ability to make and report measurements on the radio environment. A "Network Management Entity" would be able to obtain configuration and statistics information.
* Improved WLAN information. Information would be available to stations to assist in making association decisions that result in good service characteristics for the individual station, without affecting those experienced by other stations in the network, both wired and wireless. This enables efficient usage of the available resources, which is likely to be particularly important if applications with QoS requirements are being supported.
* Information to streamline roaming, i.e. information would be available to stations to assist in streamlining the roaming process.

These are exactly the problems that wireless "switch" makers are trying to solve (see Chapter 18).

802.11l

IEEE has stated that it has skipped over the "l" designation because it is easy to confuse it with the number "1."

802.11m

The 802.11m Task Group's mandate is to go back and correct any errors in any previous amendments to the 802.11 series of specifications.

802.11n

The 802.11n Task Group is working on a high-speed variant of the existing 802.11 standards. This revision of the 802.11 specifications would double the speed of existing 802.11 standards. The body working on this standard is called the "High Throughput Study Group."

This group is trying to increase the "throughput" of 802.11, i.e. the actual amount of data that is transferred over a wireless link in a set period of time, rather than just upping the data transfer rates. As the reader should know by now, the data transfer rate is entirely different from the actual data throughput rate. That's because once a wireless link is established, the connection can be subject to interference, lost packets, and all sorts of problems, causing the "actual" transfer rate to be much lower than advertised.

"We're talking true throughput here," says Stuart Kelly, chairman of the new Task Group. "We've had proposals running at 108 Mbps and on up to 320 Mbps." Kerry also adds, "We want to harmonize European, Japanese and North American protocols right from the beginning." However, don't expect to see these newer, faster standards for a number of years.

CONCLUSION

Although there are often complaints about market confusion resulting from the "alphabet soup" nature of the 802.11 series, it's important to bear in mind that the formation of the various Task Groups represents an ongoing effort to modernize and improve the standard. The IEEE's goal is to identify the best technical solutions available, not to develop clear and simple marketing tag lines to promote the resulting products.

Wireless LANs have been with us for a few years now, but enabling wireless equipment from one vendor to communicate with equipment from another ranks as a relatively new development. Yet, this stumbling block must be overcome if wireless LANs are to become predominant. Service must allow a seamless experience to the end-user, whether they are communicating via their home network or not. With wireless LANs finding wider acceptance in the business community, the need for greater compatibility grows.

So, while the road to Wi-Fi nirvana may be a long and winding one, it is clear that with the courageous work of the IEEE and its 802.11 Working Group (and its growing number of Task Groups), the 802.11 series of standards are on the fast track toward dominating wireless LAN deployment worldwide.

Section III: Practical Deployments

Chapter 7: A WLAN Primer

Take a moment to look around you. Unless you are outside, you should see walls and ceilings decked out in technological artifacts: lighting fixtures, jacks (telephone and data), power outlets, and so forth. They are so common they've become a part of the landscape. Now an invisible artifact is joining this landscape. It provides high-speed, low-cost, low-power, wireless access to the Internet and other networks over a short range. It is Wi-Fi, which is a marketing term that has grown to encompass technology built upon the three latest 802.11 Physical Layer specifications: 802.11a, 802.11b and 802.11g.

In its base set of applications, Wi-Fi provides wireless connectivity via portable devices, including laptops, tablet PCs, handheld computers, personal digital assistants (PDAs), digital cameras, audio and video players, cell phones, headsets, and even portable devices that people can wear on their clothing. Wi-Fi also provides end-users wireless access to a host of new services through a topology referred to as a "wireless local area network" (WLAN), which usually consists of a wired Internet connection and more traditional wired voice and data connections.

Wireless LAN technologies (i.e. the 802.11 series of specifications) emphasize a high data speed and a range that makes using these technologies in a networked environment feasible. Typically, WLANs provide wireless links from portable computing devices to a wired LAN via access points (essentially a wireless hub), but there are also a growing number of stand-alone WLANs.

Now, what many people don't realize is that the wireless LAN's foundation is built upon wired technology. Let me explain.

Computers have shared information across a network via a wired connection since the 1980's. Initially, the physical path over which computers shared information was a wire, usually a coaxial cable or a Category 3 (and later Category 5 and 5e) cable. As users began to connect computers together to share information and common resources, it became essential that each computer agree on how to send that information across the networked computers. This need for different computers, built by different manufacturers, to share information eventually led to the development of a standard set of rules that each computer obeyed in order to communicate with one another.

The Institute of Electrical and Electronics Engineers (IEEE) formed the 802 Executive Committee to design specifications to standardize the physical path over which computers communicated. The result was an Ethernet-like standard published in 1985 by the 802.3 Working Group. This first publication was named "IEEE 802.3 Carrier Sense Multiple Access with Collision Detection Access Method and Physical Layer Specifications." The local area network (LAN) standard defined in that document is more commonly known as "Ethernet," although IEEE doesn't refer to 802.3 as "Ethernet," because Ethernet is a specific product trademarked by Xerox, whereas 802.3 is a set of standards. The 802.3 standard, albeit with many extensions, still governs how computers communicate today.

As the need for user mobility increased along with the cost of installing a network's cables, the business community demanded alternative methods to network its computers. The result was the development of wireless connections that use radio frequencies to transmit information.

In 1990, the IEEE 802 Executive Committee established the 802.11 Working Group to create a wireless local area network standard to govern how computers communicate over a wireless connection. The goal of that Working Group was to describe a wireless LAN that delivers services commonly found in wired networks, i.e. high-speed throughput, reliable data delivery, and continuous network connections.

The resulting 802.11 series of standards (which are in many ways extensions of the 802.3 standard) define a Medium Access Control (MAC) sublayer (of the Data Link Layer) and three Physical (PHY) Layers. The MAC sublayer is mostly made up of software-based protocols that enable devices to talk to each other and to wired local area networks. The PHY Layer defines the physical characteristics of the radio signal, i.e. the frequency, power levels, and type of modulation.

All Wi-Fi networks have an architecture that is specifically designed to support a network where all decision-making is distributed across the network's stations (i.e. the components, which may be mobile, portable, or stationary, that connects to the wireless medium). The building blocks of all Wi-Fi networks include:

* Support of all station (computer, printers, scanners, etc.) services including authentication, de-authentication, privacy, and delivery of the data (MAC service data unit).
* Basic Service Set (BSS), a set of stations (e.g. computing devices) that communicate with one another. When all the stations in the BSS communicate directly with each other and there is no connection to a wired network, the BSS is called an "Independent BSS" (IBSS), although it is more commonly known as an "ad hoc network" (i.e. it is typically a short-lived network with a small number of stations in direct communication range). When a BSS includes an access point (AP), the BSS is no longer independent and is called an "infrastructure BSS" or simply "BSS." In an infrastructure BSS, all computing devices (stations) communicate with the AP. The AP provides both the connection to the wired LAN (if there is one) and the local relay function within the BSS.

* Extended Service Set (ESS), a set of Infrastructure BSSs, where the APs communicate among themselves to forward traffic from one BSS to another. The APs perform this communication via what is referred to as a Distribution System (DS). The DS is the backbone of the WLAN and may be constructed of either wired or wireless networks.

(See Fig. 7.7 for a visual explanation of each of the above-referenced Basic Service Sets.)

As explained in previous chapters, the original 802.11 standard defines the OSI's (Open System Interconnection) Data Link Layer's MAC (Medium Access Control) sublayer and PHY (Physical) Layer. And, to date, all of the ensuing amendments and supplemental standards either enhance the original MAC for QoS (Quality of Service) and security, or extends the original PHY for high-speed data transmission.

The design of 802.11's MAC interface means it is compatible with 802.3 networks. But the 802.11 MAC offers many other functions as well, including:

* Providing a reliable delivery mechanism for user data over wireless media, which is achieved through a frame exchange protocol at the MAC level.
* Controlling access of the shared wireless media via two different access mechanisms: the contention-based mechanism, called the distributed coordination function (DCF), and a centrally controlled access mechanism, called the point coordination function (PCF).
* Protecting the data it delivers, which is done through a privacy service, called Wired Equivalent Privacy (WEP) that encrypts the data sent over the wireless medium.

Furthermore, Physical Layer dependent parameters (like timing intervals and backoff slots) are also modeled in the MAC.

ACCOMMODATING DIFFERENT APPLICATIONS

As applications using the 802.11 specifications broaden, e.g. HotSpots, integration into 2.5G and 3G cellular phone devices, support for streaming media, and more, a different implementation focus will be needed. All 802.11 implementations are comprised of four key components: the PHY, the MAC (both are normally delivered via a silicon chip), the Distribution Services, and the Management Services (the latter two are software components that enable 802.11 Access Points and Gateways to be created). And while all implementations of 802.11 require a MAC and PHY to operate, Management Services and Distribution Services are only required in certain applications. For instance:

The MAC and PHY components implement the 802.11 standards for MAC and PHY.

The Distribution Services provide the functionality to enable communication with a wired and wireless infrastructure via routing of frames that provide scalable, reliable and secure networks.

The Management Services provide the ability to manage a large-scale wireless infrastructure, to monitor performance, to determine the best method to tune network performance, and to modify operational parameters such as security and Quality of Service.

a OR b?

Wireless LANs can provide network flexibility and make it easy to support a roving workforce, but pitfalls abound. For instance, you must decide which flavor of 802.11 technology to select. Not long after the original 802.11 specification was ratified in 1997, the 802.11 Working Group formed two Task Groups, 802.11a and 802.11b, to work on extensions to the PHY Layer of the 802.11 specification and to solve some of the shortcomings of that original specification. The resulting 802.11a and 11b specifications also were designed to work with the original 802.11's MAC sublayer. But that's where much of the resemblance between these two standards ends.

To help in your decision-making process, we will now consider some of the options, security concerns, and the pieces of the puzzle you'll need to consider when evaluating a WLAN deployment, at least as things stand at this writing.

Once 802.11b gear hit the marketplace in 1999, network managers chose it as their preferred WLAN standard, mainly because there were no significant contenders since manufacturers were slow to produce any quantity of 802.11a gear. But by late 2001 everything changed as vendors released 802.11a access points and radio network interface cards. This change meant that anyone considering the deployment of a WLAN had another viable wireless technology to consider. Then in 2003 the manufacturers "jumped the gun" by releasing 802.11g gear in advance of a finalized set of specifications upon which to build that gear. Of course, now that 802.11g has been finalized expect also to find Wi-Fi certified 802.11g products all over the place.

Nonetheless, 802.11b is still the technology in current favor. And while some industry experts believe that 11a might be the anointed successor to 802.11b, others believe that since the 802.11g specification has been finalized, 11a may find itself out in the cold. So the decision on which specification to use when deploying a new WLAN can get a bit complicated.

Since 802.11g is spanking new and thus products built upon that standard do not have the experience behind them that only an installed base can provide, we will consider the major points of differentiation between 11a and 11b. However, there is an extensive dis-

	802.11a	**802.11b**
Frequency :	UNII Band	ISM Band
Frequency Band:	5 GHz range (300MHz)	2.4 GHz range (83MHz)
Modulation:	OFDM	DSSS
Channel Bandwidth:	20MHz (8 usable channels) non overlapping channel	22MHz (3 channels) Overlapping channel : 14
Data rate:	54Mbps (Layer3 -> 36Mbps) -PHY rate: 6, 9, 12, 18, 24, 36, 48, 54Mbps	11Mbps (Layer3 -> 5Mbps) -PHY rate: 1, 2, 5.5, 11Mbps
Coverage:	indoor (30m) outdoor (100m)	indoor (50 m) outdoor (150m)
Max power:	EIRP 200mW, 1W, 4W	EIRP 1mW/MHz
AP simultaneous Users:	100+ users	20-30 users

Figure 7.1 Technical differences between 802.11a and 802.11b.

cussion of the new specification in Chapter 6 and a discourse on the challenges of adding 802.11g to an existing 802.11b network in Chapter 10.

It's only through the examination of the main technical differences between 802.11a and 802.11b specifications that it is possible formulate a choice between 802.11b (and its 802.11g extension) and 802.11a. (Some may even opt for adopting a mixed environment.)

In 802.11b's favor: unlike 802.11a, 11b's products were introduced into the marketplace in 1999. Next, 802.11b's throughput is adequate for many network scenarios. Since the technology has evolved through several generations and has been used in numerous real-world situations, it has had most of the kinks worked out and its networking gear has come down to near-commodity prices. Finally, almost all HotSpot technology is based on 802.11b.

As Fig. 7.1 indicates, products with the 802.11a designation offer higher bandwidth and more channels than 802.11b products. But before you upgrade (or before you go wireless for the first time), there are some traps to be aware of.

***Note:** This analysis uses published information, published data and the system performance parameters readily available from the FCC, the IEEE and the International Telecommunications Union (ITU). It also tries to clear up some of the confusion that has been perpetuated through the information that's been available to date, much of which, in the author's opinion, misinforms about the actual performance and network capacity of both 11b and 11a technologies.*

Everyone who compares these two technologies finds major points of differentiation, but when the two are examined more closely there is no clear "winner." They each have different operational criteria.

To assess the network capacity of a wireless LAN built upon each specification, you must first understand that these two standards have different physics and operational characteristics such as:

- The possibility of interference—although the scale may tilt a bit in favor of 802.11a there is little difference between the two.
- Throughput—802.11a is much higher than 802.11b's throughput, but in many instances the throughput difference isn't that important because of the applications that will be run over the WLAN.
- Decipherable Signal Range, i.e. achievable communication range between the AP and the station, and the corresponding service coverage area—802.11b wins this contest; but in some instances, such as an area that routinely hosts a multitude of wireless users, 802.11a would be more desirable than 802.11b.

Then you must consider the different value propositions that these standards offer. They touch on many different areas including:

- Different end-user values by segments (e.g. corporate offices have different requirements than R&D departments).
- Usage models (e.g. a large, but busy, corporate conference room has different networking requirements than a warehouse).
- Applications (word processing has different bandwidth requirements than videoconferencing).
- Existing WLAN equipment (e.g. is there an existing 802.11b network?).

Now let's examine the differences between these two technologies as they pertain to the average corporate network's needs.

Interference: There are two "givens" when considering the possibility of interference in the operation of an 802.11a or b WLAN. The first concern is the unlicensed bands that 11a and 11b use, and the interference that might exist between all of the different devices that use that spectrum. The second is that given the cost of licensed spectrum, free spectrum is, and always will be, very attractive. Slivers of frequency that may not now be overcrowded will, in the near future, become so. Therefore, the claim that a technology such as 802.11a utilizes "un-crowded spectrum" is not a relevant buying consideration. Both the 2.4 GHz and the 5 GHz bands are subject to overcrowding and interference, and are used by many devices outside the WLAN arena.

The 2.4 GHz band (2.400-2.4835) is used by, not only, 802.11 and 802.11b devices, but also another WLAN technology, HomeRF. In addition, this ISM band is also used by communication systems such as Bluetooth, proprietary cordless phones, and non-communication devices, e.g. microwaves and lighting devices.

The 5 GHz band (5.15-5.25 GHz, 5.25-5.35 GHz, and 5.725-5.825 GHz), which at the moment appears less crowded than the 2.4 GHz band, hosts an ever-increasing amount of traffic. In addition to 802.11a devices, communication systems such as satellite systems (mobile and earth exploration), short range wireless systems, radio location systems, and electronic news gathering systems use the 5 GHz frequencies. The main non-communication device utilizing the 5GHz band is radar.

Throughput: An important consideration when determining which WLAN technology to use is the amount of bandwidth, data rate, or throughput the technology provides to each network user, and how well that throughput can support the applications running on the network.

***Note:** For our purposes, data rate is the amount of data that can be sent from one node on the wireless network to another, within a given timeframe—usually seconds, e.g. 11 Megabits per second or 11 Mbps. The difference between data rate and throughput is the measure of raw bits traveling from one node to another, in comparison to the bits representing the message content. This difference is determined by a number of factors including the latency inherent in the PHY components of the radio, the overhead and acknowledgement information that accompany every transmission, and pauses between transmissions.*

To help put data rate and throughput numbers into perspective, consider this:

Technology	Data Rate	Actual Throughput	Shared Among Users?	Est. Time to Download 100 MB File (actual throughput)
56.6 Kbps Modem	56.6 Kbps	56.6 Kbps	No	4 hours
10/100 Ethernet	100 Mbps	100 Mbps	Yes	8 seconds
T-1 line	1.536 Mbps	1.536 Mbps	Yes	8 minutes 41 seconds
801.11b	11 Mbps	5-7 Mbps	Yes	2 minutes 8 seconds
802.11a	54 Mbps	31 Mbps	Yes	26 seconds
802.11b/g	54 Mbps	12 Mbps	Yes	1 minute 13 seconds
802.11g	54 Mbps	31 Mbps	Yes	26 seconds

Figure 7.2 Speed comparisons of various connectivity technologies.

As you know, 802.11b offers a maximum data rate of 11 Mbps, which translates into approximately 5 to 7 Mbps of actual throughput, while 802.11a offers a 54 Mbps data rate, or approximately 31 Mbps of actual throughput. But, keep in mind when considering throughput rate that:

* It is shared among all network users who use it simultaneously.
* The throughput is managed through a CSMA/CA (Carrier Sense Multiple Access/Collision Avoidance) technique modeled on its wired equivalent (i.e. Ethernet).
* Most network traffic (wired and wireless) is bursty, and there are typically only a few users on the network at any one time, so all WLAN networks (whether using 11a or 11b technology) offer their users generally very good connectivity speeds.
* An 802.11a network can easily support several simultaneous streaming media data streams *and* still have enough capacity to serve other end-users with high wireless data rates. Conversely, 802.11b networks are limited in the types of applications that they can host.

Decipherable Signal Range: One of the most fundamental and significant differences between communication systems operating at 2.4 and 5 GHz is the achievable communication range between the AP and its computing devices, and the corresponding coverage area—the decipherable signal range. As you shall learn, when comparing each specification's decipherable signal range, both 802.11a and 802.11b wireless systems are governed by the same variables.

Without delving deeply into the theories of electromagnetic wave propagation, we will look at the propagation performance (refers to the movement of the transmission signal through the airways) in the ISM and U-NII bands. Here's what you need to understand about decipherable signal range.

Because the 802.11a carrier frequency is more than twice as high as the 802.11b carrier frequency, the electromagnetic propagation through the channel should *theoretically* attenuate the signal twice as much (6 dB according to the so-called Friss equation). While this holds true in an "open field" situation, in an indoor or campus environment, additional parameters must be considered, such as reflection, wall penetration, moving vehicles, how fast the channel changes, and so forth. Thus, while the laws of physics dictate that the range of free-space (i.e. open field) radio communications decreases with higher frequencies, indoor propagation differs from free space because of absorption and reflections. Moreover, power transmit levels and the type of modulation used also affect range. The result is that it is very difficult to arrive at a generic formula to accurately describe such propagation.

Note: *The Friss equation gives the free space power received by an antenna. This equation helps in finding out the received signal strength at the receiver side when there is an unobstructed line of sight path between the transmitter and receiver. The function is of the form*

$$P_r = P_t G_r G_t \lambda\text{^}2 / (16*\pi*\pi*d*d*L)$$

where P_r is the power of the signal at the receiver, P_t is the power of the signal at the transmitter, G_t is the transmitter antenna gain, G_r is the receiver antenna gain, d is the separation between the antenna and the receiver, λ is the wavelength and L is the system loss factor which is assumed to be 1 for practical purposes.

Various models, techniques and analyses have been created to give some insight into the propagation question, including:

* The Motley-Keenan model (J.M. Keenan, A.J. Motley, "Radio Coverage in Buildings," British Telecom Technology Journal, Vol. 8, No.1, January 1990, 19-24).
* The Kamerman path loss model (A. Kamerman, "Coexistence between Bluetooth and IEEE 802.11 CCK Solutions to Avoid Mutual Interference" Lucent Technologies Bell Laboratories, Jan. 1999, also available as IEEE 802.11-00/162, July 2000).
* Long-distance models (T. Rappaport, *Wireless Communications*, Prentice Hall, New Jersey, 1996).
* Multi-breakpoint models (D. Akerberg, "Properties of a TDMA Picocellular Office Communication System," IEEE Globecom, December 1988, 1343-1349).
* Sophisticated ray-tracing techniques (K. Pahlavan, A. Levesque, *Wireless Information Networks*, J. Wiley & Sons, Inc., New York, 1995).

Unfortunately, some of the indoor models suggest that a carrier frequency that is twice as high will attenuate the signal by 6 dB, whereas other models reach an entirely different conclusion. While there are many reasons for such contradictory conclusions, the primary explanation is that there are a number of variables that govern the decipherable signal range of every wireless system, whether 802.11-based or not. Some considerations are

* RF power transmit level, the power at which the signal is transmitted.
* Required E_s/N_0, the signal energy required to recover the transmitted symbol (the technical wireless term for the information contained in a message) compared to the environmental noise. (Because symbols are shorter at greater throughput levels, they require more energy in the symbol to recover it for the same error rate. This is one reason a WLAN system's throughput decreases as the distance between the AP and its computing devices grows.)
* Environment, the physical characteristics of the radio's environmental surroundings affect the path loss.
* Signal propagation, the physics of the radio spectrum and frequency in which the radio operates.

But the *core* difference between communication systems operating at 2.4 and 5 GHz is the achievable communication range between the AP and the computing device, and the corresponding service coverage area. Assuming common environments and system operating parameters, systems operating in the 2.4 GHz frequency band offer roughly double the range of those operating in the 5 GHz band, again holding power and throughput constant. This doubled range is explained by radio wave propagation physics, which dictate that, all other things being equal, a higher frequency signal will have a lesser range than a lower frequency signal. (Of course, this assumes a conservative 5 GHz path loss.)

Recently published papers based on varying models and techniques have compared the network capacity of 802.11a and 802.11b WLANs, but many reached contradictory conclusions. Here's why: one model can indicate that 802.11b networks offer superior performance to 802.11a, but only under the assumption that all 802.11a solutions transmit at 15 dBm, which isn't always the case. Another model can show clear network capacity advantages of 802.11a over 802.11b; however, the model is based upon a system that does not meet the full specifications of 802.11a, so the model does not provide a truly accurate representation of the full performance of optimized 802.11a systems.

Let's now assess the network capacity of wireless LANs by first looking at the network capacity offered by current 802.11a solutions and compare those to 802.11b solutions.

3COM TESTS 2.4 AND 5 GHZ PROPAGATION IN TYPICAL OFFICE ENVIRONMENT

The authors (Niels van Erven, Robert Yarbrough, and Lloyd Sarsoza) of a 3Com document entitled "2.4 and 5 GHz propagation measurements for WLAN," measured 2.4 and 5 GHz propagation in a common U.S. office environment. In other words, the test was conducted in a building with metal floor and metal roof, cubicle walls with metal sheets inside, and some drywall in the line-of-sight (LOS) path. In the case of a LOS path, where the Tx (transmit) and Rx (receiver) vertical polarized antennae were above the cubicles, the average path loss difference between 2.4 GHz (11b) and 5.2 GHz (11a) was around 7 dB. When the antennae had a non-LOS path (e.g., one antenna inside a cubicle), the difference was 2 to 3 dB.

The test results indicate that the more reflection there is in the signal's path, the less difference there is in propagation between the two frequencies. The test's results also suggest that with proper antenna positioning, 11b will have an additional link budget of about 6 to 7 dB compared to 11a. (However, for less optimal antenna positions, this difference is closer to 2 to 3 dB.)

It was also discovered that additional factors can play an important role in the actual range difference between 802.11b and 802.11a, e.g. the achievable Tx power (the transmit power of the radio) and Rx sensitivity (the receiver sensitivity of the radio), use of antenna diversity schemes, use of sophisticated time equalizers for 11b, etc. Current 802.11b implementations show a high level of optimization with respect to these parameters. (Note that for 802.11a, it is very difficult to get high Tx power with commercially available power amplifiers with linear performance to guarantee proper receiver sensitivity for the high bit rates.)

In another test, 3Com found that when 802.11b (running at 11 Mbps) and 802.11a (running at 6 Mbps) are compared, the data throughputs are similar. This is documented in a 3Com paper entitled "Comparing Performance of 802.11b and 802.1a Wireless Technologies," which states, "The available link budgets are also about the same." 3Com also found that the propagation is the only major difference. In fact, 3Com concluded, "In the best case, 802.11b can go roughly 50 to 70 percent farther than 802.11a. In the worst case the range is nearly the same." 3Com's overall conclusion based on its test was, "the high rate ranges of 802.11b are very similar to the 'low rate' ranges of 802.11a."

The importance of such tests is that they can help a network manager create a wireless network that can transition between 802.11b and 802.11a with respect to access point placement.

Measuring Performance

The primary industry-standard performance metrics used to define connectivity for a WLAN deployment include range, coverage and rate-weighted coverage.

* Range is the greatest distance from an access point (AP) at which the minimum data rate can be demodulated with an acceptable packet error rate or probability of error per bit (commonly known as "bit error rate" or "BER"), where it is assumed that there are no co-channel or adjacent-channel radiators in the vicinity.
* Coverage applies to moderate-size or large cellular deployments and is a measurement of the resulting cell size, or square meters per AP.
* Rate-weighted coverage is the integral of the bit rate with respect to area covered (expressed as megabits/second times square meters).

Range, coverage, and rate-weighted coverage are strongly influenced by not only the physical environment, but also by transmit power, receiver sensitivity, noise and interference. By analyzing, understanding and managing such parameters, WLAN system designers can have a significant effect on the overall performance of the system.

Transmit (Tx) power: Refers to the transmit power of the radio. FCC regulatory standards set upper bounds on transmitted power for 802.11a systems operating in the United States. The limit in the 5.15 to 5.25 GHz band is +16.02 dBm. In the 5.25 to 5.35 GHz band it is +23.01 dBm. The maximum upper bound on transmit power for 802.11b transmissions is +30 dBm. (Other national regulatory agencies also have set limits.)

Receiver (Rx) sensitivity: Refers to the receiver's sensitivity to the radio. According to the IEEE, an 802.11b receiver should be able to detect a -76-dBm signal and demodulate it with a bit error rate of less than or equal to 10e-5 in the absence of adjacent-channel interference (ACI). If ACI is present, the receiver sensitivity figure is specified at -70 dBm. In comparison, Fig. 7.3 shows the minimum Rx sensitivity for various data rates with and without ACI for 802.11a.

Noise and interference: Interference can dramatically affect the performance of any WLAN. In general, interference is either caused by radio devices operating in the same bands or by thermal noise, or both. Thermal noise is the only source of interference for a single AP. For multiple APs (i.e. multiple cells), however, there is also interference from adjacent channels and co-channels. The overall impact of such interference is heavily dependent upon the number of available frequency channels and cell deployment. Careful cell deployment and management of the number of available channels can mitigate its effects.

Bit error rate: A bit error rate of better than 10e-5 is considered acceptable in WLAN applications. By using standard graphs of BER vs. E_b/N_o (i.e. the bit error rate versus the ratio of Energy per Bit (E_b) to the Spec-

Modulation (Mbits/s)	RX sensitivity dBm (no ACI)	RX sensitivity dBm (with ACI)
54	-65	-62
48	-66	-63
36	-70	-67
24	-74	-71
18	-77	-74
12	-79	-76
9	-81	-78
6	-82	-79

Figure 7.3 For bit error rates less than or equal to 1e-5, the 802.11a standard specifies minimum receiver sensitivity with and without adjacent-channel interference (ACI).

Figure 7.4 For bit error rates less than or equal to 1e-5, there are defined minimum signal-to-noise ratios for 802.1b and 802.1a that will allow the required data rates to be met.

802.11b

Rates (Mbits/s)	Signal-to-noise (dB)
11	6.99
5.5	5.98
2	1.59
1	-2.92

802.11a

Rates (Mbits/s)	Signal-to-noise (dB)
54	24.56
48	24.05
36	18.80
24	17.04
18	10.79
12	9.03
9	7.78
6	6.02

tral Noise Density (N_o)) for the different modulation schemes, it's possible to then calculate the minimum required signal-to-noise ratio (S/N) values in decibels.

Physical environment: By using indoor "path-loss models," designers can quantify ambient impairments and achieve an understanding of how a system will operate when deployed. For instance, in an indoor environment, the signal power at the receiver SRx is related to the transmit power STx as shown in the following equation (Equation 1).

$$10\log_{10}(S_{RX}) = 10\log_{10}(S_{rx}) - 20\log_{10}\left(\frac{4\pi f}{C}\right) - 10N\log_{10} r \quad \text{for } r \geq 1$$

Equation 1 Here, C = the speed of light (m/s), f(Hz) = the center frequency and N = the path-loss coefficient (dimensionless). The ITU recommends using N = 3.1 for 5-GHz and N = 3 for 2.4-GHz applications.

Range Comparisons

Some published analyses argue that 802.11b provides superior range when compared with 802.11a. However, those conclusions are based on an incorrect assumption that both 802.11b and 802.11a radiate at the same transmit power, which may not necessarily be the case. Optimized 802.11a solutions can transmit at +23 dBm as compared with practicable (meaning real-life devices) 802.11b effective isotropic-radiated-power (EIRP) values of +15 to +19 dBm.

Another possible error arises from choosing an incorrect path-loss coefficient (N). This error can lead to an assumption that 802.11b solutions have a range of 100 meters with +15-dBm EIRP. Using those numbers, we can calculate that the N used in the 802.11b calculations was approximately 2.535 (see Equation 2). This is less than the ITU's recommended figure of three.

$$15 - 40.34 - 10N\log_{10}(100) \geq -76$$

Equation 2

However, by using the EIRP values quoted from published papers, together with the ITU reference model and path-loss coefficient N = 3, the maximum theoretical range of an 802.11b network operating at the maximum EIRP of 30 dBm is 154 meters, declining significantly at an EIRP of +19 dBm to 66.4 meters, then to 48.4 meters at +15 dBm.

The same analysis also uses N = 3 in its 802.11a calculations, resulting in an 802.11a range calculation that appears to be very small next to that of 802.11b. However, as mentioned above, the ITU recommends N = 3.1 for 802.11a calculations. When those ITU-recommended figures are used, the range of 802.11a improves, relative to 802.11b, giving 54 Mbps at a distance of 14 meters, down to 6 Mbps at 51 meters.

In addition, conventional 802.11a WLANs transmit at +18-dBm EIRP, while an optimized solution can transmit at +23 dBm, thereby extending the range to 30 meters at 54 Mbps and 108 meters at 6 Mbps.

***Note:** If you have a WLAN running at a full 11 or 56 Mbps (sending, receiving) and there is only one user connected, then that user will get the whole 11 or 56 Mbps; two or more simultaneous users must share that bandwidth. Note that this is not the number of users connected to the WLAN—you could have 50 WLAN users within range of the AP and if none of them are sending or receiving data then very little bandwidth will be utilized; they must be sending or receiving data at the same time to slow the resource down. Also, the further a WLAN user gets from the AP, or if there are physical structures in the way, the connection speed drops for "falls back." In the case of 11b APs, the speed falls back from 11 Mbps to 5.5 Mbps to 2 Mbps. A similar fall back occurs with 11a APs. Since there is also a certain amount of overhead required to process the data traversing the airwaves (e.g. encrypting, sending, receiving, and decrypting), an 11b WLAN's speed could top out at 5 or 6 Mbps and an 11a network's speed could be limited to less than 32 Mbps.*

Interoperability

Interoperability provides the assurance of backward compatibility with existing equipment, as well as a future migration path to any new technology that might be purchased in the future. Thanks to IEEE and the Wi-Fi Alliance, interoperability is a non-issue. Without these two organizations and the participation of the wireless networking industry en mass, 802.11 technology would probably never have taken off. As long as you buy gear that carries the Wi-Fi Alliance's "Wi-Fi" brand, you are golden as far as interoperability is concerned. The vast majority of each specification's WLAN gear is capable of interoperating with products from competing vendors, without the need of any special engineering support to make them work together. (Of course, 802.11b products can't interoperate with 802.11a products or vice versa.)

There are exceptions, of course, especially in the "Smart" AP arena because of the proprietary technology used to provided additional features and functionality such as Quality of Service, security, management, etc.

Global Spectrum Unification

Although both 802.11b and 802.11a have been on the books since September 1999, only 802.11b has been accepted globally. 802.11a has run into 5 GHz band allocation and regulation disputes in several countries. Multi-national companies and international travelers must take these issues under consideration when considering which wireless technology to purchase. (802.11g also is expected to run into quite a few roadblocks on its way to global acceptance.) This means that only 802.11b devices comply with virtually all local regulatory standards, and thus 11b devices are the only devices that are legal to operate throughout most of the world.

Such global spectrum unification issues are not a problem for solutions contained within a single country. But if a multi-national company or international traveler opts

5 GHz Band	USA	Japan	Europe	Usage
4.90 – 5.00		TBD		Authorized in Japan only
5.15 – 5.25	50mW	200mW	50mW	Indoor only
5.25 – 5.35	250mW		250mW	Indoor/Outdoor
5.47 – 5.725			1 W	Not allowed in US
5.725 – 5.875	1 W		25mW	ISM band in US

Figure 7.5 A general view of the 5 GHz worldwide spectrum allocation and authorized transmit power as of 4/01/02.

Country	2.4 GHz	5 GHz
Australia	AS/NZS 4771	AS/NZS 4771
Canada	RSS-210	RSS-210
European Union	EN 301.328	EN 301.893
Japan	Std 33A & Std 66	
New Zealand	AS/NZS 4771	AS/NZS 4771
United States	FCC Part 15.247	FCC Part 15.401

Figure 7.6 The international regulatory standards to which all 802.11 devices must adhere.

for 802.11a (or 802.11g) for their wireless networking technology, they may be required to purchase different devices or maintain different networks in order to adhere to individual country's regulations.

For more information on Wi-Fi's regulatory restrictions see Appendix III: Regulatory Specifics re Wi-Fi.

The Cost Differential

Current list prices have 802.11a components at about 25% higher than 802.11b equipment. Furthermore, in the typical WLAN layout, 802.11a networks require many more access points than an 802.11b network. Thus, if costs are a concern, and your network isn't hosting bandwidth intensive applications such as large schematics or streaming media, you should probably look into 802.11b. However, 802.11a can provide a WLAN with enough capacity for growth—not only in the user base, but also in the inevitable bandwidth-hungry applications of the future.

Eventually, the price gap between 802.11a and 802.11b will shrink with economies of scale. In addition, many wireless chipset suppliers have announced dual-mode 802.11a/b chipsets, and wireless NIC vendors are releasing dual-mode NICs. Thus, similar to the 10/100Mbps Ethernet gear, single-priced, dual 802.11a/b gear should become commonplace, with one set of gear being used for both types of networks—802.11a and 802.11b.

Existing Installations

Because there is no provision for interoperability between 802.11a and 802.11b, it's often not cost-effective for companies with an existing 802.11b WLAN to migrate to 802.11a.

Also, some industry experts feel that since 802.11g has been finalized, for those that need to upgrade an 11b WLAN, 11g will be the better upgrade path. In such an instance, 802.11b vendors will provide a simple firmware upgrade so that 802.11b gear can become 802.11g gear. But remember that 802.11g is still limited by many of the same issues plaguing 802.11b, e.g. three nonoverlapping channels and frequency interference.

THE DUAL-BAND OPTION

By designing a network around a dual-band solution, you can increase the WLAN's flexibility and provide additional user bandwidth by simultaneously supporting both 802.11b and 802.11a wireless networks. Going the dual-band route means that a network manager can deploy additional wireless PC cards and access points (dual-mode, of course) into the WLAN infrastructure without forcing a complete and immediate turnover of technology. Thus, a dual-band solution can protect any investment in existing 802.11b technology, while providing a gradual upgrade path to the higher-throughput 802.11a (or 802.11g) standard when the organization is ready.

Most 802.11b radio cards come in a PC card form factor, so they can be easily swapped out of client devices or access points. This can be an important feature if and when the WLAN is upgraded in the future to higher speed technologies. However, upgrading single-mode access points requires an upgrade of the entire wireless system at the same time. This can be costly; it also requires a potentially lengthy shut down period. Using a dual-band system to deploy a new WLAN obviates these problems.

The benefits of a dual-band network include 802.11b's range and sustainable 11 Mbps data rate being complemented by 802.11a's space-concentrated, 54 Mbps data rate and relief for network executives grappling with long-term migration issues. But 802.11b and 802.11a are not a one-for-one trade off. Mixing these technologies engenders a complexity that requires strict attention to the network's design. This is mainly due to the following factors:

- The propagation characteristics among the radios are very different—802.11b's coverage is far greater than that of 802.11a (although there are a few exceptions).
- In a dual-AP environment, a network that's designed to accentuate the properties of 802.11b cells could leave large dead zones between 802.11a cells. You might be able to address the problem by lowering the output power of the 11b radio (if the access point allows) so that the coverage areas are concentric, or by supplementing the patchy 11a coverage with single-mode 11a access points.
- For the end-user, there is still some question as to how smoothly different computing devices can transition from one band to the other, especially smaller devices such as PDAs and handheld computers.
- Seamless roaming also might be quite a technical challenge when a network is designed around dual-band equipment. However, this is a problem primarily for applications such as voice over IP, where retransmission is not an acceptable solution to data loss.

802.11g

This new PHY extension to 802.11b will operate in the 2.4 GHz band and specify three available radio channels providing a maximum link rate of 54 Mbps per channel, compared with 11 Mbps for 802.11b. Even though the charter of the 802.11g Task Group is to develop a higher speed extension (up to 54 Mbps) to the 802.11b PHY while operating in the 2.4 GHz band and implementing all mandatory elements of the IEEE 802.11b PHY standard, the Task Group opted not to use 802.11b's DSSS. Instead, OFDM provides the higher data rate extensions. But, to provide backwards compatibility with 802.11b, the specification supports complementary code keying (CCK) modulation and, as an option for faster link rates, allows Packet Binary Convolutional Coding (PBCC) modulation.

The IEEE standards board finalized the 802.11g specification in mid-June 2003. The next step is for the Federal Communications Commission (FCC) to approve the use of OFDM in the 2.4 GHz band (a generally necessary action when messing with the PHY). Also, other national regulatory bodies must approve the operation of 802.11g equipment within their domain. It could take awhile for that to occur.

Despite all of these constraints, leading vendors are racing to introduce new 802.11g compliant gear. But don't jump on the 802.11g bandwagon until products based on a *final* version of the specification *and* sport a Wi-Fi certification label are available. Buyers of 802.11g gear that doesn't meet such criteria risk interoperability and performance problems in a multi-vendor environment, particularly with certified 802.11b products installed in PCs in a mixed 802.11b and 802.11g operating environment.

The highly respected market research group, Gartner, Inc. suggests that potential users wait until 802.11g products have been certified by the Wi-Fi Alliance before adding such equipment to their networking environment. (The Wi-Fi Alliance has begun the certification process.)

WI-FI AND ETHERNET

The fundamental components of a WLAN are clear-cut. And it's relatively easy to describe the 802.11 suite in basic networking terms—a series of specifications that provide an open asynchronous networking environment that requires a distributed control function. But the technology isn't as simple as it sounds.

Wired LANs use a physical medium to interconnect their terminals. Access nodes are provided at various points on the physical medium to allow for the connection of terminals to the medium. LANs may be interconnected by means of bridges or switches.

The most popular set of LAN specifications is 802.3 (commonly referred to as "Ethernet") with all of its evolutionary variations. These specifications provide for message structuring rules, station naming, allocation of resources, and other housekeeping functions.

Most wireless LANs extend their access to a wired LAN's medium via an Access Point (AP) that attaches as a bridge to the wired LAN. (I say *most* because you can set up a WLAN that stands alone, i.e. no wired access whatsoever.) The AP uses radio spectrum to extend the LAN's medium to radio equipped network devices within the AP's range.

***Note:** Although many LAN protocols exist, and many wireless LANs can work with these other protocols, this chapter assumes the LAN referenced uses 802.11's wired successor, the 802.3 series. It is also noted that the 802.3 suite of protocols is commonly referred to as "Ethernet," but in actuality, Ethernet is a LAN architecture developed by Xerox Corporation in cooperation with DEC and Intel. Xerox's Ethernet, however, did serve as the basis for the IEEE 802.3 standard.*

In a mixed wired/wireless environment, the most common WLAN configuration uses a single AP to provide service to all terminals within its coverage area. In such situations, the AP is analogous to a wired network's hub in that it supports the shared usage of the medium by its active computing devices. A wireless computing device becomes a full-fledged member of the wired LAN, with all assigned LAN privileges, after it is associated with an AP. Wireless stations (e.g. computing devices) may talk to each other, whether they are on the same or different APs, and they can also communicate with devices on the wired LAN.

Mobile computing devices may roam among multiple access points without losing their connection. However, if the access points are in separate subnets within the wired LAN and there is no specific roaming technology in residence, it may not be possible for a station to maintain session persistence when roaming outside its subnet due to routing constraints within the LAN.

The 802.11 series of standards that are spreading throughout today's networking landscape are sometimes being described as wireless Ethernet. This is because they use similar modulation techniques and are simple and flexible. And perhaps also because Bob Metcalfe (he is considered the "inventor" of Ethernet) has stated that he chose the word "Ether" because he didn't want Ethernet to be associated with a particular media. So there is some logic behind the term wireless Ethernet.

But although the 802.11 standards share some common aspects of an Ethernet network, in reality, the 802.11 specifications provide for a CSMA/CA (Carrier Sense Multiple Access/Collision Avoidance) network while 802.3 provides for a CSMA/CD (Carrier Sense Multiple Access/Collision Detection) network. To understand, let's consider the more established 802.11b standard since it is still the most widely implemented WLAN technology.

The 802.11b specification is designed to use a variant of Ethernet's CSMA/CD—CSMA/CA. CSMA/CA was chosen because CSMA/CD would require that the wireless radios be able to send and receive at the same time. Not only would that serve to substantially increase 802.11b product price and complexity, but also, in a wireless networking environment devices are not always in a position to hear all of the other wireless devices on the network.

CSMA/CA utilizes the RTS/CTS (request to send/clear to send) protocol to notify other workstations that a transmission is about to take place. This four-way handshaking minimizes the number of collisions and makes sure that hidden nodes are aware of transmissions across the entire wireless segment; however, this method introduces significant overhead on the network. 802.11's MAC is also significantly more complex than a typical 802.3 MAC, because the wireless specification calls for four MAC addresses instead of the two found in an Ethernet header.

Finally, to maintain backward compatibility with the original 802.11 specification devices, 802.11b wireless device transmits the preamble and a portion of the packet header at 1 Mbps. This accounts for significant additional overhead, as that preamble is significantly longer than an Ethernet preamble. The overall result is a network that can be at most 70 percent efficient (allowing a maximum data throughput rate of about 7.7 Mbps).

Losses in effective transmission rates can be reduced by improving the strength of the primary signal, which, in turn, reduces the time it takes to discern ghost signals from the true signal and the amount of time it takes to sample diverse antennae. Furthermore, many of the 802.11b WLAN products on the market today push throughput at the cost of interoperability. In the case of point-to-point devices, eliminating the 1 Mbps 802.11 legacy transmit speed can significantly improve performance. Likewise, if a network is known to be point-to-point, the Random Backoff algorithm, interframe gap, and preamble can all be minimized to maximize throughput.

This adds up to a wireless standard that really isn't Ethernet at all. When Xerox first licensed Ethernet, it charged a pittance, but in exchange it stipulated that the technology couldn't be changed; it had to interoperate with *all other* Ethernet implementations. The 802.11 series doesn't meet that directive.

A WLAN'S ARCHITECTURE

Typically a WLAN is built upon "plug and play" equipment, an open architecture, and a wired LAN. The WLAN's architecture also should be customized to facilitate the efficient and effective coordination of the organization's common business processes, information flow, and systems. Thus, a proper WLAN architecture provides a framework around which it is possible to develop, maintain, and implement an excellent operation environment.

All 802.11 specifications require that the network be built on a cellular architecture where the system is subdivided into cells. An access point controls each cell with coverage generally mapped with cells overlapping by 30 percent to support continuous communications as end-users move around the facility. Three different kinds of cellular architectures are specified for a WLAN implementation:

Ad Hoc: The first is an ad-hoc architecture, i.e. no access points. Computing devices communicate directly via the antennae built into their PC card. This type of architecture, which is also known as an Independent Basic Service Set (IBSS), has limited usage. This form of wireless networking can be applicable in situations where end-users spontaneously form a wireless network (e.g. conference rooms, demonstrations, small informal work groups such as a traveling business development team, or a group of people gathered for a business meeting) to share documents such as presentation charts, spreadsheets or data files. But since an ad-hoc network doesn't provide connection to the entire network ecosystem, its capabilities are limited. (See Fig. 7.7.)

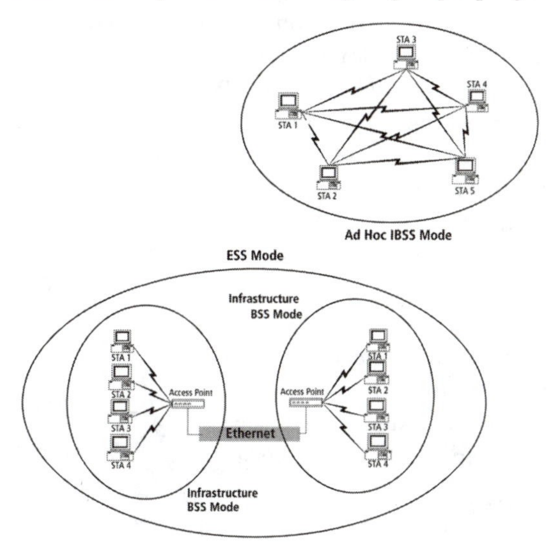

Figure 7.7 The various Basic Service Set (BSS) modes that can be deployed in a wireless networking environment.

Infrastructure or BSS: In this architecture where there is at least one access point connected to a wired LAN and a number of wireless computing devices. This type of network architecture is referred to as a Basic Service Set (BSS).

Extended Service Set: ESS refers to a network designed with two or more BSSs that are linked together to form a subnetwork.

To pull all of these different Service Sets together sometimes requires not only a wired Distribution System, but also a Wireless Distribution System (WDS). A WDS allows a network manager to deploy a completely wireless infrastructure.

Wireless Distribution System

A wireless system is only as good as its backbone, since achieving optimal transmission rates and coverage depends more on the quality and installation of that backbone than on any other component. A good wireless backbone should be reliable, easy to install and administer, and scalable. It is the crucial component not only to future migration, but also to good radio coverage and optimal data throughput.

As mentioned at the beginning of this chapter, this backbone is typically referred to as the "Distribution System," in that it connects all Basic Service Sets (BSSs). More often than not, the backbone is made up of a wired network (or the wired components of the system) and access point(s). Ideally, the WLAN will be invisible (or transparent) from an IT management perspective. The wireless system should be seamless to the existing wired system and should not require additional expertise to manage; although as you should already understand, this is not always the case.

A Distribution System is usually Ethernet-based and the networks' APs are connected to the LAN while creating cells to allow wireless connections. A Wireless Distribution System, on the other hand, allows the APs to be wirelessly connected–the connection between the APs is established using the AP's PC card. A WDS is useful in a situation where the coverage area is too large for one AP and a second AP is needed to act as a wireless repeater.

A common WDS deployment is to use two APs to create a wireless bridge between two wired networks with one AP configured to forward all data to the other AP and vice versa. To communicate effectively, both APs apply the same wireless parameters. Since they are acting as a bridge, the APs learn the network devices that are connected to their respective Ethernet ports to limit the amount of data forwarded. Data destined for stations that are known (1) to reside on the peer Ethernet, (2) to be multicast data, or (3) to be for an unknown destination, must be forwarded to the companion or peer AP. The fact that the data is being wirelessly bridged is completely transparent to the LAN, its components, and the end-users.

A Wireless Distribution System can also be designed to have forwarding functionality. By setting up an Extended Service Set (ESS) between APs and manually configuring the WDS peers, sta-

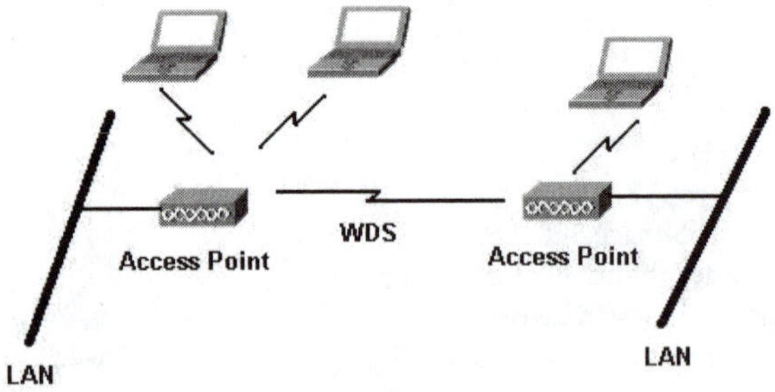

Figure 7.8 A WDS bridge setup.

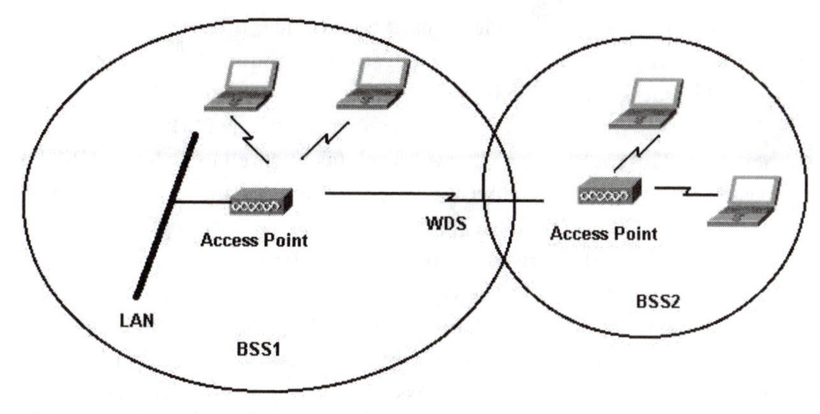

Figure 7.9 A WDS can have forwarding functionality by setting up APs as repeaters.

tions can intersect with any AP within the ESS and move between the coverage of both APs while the higher layer network connection remains intact. This is similar to the mobility provided in an ESS environment with a wired Distribution System. For instance, a WDS might be used when expanding an existing wired infrastructure network to provide coverage for office space that is not adjacent, but perhaps located across the street. WDSs also are a good solution when creating a roaming network in an area where wired connections between the APs cannot be installed, such as for a trade show where a large area needs to be covered requiring multiple APs.

Wired or wireless, when a WLAN is part of the network mix, a Distribution System connects cells in order to build a premises-wide network that allows users of mobile equipment to roam *and* stay connected to the available network resources. The whole interconnected WLAN, including the different cells, their respective access points, and the Distribution System, is seen as a single 802 network to the upper layers of the OSI (Open System Interconnection) model.

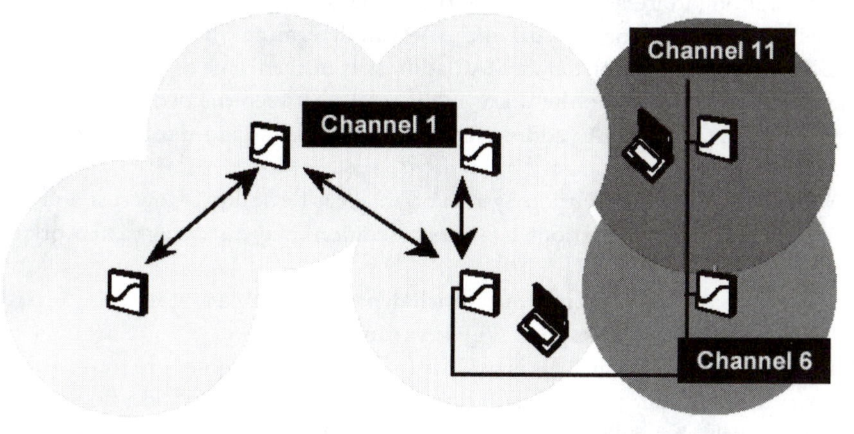

Figure 7.10 The three access points on the right hand side of this graphic are connected by Ethernet cable and hence use a wired Distribution System, while the four access points to the left are wirelessly connected, and are said to use a Wireless Distribution System. *Graphic courtesy of Agere Systems, Inc.*

WDS can use an AP with its single PC card to assume multiple roles simultaneously. But for an AP to take on such roles, the operational (frequency) channel must be the same for the cell that is controlled by the AP and for the wireless links to the other APs. In Fig. 7.10, this is illustrated by the four cells on the left side of the graphic all operating on channel 1. Thus, a WDS can "drive" a cell (as in wired connected APs), can connect wireless clients to the infrastructure, and can maintain up to six different wireless connections to other APs. (This is in contrast to other existing wireless AP-to-AP connection schemes such as those used in outdoor installations.)

To fully comprehend how a Wireless Distribution System works, we must also consider MAC addressing issues that are inherent to all wireless distribution systems. All LAN devices (including WLAN devices) communicate with each other by using MAC addresses (hardware addresses uniquely assigned in the factory to each device). Each wireless network interface device, whether an adapter, NIC, PC Card, or Compact Flash has a unique MAC address that is used by the system to send data frames to it. If a LAN device transmits data, it will add its own MAC address to the frame in order to indicate to the recipient where the frame originated. Thus, all data frames transmitted over a LAN contain a Destination and a Source MAC address as part of the frame header. When a data frame is transmitted over an Ethernet cable, just those two MAC addresses are required. But when data frames are transmitted between LAN end-stations that are not connected to the same LAN segment, an intermediate device is required to "bridge" the frame from one segment to another. An AP can act as that bridge. To relay traffic from one segment to another the AP uses a "bridge learn table," where MAC addresses are stored in association with the LAN segment (or physical interface) where they reside (from the perspective of the bridge).

However, instead of the two above-referenced addresses, traffic between 802.11 WLAN devices requires four MAC addresses. That's because when a wireless device is associated with an AP it will always direct its traffic to that AP by using the MAC address (address one) of the AP's PC card as its direct destination address. The MAC address of the end-station (address two) to which the frame is being sent is also included in the frame header, so that the PC card in the AP can determine where to relay the frame. Finally the sending station's own MAC address is in the frame as the source address (address three). Then when the WDS link is set up between the two relevant APs, the receiving AP's PC card's MAC address (address four) must be added to the address fields in the MAC header.

While a Wireless Distribution System offers great flexibility at low cost and can be applied in many useful situations, a few considerations may discourage a network manager from using a WDS:

- It is not possible to use encryption with dynamic assigned and rotating keys, on a WDS link. Only fixed assigned WEP keys can be used.
- Performance takes a hit. This is because (1) the frame is required to traverse the airwaves three separate times, and (2) since CSMA/CA technology and a single PC Card (and a single channel) are used, the end-to-end throughput will be about one third of the maximum attainable value. Obviously using a second PC card can improve this situation but in that case the expense of a second card is incurred.
- WDS allows creation of point-to-point connections. This would suggest that a WDS

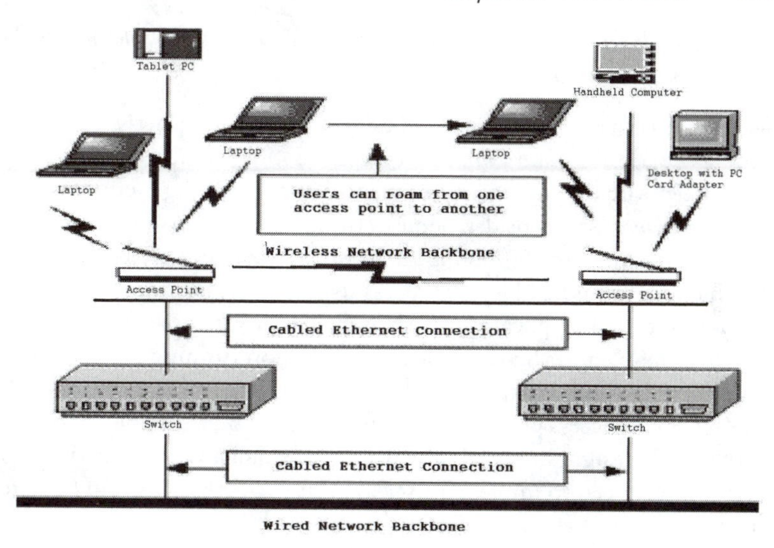

Figure 7.11 A typical corporate network architecture that uses both wired and wireless networking technologies.

could be applied to outdoor installations, but in outdoor situations (especially long distances and point-to-multipoint configurations) additional provisions must be implemented.

* Data could potentially be forwarded and duplicated endlessly, although there are solutions that can prevent such looping (such as spanning tree algorithm), but using such solutions add complexity to the network's management.

Supporting Mobility and Roaming

A key demand of untethered employees is to maintain a network connection as they roam throughout an organization's facilities whether from office to office, the first floor to the 20th floor, or from the distribution center to the customer service area. Changing a device's network attachment point, however, causes topology and identity changes at the IP level. Traditional end-user applications usually react badly to having the underlying host's identity changed in mid-session.

There are a variety of methods that can be used to decouple a host's identity from its location on the IP network, so that identity can remain the same while attachment points change. In order to accomplish this, however, the network must have a backbone that can track mobile workers wherever they go.

Many limitations are posed when roaming workers cross over subnets, including issues that do not permit users to cross over subnets or even to leave a specific coverage area. Consequently, when evaluating the hardware and software to build a WLAN, both standards-based and vendor-specific roaming capabilities must be closely examined if the WLAN being deployed is to support mobility.

Note: *A subnet (short for subnetwork) is a logical portion of a network that has a contiguous string of IP addresses. Addresses in a subnet are reachable without going through a router, and thus can be reached by broadcasts. To reach addresses outside of a particular subnet, you must transmit through a*

router. Typically, a subnet will consist of all networked devices in one geographic location, or in one area of a facility, or on the same network.

When connecting a WLAN to a wired network, map out exactly how the connection(s) will take place. End-users will be on the move. One minute the computing device will be associated with one subnet, the next another. Supporting mobile devices means there are new challenges to consider. Some of the most common operating systems (e.g. Windows XP and Windows 2000) support automatic Dynamic Host Control Protocol (DHCP), which is a release and renew process to obtain the IP address for the new subnet. However, certain IP applications such as virtual private networks (VPNs) will fail when DHCP is enabled. If this is an issue you envision running up against once the WLAN is deployed, then perhaps it would be best to deploy a flat network design for the WLAN, where all access points in a roaming area are on the same segment.

A flat network design, however, only works for small and static networks. Larger organizations may need to implement several flat networks. This can be done if the designer first determines where the end-users will roam (e.g. from their office to a conference room, or from their office to the warehouse, or from the warehouse to the cafeteria). Then segment the wireless network based on coverage areas with a minimum number of users roaming between them.

Note: Roaming between cells that are interconnected by a WDS link works exactly the same as for cells that are interconnected via Ethernet. When a station is relocated from one cell to another, the access point's "bridge learn tables" is updated via a hand-over request message that is part of the Inter Access Point Protocol (IAPP) to reflect the station's new location.

Routers and Switches. These devices provide end-users with the ability to roam the confines of the wired LAN without interruption. But this requires the wireless network to integrate seamlessly with the wired network's routers and switches.

The main function of a router is to keep network traffic at a manageable level, ideally five percent of network capacity, but again we run into the subnet issue—routers segment data to control traffic flow by dividing the network into subnets. (Routers can act as filters to the data as well.) However, since a router can be configured, its functions can be customized to meet a network's demands, including a roaming user-base.

A switch's tasks are usually a bit different than a router's—they are normally responsible for segmenting data to various ports, i.e. acting as subnets within the network. Switches are more affected by mobile devices roaming the network than routers. This means a switch is usually the roadblock when end-users who roam have problems maintaining a consistent connection. For instance a mobile user might be dropped from the network if the switch cannot keep up with the roaming computing device's hops from subnet to subnet.

Seamless IP Mobility

Many readers might wonder at the fuss over network mobility. After all there is a certain degree of mobility in IP networks today. Just look at the Internet, the king of IP networks. An Internet user can move from one city to another and essentially have connectivity and the same set of services available everywhere. But this type of nomadic mobility means that end-users have to shut down an application or a session and restart it when they connect at the new point of attachment.

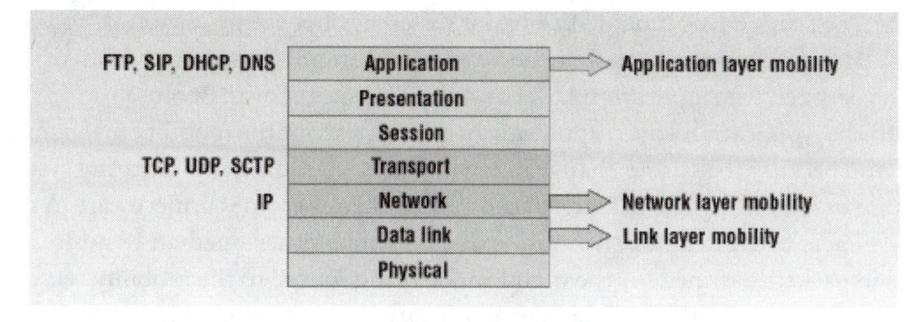

Figure 7.12 Various solutions have been proposed to solve the problem of seamless continuity of IP sessions and applications. They can be classified according to the layer of the OSI model at which they are implemented.

While for many users, this type of mobility is sufficient, for others it's not—especially the computing nomads who are the most frequent users of wireless networks. This group demands seamless mobility where session continuity is maintained even as the mobile device changes its network point of attachment or interface type. One method that can be used to provision a wireless network to support employee mobility is to create a separate sub-network within the existing wired network, where all wireless access points are wired back to a single hub. While this method can simplify network administration, it requires extra cabling, upping the costs of the WLAN.

***Note:** When a WLAN or WLAN/LAN Distributed System is correctly designed for roaming, a roaming computing device can move from subnet to subnet and in and out of a fixed Ethernet 802.3 connection or interface to an 802.11 WLAN interface without even a blip on the computer screen.*

Various solutions to the seamless-mobility problem have been proposed. These can be classified according to the layer of the OSI model at which they're implemented. The approaches may vary, but the end result is always the same: seamless continuity of applications or sessions. Hence, mobility can be solved at the Data Link Layer, the Network Layer or the Application Layer, as shown in Figure 7.12.

Application Layer: Application Layer mobility essentially moves the burden of managing the session and the underlying changes at the Network Layer's IP layer to the Application Layer protocol itself. For example, File Transport Protocol (FTP), which is commonly used for downloading files, music or video, would have to be enhanced to support mobility. What happens to other applications if FTP is extended? Mobility would have to be added to Simple Management Transfer Protocol (SMTP), Internet Message Access Protocol (IMAP), Session Initiation Protocol (SIP), HyperText Transfer Protocol (HTTP) and every other Application Layer protocol used. Applications would have to be rebuilt to support mobility. Such an approach is not viable. The impact is too drastic and backward compatibility would be a major issue. A shim could be developed, however, to sit between the application and the transport layers to perform the mobility task. NetMotion Wireless Inc. takes that approach.

The concepts behind the NetMotion Wireless solution are analogous to the Internet Engineering Task Force (IETF)-defined Mobile IP, which refers to protocol enhancements that allow transparent routing of IP datagrams to mobile nodes within an IP extended network environment. The major difference between NetMotion's approach and the Mobile IP approach is that the mobility solution is based on a shim, or driver, that sits between the Application Layer and the Transport Layer. Because the driver sits beneath the Application Layer, applications are unaware of the mobility mechanism in place. And because there is no change in the IP stack, rebuilding the operating system or replacing or enhancing the IP stack of the mobile client becomes unnecessary. A mobility server acts as a proxy for the mobile device, which is assigned an IP address that results in packets destined for the mobile node being routed to the mobility server. The mobility server knows the mobile's current location and care-of address and is able to forward the packets.

This means that the NetMotion solution requires a mobility server as well as the installation of proprietary software on the client. The same is true for Mobile IPv4. However, since NetMotion's solution does not include the concept of a foreign agent, there are no agent advertisements, as required with Mobile IPv4. The motion-detection mechanism is based on either Data Link Layer triggers provided by the interface card driver or by DHCP discover broadcasts. Triggers from the Data Link Layer may be available for certain types of wireless technologies like 802.11, but getting such triggers from other wireless interfaces is a highly complex task.

Network Layer: Mobile IP (which we examine in detail later in this section) solves the mobility problem at the Network Layer. Network Layer mobility hides the changes in IP address and network attachments from the upper layers, thus applications are essentially unaware of mobility enhancements. It also provides mobility to all applications, rather than dealing with applications individually. The mobile IP scheme is derived from work done by the IETF, which defines Internet protocols and standards, and maintains the most developed and deployed model today.

Data Link Layer: Drivers at the Data Link Layer can be developed to handle IP mobility. To understand, one way to think about Data Link Layer mobility is that the access technology handles all the mobility and the IP/Network layer is unaware of changes in the points of attachment. A device moving across 802.11 access points within the same Distribution System continues to maintain its sessions uninterrupted. Data Link Layer mobility solutions for seamless mobility across heterogeneous access media are extremely complex, so it is generally considered easier to instead develop and deploy a Network Layer solution.

In a network (wired or wireless), IP routing depends on a well-ordered hierarchy. At a network's core is the router. This device isn't concerned with individual users. It looks only at the first few bits of an IP address (the prefix) and forwards the packet to the correct network. Routers further out look at the next few bits, sending the packet to a subnet. At the edge, access routers look at the final parts of an address and send the packet to a specific networked device.

The hierarchy depends on devices that remain fixed to one network or subnetwork (subnet) and move between networks or subnets. In the case of an organization's networking environment, whether wholly wireless or a mixture of wired and wireless, when a computing device moves from one subnet and connects to another, its IP address must be altered. The result is that most computing devices don't have a permanent IP address, but acquire a new one each time they log on to a network. Most laptops, for example, have an IP address on the employer's network while docked at the office, but another one when accessing the employee's ISP while at home.

Mobile IP

IP addressing in a WLAN environment is not a problem if users don't often switch between subnets and if they are willing to log off and on again whenever they do. However, it is a problem if users need to stay connected while on the move, because it entails moving connectivity between subnets. Higher-level protocols, such as TCP, use the IP address to identify users, so a user can't maintain a TCP connection if the IP address changes. The solution to this is mobile IP, an IETF standard enabling users to keep the same permanent IP address no matter how they're connected.

One of the most popular methods for providing mobility within a networking environment is mobile IP with IP tunneling. The Internet Protocol (IP) is a connectionless protocol that operates at the OSI's Network Layer, meaning it avoids failures in intermediate networks by rerouting packets, an activity that brings into play the Transport Layer. The OSI's Transport Layer, which supports the majority of applications including the World Wide Web, uses the workhorse of the Transport Layer, the connection-oriented protocol, Transmission Control Protocol (TCP).

The two end points of a session or application use the IP address and the TCP port number at each end point as a tuple (an ordered sequence of fixed length of values of arbitrary types) to form a connection. Any change in those identifiers tears down the connection and breaks the session continuity. When a mobile node, such as a laptop or PDA moves from one point of attachment to another point of attachment, that node may be assigned a new IP address. This change in IP address will usually break an ongoing session. The relevance of a node moving to different points of attachment is especially high in wireless networks due to the mobility factor. Thus arises the need for IP mobility to support seamless session continuity even as the address of the node (one or both of the tuples in the connection, depending on whether both ends are mobile or one is static) itself changes.

The Network Layer's mobile IP is used rarely, partly because, until the WLAN explosion, there was little need for it and partly because present implementations (for IP version 4) waste bandwidth and require at least two precious IP addresses per user. But with wireless networking becoming ever more popular, mobile IP will become a blip on every WLAN manager's radar.

Every type of mobile IP depends on giving the mobile node two IP addresses: a permanent address on its home subnet, and a care-of address on another subnet. The permanent address is the one that higher-level protocols use, while the care-of address signifies the node's actual location within a network and its subnets.

At the moment there are two official versions of mobile IP, one for Internet Protocol version 4 (IPv4) and the other for use with the new IP version, 6 (IPv6). The design of these two mobile IPs varies to a certain extent.

Mobile IPv4 is not an inherent part of the IP stack and is an add-on that is built into nodes that require it. As a result, it is not universally used. Mobile IPv4 uses the basic concept of a home agent and a foreign agent, but because address space in IPv4 is a concern, many nodes share a single care-of address that is advertised by the foreign agent.

As the names suggest, the home agent is a router in the mobile's home subnet and the foreign agent resides on visited links. In most cases, the foreign agent assigns the care-of address. (A mobile node can also obtain a care-of address for its own interface, called a co-located care-of address, but this is not the general model.)

Mobile IPv6 is still a work in progress. The IETF Mobile IP Working Group is in the process of developing routing support to permit IP nodes (hosts and routers) using either IPv4 or IPv6 to seamlessly "roam" among IP subnetworks and media types. This will support transparency above the IP layer within the OSI Network Layer, including the maintenance of active TCP connections and UDP (User Datagram Protocol) port bindings. Where this level of transparency is not required, Mobile IP will not be needed. In such instances, the Working Group assumes that solutions such as DHCP and Domain Name Service (DNS) updates will be adequate.

Mobile IPv6 provides many advantages over the mobility support provided for IPv4. This includes the fact that IPv6 is designed from the ground up to include route optimization. This means that the mobile and correspondent node communicate with each other without the support of a home agent. Hence, routing of packets between the session end points is optimal. Since the mobility support is a standard feature of IPv6, every IPv6 node is expected to support IP mobility. Therefore, the deployment and support of true IP mobility are expected only when IPv6 networks are widely built out and begin to replace the current IPv4 networks.

Let's look a little closer at how Mobile for IPv4 and IPv6 differ.

Whenever a computing device accesses a new subnet, it must acquire a new care-of address on the subnet it's visiting. In IPv4, this means requesting an address from a special mobility agent—essentially a DHCP server, with some authentication, authorization, and accounting (AAA) functionality added-on. On the other hand, IPv6 has so many addresses available that the mobile node can make up its own by combining the visited subnet's prefix with an identifier unique to the device, such as its MAC address. This eliminates the need for a mobility agent, which, in turn speeds up the process, and ensures that a care-of address is always available.

Back at the home subnet, another mobility agent, usually an edge router with some AAA functions keeps track of all the mobile nodes with permanent addresses on that network, associating each with its care-of address. The mobile node keeps the home agent informed of its whereabouts by sending a binding update whenever its care-of address changes.

When a computing device on another subnet within the network needs to correspond with the mobile node, it sends packets via the home subnet. The home agent must intercept these packets and forward them to the visited subnet via a process known as "tunneling." This allows correspondent nodes to use the permanent address and remain unaware of the mobile node's movements.

The next step depends on which type of mobile IP you're using. In IPv4, all packets intended for the mobile node are tunneled via the home subnet, where the home agent intercepts and forwards them to the care-of address. This is the simplest way to enable mobility, but it adds extra routing hops, uses more bandwidth, and increases latency. The latter is particularly important for wireless networks, where latency is already high and unpredictable.

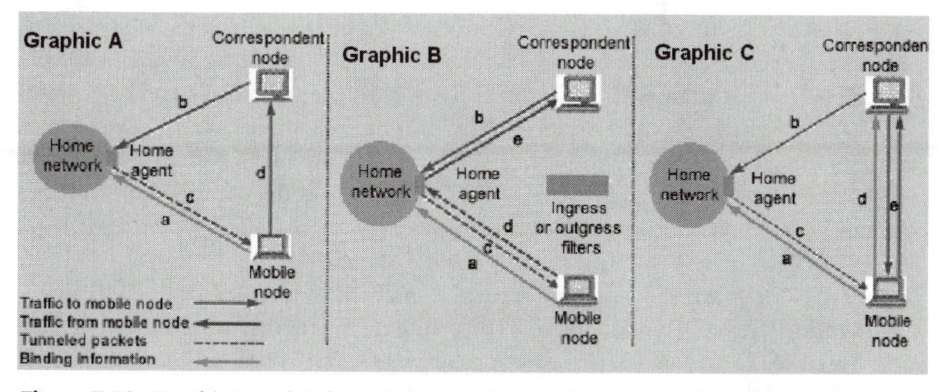

Figure 7.13 Graphic A and B depict how mobility for IPv4 works, whereas Graphic C depicts mobility for Ipv6.
Graphic A. When a mobile node first moves into a visited subnet (or network), it acquires a temporary care-of address and informs an agent on its home network of this (a). Packets intended for the mobile node are sent to its permanent address on the home subnet (b), they are then forwarded by the home agent and to the care-of address (c). The mobile agent sends replies directly to the correspondent node (d).
Graphic B. Reverse tunneling works in the same way as triangular routing (as shown in Graphic A), except that replies are tunneled back via the home agent (d) so the correspondent node sees them arriving from the mobile node's permanent address (e).
Graphic C. When mobility for IPv6 is used, the correspondent node initially sends packets to the mobile node's permanent home address (b), where they're tunneled to the care-of address (c) as in IPv4's mobile IP solution. However, in IPv6, the mobile node then informs the correspondent node of the care-of address (d), so further packets can be exchanged directly (d).

In the original version of mobile IPv4, mobile nodes sent replies directly to correspondents. For compatibility with higher-level protocols, the "source" address field in these packets had to be the permanent address on the home network, even though routers on the Internet would see that the packets were actually coming from the care-of address on the visited subnet or network. This wasn't a problem in 1996, but it is now.

Thanks to Denial of Service (DoS) attacks, where malicious packets often claim to be from fake IP addresses, routers have begun to incorporate ingress and egress filtering whereby routers only allow a packet through if its source address field is consistent with its origin. To get around these filters, mobile IPv4 was updated in 2002 to include reverse tunneling. Instead of taking a triangular path, all packets travel via the home subnet in both directions. Unfortunately, this step wastes even more bandwidth and adds further latency, making it unsuitable for any wireless network running bandwidth intensive applications.

Mobile IPv6 reduces the bandwidth and latency problems by avoiding tunneling as much as possible. Though the first few packets of every session are still tunneled via the home agent, the mobile node also sends binding updates to every correspondent. Future packets can be sent directly, just as if the mobile node belonged on the network it was visiting. You can apply the same principle to entire mobile subnets, such as a WLAN inside a moving vehicle.

You can accomplish this with extensible headers, a feature allowing IPv6 packets to contain extra protocol information to deal with issues such as QoS and prioritization. Extensible headers allow each packet to contain both the permanent and the care-of address, satisfying both higher-level protocols and routers. However, the extra bandwidth taken up by this information can be significant, especially for small packets such as those used in VoIP. That is why IPv6 uses the robust header compression (ROHC) standard. By taking advantage of the fact that consecutive packets often have identical headers, ROHC can reduce header size by around 95 percent.

Because a mobile node can move rapidly, it might have several care-of addresses at any one time. These addresses include the primary one, representing the subnet the node is attached to, and several older ones on subnets the node previously passed through. Packets sent to these older care-of addresses must be tunneled by agents on the previously visited subnet, just as if they were sent through the home subnet. To prevent a node from accumulating too many old care-of addresses, mobile IPv6 provides binding updates, which always include an expiry time for a care-of address.

The mobile node registers its care-of address with its home agent, and the home agent forwards packets destined for the mobile node to the care-of address via an IP-in-IP tunnel. The tunneled packets are stripped out of the outer header and the inner packet is delivered to the mobile node. Because the application/session is using the home address, session continuity is maintained. As the mobile node moves, it obtains a new care-of address and performs a re-registration with the home agent to indicate its new care-of address, and the tunnel end point is changed to the new care-of address.

All of this would seem to require extra functionality within every device connected to the network; edge routers must be able to tunnel packets not just to their own mobile nodes, but also to other nodes that have previously used a care-of address on their network; TCP/IP stacks on individual devices must be able to understand the difference between a permanent and a care-of address. However, such functionality is standard in the IPv6 specification, whereas the ability to act as a home or foreign agent has to be retrofitted to IPv4 devices. These facts, rather than the larger address space, is why the wireless industry is so keen to promote IPv6 adoption.

IP Mobility Solutions

While a few proprietary IP-mobility solutions have been developed, the emphasis remains on developing Mobile IP standards-based solutions with enhancements to handoffs, security and tunneling. A few of the companies developing solutions based on Mobile IP include not only the above-mentioned NetMotion, but also Flarion Technologies, a company that has developed an all-IP mobile network based on FlashOFDM as the air interface and uses Mobile IP for roaming and handoff support. The Flarion architecture includes a RadioRouter base station, a concept that puts the access router at the very edge of the network. Here, the access router that terminates IP between the mobile and the network is located at the base station instead of at a distant point deep in the network.

The RadioRouter base station is connected to a packet data network. Mobility is accomplished using Mobile IPv4 albeit with enhancements to support fast handoffs at up to vehicular speeds for real-time services like VoIP and video streaming. Flarion has also demonstrated seamless handoffs between 802.11-based WLAN networks and the Flarion FlashOFDM-based network, using Mobile IP as the IP mobility glue.

Other Mobile IP-based solutions are available from Birdstep Technology, ipUnplugged and Airvana (for 3GPP2). Though standards based, almost all have their own enhancements in terms of security and handoff speed. Also, ipUnplugged, in conjunction with Airvana Networks, has demonstrated seamless mobility between 802.11 WLAN and cdma2000 1xEV-DO networks.

Mobile IP's Downside

While mobile IP is widely used, some drawbacks should be considered before implementing this mobility solution.

- Mobile IP relies on dedicated servers to maintain tables of remote and mobile clients.
- In most instances the "anticipated" location of each mobile client must be manually entered into the system. This is a tedious process. Furthermore, if a client roams outside its pre-assigned "territory" of defined subnets, the Mobile IP server will drop the network connection, forcing the client to re-boot to reestablish a link. However, Intermec Technologies Corp. has develop a "smart" wireless access point that can automatically track mobile clients without this administrative burden.
- Mobile IP requires additional software for every computing device that accesses the wireless network. This means that licenses must be purchased for each device, which can be an expensive proposition.
- Mobile IP servers represent a single point of failure in a network, i.e. if a Mobile IP server fails there is no support for roaming. But the Intermec solution, mentioned in an earlier point, lessens the single point of failure aspect of Mobile IP. Intermec does this by using IP tunneling support to provide robust, seamless connections for all mobile devices. The Intermec solution works with any IP-based networked device and does not require any client-side modifications. Any computing device that attaches to the Intermec smart AP has the capability of roaming across subnets.

SECURITY AND QUALITY OF SERVICE

IT managers are fighting a rear-guard action to secure and manage the quality of their WLAN networks. Most primary wireless LAN security mechanisms, such as Wired Equivalent Privacy (WEP) and 802.1X, are seated in the Data Link's MAC sublayer, which is neutral regarding 802.11a, b, and g. Because all three 802.11 WLAN PHY specifications suffer from the same security and QoS maladies, Chapters 16 and 17 are devoted exclusively to Quality of Service and security, respectively.

That being said, for continuity's sake, here is a brief discussion concerning both issues. First, let's look at Quality of Service. The quality of a WLAN's management and maintenance programs can have a tremendous effect on a network's QoS. But also know that the IEEE is diligently working on the 802.11e specification, which will address both asynchronous data traffic and data traffic that is time controlled, such as voice or video. It will lay down rules on how each traffic stream can employ different policies. For example, a video stream that is time sensitive could employ forward error correction instead of packet retransmission.

QoS is an essential capability for voice and video support, but if and when the QoS standard is finalized, its mechanisms will need to be integrated with QoS mechanisms in infrastructure networks at large, and this will take some time. So while exciting, it may be years before applications in corporate environments can truly take advantage of any new QoS standard. Home use of integrated voice/video/data networks will happen much faster. However, there is no reason to wait for the more exotic features that should become available once 802.11e is ratified: Today's products offer more than sufficient capabilities for many applications. And as long as you put some hard questions to your vendors about their upgrade paths, you can safely deploy a network that you can enhance, as needed, over time.

Now to security. While implementing a fail-safe security policy is very important, it also must be pointed out that malfeasance from unauthorized access to a wireless network is greatly exaggerated; an individual or organization is as likely to lose valuable information via a stolen or lost computing device as from someone maliciously breaking into its WLAN. If you take the necessary steps to protect your computing devices by password protection and by encrypting files on the hard drives and generally exercising caution, you've taken a big step toward decreasing the likelihood that your valuable data will wind up in the wrong hands.

Currently, there is no single blanket solution that addresses all of the security issues that crop up in a wireless networking environment. But there are a variety of steps that can be taken to increase a WLAN's security. As with wired networks, effective security requires a multi-layered strategy, as discussed in Chapter 17.

NETWORK MANAGEMENT

The first WLANs were primarily small networks, making it relatively easy to manually operate the network and manage radio coverage by trial and error. End-users would detect failures and report them, there were few eavesdroppers, and hackers were still a plague of the future. This changed as the number and size of WLAN installations grew. Wireless network management tools are invaluable for anyone deploying a WLAN supporting more than 20 stations. Network managers of larger networks are reluctant to deploy a WLAN unless they can provide the same level of security, manageability, and scalability as their wired LANs.

While tools are incorporated into Ethernet networks to support the operation phase of a network, these tools are of minimal use for development and deployment of a WLAN; furthermore, many do not adequately address the dynamic nature of a wireless environment. A WLAN's traffic demands and radio performance change dynamically, and network performance, as seen by users, can suffer if the WLAN's parameters are not optimized and reviewed regularly. Long-term throughput degradation resulting from poor performance can be even more costly to an organization than short-term outages.

WLAN managers realize that if they are to maintain the kind of performance that an organization expects from a wireless network, they must seek tools that can regularly monitor performance and allow easy identification of the root cause of performance problems.

Currently, the typical AP vendor provides only device-configuration tools or web-based device management, and very little in the way of capacity management or networking management. But this segment of the WLAN marketplace is evolving rapidly; it is now possible to find vendors that offer network management tools, although they vary greatly in usability, performance, interoperability and manageability. These vendors include 3Com, AirWave, Cisco Systems, Computer Associates, Lucent Technologies, Sniffer Technologies, Symbol Technologies, Vernier Networks, Wavelinke, WildPackets, Wireless Valley Communications, and others.

Many managers also find wireless network analyzers to be useful as a WLAN management tool. While not scalable for large WLANs, managers of smaller wireless networks can use a wireless network analyzer for site surveying, troubleshooting, intrusion detection, and logging.because they can observe each of the layers of the stack and quickly reveal problems. This can greatly reduce the amount of time spent diagnosing or pinpointing, whether the issue is wired or wireless in nature.

THE CAVEATS

A wireless LAN is an ideal solution for any business or organization that needs to provide its employees with mobility. Wireless LANs empower mobile workers, enabling them to stay in touch with easy access to real-time information. Wireless LANs also can provide flexibility in a networking environment that requires frequent LAN wiring changes. And, of course, WLANs are a godsend when deploying a networking solution in a difficult wiring situation.

But, a WLAN will never behave or perform precisely like a wired network. Thus, when considering the deployment of a wireless network, the first item everyone needs to attend to is the development of a thorough understanding of how a WLAN and its clients (i.e. computing devices, printers, etc.) perform in various networking environments.

Second, if you are introducing wireless into an existing wired network or in anyway mixing wired and wireless, you need to understand that the two network types differ in their access methods—wireless uses CSMA/CA and wired needs CSMA/CD. Consequently, TCP/IP, with its inherent back-off algorithms, will actually degrade the performance of wireless clients when it attempts to retransmit lost packets.

Finally, while the appeal of WLANs is their ability to augment or supplement existing wired LANs in difficult-to-wire locations, wiring remains a consideration when deploying WLANs, as the wired infrastructure must be extended to the WLAN's access points.

The combination of the proper design and a good deployment plan that can implement the proper extension of existing technologies and corporate policies, accommodate the WLAN's management and maintenance needs, and plan around any performance issues, will enable a WLAN that is a part of a strong, viable networking system.

CONCLUSION

Before the introduction of the 802.11 series of standards, wireless local area networks (WLANs) were slow, expensive, proprietary, and were only deployed to support vertical market applications, i.e. sold as components of a specific solution to a specific problem where mobility was required. Shipping companies were the first to see the value of WLANs in warehouses; today, FedEx, which has extensive WLAN deployment throughout its global infrastructure, estimates its Wi-Fi-enabled workers are 30% more productive since they've been unleashed. The manufacturing sector also was among the first adopter group. This sector immediately saw WLANs as a way to help with inventory management; Wal-Mart uses WLANs for inventory and to connect pricing terminals. Hospitals and college campuses came next and it wasn't long before car rental companies used WLANs to connect portable check-in terminals to their back-office systems.

In each of these applications, the organization deployed the WLAN to support a specific application, not as a part of the general-purpose LAN infrastructure. But today's 802.11 (a, b, and g) networks are flexible wireless data communication systems that can be implemented as an extension to, or as an alternative for, a wired network. The current wireless landscape looks significantly different from that of yesteryear. WLANs are practical and cost-effective for almost any networking environment because a WLAN combines data connectivity with mobility delivering powerful results: the ability for end-users to communicate with a network at the point of activity.

At the beginning of 2003, almost 60% of all U.S. corporations, including all of the FORTUNE 1000, had at least a small-scale Wi-Fi network, and a few tech-savvy firms like Microsoft, Novell, and Qualcomm have deployed WLANs company-wide.

The speed and performance of 802.11b products make WLAN technology attractive, especially where a wired Ethernet installation is not feasible. And thanks to the recent dramatic drop in price for WLAN gear, SOHO (small office/home office) and mobile applications are also a possibility.

Another exciting event in the WLAN arena is that vendors across the board are designing and shipping innovative products. All the major enterprise infrastructure providers—Cisco, Enterasys Networks, Lucent Technologies and Nortel Networks—have a wireless story to sell. Even more compelling, these products (1) are truly interoperable, (2) cost less, and (3) provide better performance than their predecessors. The quantum leaps made in all three areas make wireless LANs a viable technology for the masses. Never before has there been such a strong backing of wireless technology, feeding a growing momentum for everything wireless.

The Wi-Fi craze is also aided by new and better designed computing products, e.g. lighter, more portable notebooks, tablet computers, and PDAs like HP's iPAQ, to name a few. For stability and continuity, there is the Wi-Fi Alliance, a highly active group of vendors with a mission to guarantee interoperability across all 802.11 products. As for the broadband revolution, it's enabling the WISP (Wireless Internet Service providers) industry to provide low-cost wireless Internet access to the general public the world over through a new business model known as "HotSpots."

Chapter 8: A Practical WLAN Deployment Plan

A wireless local area network (WLAN) that is based on the 802.11 series of standards provides a local data transmission network environment that offers high-speed connectivity to any compatible wireless-enabled networked device. A WLAN solution can add actual benefits to the organization that it serves. Benefits that include: cost reduction in areas where network cabling is difficult or expensive; flexibility to add to or make changes in the networking ecosystem with no extra cabling costs; duplicate networks, i.e. you can have mobile networks with no need of new cabling structures for new work lines; and efficiency and productivity increase by allowing employee mobility within the work facilities.

As the reader has learned, the 802.11 specifications use electromagnetic waves to transmit and receive data over the air. They offer a basis for WLAN technology that provides flexible data communication systems that can be implemented either as an extension to or as an alternative for a wired network.

The beauty of any 802.11 WLAN is that it can combine data connectivity with a mobile workforce to provide a powerful result–the ability to communicate and collect data at the point of activity. WLAN deployments are growing by leaps and bounds. There are a number of reasons for this phenomenal growth:

First, the Institute of Electrical and Electronic Engineers (IEEE), the same group that established the 802.3 wired network standards, drafted, finalized and continues to improve upon the 802.11 suite–the industry-wide, vendor-independent WLAN standard.

Second, the International Organization of Standardization (commonly known as the "ISO") adopted the IEEE 802.11 as a worldwide WLAN standard.

Third, the Wi-Fi Alliance set up a program to certify the interoperability of the 802.11 WLAN variants, e.g. 802.11a, 802.11b, and 802.11g. The goal of the Alliance is to ensure interoperability among a wide variety of wireless systems and products and to promote the term "Wi-Fi" as the global WLAN standard across all market segments.

Fourth, the actions taken by the IEEE, the ISO and the Wi-Fi Alliance assure users of a stable technology and competitive pricing. In fact, WLAN infrastructure prices have decreased dramatically over the last few years.

Fifth, we see a shift to mobile computing platforms in both the workplace and at home. Workers and consumers use laptops, PDAs and tablet computers as their primary computing devices, and all of these mobile devices have gained acceptance in the workplace.

Finally, many organizations are evaluating the benefits of improved productivity and efficiency that comes from providing individuals with the flexibility to move freely within their work environment while maintaining their computing connectivity.

THE TASK AT HAND

Just when most people understand how Wi-Fi technology can improve their business, the technology evolves again. Now, there are wireless adjuncts such as wireless voice over IP and Power over Ethernet that also should be considered. This fast evolutionary pace makes it difficult to assess the proper timing for adoption of a wireless networking strategy. Nonetheless, there are many organizations developing wireless strategies to streamline productivity and increase their data throughput.

But organizations often underestimate the effort required to deploy a wireless networking system. To get the full benefit that automation and wireless technologies offer, a plan is needed—a strategic plan that matches the end-users' needs and wireless technology capabilities with how those needs are expected to evolve over the next five or so years. Designing a WLAN, especially an enterprise WLAN, can be a daunting task. It requires an understanding of how 802.11 radios work, the differences between vendor implementations, and how varying building structure and outdoor elements and sources of external interference affect a WLAN's performance.

Before the initial installation, invest time in the plan and design of your system. While a wireless system will enable worker mobility, the supporting infrastructure is not mobile and its placement for performance and coverage must be carefully planned. The exact design steps needed depend on the organization's critical requirements for performance coverage and future growth.

The designer also will need to address core network services—IP address management, authentication, encryption, access control, accounting, and maybe even Quality of Service—that must be delivered to wireless users.

Once you've assessed the WLAN's requirements, you need to develop a plan for day-to-day management, service, and support to ensure your wireless network investment will continue to meet the changing demands of the organization.

THE ASSESSMENT PROCESS

The first step in planning a wireless strategy is to evaluate the existing corporate culture, networking environment, and users' needs. By talking to potential end-users you can discover their expectations, how they would like to utilize a wireless network, and a bit about what benefits a wireless network would provide them. In a corporation, for example, one group of users may be delighted to take a mobile computing device with them as they check inventory in the adjacent distribution center. Another group may be eager to use wireless networking capabilities during conference sessions. And still another group may be anxious to offer wireless networking to important corporate visitors.

In a corporation, you could classify the employees according to their job functions, since based on their job functions some end-users will derive more benefit from a WLAN than others. You can segment the WLAN user base into categories, such as executive management, engineering/product management, manufacturing, sales, marketing, and support.

Survey the potential user base about how they do their jobs. Have them provide an outline of their daily routine, and ask them to identify the difficulties they feel they encounter without wireless connectivity. Once that information is in hand, you will have a good foundation for determining whether a WLAN is right for your organization and if it can deliver an acceptable return on investment (ROI).

Then take a hard look at the impact of adding wireless networking to the current network environment. Make a list of the resources and people that will be affected by a wireless networking environment. This list should include WLAN technicians, network management software, upgrading and/or purchase of wireless computing devices, and more.

Also, before taking on a WLAN project, answer this question: Why is your organization considering wireless? Is there a need to implement (or expand) a mobile workforce environment?

Here is a *suggested* "wireless assessment check list" to help the reader plan his or her assessment strategy.

- Can wireless devices operate within the current IT infrastructure? For example, does the existing backbone have sufficient bandwidth to handle the additional traffic of a WLAN? Are there enough ports available for the necessary access points? Will the access points (APs) interoperate correctly with the existing local area network's (LAN) routers and switches? While it's not strictly necessarily to use the same vendor for wireless as you do for your existing wired infrastructure, if the wireless components do not work properly with the existing network components, you're in trouble.
- Is the technology consistent throughout your wired network?
- How many users require mobility, and where do they need to go?
- Will the WLAN support subscription-based users?
- What user applications will the WLAN support and what performance levels do these applications require?
- What is the corporate culture, and what are the goals for the WLAN? For instance, is it to bring more business to your location or to add an additional revenue stream? Or is it to gradually increase mobile employee support? If the organization has a substantial mobile workforce (e.g. a distribution center), supporting such a mobile force can have a significant impact on the company's information technology (IT) infrastructure. The mobile employee may need, for example, access to inventory management systems, corporate databases, company policies and procedures, the organization's email server, and customer support. If so, a WLAN is definitely worth investigating.
- Does management want to support a mobile workforce to increase communication within the organization? If so, a WLAN is not a one-size-fits-all solution on how to improve communication within an organization. If the there is a lack of enterprise communication to begin with, no device in the world will fix the problem.
- Is the push for a WLAN an effort to increase employee productivity? If so, how will the employees respond? Some employees are pleased that they are able to connect to the office 24/7. Others want to leave work behind when they leave the office. Forc-

ing employees to maintain contact and availability can backfire by causing resentment or burnout. Evaluate the employee situation and formulate a strategy that respects the working environment as a whole.

* Does the organization have both the financial and human resources to support a WLAN environment? The spending doesn't stop with the purchase and installation of the network and its components and computing devices. Resources will be needed for such things as training, employee technical support (24x7), lost devices, just to name a few. Supporting a WLAN environment can tax an IT department, particularly if it's already operating with minimal resources. Talk to the IT department and get the staff's opinion—the more they contribute to the decision process, the less likely they will be to protest later.

The assessment process lays the groundwork for the initial preparation for range, capacity, and coverage planning.

Range: You now know that every WLAN device is a transceiver, i.e. capable of both transmitting and receiving radio signals. This can, unfortunately, make predicting the range of a specific WLAN system in a specific environment challenging. Using identical components, effective system range may be well over 300 feet in one location and less than 50 feet in another.

A number of variables, including building layout, construction materials and noise sources can affect transmission range. While some might consider the range limitations of radio to be its downfall, others understand that this very limitation is the main ally of a WLAN designer. That's because range limitations let you reuse frequencies (i.e. the same channel can be used in more than one cell, as long as the cells don't overlap in their coverage area), just like you do with conventional wireless services like FM (Frequency Modulated) radio. For example, say that WXXX FM in New York City is operating at 90.6 MHz. In all likelihood, careful design and technology has enabled at least one other FM station to use that same frequency in another locale within the state of New York. The FCC just requires that 100 miles or so separate FM stations using the same frequency.

Capacity: The same holds true for capacity. WLANs and 802.3 networks are shared-medium technology. With a WLAN, an access point (AP) is used to establish a coverage area within which an aggregate amount of throughput is shared by all wireless computing devices that access the access point(s). However, unlike a wired network, where the number of computing devices can be defined by choosing how many ports will be used on any one wire hub, a WLAN has no physical ports; only the size and shape of the coverage area defines the limits on the number of users accessing the network. So with a wired LAN, capacity planning can be absolute, whereas with a WLAN the user numbers vary. Not only are WLAN users mobile, entering and exiting the coverage area at will, but a WLAN's throughput is subject to variation as factors such as radio wave interference can decrease transmission rates.

When planning a WLAN's level of service, the goal is to provide end-user groups with the throughput rate they need to work efficiently and effectively. To reach this goal the deployment team must determine what are the average throughput requirements for each user group, and then how much throughput to provide to each group at any given time. Different groups have different throughput requirements. R&D departments, engineers in a manufacturing plant, and a graphical design group will required a higher throughput rate than a typical corporate office, which can get by with lesser throughput speeds. Distribution center employees and retail workers whose job requires that they use a WLAN to gather data via barcode scanners will need very modest throughput rates.

Coverage: The assessment process should provide the deployment team with the information needed to determine exactly where and what type of wireless connectivity is needed. For example, after the assessment process is completed it might be determined that the WLAN should provide connectivity in all conference rooms except those located in the accounting and human resources floors. Or that the WLAN should extend to the distribution center but not the loading docks.

When the requirements of the proposed WLAN can be clearly defined, it's much easier to choose the right Wi-Fi standard, plan a viable site survey process, and develop a realistic deployment plan. For instance:

- Identifying the proposed WLAN users and their application requirements helps to define the network's technology and coverage areas, eliminating not only wasted money, but also security risks (by not sending signals beyond the necessary areas).
- Listing the applications required by users will go a long way in helping to determine the WLAN's minimum bandwidth requirements and identify WLAN candidates. Keep in mind, however, that wireless is a shared medium, thus while most mainstream networked applications can be migrated to a shared WLAN, it's not necessarily appropriate for all applications (e.g. sensitive documents such as R&D, and human resources).
- Once you know how many users the WLAN is expected to support and who these users are, it is relatively easy to determine which end devices the WLAN will serve directly. Then it's a simple task to establish which devices are already equipped with the correct radio NIC, and which need to be upgraded.
- Doing a thorough assessment enables you to do a realistic *estimate* of what it will take—hardware, software, personnel—to deploy a WLAN in your organization and the estimated costs for the same.

You may use the results of the assessment study to compare a WLAN with an Ethernet alternative, perform a feasibility study for a specific mobile application, or provide the basis for a budget to present to upper management. Thus, an assessment study and determination of the WLAN's technology criteria not only saves networking dollars, but also can serve as the starting point in an increase in corporate productivity and end-user satisfaction.

DEPLOYMENT TEAM

If the assessment process indicates that a WLAN is feasible, the next step is to form a deployment team. A deployment team considers the company's needs, looks at wireless options, assesses barriers, determines integration issues, and more. The team must effectively understand a WLAN's functionality and usability, as well as deployment management. It can then write a deployment strategy, and afterward methodically maintain a project management plan to which all stakeholders have access.

The size of the deployment team is dependent upon the size of the WLAN. For a small WLAN, the team may consist of one person, while an enterprise WLAN team could consist of 35 people or even more. Whatever the number, the team must provide all of the core competencies needed to deploy the WLAN. This includes not only IT skills, but also radio frequency (RF) expertise, a good background in finance (for ROI), project management experience (if a large WLAN is being deployed), end-user training experience, and more. The team should also have at least one member from every department affected by the WLAN's deployment, along with a member of the organization's executive staff.

Many companies struggle to determine the best way to staff new projects. This is especially difficult when technical skills simply are not available within the organization, when the project needs them. Using outside professionals can close that gap and allow the WLAN project to move forward.

To overcome staff deficiencies, bring in outside experts who understand site surveys, wireless specifications, access point infrastructure, network configuration and integration, device deployment, data traffic, and wireless security. Many of these experts have the experience, training, and access to the newest technology, so they can intelligently discuss what is right for the specific needs of your organization.

Project Manager

If the WLAN project encompasses campus coverage, or involves a full-scale WLAN deployment for a large organization, consider using a professional project manager who can assist in managing the organization's IT resources, and/or oversee the work of outsourced contractors and vendors. These professionals are experienced in the design, implementation, and management of system solutions and rollouts.

The consulting project manager's first job is to gain a thorough understanding of the goals and objectives of the WLAN project. Once the goals are understood, the project manager prepares a Statement of Work (SOW) that clearly documents objectives, work scope, expectations, assumptions, deliverables, cost, schedule and schedule of payment, and outlines the acceptance criteria of the job that is to be completed. A project manager then begins a detailed analysis to better understand potential trouble spots.

An outside project manager also can offer objectivity, and can sometimes influence decision-making parties or resolve conflicts without in-house political ramifications. In addition, an outside project manager handles one project at a time without additional daily responsibilities. This helps the WLAN deployment to quickly move forward.

Finally, because outside project managers typically have a wealth of experience, they should avoid easily overlooked pitfalls and mistakes.

Outsourcing

Too often, because of budget constraints, internal IT staff is expected to research, manage, configure, and install a new wireless system while still handling daily responsibilities. Relying on current employees who don't have the proper skills or enough time can increase the risk that the WLAN project will take longer and not meet the quality standards expected. Not only is there a risk of shattering IT budgets, but also of staff burnout and costly turnovers. Professional service firms or large vendor can help with your wireless strategy and WLAN deployment. Although each firm will have its own unique focus, when selectively used, these firms can help an organization move a wireless project from conception through deployment and beyond.

It's up to the WLAN project manager to decide what balance of services is right for the project, staff, and budget. For example, you may want to outsource the site survey and hardware implementation, but handle the post-deployment tasks in-house.

While there are many reasons that businesses decide not to use outside services, especially now with the short-term focus on decreasing IT expenditures, this environment can make it difficult for IT departments to meet business demands while fulfilling critical IT initiatives. When the deployment team contracts with a professional services firm for some or all of its WLAN needs, they buy peace of mind from somebody that understands the wireless environment. Reputable vendors and professional services firms stand behind their services, and will work with a deployment team until the WLAN is operating to the team's specifications. Furthermore, outside service professionals provide an objective view to the project, as well as becoming onsite educators for IT management.

Whether in-house or outsourced, as a member of the deployment team, the project manager ensures the project stays within budget, on schedule, and maintains acceptable performance and specification criteria. He or she gives consistency to a project by providing a single point of contact. And after the deployment is complete, the project manager reviews the project with the stakeholders.

The Deployment Plan

Once the deployment team is on-board, it's time to create the deployment plan. This document outlines the scope of the WLAN project, its objectives, the work breakdown structure, roles and responsibilities, detailed task schedule, budget information, risks and issues, change control information, and all communications between the deployment team and/or project manager and the stakeholders. Next step—technology and hardware selection.

The key to making the right technology decisions, as well as a smooth deployment, is to have a strong strategy spelled out in a detailed plan before deployment begins. Preparation is vital to the build-out phase because wireless products must, by design, touch many different parts of an organization's environment. Strict attention to detail and unfailing diligence offers the best chance for success.

TECHNOLOGY SELECTION

In the decision-making process, management and IT analysts need to evaluate the WLAN technology alternatives from the perspective of their organizations' short-term and long-term communication objectives. Here's a suggested check list to use during the evaluation process:

- Number of concurrent users—how many people do you expect to access the WLAN on a daily basis? What percentage of that group is expected to be accessing the WLAN at any one time?
- Number of stations—this refers to the total number of computing devices, wireless printers, perimeter surveillance systems, and so forth that is expected to be used to access the WLAN.

* Type of use—to provide wireless Internet access to the public, to support distribution center activities only, to enable visitors to access the network with ease, or perhaps to enable widespread use of wireless connectivity.
* Short-term and long-term objectives for the WLAN.
* If the wireless technology is to be deployed to support a business or home network, what applications—email, instant messaging, word processing, PowerPoint presentations, instant sharing of design schematics, video streaming—will regularly traverse the WLAN system?
* If the WLAN isn't to serve a HotSpot—determine the type of computing devices (e.g. laptops, PDAs, printers) that will access the WLAN. If certain end-users will use the WLAN to access data-rich applications and Internet content, including streaming audio and video, laptops are a better choice for those users than a smaller PDA or other type of handheld computing device.
* Expandability—is the WLAN deployment designed for only a small group of end-users and if so, are there plans to expand the WLAN in order to bring wireless connectivity to others in the future?
* Software and hardware—if the network will be built with gear from various vendors, stage a trial run to ensure compatibility.
* Vendor support—what do they offer? As with any technology, good vendor support is crucial.
* Mobility and flexibility—what degree of mobility will the end-users require? Moving about within a single building, from building to building, throughout a large campus or what? Will all of the end-users be using the same type of computing device (e.g. all using laptops or will some be using PDAs or vice versa)? For instance, a HotSpot must be designed to support a variety of devices, but a corporate WLAN also may be expected to support a mix of computing devices. In that case, you need to determine if the mix is throughout the organization or only within groups (e.g. only warehouse personnel will be using handheld computers, everyone else will be using laptops)?
* Environment—will the WLAN provide wireless connectivity to a warehouse, office space, medical facilities, manufacturing plant, campus, or some mixture of these?
* Maintenance—how will the WLAN be maintained so that end-users won't suffer from dropped signals or interference?
* Life cycle—are you building the WLAN to accommodate end-users while attending a conference, or a training session, or are you building the WLAN as part of the organization's permanent network infrastructure?
* Speed—applications and number of concurrent users—bandwidth intensive applications will be intolerant of speed fall backs due to interference, distance or heavy user activity.
* Range—here you need to be specific about how far the signal should reach, if only for security purposes.
* Frequency—there are some differences between 11a, which uses the 5 GHz frequencies, and 11b and 11g, which use the 2.4 GHz band.
* Equipment connectivity—a clear line of sight is important for quality connectivity, but also consider network equipment placement insofar as electrical power and cabling are concerned.

* Manageability—this mainly refers to training the IT department in the complexities of managing a WLAN. But you also need to consider the overall WLAN architecture in light of possible manageability problems that could crop up or could be avoided with more attention to design details.
* Security—there are many other aspects of securing a WLAN such as availability, survivability, intrusion detection, and user privacy. Even with the 802.11's flaws, it is still possible to secure a WLAN to an acceptable level by following a few basic principles.
* Connectivity with other wired and wireless networks—is there a wired network that will be accessed via the WLAN? If so, is its technology compatible, i.e. all Ethernet, all Token Ring? Are there any other wireless networks in operation? If so, are they compatible with the proposed WLAN?
* Location—this refers to the country where the WLAN is in operation. Does it meet that country's regulatory conditions? For example, are the antennae in compliance with FCC regulations?

Filling in the "blanks" for each of these items puts you in a much better position to make the right technology choices. Use the information in Chapters 7 and 10 where we examine the different characteristics of 802.11a, b and g. This includes the particular networking environment issues relevant to making a decision as to the make-up of a wireless network's architecture compared to the technology to be used.

Read Section VII: The Hardware for an in-depth discussion of the wireless networking gear available for deploying a wireless network. Only the WLAN deployment team will know which components (including their various permutations) are necessary to enable the end-users to obtain an optimal wireless networking experience.

THE VENDOR SELECTION PROCESS

During the decision-making process, you will consider many vendors' products. And while the final product selection might not be of vital importance to a small WLAN in a small company, it will be of *utmost importance* for *all* WLAN projects (large or small) in a large organization. When you purchase the access point, wireless network card and other hardware, you not only buy the device, you also acquire add-on utilities, such as software that can be used during the site survey, for network management, and even for a WLAN's on-going maintenance. These add-on utilities are key to the overall success of a WLAN project. But since these utilities are typically provided at no additional cost, their strengths, features and usefulness can vary greatly from vendor to vendor. Also don't underestimate the value of pre/post sales support. The specialized expertise and skillsets that a vendor's organization can provide during the deployment and post-deployment stages are invaluable.

Thus, when deploying a WLAN, look to partner with a vendor that can support the entire wireless infrastructure, including 24/7 help desk support, remote administration capabilities, and RF diagnostics. Use only vendors who continually focus on a compelling industrial design at competitive prices. The vendors chosen also should offer business integration and flexible solutions, and a broad product/service mix, as well as flexible contract terms.

Only consider vendors who offer suggested systems and approaches that promote a rapid ROI. Ask the vendors to issue Statements of Work (SOW). Finally, if needed, ask vendors if they can assist in preparing a needs analysis and feasibility study; many vendors servicing the enterprise market will readily provide such services.

When choosing your technology vendor, keep the following items and issues in mind:

- Approach the vendor selection process recognizing its strategic importance in the planning of your wireless system.
- Consider all of the important characteristics that will determine the WLAN's success, including robustness, manageability, scalability, value, and vendor capabilities.
- Think carefully about the speeds and applications you will need now and into the future. High-speed systems are best in limited spaces where wide bandwidth applications are a priority; lower-speed systems provide better coverage and immunity. Select your networking gear to complement each other and to optimize their individual strengths.
- Take into account upgrade planning, which should be approached with accurate knowledge of differences in coverage range; simplistic schemes that talk about one-for-one access point upgrade or replacement will not yield the desired results. Higher speeds may require more access points.
- Insist on standards compliance from everyone—network designer, vendors and during implementation.
- Consider the capabilities, capacity and availability of your vendor to work with you throughout the life of your wireless system—from the initial planning, network analysis, design and site survey through integration and installation of the system to ongoing service and support.

Once you've settled on a short list of vendors, ask those vendors to provide you with an analysis of the technical advantages of their products (heightened security, performance), and the business benefits (scalability, lower cost of management), and have them prove how their products deliver these advantages. Use those analyses to compare each vendor's offering. Next, issue requests for proposals and perform the proper due diligence. For instance:

- Assess the vendor's economic health—an important consideration because the IT department will want to know the vendor will be around, if needed.
- Determine the number of installs the vendor has completed. How many of those were in your organization's industry?
- Ask the vendor to provide at least five customer references.
- Follow up on those customer references and ask how well the vendor backs its services and products, i.e. does the vendor have a track record of being present, not only before, but also after the sale?
- Find out if the vendor has a product support group. To determine how responsive that group is to a customer's needs, ask for customer references.
- Ask the vendor if the system's information is stored on a database, so the IT department doesn't have to bring the product support person up to speed every time it has a question or service request.
- Find out how has the product evolved over the years. Is the vendor recommending a relatively new product without a track record, or one that is approaching end-of-life?

A good vendor will be knowledgeable about the organization's industry and business models, as well as hardware and software integration.

Consider your WLAN's specific needs. In certain deployments vendor specific utilities and specifications may not be applicable. For example, when deploying a WLAN in a public area such as an airport, or a HotSpot, vendor-specific utility may not work for all end-users because every user will not be accessing the WLAN via the same product. Consequently, generic wireless characteristics need to be retrievable and monitored.

Be careful that you do not to create a "closed system," thereby locking the WLAN into a vendor-specific solution. While such a "closed system" might not be a problem in a corporate environment, it can be particularly problematic in public areas such as HotSpots, where various radios will be present.

If the WLAN project is large or planned around a complex WLAN deployment, consider using only products from an established vendor. Large, reputable vendors are better able to fulfill the project commitments through the utilization of its staff experts and in-depth resources than are smaller vendors with limited staff and experience. Also, since the entire IT industry faces attrition and vendor consolidation, a good percentage of IT suppliers in business today will disappear from the competitive landscape in the coming years. Thus, contracting with an established vendor offers the best chance for stability.

Finally, bear in mind that for the most part, products that have obtained the Wi-Fi Alliance's certification will, at least, guarantee a basic level of interoperability.

PRE-DEPLOYMENT PREPARATION

The next step is to plan for the actual wireless deployment. When planning the deployment of a wireless LAN, you need to carefully assess and resolve risks. Otherwise unforeseen implications, such as RF interference, poor performance, and security holes will

CHECK LIST FOR A SUCCESSFUL WLAN DEPLOYMENT

- Match the needs of the organization and end-users to the WLAN's final design.
- Identify the area the WLAN should cover to provide the optimal wireless/mobile activity.
- Measure radio characteristics of site.
- Survey and identify power options.
- Verify requirement of host connectivity.
- Survey existing network connections and existing equipment.
- Analyze results.
- Design the WLAN to provide optimal coverage for throughput and mobility.
- Design network as "legal" segment of any existing wired network.
- Be pedantic during the vendor selection process.
- Carefully plan integration into any existing network(s).
- Document equipment placement, power considerations, and wiring.
- Carefully plan for post-deployment needs, e.g. upgrades, network management and maintenance, user training and support, etc.

wreak havoc. By handling risks during the early phases of the deployment, you'll significantly increase the success of your new wireless network.

The Site Survey

With the assessment completed and the budget for at least a site survey in hand, the next step is to conduct a comprehensive site survey. Site surveys differ in their complexity and level of effort, based on technology and space.

Never deploy a WLAN without conducting a site survey. Even a small WLAN, serving only a limited number of end-users, should be deployed only after a site survey. And for larger WLAN systems a site survey is vital—unlike wired networks, WLANs have many variables and few fixed rules.

A site survey not only provides detailed specifications that address coverage, equipment placement, power considerations and wiring requirements, but it also gives the installers a realistic understanding of the wireless installation. A site survey also helps to determine whether a site has unusually high interference. Furthermore, it guides financial decision-making since it can be used to obtain accurate quotes from vendors, consultants, and installers. Without an effective understanding of your site's requirements, your wireless LAN installation *will* be more problematic, expensive, and time-consuming than necessary.

***Note:** Every WLAN will have different goals and requirements so take your own organization's wireless connectivity needs into account when reading the following discussion.*

Every WLAN design will be addressed differently. For example, one WLAN may need to provide for coverage that is isolated to a warehouse, and another will provide perva-

PLANNING FOR GROWTH

Plan the initial installation with the future in mind. Requirements that need to be understood before beginning a site survey include:

- How many users will the WLAN support at the outset? Will there be user growth? If so, how much and when?
- Where are the potential users located? Where are the potential additional users located?
- What sort of applications will be run over the WLAN? You must plan your network layout so that data-rate "fall backs" do not interfere with the needs of more bandwidth intensive applications. For example, you may want to take advantage of zones with higher speed coverage to meet certain application requirements (e.g. the R&D department or the marketing department), while optimizing longer range coverage elsewhere. High data rate sites require a very thorough site survey.
- What types of computing devices are to be used by various user groups (PDAs, laptops, tablets, a mixture of both)? Is the make-up of these computing devices expected to change over time?
- If you use antennae to "reshape" rather than extend an AP's signal, beware that on top of their tricky installation, these antennae may or may not work with next generation technology.

sive connectivity throughout a three-story office building, another will be set up to allow hotel guests to enjoy wireless Internet access.

The site survey either can be performed in-house using tools and equipment provided by wireless vendors, or it can be outsourced to an expert in site surveys. If this is an organization's first WLAN, outsourcing may be the preferable route, but later, as the WLAN is expanded to serve more end-users and to cover larger areas of the facility, in-house resources can usually do the job.

The first phase of a site survey is research.

- Gather as much information about the site as possible. It is vital that you obtain a copy of the site's floor plans and/or layouts of all areas that the wireless network is expected to cover. These are the documents you will use to create a diagram of the coverage area.
- Test different access points by measuring each access point's signal strength. (Access point and antenna ranges differ by vendor specification.) Potential vendors should loan you the gear for these tests.

SITE SURVEY CHECK LIST

Make sure your site survey includes:

- A detailed description of the desired coverage areas, along with a description of areas that do not need coverage.
- A detailed layout of the coverage area. Use that as your site map. Consult facilities drawings, blueprints and wiring documents; a topographical map of the campus; an architect's floor plans; and anything that can show the location of host systems, power outlets, passageways, structural elements such as metal firebreaks and walls, and doorways. Make document copies so that you can mark them up.
- A close estimate of the total number of WLAN users, descriptions of applications to be accessed through the WLAN, and data rates needed. This will help determine how to properly provision your collision domains through access point placement.
- A set of the same brand and model of WLAN equipment that will eventually be deployed. (If possible, test more than one brand and/or model.)
- Antennae. Consider trying out more than one kind of antenna (different antennae provide different performance ratios in different coverage areas). Understand that the actual pattern of an antenna can vary, and be careful of antenna polarity.
- A portable battery pack or other method of powering access points as they are being tested.
- Tie wraps, duct tape, or some other method for temporarily mounting access points and antennae.
- Pictures (preferably digital). Use a camera to provide pictorial evidence of each step of the site survey. Such pictures will prove to be useful throughout the installation process.
- A small wheeled cart with an UPS (uninterruptible/universal power supply) and extension cord at the bottom, and an extendable pole attached so you can mount an AP and antenna.
- A small flashlight for seeing under ceiling tiles and the like.

* Note the location of such items as microwave ovens, cordless phones, satellite systems, RF lighting systems and neighboring WLANs—all can be sources of RF interference. Some or all of these items can seriously affect a WLAN's performance, depending on the item and the flavor of WLAN being deployed.

The Four Major Considerations

You are now at the phase where you can do a detailed examination of the site where the WLAN will be used. There are four major areas of consideration when designing a WLAN. They are:

1. Range and Coverage
2. Data Rate and Capacity
3. Interference Immunity
4. Connectivity and Power Requirements.

As you examine each of these areas to determine the WLAN's current needs, keep in mind possible future expansion needs. It will save a lot of headaches down the road. Let's now look at each area of consideration, individually.

Range and Coverage: Define the physical space and environment that the WLAN should encompass.

* Is it open field, in-building, or mixed?
* Does the WLAN coverage need to be isolated to only serve certain specific areas?

SITE SURVEY TIPS

- Since materials such as wooden floors can cause floor-to-floor interaction between access points, be sure that your channel selections are appropriate for vertically adjacent access points.
- Concrete and steel rebar will bounce a signal, while single brick walls and sheetrock will allow for greater signal penetration.
- Untreated windows will allow for greater penetration, while treated windows can (depending on treatment) cause tremendous problems. The same is true for HVAC ducting and elevator shafts.
- Close all office and room doors before beginning the survey, in order to assess reception at its lowest, everyday level.
- But also test the environment at its busiest. Water can absorb radio signals and, since the human body is 55 to 65 percent water, there can be a difference in range as people move about during the workday. Vehicle movement can also affect a signal's strength. If any radio signal's path crosses a parking lot, transportation area, or docking area, there can be a difference in the range during busy periods.
- Consider redundancy for conference rooms (including spaces with multiple conference rooms in close proximity), cafeterias, and other multi-use spaces to ensure satisfactory throughput.
- In high-security, limited-access areas, consider placing access points adjacent to or straddling the area.

* Since an access point's coverage area is typically spherical in shape, the size, shape, use and nature of the physical make-up (e.g. construction material, lighting, geographical features, etc.) of the coverage area will determine the number of access points that will be needed to provide adequate signal coverage.

Data Rate and Capacity: Next, determine the required data rate and capacity. Again, this data is essential in determining the number of access points needed for the WLAN to provide end-users with quality connectivity. This requires that you establish:

* How many users the WLAN is expected to serve at completion.
* What type of applications will they be running?
* What kind of computing devices will they use?
* Is it expected that there will be substantial changes over the next couple of years? If so, what are the changes and how will they occur?

Interference Immunity: Now it's time to assess potential sources of interference. This includes other WLANs, RF lighting, cordless phones, microwave devices, and satellite systems, etc., along with adjacent channel interferers as discussed in Chapter 10. A site survey will not only be used to determine how interference affects AP placement, but, in many cases, it may also be used to examine potential antenna performance patterns for antenna selection.

Connectivity and Power Requirements: Existing networking constraints often are overlooked during a site survey. For example, all of the organization's networks (wired and wireless) should be homogeneous. Care must be taken as to where the APs are placed, since many times they are installed in areas where there is no access to AC power or network cabling. But also review access point electrical installation alternatives that will prevent performance degradation from inherent or random electrical problems.

Now that you have catalogued your site considerations, you are ready for the actual survey. Mark all coverage areas on your site map. Next walk around the site and identify all possible *obstructions* for RF (e.g. freezers, coolers, X-ray rooms, elevators, etc.) marking their locations on the site map. Since metal is highly reflective for RF signals, collections of metal bookshelves and cabinets can constitute potential RF problem areas, so note their locations as possible obstructions. Also keep in mind that moving vehicles such as trucks, forklifts, and other equipment temporarily block signals. Therefore, if your site includes a docking bay, a busy transportation yard or a parking lot, conduct the survey when there is heavy activity in those areas.

***Note:** Placing access points in very high locations (but not too high), or at opposite ends of the high traffic areas, can probably overcome the temporary blocked signal syndrome caused by activity in heavily trafficked areas.*

Because of an access point's typical spherical coverage, the identification of areas where RF signals extend beyond the wireless network's intended coverage area is as important as identifying RF problem areas. Successful WLANs provide adequate coverage where needed, but take pains to minimize or eliminate coverage that extends beyond the organization's physical campus. Failing to limit RF coverage can expose the organization's network to unauthorized access.

Obstruction	Degree of Attenuation	Example
Open Space	None	Cafeteria, courtyard
Wood	Low	Inner wall, office partition, door, floor
Plaster	Low	Inner wall (old plaster lower than new plaster)
Synthetic Materials	Low	Office partition
Cinder block	Low	Inner wall, outer wall
Asbestos	Low	Ceiling
Glass	Low	Non-tinted window
Wire Mesh in Glass	Medium	Door, partition
Metal Tinted Glass	Medium	Tinted window
Human Body	Medium	Large group of people
Water	Medium	Damp wood, aquarium, organic inventory
Bricks	Medium	Inner wall, outer wall, floor
Marble	Medium	Inner wall, outer wall, floor
Ceramic (Metal Content or Backing)	High	Ceramic tile, ceiling, floor
Paper	High	Roll or stack of paper stock
Concrete	High	Floor, outer wall, support pillar
Bulletproof Glass	High	Security booth
Silvering	Very High	Mirror
Metal	Very High	Desk, office partition, reinforced concrete, elevator shaft, filing cabinet, sprinkler system, ventilator

Figure 8.1 Relative attenuation of RF Obstacles. The ability of radio waves to transmit and receive information, as well as the speed of transmission, is affected by the nature of any obstructions in the signal path. This table shows the relative degree of attenuation for common obstructions. *Graphic Courtesy of Intel Corp.*

To help ensure that the wireless network's coverage area does not extend beyond the physical campus, survey the area from "the outside in." Begin by placing an access point in the far outside corner of the *proposed* coverage area (position A in Fig. 8.2). Moving inward, locate the edge of your *actual* coverage area (position B) using the aforementioned survey methods.

Now move the *access point* to position B. Since position B represents the coverage edge when the access point was in position A, with the access point at position B, the reverse is true—position A is now the edge and it should be able to receive an adequate signal without RF spilling beyond the physical campus area. By performing another survey at position B, you can determine if your maximum coverage area includes more users than desired. If so, reduce the coverage area by using a smaller antenna or lowering the power level of the access point.

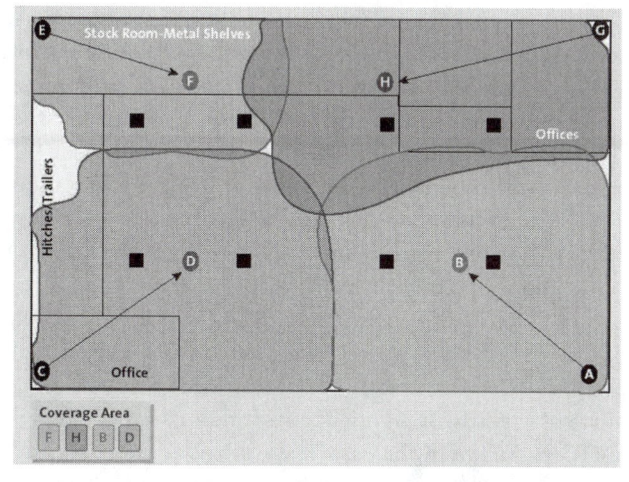

Figure 8.2 Example of an "outside in" survey method. This method helps to ensure that the WLAN's coverage area doesn't extend needlessly beyond the physical plant. *Graphic Courtesy of Cisco Systems, Inc.*

Surveying a site for the "weakest link" is another important activity. This requires consideration of different radio cards (i.e. wireless network interface cards), as well as knowledge of the computing devices themselves and how they house the transmitter/receiver. For example, surveying with a laptop with an exposed radio will not accurately indicate the coverage experienced by a traditional Automatic Identification and Data Collection (AIDC) terminal that reads barcodes and RF tags. The same holds true if the WLAN design calls for antenna diversity.

When the site survey is finished, provide a complete report to the deployment team, including the marked-up site map showing access point placement, and possible interference sources, along with cell structures, antenna choices, configuration parameters, power requirements, and photographs.

Professional Assistance

Site surveys are critical to a successful WLAN deployment. If not implemented correctly, the WLAN can end up costing more than just money and result in employee frustration and/or customer dissatisfaction. When planning the site survey, the in-house staff's knowledge and expertise with the technology and installation process must be considered.

A great deal of information can be obtained from a site survey, and while you may feel your in-house staff has the skills to perform the survey, they might not be able to adequately perform the necessary analysis. How the end product is analyzed is just as important as the survey itself. A good, informative analysis can support cell planning; cell search threshold; range and throughput; interference/delay spread; and bandwidth management for bandwidth-intensive applications, access point density, and load balancing.

To perform site surveys, vendors who cater to the enterprise market employ wireless experts, often engineers with many years of expertise. Using these specially trained professionals is like taking out a WLAN insurance policy. If the coverage isn't what was agreed to up front, the professional will come back and make it right (usually at no extra cost) because a reputable vendor will stand behind its expert's recommendations.

As mentioned previously, outsourcing to WLAN experts may be the wisest course, especially if it is the organization's first WLAN installation or if the WLAN is to serve a large organization or operate in a campus environment. Such assistance is available from wireless vendors and value-added resellers (VARs). Sometimes these wireless resellers charge a small fee for the site survey; sometimes the costs are the same as hiring an independent consultant, so shop around. An installation company or consultant can perform the site survey at a cost of approximately $1000 per day. The average survey will take a couple of days for a typical two- or three-story office building and close to a week for, say, a hospital or corporate campus.

Another source of expertise is an outside systems analyst who has worked with the organization's IT department in the past. The systems analyst can perform a series of on-site tests to help design and implement a wireless network. The survey determines the optimal number, placement, and configuration of access points and antennae for the required radio frequency (RF) coverage in a facility. Industry experts use tools to analyze frequencies, power capabilities, network loads, and coverage on the site. During this process, specifications are carefully developed for the required network layout and cabling, as well as for the necessary bridges, routers, and hubs.

Many environmental factors can disrupt WLAN coverage. Building construction materials such as paper, cardboard, and fabric absorb a lot of signal strength; whereas sheet rock and metal cause the signal to bounce around. And as previously mentioned, RF interference can come from a variety of sources. In such situations, an outside expert may be the best solution. Testing the proliferation of these frequencies, and knowing where to place access points and directional antennae, can eliminate these problems. Depending on the organization's needs, the expert can also consider data rates. End-users may have different data needs and may require more than one rate. Site surveyors will take those needs into consideration and recommend an appropriate solution.

Once all tests are completed, the outside expert prepares and presents a detailed report of findings and recommendations. After an internal assessment of needs and budget, the organization can move forward with the expert's wireless recommendations.

If the decision is to use a vendor's professional services or to hire a consultant, ask for references—particularly of customers that are in the organization's industry and with a similar facility. Use someone who has received certification from an independent organization such as a Certified Wireless Networking Professional (www.cwne.com). This holds true whether using a vendor's expert or hiring your own.

Note:* *Some vendors offer RF Site Analysis training, which can be useful for any in-house personnel who might have the responsibility of conducting a site survey.

Site Survey Tools

A WLAN, while relatively inexpensive when compared with a wired network, a large WLAN is more complex and far more difficult to engineer than the conventional LAN design. Without good tools to help in the planning and deployment's troubleshooting process, the deployment team is operating blind. For example, rogue wireless devices and departmental systems can show up just about anywhere; a pilot WLAN that was not originally intended for production may still be in operation.

Experienced WLAN designers can walk into a facility, be it an office, warehouse or campus environment, give it a once-over and make educated guesses about how the system should be designed. But even they lean on site survey tools to ensure the final design is problem-free. For the average network manager, the site survey will be a trial and error process, and thus a good site survey tool will be invaluable.

Site survey tools are software that can be installed on a portable computing device to assist in the site survey process. This software establishes a two-way data network using both stationary and mobile devices at various points within the proposed radio coverage area. Once installed, these tools provide signal strength, throughput, best channel, and address information, which can be invaluable in making deployment decisions for the system.

Since site survey tools measure performance between access points and identify sources of interference, they help in determining effective operating range (i.e. coverage area) between end-users and access points, and in formulating optimal access point placement. For example, after placing the estimated number of APs in the locations you feel will provide the best coverage and overlap, verify the placements by walking around the WLAN's proposed coverage area with the site survey tool monitoring the signal levels of each AP. This allows you to verify the maximum distances that will maintain adequate signal levels (e.g. 4 or 6 Mbps). If a specific AP position doesn't provide the coverage needed, reposition the AP or add additional APs, and repeat the test.

***Note:** Use site survey tools carefully. To give you an example, Intel reported that when one of its WLAN deployment teams was searching for other wireless devices on an IP address range that was fairly large (meaning the process involved probing every address within a specific range for information), the probed addresses ran through the server farm in Intel's CIS department, setting off alarms. The IT center's administrator, convinced that a major hacker attack was under way, later discovered that Intel was just testing to see whether its large SP2 data center had hidden wireless network ports. So unless you want to test the efficacy of some of the IT administrator's blood pressure medication, the author (and Intel) recommend carefully limiting the IP range on discovery probes.*

Vendor-provided Tools. Most vendors include simple site survey tools with their wireless card and AP installation disks, although features vary greatly. For example, you can find vendor-provided survey tools that measure signal strength and quality, data rate, or a host of other relevant information about a WLAN's performance and efficiency. All of Cisco's wireless client adapters include the Cisco Aironet Client Utility (ACU) that helps in the configuration, monitoring, and management of the adapter. The site survey functions also produce easy-to-understand, detailed graphical information, including signal strength, to assist in the correct placement of APs. Moreover, the ACU also provides signal-to-noise ratio measured in decibels (dB), and signal level and noise level measured in decibels per milliwatt (dBm).

The survey tools that come with the today's APs are generally web- or command line-based, and provide the ability to configure APs individually. Vendor-provided survey tools vary widely in functionality and usability. In the author's view, the only site survey functions shared by most vendor-provided tools are displays showing the strength and quality of the signal emanating from the AP. Fortunately, the tools available from most enterprise-oriented vendors have improved significantly over the past several years and are expected to continue to improve. Industry experts expect that by 2004, most major wireless LAN vendors will offer a comprehensive suite of products (primarily software) focused on site surveying and RF troubleshooting.

Advanced Site Survey Tools. Some WLAN deployment teams may feel that the site survey tools provided by the vendor are insufficient for their needs. In that case, you might want to look into advance site survey tools. A spectrum analysis tool can provide the "eyes" and "ears" for a deployment team to understand the affects of the environment on the transmission of 802.11 signals. Some tools can graphically illustrate the amplitude of all signals falling within a chosen 22 MHz channel, enabling the deployment team to distinguish 802.11 signals from other RF sources that may cause interference. Thus the team can locate and eliminate the source of interference, or use additional access points to resolve the problem.

Another key spectrum analysis feature that a deployment team might find useful is the ability to monitor channel usage and overlap. For example, 802.11b accommodates only three access points in the same general area; if there are more, the APs experience interference and corresponding performance problems. This limitation creates difficulties when planning the location and assignment of channels in large networks. Advanced site survey tools can display these channels, enabling the team to make better decisions on locating and assigning channels to access points.

Many may question if the value of advanced survey tools is worth the expense. Others understand that such tools not only help them to spot potential design flaws, but also can help to avoid serious problems down the line. Due to the high cost of these advanced tools, you should only consider them if you are designing multiple WLANs or a WLAN whose environment is complex, e.g. warehouses with lots of high metal racks, manufacturing plants full of machinery, a large campus–any of these WLAN designs easily warrant the additional costs of using advanced tools. If you want to investigate the possibility of procuring advanced site survey tools, consider those offered by such companies as AirMagnet, Berkeley Varitronics Systems, and Fluke Networks.

Once the WLAN is up and running, these same tools can be used to monitor the installed access points' signal strength to (1) ensure they are providing adequate cover, and (2) to identify failure zones.

***Note:** You also might want to check out a freeware product—Netstumbler (www.netstumbler.com)—in many cases this tool is far superior to the site survey tools the wireless equipment vendors provide. Network Stumbler not only captures signal strength and signal-to-noise statistics, but even more importantly, it helps network administrators identify and locate rogue access points, including any WLAN that an employee or group of employees may have installed without the IT department's permission.*

DETERMINING THE COSTS

Before arriving at the cost of your WLAN, you must know the network's specific elements. To know which components will be necessary for a WLAN project, draw up a preliminary sketch of the requirements and design. For example, the cost structure will differ radically between a network that provides Internet access in a small HotSpot, a

WLAN that services a small business, and a WLAN that accommodates a large organization with a variety of applications.

The components:

Wireless NICs: Each WLAN computing device will need the ability to connect to the wireless network, this includes devices such as PDAs, laptops, PCs, printers and so forth. Thus these devices will need to be equipped with the correct network interface card, typically a PC card. (Unless you have unique application requirements, the user devices will not require additional components.) However, since more and more computing devices ship with wireless ability embedded, there may be no need to purchase a wireless NIC for each user device. But, for each wireless NIC that must be purchased, plan on paying on the average of $65 for each 802.11b card, and figure an additional 25 percent for 802.11a cards (although 802.11a prices could come down as usage of this technology grows). It's too early to predict the average cost of an 802.11g card.

Access Points: Most WLAN deployments will be some type of infrastructure network, which means you will need at least one access point. (Of course, an access point isn't needed for an ad hoc WLAN.) Access points vary widely in price, ranging from less than $100 (for a home or SOHO WLAN) to $2000 for APs that serve enterprise WLANs. The cost depends on the features the device provides. (The benefits gained from an access point's advanced features will likely make the difference cost-effective.)

Since an access point has limited range and capacity, most corporate WLANs will need more than one AP to provide adequate coverage. The site survey will determine the number of access points needed. However, if you are trying to determine the WLAN's preliminary costs before performing a site survey, plan on approximately one AP per 70,000 square feet (6503 square meters) of coverage area. This assumes a range of about 150 feet (45.7 meters) without any limitations due to attenuation. For the WLAN to provide a consistently high data throughput rate, a single AP's capacity also must be limited, typically to somewhere between 30 to 50 end-users. So even though a single AP might be able to provide adequate geographical coverage, additional APs might be needed to provide adequate data throughput rates.

Also consider the costs for mounting the APs and running the necessary network cabling to each access point from the Ethernet switch or hub. If you use in-house labor, calculate the time and cost of (1) running cable for each AP, (2) installing and mounting the APs, and (3) installing network cards in all user devices, and (4) making applicable configuration settings.

Professional Fees: Besides having a professional to perform the site survey, a large organization may want to consider hiring a company specializing in wireless LAN installations. If so, the contract should cover the costs of the site survey and the equipment installation. Here's an example of what to expect for these types of professional services. Cable runs cost approximately $100 for each AP (depending on the scope of your network), then add another $150 or so per AP for installation and mounting hardware. If you also want the contractor to install the network cards in user devices and make applicable configuration settings, you will need to include some additional costs for those services.

There are firms that will handle the entire WLAN project—feasibility, assessment, site survey, financials, and the business case. The fees for the "project" approach, which can be relatively high, may be well worth the expense, especially if the IT staff is already overburdened with its current workload.

Power: A wireless LAN installation may require the addition of electrical wiring and outlets to power the access points. If you don't have an outlet within a couple yards from an access point, you need to install new outlets. The cost can run upward of $300 per outlet (depending on the facility's geographical locations). Or you can sidestep the "power" issue altogether by using power-over-Ethernet (PoE) technology. (Do not use extension cords.)

Switches and Hubs: If you're designing a WLAN to link to an existing wired network, determine if there are enough ports open on the wired network's switches/hubs to connect the WLAN's APs. If not, be sure to include in the WLAN financials the costs for additional switches/hubs to interconnect the APs. The cost of these hardware devices varies widely—from $50 to several hundred dollars—depending on features.

Redundancy: I bet you didn't plan for redundancy, i.e. having spare hardware on-site to replace components that may be defective. This would apply to only the APs and the antennae since they are the only components that offer a single point of failure. If you have a spare on hand, you can replace an access point within an hour or less, but it could easily take a couple of days if you have to order the AP. Only you can assess how important the availability of wireless networking is to your organization and how long the network can be down. My thoughts are that it is advisable to have, at a minimum, one spare access point.

Servers: In addition to the network hardware components, authentication and encryption are required for securing WLAN communications. These capabilities are embedded in the wireless NIC, the AP, and in a back-end wireless authentication server that uses the same protocol as the other components. The wireless server can be a Remote Authentication Dial-In User Service (RADIUS), Lightweight Directory Access Protocol (LDAP), or other server already in place to enforce user access rights. Or it can be a separate server that connects to these other servers specifically to enforce the access rights of wireless users. The costs vary widely, but plan on spending at least $1000 if an extra server is needed for authentication. (Note that some access point vendors have built access control functions into their access point.)

Internet connection: This cost, ranging from $30 for a DSL connection up to $1000 or more for a T-1 line, pertains almost exclusively to the WISP business models, since most corporations and other organizations will already have Internet access.

Additional security measures: This includes extra firewalls, additional security software, etc. The costs for these extra security measures can run from a few hundred dollars to thousands of dollars.

Network management and maintenance: Operational support such as network monitoring and problem resolution and the tools to aid in these exercises should be included in a WLAN's overall cost figures. Such tools usually cost anywhere from $5000 on up.

Intelligent boxes: Refers to catchall switching in combination with dumb access points. Some enterprise-size WLANs may be built upon a thin AP/intelligent box architecture. The prices vary, but the average combined price of an access point and switch port is just under $900 (a price that is similar to an enterprise-class access point). While this type of solution is expensive, the more APs deployed the less the overall cost.

That pretty much sums up the hardware. Of course, there will be additional costs, e.g. purchase of mobile computing devices, project management (in the enterprise environment), user training and support, and IT staff training.

DEPLOYING THE WLAN

Now you're ready for an actual WLAN deployment, but—take it easy, start with a trial or pilot program. If possible, borrow the necessary components from the proposed vendor(s), if not, make a small initial purchase for the trial run. Choose a test group consisting of both experienced wireless users (e.g. they have a WLAN at home) and neophytes. (For more details on planning and implementing a pilot project, see Chapter 14.)

A well-run trial or pilot program will reveal design flaws and possible incompatibilities within the overall networking environment. While you should use the test program for normal daily business processes, also introduce intentional failures so you can see how end-users and IT staff react when the network goes down.

Once everyone's satisfied that the design functions well, the components are compatible, and a WLAN is beneficial to the organization's bottom line, it's time to train the IT support staff. The training program should include support for the wireless computing devices, support of the end-users, and competency in the use of any new applications that may be needed for optimal network management, maintenance and security.

The most efficient way to rollout an enterprise WLAN system is to do it in phases, one department or locale at time. This allows the deployment team to deal more easily with problems that might crop up along the way.

Check List

The final step before deployment is to establish a check list. Use the check list to determine that everything is ready for deployment. Items that should be on the check list include:

- Determine that you have the right amount and the right kind of wireless adapters to equip all mobile participants and, if necessary, create auto-installers to install all the WLAN drivers, VPN software (if any), etc., in participants' computing devices.
- Make sure to quality-test the drivers and software in the computing devices as an integrated solution set (not only as individual units) prior to deployment.
- Consider what kind of security will be implemented, e.g. WEP, installing switches to route traffic from access points to a network demilitarized zone in only one direction, using RADIUS servers for secondary authentication, VPNs, etc.
- Determine whether a VLAN architecture is necessary, especially if the WLAN is expected to support roaming or bandwidth-intensive applications.
- Decide what type of network management tools will be used to manage the network after it is operational.
- Decide the end-user support to be provided, including training and ongoing help desk support.
- Equip the deployment team with the necessary tools to deploy the equipment, e.g. blueprint, photos taken during the site survey, flashlight, etc.

Access Point Placement and Power

An access point's functionality can be equated to a media-independent, multi-port bridge that provides bridging from a wired media to a wireless media. For the best access point placement, management, and performance, the designer, installer and IT staff need to understand basic AP functionality and configuration options. Most APs include features for different interface connections and network management, and all provide MAC bridging between their interfaces.

APs monitor traffic from their interfaces and, based on frame address, forward the frames to the proper destination. APs track the frames' sources and destinations to provide intelligent bridging as mobile computing devices roam or network topologies change. APs also handle broadcast and multicast message initiations, and respond to mobile computing devices' association requests.

An access point's radio frequency signals may have to pass through ceilings, floors, walls, and other objects–all of which can cause signal degradation. In a typical corporate office, education, or healthcare installation, the access points will be mounted at ceiling height, whereas in distribution centers, manufacturing plants, warehouses and other facilities with high ceilings, the access points should be mounted in a location that is no more than 15 to 25 feet from the floor. Of course, mounting access points in such locations can create the additional problem of getting power to the unit. If your installation requires such AP placements, you should consider purchasing units that utilize PoE (i.e. can access power over the Category 5e Ethernet cable). And in fact, in all but the smallest WLAN installations, using PoE can drastically reduce the cost and complication of installation.

Unusual access point placement can jeopardize proper signal distribution, whereas creative antenna placement can provide a solution. For example, if an AP needs to be placed above ceiling panels, its antenna will need to be positioned below the tiles. Such a configuration requires that the access points used should (1) have remote antenna capability, and (2) be plenum-rated, since many regions legally require that devices installed above a ceiling have a metal casing that meets the specific fire code requirements of your area. Outdoor situations bring up another host of challenges, such as how to provide power and bring cabling to the devices, and how to harden the device(s) to inclement weather.

The best way to determine optimal access point placement is the site survey. Of course you must use the same model of antennae you intend to use in the final deployment stage. Using different models produces different wave propagation patterns. This is the only way to determine how a building's construction materials or campus's topographical features will block or absorb signals.

Too Much of a Good Thing

While you must install enough access points to support the end-users and their applications, too many access points can be as bad as having too few. When more than one access point sends at the same signal strength to a specific location, an accessing client can become confused when forced to constantly evaluate which access point it should utilize. However, if only one access point sends at a strong signal and another is far enough away that there is a 20-decibel or so difference, there is no problem, the accessing client knows to go for the stronger signal.

Once an organization determines where to place the access points, the rest is easy. In fact, installing and configuring the access points isn't much harder than just hooking up the wires and turning them on. Windows 2000 and XP, as well as many handheld operating systems, are designed to automatically locate Wi-Fi signals.

The Importance of Antennae

The antenna is a significant component in your wireless network. Understanding antenna technology can make a difference not only in the overall performance, but also in the total cost of deploying and maintaining a wireless system. The right antenna technology can mean the difference between meeting your organization's specific wireless connectivity needs in the most cost effective manner possible, or incurring added costs from installing more access points than are truly necessary.

The antenna directs radio frequency from the radio to the coverage area. Different antennae produce different coverage patterns, and thus need to be selected and positioned according to a specific site's coverage requirements. Base antenna selection upon regulatory requirements, size and shape of the area requiring coverage, antenna mounting options, and aesthetics.

While the size of the coverage area is the most important determining factor for antenna selection and placement, it isn't the sole criterion. Building construction, ceiling height, internal obstructions, available mounting locations, and physical appearance all must be considered. External considerations, such as public locations that prohibit the use of larger antennae, and other areas (e.g. executive offices), may require creative antenna placement so that they are unobtrusive and blend well with the surroundings.

***Note:** Access points designed for the 802.11a U-NII 1 indoor band are required to have an antenna that is permanently attached. Other APs have antennae that can be placed remotely via an antenna cable, but even those antennae can't be mounted more than a few feet from the AP's position, because the coax cable used for RF has a high signal loss.*

Antennae deliver flexibility and robustness to any WLAN, but for some reason the complexity of antenna technology, placement, etc., is hardly referenced in most WLAN documentation. Nevertheless, antennae can optimize certain applications, e.g. building-to-building bridging. Furthermore, because wireless is a dynamic medium, by using high or low gain antennae, it is possible to alter how signals propagate. For example, by understanding how antennae work, it is possible for a designer to focus an RF pattern and energy down a long narrow hallway, avoiding wasted energy and/or multipath interference.

Antenna radiation patterns are affected by polarization, free space loss, and propagation in solids. The transmission loss between a transmitting and a receiving antenna is a function of the antenna gain, the distance, and the frequency. For best performance, the transmitting and receiving antennae must have similar polarization alignments. Additional "free space loss" occurs because of signal spreading: as a signal radiates outward from an antenna, its radiated power is spread across an expanding spherical surface, with the power level inversely proportional to the distance from the source antenna. This antenna radiation must pass by and through solid objects, so it will be subject to losses from reflection and absorption. For example, oxygen atoms in the atmosphere cause a prominent peak in the attenuation effect of the atmosphere. Clandestine inter-satellite communications are performed on this frequency so that the signals will not reach earth. The atmosphere's attenuation almost vanishes at 94 GHz (the W-band), which is why many radar systems operate around this frequency.

Water, however, absorbs signals above 2 GHz. Fog, rain, the leaves of trees, people, etc. can rob energy from a microwave signal. The reason microwave ovens use the 2.4 GHz range is that 2.4 GHz penetrates food very well, but since water molecules cannot vibrate as fast as the microwaves push them, the molecules absorb the microwave energy and release it as heat. And that dear reader is how a microwave oven delivers piping hot food for our dining pleasure.

Reflections from objects (metal objects in particular) give rise to multipath distortion (fading). Some paths converge and become constructive (adding to signal strength) or destructive (fading). The selection of an antenna's azimuth pattern is driven by the shape of the coverage area and the location of the items within that area. Elevation pattern shapes are controlled to keep the maximum response at or slightly below the horizon for best far and near field coverage. The dimensions and height of a communication sector will determine an antenna's azimuth and elevation beamwidth requirements. These can be derived from the established beam area formula for the approximate gain needed:

G(dBi) = 10 log 10 / 29,000 / antenna azimuth * the antenna elevation

Spacing in excess of 0.75 wavelengths from a large conducting surface leads to deep nulls. And, antennae placed close to the ceiling will be prone to diffraction loss if they can't clear the doorjambs.

Radio Propagation

The WLAN designer's job is to deploy a WLAN that provides the best performance possible given the constraints of cost and power. The variables a WLAN designer considers include:

- Radio design variables, such as RF Tx power, antenna gain (Tx/Rx), receiver noise figure, required data rate, required packet error rate, E_b/N_o (waveform PSK, CCK).
- External link variables, such as propagation conditions (range, multipath, environment), interference from outside sources (light fixtures, cordless phones, microwave ovens), and co-channel interference.

Some variables are under the control of the designer and others are not. The external link variables can be managed by careful site analysis, planning, and installation. Access point placement and antenna pointing can help, as can better design and installation of the station antennae along with antenna diversity. The final key to good performance is equalization in the receiver processor.

***Note:** This might be a good time to review the detailed discussion on propagation provided in Chapter 7*

If you've ever experienced bad reception on a car radio when driving through a downtown area or through terrain with hills and valleys, you can appreciate the importance of setting up a WLAN so that it can provide all users with adequate coverage. Appropriately placed access points, paired with the correct antennae positioned relative to environmental obstructions or competing radio signals, are mandatory for good coverage and data throughput.

To that end, a site survey helps to define the wireless network's coverage and bandwidth performance at different locations within a cell, and to indicate where the "fall back" rates will occur. It also allows the designer to determine the exact number of access points needed and their optimal placement, and whether a special antenna system will be required.

Security

Does the deployment team know the WLAN's vulnerabilities, and/or how to test for them? Do they know what else the wireless network is doing other than talking to its clients? Do they know what is being sent out over the airwaves? If not, it's not too difficult to find an expert who can.

Wireless security experts can perform vulnerability studies that estimate how long it would take someone to break into a network, how they would do it, and what type of data is vulnerable. They can figure out ways to maximize internal wireless coverage without allowing too much data to seep outside the corporate facilities. An expert also can suggest security tools that can be used to reduce the risk of network penetration, as well as other means to make a network more secure.

Furthermore, a wireless security expert can perform an equipment audit that looks for rogue employee devices that make networks more vulnerable to attack.

There are a number of security measures that are easy to implement and that can help to ensure a WLAN's protection. They are discussed in detail in Chapter 17.

VLANS

Will the WLAN you're about to implement support roaming and/or bandwidth-intensive applications? If so, consider building the WLAN upon a virtual local area network (VLAN) architecture. Adding VLAN capabilities to an existing WLAN, or deploying a new WLAN using VLAN technology, enables wireless clients to roam without dropping a connection, and segments "chatty" applications so that other WLAN users don't experience degraded service. For example, if the WLAN is to support voice over IP, videoconferencing, or other bandwidth hogging applications, a VLAN can be used within the WLAN to segment this bandwidth intensive traffic. A VLAN can also logically segment WLAN users on an organizational basis, i.e. by functions, departments, project teams, applications, etc., rather than on a physical or geographical basis. For example, a WLAN built upon a VLAN architecture, can segment all computing devices and servers used by a particular segment so that they connect to the same VLAN, regardless of their physical connection to the network. (Marketing personnel in different physical locations and on different network topologies could be part of one VLAN, while engineering personnel could form another distinct VLAN.)

When VLANs were introduced in the late 1990s, they were touted as a way to simplify address management by letting IT departments physically deploy servers and PCs anywhere on a network, and then associate the machines into virtual groups. Software on most managed network equipment can be used to associate client MAC addresses with VLANs, letting the computing device automatically connect to its network when moved from one port to another.

Most of today's VLANs are based on the IEEE 802.1Q and 802.1P standards. 802.1Q provides a standard method for inserting VLAN membership information into Ethernet frames, and 802.1P gives Layer 2 switches the ability to prioritize traffic and perform dynamic multicast filtering.

A VLAN architecture, whether in a wired, wireless, or mixed network environment, can provide the following benefits:

Improved scalability: This is especially true in LAN environments that support protocols and applications that can flood packets throughout a network.

Network organization based on function: VLANs allow logical network topologies to overlay the physical switched infrastructure, such that any arbitrary collection of LAN ports can be combined into an autonomous user group. Logically segmented portions of the network are separated into Layer 2 broadcast domains, whereby packets can be switched between ports that are designated as being within the same VLAN. By containing traffic originating on a particular LAN only to other LANs in the same VLAN, the virtual networks avoid wasting bandwidth. (This is a drawback inherent to traditional bridged and switched networks in which packets are often forwarded to LANs with no need for them.)

Common broadcast domains: By offering common broadcast domains, there can be complete isolation between VLANs. Just as switches isolate collision domains for attached hosts and only forward appropriate traffic out a particular port, VLANs can provide isolation between VLANs.

Improved security: Not only can firewall protection be provided for individual VLANs, high-security users can be grouped into a VLAN where no users outside that VLAN can communicate with them.

Improved performance: A VLAN supports logical grouping of users. Thus it is possible to improve general network performance through traffic segmentation, e.g. isolating groups with high bandwidth usage that can slow down other users sharing the network. This in turn allows the intensive use of a specific application (e.g. an accounting suite sitting on a group of servers), or the use of a bandwidth intensive application (e.g. videoconferencing), to be assigned to a VLAN that contains one specific group and its servers (such as the R&D department), or that specific application server and its users (videoconferencing application and its local attendees). Thus that group's work or application's usage will not affect other users on the same network.

Wireless VLANs

The majority of WLAN products and technology available today focus only on the Data Link Layer and on solving discrete security and mobility problems. None provide the comprehensive capabilities needed to help enterprises centralize control and to easily and cost-effectively deploy, secure and scale wireless LANs. However, it is possible to extend the concept of Layer 2 wired VLANs to wireless LANs, using wireless VLANs. As with wired VLANS, wireless VLANs define broadcast domains and segregate broadcast and multicast traffic between VLANs. Without a VLAN, the deployment team must install additional wireless LAN infrastructure to segment traffic between user groups or device groups. For example, to segment traffic between employee and guest VLANs requires two APs to be installed at every location where both are expected to access the WLAN. (See Fig. 8.3.)

However, with the use of wireless VLANs, one access point at each location can provide access to both groups. VLAN architecture can also be of great help in a WLAN infrastructure where there are roaming difficulties. VLANs can provide a means for wireless clients to roam among 802.11(a, b or g) access points without losing connectivity.

Consider Bridgewater State College in Massachusetts, which deployed more than 100 Enterasys Networks RoamAbout 802.11 dual-mode APs, along with a mix of Cisco and Enterasys stackable switches throughout the campus so that students could access the Web. As long as the students were on campus, they could get online. According to Pat Cronin, telecommunications director at Bridgewater State, "After getting over a few cross-vendor VLAN configuration issues, students and faculty now can roam anywhere and stay connected on the same VLAN, whether they are connected to a Cisco or Enterasys switch." Cronin goes on to say, "If we didn't segregate the wireless traffic into its own VLAN, it would be a nightmare with people moving around campus who want to stay connected. Putting all wireless traffic on a VLAN ensures that no one drops off the network as they move from access point to access point."

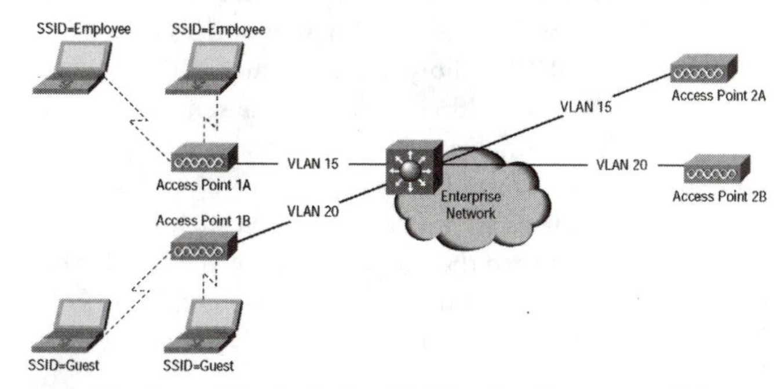

Figure 8.3 User segmentation without wireless VLANs. *Graphic courtesy of Cisco Systems, Inc.*

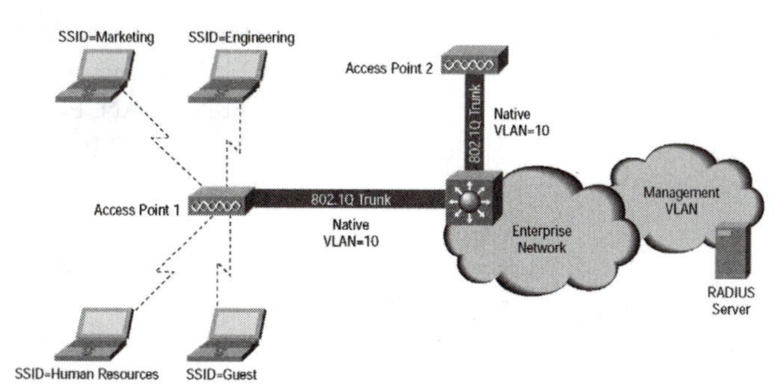

Figure 8.4 An indoor wireless VLAN deployment where four wireless VLANs are provisioned across a campus to provide WLAN access to full-time employees (segmented into engineering, marketing, and human resources user groups) and guests. *Graphic courtesy of Cisco Systems, Inc.*

Criteria for a VLAN in a Wireless Network

To properly deploy wireless VLANs, the deployment team must evaluate the need for VLANs within the existing or proposed WLAN architecture. The evaluation should include, but not be limited to:

- A review of any existing wired VLAN deployment rules and policies, since existing wired VLAN policies can be used as the basis for wireless VLAN deployment policies.
- Identification of the common applications used by all WLAN users, e.g. wired network resources (such as servers). Then determine the Quality of Service (QoS) level needed for each application.
- A list of the common devices used to access the wireless LAN.

Once the data is gathered, determine (1) what security mechanisms, e.g. static WEP, MAC authentication, Extensible Authentication Protocol (EAP) authentication (LEAP, EAP-TLS or PEAP), virtual private networking, and so forth, are supported by each device; (2) which wired network resources (such as servers), are accessed by each WLAN device group; and (3) the QoS level needed to support each device group.

After completing the evaluation, determine the VLAN deployment strategy for your WLAN. There are two standard deployment strategies. One is segmentation by user groups. For example, three separate wired and wireless VLANs could be created—one for R&D, another for accounting, and a third for guest access. The other is segmentation by device type. This allows a variety of different devices with different access-security "levels" to access the WLAN. For example, handheld computers that support only 40/128-bit static-WEP shouldn't coexist in the same VLAN with WLAN client devices that support 802.1X with dynamic WEP. Instead, group and isolate these devices by their different "levels" of access security into separate VLANs.

The next step is to define implementation criteria such as use of policy group (set of filters) to map wired policies to the wireless side; use of 802.1X to control user access to VLANs using either RADIUS-based VLAN assignment or RADIUS-based SSID access control; and use of separate VLANs to implement different quality of service or class of service.

Deploying a Wireless VLAN

The criteria for a wireless VLAN deployment differs according to whether the VLAN is operating in an indoor or outdoor environment. For indoor deployments, the APs are generally configured to map several wired VLANs to the wireless LAN. For outdoor environments, 802.1Q trunks are deployed between bridges with each bridge terminating and extending as an 802.1Q trunk, participating in the 802.1D-based Spanning Tree Protocol process.

Each wireless VLAN is configured with appropriate network policies and mapped to a wired VLAN. A network manager enforces the appropriate network policies within the wired network for each different user group.

VLANs and Security

A VLAN architecture ups the WLAN's security quotient by allowing network managers to define appropriate restrictions per VLAN, using the APs as the security policy vehicles.

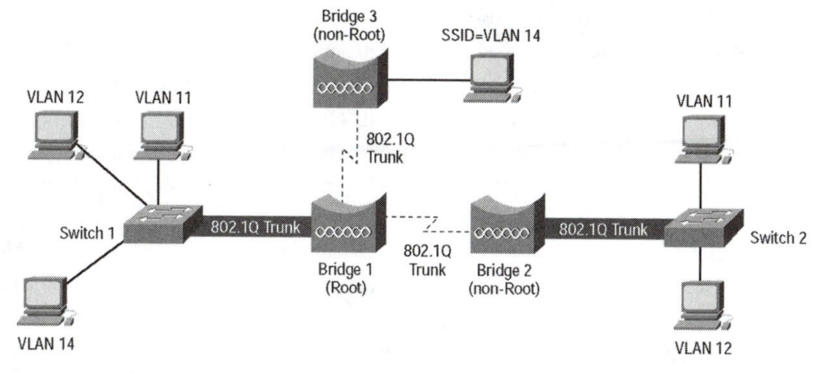

Figure 8.5 An outdoor wireless VLAN deployment scenario. Wireless trunking connects the root bridge to the non-root bridges. The root and non-root bridges terminate the 802.1Q trunk and participate in the Spanning-Tree Protocol process of bridging the networks together. *Graphic courtesy of Cisco Systems, Inc.*

Cisco Systems offers an example of configurable parameters on a SSID wireless VLAN and on the wired VLAN side. SSID wireless VLAN parameters include:

* SSID name—configures a unique name per wireless VLAN.
* Default VLAN-ID mapping on the wired side.
* Authentication types—open, shared, and network-Extensible Authentication Protocol (EAP) types.
* MAC authentication—under open, shared, and network-EAP.
* EAP authentication—under open and shared authentication types.
* Maximum number of associations—ability to limit maximum number of WLAN clients per SSID.

SYMBOL'S MOBIUS AXON WIRELESS SWITCH

Symbol Technologies Inc., a big provider of wireless and wireless LAN products, offers a Mobius Axon Wireless Switch that delivers centralized wireless connectivity through Mobius Axon Access Ports. The switch supports 802.11b, 802.11a, 802.11g, and legacy Symbol wireless protocols. Existing wireless LAN products require customers to integrate and manage separate products for wireless connectivity, security, and management.

The product uses a virtual LAN architecture and policy-based networking to deliver bandwidth, security, and networking services by device, by user, by application, and by location, all from a single access port. It performs traditional Layer 2, Layer 3, and Layer 4 switching, all of which is managed through an XML-based or command-line user interface.

There are two hardware components: the Mobius Axon Wireless Switch and Mobius Axon Access Ports. Software components include the Symbol MobiusGuard security portfolio, authentication features and virtual private network capabilities, and wireless network management.

The wired VLAN parameters are:

* Encryption key—this key is used for broadcast and multicast traffic segmentation per VLAN. (It is also used for static WEP clients.) Network managers must define a unique encryption key per VLAN. With an encryption key configured, the VLAN supports standardized WEP.
* Enhanced Message Integrity Check (MIC) verification for WEP—enables MIC per VLAN.
* Temporal Key Integrity Protocol (TKIP)—enables per-packet key hashing per VLAN.
* WEP (broadcast) key rotation interval—enables broadcast WEP key rotation per VLAN. This is only supported for wireless VLANs with IEEE 802.1X EAP protocols enabled (such as EAP Cisco Wireless [LEAP], EAP-Transport Layer Security [EAP-TLS], Protected Extensible Authentication Protocol [PEAP], and EAP-Subscriber Identity Module [EAP-SIM]).
* Default policy group—applies policy group (set of Layer 2, 3, and 4 filters) per VLAN. Each filter (within a policy group) is configurable to allow or deny certain types of traffic.
* Default priority—applies default class of service (CoS) priority per VLAN.

END-USER TRAINING

The WLAN has been deployed, but does everyone know how to use it? Achieving the desired results with any technology requires trainers with knowledge and expertise. Too often a project is completed and the end-user is given a quick download, handed documentation, and sent on his or her "confused" way. Set up a training program to teach the end-user how to make the most of this wondrous technology. The training program can be handled in-house or outsourced.

Professional trainers can develop a program that ranges from a simple quick-reference guide to complete multi-lingual training packages. Or perhaps all that's needed is a professional trainer to train the in-house instructor on the ins and outs of wireless connectivity.

Intermec Technologies Corp., one of the wireless industry's "big guns," perhaps best explains the benefits of using an expert training program. According to Intermec, "By using an expert training program, employee productivity, system performance, and reliability are increased. Excellent training programs are developed to fit specific industries and cultures, often garnering better response from the end-user. It also expedites a project's ROI because the user can operate the system properly and more quickly, thus making it possible to gather data faster and make fewer data entry errors.

"Employing an expert trainer can also reduce rollout times. Professionals understand how to best communicate technology at the user's level of understanding. Sometimes IT staff members are too technical to be clearly understood when describing the system and the new processes. Finally, investing in employee training often reduces resistance to the new technology."

FOUR-NINES RELIABILITY

Many organizations excel in their WLAN design, but don't give enough thought to a wireless network's management, maintenance, monitoring and troubleshooting, all of which are as much a part of the deployment process as the implementation itself. It is very important to implement a cohesive network management strategy, and that industry standard tools, platforms and applications be used to manage the wired and wireless infrastructure seamlessly.

To ensure that a network management platform incorporates all of a WLAN's components requires that adequate planning and staff training be put in place prior to deployment. Although a WLAN's components support the standard network protocols, TCP/IP, NetBEUI, and IPX, and they can be managed with existing network management tools, often additional tools are needed to ensure the network management platform is monitoring the wireless network fully, and keeping track of performance and problems.

First, think about how you're going to manage the access points. The process of configuring and upgrading the large number of APs serving the average enterprise WLAN needs to be automated. Some AP vendors, including Proxim, design their systems with this in mind. Others, including Agere and Symbol, provide management software to accomplish that goal. In some cases, you may find it valuable to turn to third-party systems, such as Wavelink's Mobile Manager, for added functionality or to integrate APs from multiple vendors under a single management framework.

If the WLAN can't be managed remotely, whenever there is an upgrade, a "bug" fix, or a firmware change, a labor-intensive effort is required to physically disconnect and individually program each access point and then reinstall each unit in its original location. But if these same tasks can be done remotely, the access points need never be removed.

Annual site re-surveys can help to guarantee a high level of system performance and RF coverage. An organization's facilities and inventory layouts can change over time, so performing an annual site survey is invaluable. For example, a warehouse's inventory during tough economic times is a very different RF environment than during boom times. A university campus RF environment will evolve as the school year changes over to a summer classes.

Wireless Network Management Systems

Network-accessible data and mission-critical applications are essential to the operations of most organizations. They all rely on the availability of their network. The introduction of wireless into the networking environment adds additional complexities to the already complicated world of network management. You should implement a network management system that can meet the following goals:

- Improve network availability (up time) and service.
- Centralize control of network components.
- Reduce complexity.
- Reduce operational and maintenance costs.

A network management system can effectively reduce the cost and complexity of a network's maintenance by providing a set of integrated tools that allows a network manager or support staff to quickly isolate and diagnose network issues. The ability to analyze and correct network problems from a central location is critical to the management of both network and personnel resources.

In fact, network management is so critical that the general requirements of a network management system have been defined by the International Organization for Standardization (but commonly known as the "ISO") and categorized as part of the OSI specification for systems management. These general requirements are known as "The Open Systems Interconnect (OSI) Management Functional Areas," and are used as a base line for the key functional areas of network management on any system. The acronym FCAPS is used to represent the key elements of the ISO definition:

Fault management
Configuration management
Accounting management
Performance management
Security management

Fault management: This encompasses the activities of detection, isolation, and correction of abnormal network operation. Fault management provides the means to receive and present fault indication, determine the cause of a network fault, isolate the fault, and perform a corrective action.

Configuration management: Activities include the configuration, maintenance, and updating of network components. Configuration management also includes notification to network users of pending and performed configuration changes.

Accounting management: The ability to track network usage to detect inefficient network use, or the abuse of network privileges or usage patterns, is included in accounting management, a key component for planning network growth.

Performance management: Activities include the monitoring and maintenance of acceptable network performance, and the collection and analysis of statistics critical to network performance. Tools are used to recognize current or impending performance issues that can cause problems for network users.

Security management: This encompasses controlling and monitoring the access to the network and associated network management information, including the control of passwords and user authorization, and collecting and analyzing security or access logs.

The goal of a network management system is to provide the above functionality in a concise manner that views the entire network as one homogeneous entity.

Network management systems rely on defined standards to interface with network devices for monitoring and controlling their configuration, performance and functionality. The Simple Network Management Protocol (SNMP) is the basis for most modern network management, since it provides multi-vendor network management systems with the ability to manage network devices from a central location. SNMP includes standard protocols, databases, and procedures to monitor and manage devices connected to the network. Nearly all vendors of network-based components, computers, bridges, routers, switches, etc., offer SNMP. Basic SNMP components are:

- A management station or console, which is the user interface component of the network management system. It provides the applications to configure, monitor, analyze, and control the various components that comprise the network.
- A management agent program that resides on a given network device that responds

to requests from the management console or generates events (traps) based on configured parameters.

* The management information base (MIB), which is the management database for a given network component. There is a standard definition of a MIB for every device that is supported by SNMP. The management station monitors and updates the values in the MIB, via the agent. SNMP provides three main functions, GET, SET, TRAP, which retrieve, set device values and receive notification of network events.
* A proxy agent that supports devices that do not have an SNMP implementation available. The proxy is an SNMP management agent that services requests from the management console, on behalf of one or a number of non-SNMP devices.
* Remote Network Monitoring (RMON) is a specification that was developed to provide a standard interface between a management station and remote monitoring agents or probes. Remote monitoring agents are used to gather network statistical information in order to diagnose network faults and performance issues. RMON defines additional MIBs that collect this performance information.

Standard Wireless Network Management Platforms

Network management platforms can provide the management and maintenance features needed to manage a network system that provides both wired and wireless connectivity. Although they share standard elements and mechanisms, wired and wireless networks have significant differences. In addition to the conventional wired network, wireless networks have the following unique issues:

Secondary hierarchy: The wireless network environment is hierarchical, with mobile units attached or associated to a given access point. Standard network management products that are designed for wired networks cannot represent this critically important tiered topology network structure, since they represent the wireless network as a "Flat Topology."

Roaming: The wireless network environment supports dynamic cell connection or roaming, which is the process of changing the network connection of a mobile device from one access point to another. This is a unique component within the wireless environment.

Persistence of mobile computing devices. In an environment that is comprised of handheld computers, the ability to manage networked devices also becomes a factor. Unlike desktop systems or other network components that operate continuously on a daily basis, handheld computers are turned on and off frequently throughout the day, making it difficult to monitor these devices.

SNMP agents. The ability of devices utilizing wired and wireless networks to "host" an SNMP agent is another differentiating factor that impacts network management functionality. Desktop or laptop systems have adequate amounts of memory and processor power to support an SNMP agent operating as a background task handling requests from the network management station. However, on handheld terminals, many of which still run DOS, memory space and processor speed are highly limited resources, making it difficult to provide agents for these devices.

The unique features associated with wireless networks need to be addressed in order to provide a complete network management solution. Also wireless network management software is needed to continuously monitor all of the access points in the WLAN and alert the IT staff if anything strange is going on.

Due to the continually changing wireless network environment, established network management systems have not incorporated the necessary support for unique issues associated with wireless networks. Instead, they rely on the wireless vendors or third parties to provide "management applications" that can be incorporated into their enterprise network management system to manage the wireless component of the system.

A number of vendors offer wireless network management systems (WNMS), including NetMotion Wireless's NetMotion Mobility and Wavelink Corp.'s Wavelink Mobile Manager. However, let's look at Symbol's SpectrumSoft WNMS to demonstrate how a WNMS is specifically designed to solve network management issues associated with mobile computing devices in a wireless network. First off, the SpectrumSoft WNMS is seamlessly integrated into the corporate enterprise network management platform. This provides the ease of working in a single management environment and avoids the additional expense of purchasing, training, and maintaining a new management tool.

Since the SpectrumSoft WNMS is constructed using an Open Systems architecture, it utilizes proven network management standards (SNMP, MIB-2) to provide a comprehensive solution to the unique aspects of wireless networks: roaming, cell association, random connectivity, and power management.

The graphical user interface provides an intuitive tool for navigating, examining, and managing the wireless network. The products feature set provides the functionality as defined by the ISO FCAPS network management requirements, including the unique areas associated with wireless networks and mobile computing environments.

SpectrumSoft WNMS's modular design and the use of standard interfaces provide an easy integration path into a variety of enterprise network management platforms, including CA Unicenter, Cabletron Spectrum, Tivoli TME 10, IBM NetView, and SunNet Manager/Solstice. Also, a management console for SpectrumSoft WNMS has been created for wireless networks that do not require the extensive functionality provided in the above-mentioned platforms.

Monitoring and Troubleshooting

It is necessary to constantly monitor access point settings to ensure that they not only remain in compliance with the organization's current security policies, but also perform to specification. If a WNMS isn't an option, then get some monitoring and troubleshooting tools. Find software that allows the IT staff to set the performance and security thresholds at any value they wish and change them at any time. Also, be sure the software can produce detailed WLAN performance reports, so the IT staff can review the reports and fine-tune the network. You may also want to consider whether you need a software package that offers auto-repair features, which automatically return the access points to their proper settings.

Network management staff also can benefit from specialized tools, such as handheld WLAN analyzers that provide the mobility technicians and engineers need to manage a wireless network environment. If you have a protocol analyzer for your Ethernet network, you'll also need a version for your WLAN. Capable products are available from Network Instruments, Sniffer Technologies and WildPackets.

Another handy tool is a spectrum analyzer, which can troubleshoot RF problems. High-end spectrum analyzers, which may cost $20,000 or more, are available from Agilent and Tektronix. Lower-priced systems designed specifically for WLANs are available for less than $3,000 from Avcom-Ramsey. Most of these products put a ton of power into the hands of IT staffers.

Note: *Sometimes if the WLAN isn't an enterprise-size network, it makes good business sense to buy a lower-end spectrum analyzer. After all, it's not likely to be a tool you use every day, and you may not need all the advanced features of a high-end unit. But over-economizing is a bad idea when managing a large WLAN. So take the time to thoroughly understand the management issues and then select products from reputable vendors. Cutting corners may save you a few dollars today, but the cost of reduced productivity or stolen secrets will be a price you may be forced to pay in the future.*

A WLAN analyzer can help throughout the life of a WLAN with site surveys, deployment, troubleshooting, and auditing. Post-installation connectivity problems between end-user computing devices and access points can be analyzed. Traffic and signal distribution problems can be found, analyzed and corrected. WLAN analyzers can be used to sniff out rogue equipment, as well as to discover devices with inappropriate settings that could open the WLAN to unauthorized users

Some WLAN analyzers are designed for laptops and others for the smaller handheld or PDA device. The analyzer software that's designed for a handheld device is used less for protocol analysis than for WLAN-specific features, such as surveying radio channels for signal strength and device populations. The range, though, is similar to laptop-based WLAN analyzers, except that a handheld device is much easier to wave in the air while looking for a signal.

The readers may wonder if a PDA is the right form factor for advanced troubleshooting. As with many applications, sometimes a PDA can be as much a limitation as a benefit. For example, the larger screen and higher resolution of a laptop computer make high-layer protocol analysis much more convenient. Furthermore, PDA-based systems are prone to occasional system lockups that necessitate a reboot.

But when considering an analyzer remember that these products stress the capabilities of the underlying hardware and software platforms on which they operate. The painful truth is that handheld analyzers don't replace other tools; they supplement them.

Because it's a relatively new market, comparing products isn't easy. One product might offer superior Physical Layer spectrum analysis while another might have the best expert system, letting people in the field quickly identify problems—two different tools for two different problems.

Note: *Don't expect these tools to be a substitute for understanding WLANs. They're just tools.*

To choose the right handheld WLAN analyzer, consider its intended use. Which layers of the network stack do you want to analyze and who will do it? If you aim to equip your Tier 1 field technicians, perhaps the same folks who handle your wiring infrastructure, you should match the device to those workers' knowledge and experience with RF and WLANs. In many cases, this means choosing a tool that captures and analyzes WLAN traffic and provides practical expert analysis in real time. These techs also could benefit from having a tool that helps them perform site surveys, troubleshoot connection and performance problems, and identify and locate rogue devices and network attacks or intrusions.

If your needs are advanced, consider a more specialized tool. As RF-based communications systems grow, both in scope and usage, there is a greater need for precise tools that detect Physical Layer abnormalities. Remember, these are unlicensed airwaves, so the sources of interference are many. The tool of choice for serious RF engineers is the spectrum analyzer. Sometimes, a spectrum analysis tool is the only systematic way of identifying the source of WLAN performance problems, including those that may be caused by cordless phones or other unlicensed radio devices. While you can easily spend $20,000 on a capable spectrum analyzer, Berkeley Varitronics Systems' analyzer can be carried in your hand, at a dramatically lower price (less than $4000). Using a spectrum analyzer requires significant RF training, but for enterprise WLANs that must meet four-nines availability, this is a cost you may need to absorb.

Although most enterprise IT professionals appreciate the long-term business value of wireless LANs, use of this technology remains low enough to mask underlying design and management problems. The most popular applications—email and web access—don't place extraordinary demands on a WLAN's infrastructure. For this reason, buying expensive wireless troubleshooting tools to fix what isn't broken may seem like a low priority. However, if you anticipate your organization will support more wireless data services, you will eventually need specialized tools that will satisfy field technicians and WLAN designers alike.

When considering the purchase of a costly analysis tool, look for a product that can, at a minimum:

* Perform a thorough security assessment of the network.
* Provide a good site survey analysis.
* Offer simple installation and ease of use.
* Perform detailed protocol analysis.

Management Tools

Even if the project's budget can't swing a costly WNMS, the deployment team still needs to determine how to manage the access points. Although today's access points are highly reliable when supported with clean power, they still require maintenance. As previously mentioned, some AP vendors, like Proxim, design their systems with automated and/or remote management of APs in mind. Agere, Symbol and other vendors, however, provide management software to accomplish that goal.

There are also third-party tools that, while not as robust as the more costly WNMSs, can help centralized network management departments identify and solve problems before system failure. The attributes to look for when purchasing these solutions include:

* The ability to control everything from a central location—a network administrator should be able to configure and monitor infrastructure, change access point settings, and firmware upgrades from one terminal.
* System design flexibility—the management tool should support access point hardware from a variety of vendors.

* Easy upgrades—management tools that cannot change with the times are a poor investment.
* Easy integration with existing network infrastructure—it is always good when a product can integrate with a legacy system. The management tool must also be able to integrate seamlessly with other network management software that may already be in use for an existing wired network.
* The software—must be user friendly, easy to navigate, and provide intuitive help when needed.
* The ability to automatically implement changes over large groups of access points—this feature helps to eliminate the chance for human error and ensures uniform implementation of any necessary changes.

POST-DEPLOYMENT SUPPORT OPTIONS

A well-executed WLAN deployment is the first defense against overwhelming network support needs. Next is a thoroughly trained IT department that can understand and use the applications, network and hardware, and can troubleshoot the new WLAN technology. But eventually the time will come when a problem arises that the IT department can't handle. It can be as simple as a portable computing device being dropped once too often, or as troubling as an unexplained "hiccup" in the network.

***Note:** The common WLAN components usually come with a manufacturer warranty, typically for one year. But vendors also offer a variety of service contracts and/or extended warranties for resale to the customer.*

Hardware support options include vendor depot and on-site service. Traditionally, portable computing devices have been handled via depot while access points were put on on-site service contracts. But today, the price of an AP is relatively low, and the devices themselves have become so reliable, that vendors and their customers have found it more economical to keep a spare configured AP on hand and swap it for a malfunctioning one rather than going the service contract route.

Be aware that some vendors hesitate to include support costs in their initial proposal because they want to keep the bottom line low. Many vendors feel that it is easier to sell a support contract as a separate expense after they have a signed product contract in hand.

Support is often offered via a range of service levels, with different turnaround and response times. Determining the appropriate service contract level means an evaluation of the WLAN environment and whether the data and network is mission critical.

Last, but not least, is end-user support. Organizations have several options—from doing it themselves, to making themselves the first line of support with outsourced back-up, to outsourcing the whole shebang. It all depends on the size of the WLAN and the number of end-users, the strength of the IT department, help desk availability within the organization, and whether the organization currently uses outsourced help desk services.

CONCLUSION

802.11-based technology has redefined what it means to be connected. It stretches the limits of networking by providing an infrastructure that is as dynamic as it needs to be.

But wireless networking is in its infancy. The IEEE set of standards is only a few years old. With standard and interoperable WLAN products, wireless networking can reach scales unimaginable with a wired infrastructure.

In a WLAN world, end-users can roam not just within a building or a campus, but within a whole city or country, all the while maintaining a high-speed link to extranets, intranets, and the Internet itself. But the planning, design, implementation, and support of a WLAN require in-depth understanding of the special challenges of WLAN technology. The ability to identify and effectively address those challenges can mean the difference between a successful WLAN and one that fails to deliver the expected benefits and return on investment.

Still, as long as there is a good plan and proper project management, purchasing wireless resources and then deploying them can be done without too many headaches. The key is to have a plan, perform a credible site survey and use the results to guide the deployment. Begin with a small deployment, test everything, not once, but twice, then roll out the full wireless networking facilities—in phases—testing everything along the way, just as you would with any large networking infrastructure change.

Here are a few words of advice. Initiate a small pilot program to test usage, costs, and benefits of a wireless system. Invest in a qualified site survey (don't be pennywise and pound foolish). Do plenty of pre-testing in the actual production environment. Finally, be careful about mixing brands; some products may not be fully interoperable with other vendor's products due to add-on features such as those that enhance security and/or Quality of Service.

Chapter 9: Bridging Buildings

Sometimes end-users who are located in geographically separated buildings need to be connected to the same network. Wirelessly bridging using Wi-Fi technology provides a perfect way to connect two buildings. As long as the two buildings are within a few miles of each other, and there is a clear line of sight between the locations to be bridged, a wireless connection (bridge) between the buildings will allow all network users to access the same network system. Note, however, that to provide adequate signal levels, special antennae may be required.

A Wi-Fi bridge is essentially a media access control (MAC) level wireless access point that's configured so it can "bridge" or pass incoming wireless packets to one or more corresponding bridges. Wireless bridges allow network managers to quickly and inexpensively extend a wireless LAN across intervals ranging from several hundred feet/meters to a mile/kilometers or more. With specialized antennae, the distance could be up to 5 or 6 miles (8 or 9 kilometers). And new antenna technology may enable that distance to grow to more than 30 miles (48 kilometers).

***Note:** Although wireless bridges can provide line-of-sight bridging at greater and greater distances, in the U.S., the FCC limits the maximum distance a wireless bridge is allowed to span to 25 miles / 40.2 kilometers. And in European countries the maximum is 6.5 miles / 10.5 kilometers.*

Bridging is not covered, *per se*, under the 802.11 series of standard. Yet the standards do allow a wireless bridge system to be built that can span miles/kilometers at high data rates and can turn WLANs into WANs (wide area networks), if necessary. Network administrators, fed up with the high recurring costs associated with leased lines and the expense that comes with running fiber underground, especially in areas where right-of-way issues exist, eagerly exploit wireless bridge technology to extend their networks beyond the indoor network.

The quickest and least expensive bridge is the *point-to-point wireless bridge*. This type of bridge only requires a couple of access points, which are configured as either a router or a bridge. Then just attach an external double gain antenna to each, and place each in a position where the external antennae are visible to each other. This type of set-up provides a connection between networks in different buildings. It is a good choice in a campus environment, where specific areas are close to the campus yet are without fiber optic

Figure 9.1 Point-to-Point Bridging.

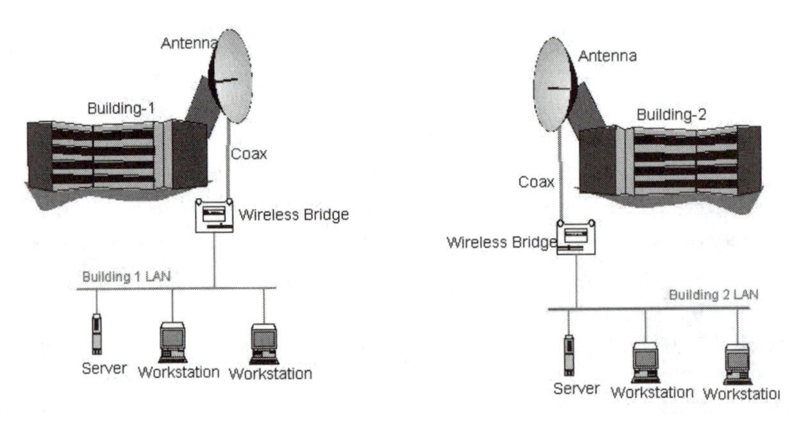

Figure 9.2 An example of how a building-to-building bridge might be used.

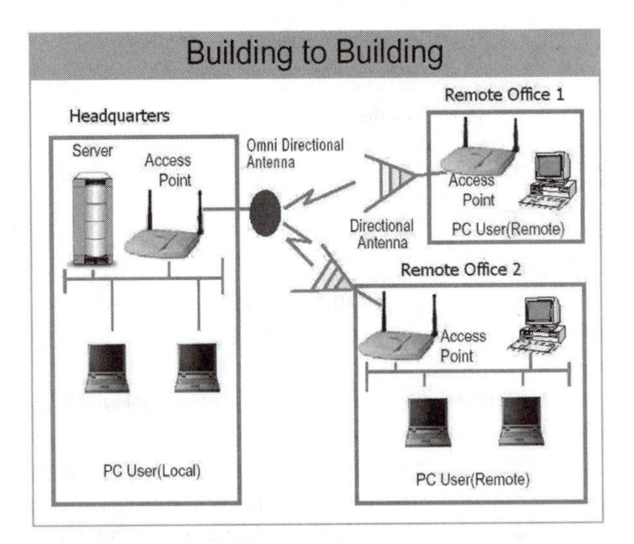

Figure 9.3 As this graphic shows, you needn't limit the bridge to only two buildings.

connectivity to the backbone. Buildings with a wired network within them can use wireless connectivity to get to the campus backbone network; but the distances for this type of set-up are limited, and you must have "clear line-of-sight" between the antennae.

To deploy a *building-to-building bridge* application, all you need is a wireless bridge at each location. Each bridge is connected to the building's network using traditional wire. Antennae are then installed on each wireless bridge to connect the two buildings' local networks.

In *multipoint bridging*, a subnet on one LAN is connected to several other subnets on another LAN via each subnet's AP. For example, if a computer on Subnet A needed to connect to computers on Subnets B, C, and D, Subnet A's AP would connect to B's, C's, and D's respective APs.

WHEN TO USE WIRELESS BRIDGING

Building-to-building bridging is ideal for organizations that:

- Are looking for a cost-effective alternative to expensive T-1 service or leased lines.
- Need an inexpensive way to connect separate wireless LANs together.
- Need a method that can provide networking capabilities to unexpected locations.
- Need a redundant, alternative path to the wired (fiber optic and T-1) network backbone.
- Don't want to invest in laying expensive and tough-to-install cables since their network requirements or location may change.
- Want a simple way to connect remote and hard-to-wire locations—between office buildings that aren't already connected to one another, portable classrooms and cross-campus office buildings, and buildings separated by other structures such as bridges or freeways.
- Want to link wireless LANs to wired networks, or broadband Internet access.

EXAMPLES OF WIRELESS BRIDGING

The following examples may give the reader some idea of how to take advantage of the benefits of wireless bridging. For additional ideas on how to use bridging technology, read the case studies at the end of this chapter.

To bridge buildings: Enterprises often use bridging to connect networks in different buildings within a corporate campus. Bridging access points are typically placed on top of buildings to achieve greater antenna reception. The typical distance over which the average AP with the typical antenna set-up can be connected wirelessly to another by means of bridging is approximately two miles (3.2 kilometers), although with the help of specialized equipment the distance can be extended to around 30 miles (48 kilometers). But even the two mile range will vary, depending on several factors including the specific receiver or transceiver being used.

3Com, one of the premier providers of Wi-Fi products including wireless bridging gear, offers a good hypothetical wireless bridging scenario. The principal of a local high school is pleased about a new addition to the school, which when completed, will encompass three computer labs, 15 classrooms, and more. But the projected completion date for the new addition is two years down the road. Classrooms are bursting at the seams now and more students are on the way. That means that the principal must find some way to network 50 new computers now, and not break the budget. The principal determines that by using wireless building-to-building bridging techniques, it is possible to turn one of the school's three new portable classrooms into a computer lab, and give the other two classrooms network access without running cables. One bridge is connected to the school network and another to a 3COM access point that was located in the portable computer lab. Now all of the new computers, including those in portable classrooms, have access to all necessary network resources, including the Internet.

Redundancy: An often-overlooked WLAN application is the use of wireless bridge links as a redundant, alternative path to the wired (fiber optic and T-1) network backbone. If the organization experiences disruption in its wired network services because of, say, a cut fiber optic cable or a failed building-to-building data T-1 transport line, a redundant wireless bridge link can take up the slack. Just remember that the redundancy is only for when the backbone service is interrupted in building-to-building connections, not when the service fails at the source or beyond the physical confines of the organization's facilities.

The benefit of wired cabling, such as fiber optics, is that it can provide data connectivity for approximately 1.25 miles (2 kilometers), whereas conventional copper cabling should not exceed 328 feet (100 meters) and three hops through repeater hubs. Thus fiber optics is often the backbone of choice for building-to-building connections. However, since fiber optic cabling often runs extensively through campus and industrial environments, it can be the victim of repeated service interruption due to construction or digging. Also, as luck would have it, redundant fiber optic link will often run in the same cable bundle as the primary link; thus, when the primary link is interrupted, it is highly likely that the redundant will also be lost. A wireless building-to-building bridge would effectively solve such a situation. Similarly, a wireless link can serve as a backup for an on-campus T-1 connection.

There are a couple of caveats to redundant wireless bridging:

- Networks using redundant bridge links must support Spanning Tree Protocol (STP) because STP constantly monitors the network for service interruptions and turns ports, switches, and bridges on and off to block and send traffic.
- Wireless bridge links require clear line of sight between the buildings connected by the redundant wireless pathway—obstructions, such as trees (which can be bare during winter months but overflowing with signal-blocking foliage other times of the year), can render such a pathway useless.

However, once the wireless bridge link is installed, some organizations may get rid of their wired building-to-building lines and install a second redundant wireless bridge. By eliminating recurring costs of, say, a T-1 line, an organization could save as much a $1500 per month. Additionally, any 802.11 (a, b or g) wireless bridge link is many times faster than a T-1's 1.5 Mbps connection.

THE DISADVANTAGES OF WIRELESS BRIDGING

While there are many advantages to implementing a wireless bridge, there are also some downsides. They include:

- More complicated bandwidth management.

* Potentially greater security risk.
* Requirement of line-of-sight path between bridges.
* At some distances, special antennae may be required to improve signal levels.

For the readers who think they might want or need to deploy a wireless bridge, here are some products that can help you in your endeavors.

BRIDGING PRODUCTS

Vendors offer specifically designed wireless bridging products, often referred to as point-to-point Ethernet bridges, which are capable of spanning distances of 30 miles (48 kilometers) or more. The sweet spot is a cost-effective 11 Mbps link that spans only a mile or so. Such products can not only provide better performance than using a T-1 line, they can also provide system reliability that approaches 99.999 percent.

It's easy to find a point-to-point bridging system for less than $1000, although most in this price range operate at 2.4 GHz and leverage commodity WLAN chipsets to deliver a maximum data rate of 11 Mbps. While these systems can be appealing, they often lack features important for enterprise implementations, including Power-over-Ethernet and flexible management and monitoring capabilities. In addition, because they're based on LAN protocols, they tend to require higher overhead, resulting in throughput that is as low as half the stated rate. Finally, the duty cycle of these products often fails to meet enterprise standards.

That's why many enterprise-grade bridging products are migrating toward the 5 GHz bands. These systems are designed typically from the ground up for fixed wireless applications. As such, their feature sets, performance, and overall reliability are better than lower-cost alternatives. The downside is that many of these systems will cost more than $10,000, though lower-cost options are emerging.

When considering total system cost, factor in not only the cost of the wireless bridges and antennae, but also the costs of components, installation charges, and maintenance required for a production system. In the past, system integrators have charged a healthy premium for installation and maintenance, though the trend these days is toward self-installation, at least in enterprise environments where adequate expertise exists.

Although all 2.4 GHz and 5 GHz bridging products discussed in this section operate in unlicensed spectrum, they still must adhere to rules defined by the FCC and other international regulatory bodies. The chart in Figure 9.4 shows some of the key regulations in those bands, but also see Appendix III: Regulatory Specifics re Wi-Fi.

	ISM 2.4-2.4835 GHz	ISM 5.725-5.850 GHz	UNII-1 5.15-5.25 GHz	UNII-2 5.25-5.35 GHz	UNII-3 5.725-5.825 GHz
Bandwidth	83.5 MHz	125 MHz	100 MHz	100 MHz	100 MHz
Max Power*	1 watt (+30dBm)	1 watt (+30 dBm)	50 milliwatts	250 milliwatts (+24 dBm)	1 watt (+30 dBm)
Max Point-to-Point EIRP**	200 Watts (+53 dBM)	200 watts (+53 dBm)	200 milliwatts (+23 dBm)	1 Watt (+30 dBm)	200 watts (+53 dBm)
Special restrictions	Requires Spread Spectrum	Requires Spread Spectrum	Indoor only with integral antenna		Overlaps with ISM 5.8 but does not require Spread Spectrum

* POWER OUTPUT AT THE RADIO **POWER OUTPUT AT THE ANTENNA

Figure 9.4 FCC Regulations for 2.4 and 5 GHZ unlicensed radio bands. (In Europe, the ETSI is the regulatory body that regulates the Antenna/Radio power output and system range).

Finally, it's worth noting that wireless point-to-point systems aren't always based on radios. Second-generation FSO (free space optical) systems based on lasers and LEDs are an option for some applications. These systems provide significantly greater bandwidth than radio systems, but some may be vulnerable to adverse weather conditions. Vendors are hard at work trying to overcome these vulnerabilities.

5 GHz Products

Why 5 GHz? First, these devices may be less vulnerable to interference than those in the 2.4 GHz band, where wireless LANs and devices such as cordless phones and microwave ovens spew RF signals in all directions. Next, 5 GHz offers more than three times as much bandwidth as 2.4 GHz. That extra bandwidth not only lets applications run at higher data rates, it offers added flexibility to deploy multiple systems at a single location and to move to alternate channels should interference occur. The downside? You may have to sacrifice a little range, and 5 GHz products tend to be more expensive than 2.4 GHz products.

Network Computing magazine tested six different 5 GHz bridging products in late 2002 (see Fig. 9.5). All were worked as wireless Layer 2 bridges. In an age where Layer 3 switches and routers dominate the market, you may wonder why these products still run at Layer 2. The positive spin is that with a point-to-point system, you don't really need the sophisticated traffic-management capabilities of Layer 3 devices; also, many point-to-point systems link to backbone routers.

So just how much geography can you cover with the 5 GHz systems? The vendors whose products *Network Computing* tested claimed ranges from 1 to 15 miles (1.6 to 24.1 kilometers), and while more expensive offerings are available from some vendors that could extend the range even farther, the products tested hit the sweet spot for most organizations. The maximum distance supported by the various bridging products vary depending on radio output power, receive sensitivity, and antenna size used. (With amplification some can send quality signals for 30 miles / 48 kilometers).

An interesting trend, visible in some of the products referenced in Fig. 9.5, involves integrating the antenna and modem into a single weatherproof enclosure. This simplifies installation and increases the effective system range by eliminating RF cabling, which is a source of significant signal loss. An alternative that yields similar benefits is to co-locate the modem and antenna on a single antenna mast; this strategy allows flexibility of antenna selection–and even more range–but it is more complex.

What about speed? All of the 5 GHz products provided enough speed to handle mainstream business applications. In the *Network Computing* list, among the six products tested, throughput ranged from a speedy 7 Mbps to a whopping 78 Mbps.

Finally, you should understand that most of today's 5 GHz bridging systems are based on custom radio implementations. However, it is likely that future products will leverage the cost-economies of commodity 802.11a chipsets. Standards are critical for WLAN products, but less important for point-to-point systems because interoperability is not a critical requirement. In fact, a lack of interoperability can be a benefit to the extent that proprietary radio designs and modulation schemes add to system security.

5-GHZ ETHERNET BRIDGE FEATURES

	Airaya AI108-1-050 Wireless Bridge	BitRage CR45-A-53 DS3 Radio and 1U45-E AC with Ethernet Bridge	Proxim Tsunami QuickBridge 60	RadioLAN Campus BridgeLink-II-P25	Wi-LAN AWE 120-58 Ultima3 RD	Young Design EX-1 Wireless Bridge
Management and monitoring						
System log	N	N	N	Y	N	N
Remote status monitoring	Y	N	Y	Y	Y	N
Firmware upgrades	Y	N	Y	Y	Y	N
Advanced features and functionality						
Multipoint support	N	N	N	N	Y	N
VLAN support	N	N	N	N	Y	N
Protocol filtering	N	N	N	IP/IPX/NetBEUI NetBIOS	N	N
MAC address filtering	N	N	N	Y	Y	N
IP address filtering	N	N	N	N	Y	N
Ease of configuration and installation						
Radio sampling port	N	N	N	N	N	Y
Antenna alignment tool	Web utility	Alignment oscillator	Java utility	Web utility	Y	None
Visible/audible signal-strength meter	N	Y	Y	Y	N	N
Automatic TX adjustment	N	N	N	N	Y	N
Power-over-Ethernet	Y	N	Y	N	Y	N
Web interface	Y	N	N	Y	Y	N
Console	N	Y	N	Y	Y	N
Telnet	N	Y	N	Y	Y	N
SNMP	N	Y	N	Y	Y	N
DIP switches	N	N	N	N	N	Y
Java application	N	N	Y	N	N	N
Technical specifications						
Receive sensitivity (dBm)	-80	-70	-77 to -89	-67	-80	-83
Transmisson power (dBm)	+20	-1.8 to +5.8	+16	+17	-10 to +21	0 to +12
Modulation	802.11a/OFDM	B3ZS	QPSK	D-PPM Pulse with CSMA/CA	MC-DSSS	BPSK
Radio channels	1	2	7	1	1	8
Operating temerature range (°C)	0 to 50	-30 to 70	-33 to 50	0 to 60	-40 to 60	-30 to 60
Range	1 mile	11 miles	2.5 miles	3.5 miles	15 miles	7 miles

Y=YES, N=NO

Figure 9.5 In late 2002, *Network Computing* magazine tested a group of 5 GHz Ethernet bridges. This chart lays out the results of those tests. The underlying radios and modulation systems varied considerably from product to product. In most cases, it's a trade-off between price and performance. Packing more bits into each clock cycle requires more sophisticated radio technology—and you'll pay for that luxury. Because regulations vary by sub-band at 5 GHz, you'll get longer range from products that operate in the 5.8-GHz UNII-3 sub-band.

2.4 GHz Products

If you don't need the higher bandwidth, quality 2.4 GHz wireless bridge products are offered by vendors such as BreezeCom, Cisco Systems, Enterasys Networks, Lucent Technologies, Pinnacle Communications, and RadioLAN. These wireless bridges are best suited to joining small to midsize networks that don't have high bandwidth requirements. That is because the throughput delivered by such products is not sufficient for networks that need to send large amounts of data over the bridged link.

***Note:** A word about wireless bridges and the 802.11 standards. These standards don't directly address wireless bridging, although they have had an impact on the wireless-bridge arena. Ironically, IEEE 802.11 can hold back a wireless bridge's performance, in that it introduces a rather large gap between data packets. Because of that gap, the actual data throughput realistically can't be close to the speed* *advertised. On average, an 11 Mbps device is able to flush out data at around 4 Mbps (half-duplex). Thus, not all bridging products strictly adhere to the IEEE 802.11 standards. So while, for example, Lucent and Pinnacle's devices, which use nearly identical management software, stay true to the 802.11 standards, other bridging products use varying methods and technologies to tackle the issue of transferring packets—many make some sort of effort to stick to the 802.11 standards, but others employ proprietary techniques.*

In general, configuration involves setting each device with an IP address via software that came with the device, or a serial connection. Nearly all of the devices offered by the above-mentioned vendors support the Simple Network Management Protocol (SNMP), and most of the software consists of a basic SNMP interface with a management information base (MIB) extension.

One of the primary differences among the above-referenced bridges is in regard to whether the device uses a PC Card radio or an internal radio. The advantage of the PC Card approach is that it affords the ability of a radio upgrade, along with the possibility of increasing the bandwidth or encryption options as budget and technology allow. In addition, use of a PC card-enabled device lets you begin with just one radio and eventually expand to take advantage of the point-to-multipoint option. With PC card radio products, however, you might notice generally lower throughput compared with those units that use an internal radio. Also, the price of the PC-card bridge usually doesn't include the PC card itself, which must be purchased separately. Prices also vary depending upon encryption options.

DEPLOYMENT

The software side of wireless bridging is fairly simple, although the physical installation of these bridges can be complicated. Choosing the right antenna and correct installation is crucial. With the wrong antenna, you'll end up with poor performance as well as inefficient use of available bandwidth.

Certain antennae are designed for straight point-to-point applications, while others are omni-directional, allowing point-to-multipoint bridging. An antenna installation consists of mounting the antenna, along with a lightning arrester, usually on the roof of the building. Consider using a professional installer to ensure line of sight, although to make the task easier, some vendors, such as Enterasys and RadioLAN, provide utilities to correctly aim the antenna.

In addition to the ability to connect two network segments, wireless bridge products offer other features; for example, performing both protocol and broadcast packet filtering. The configuration software that comes with many bridging products lets you choose which protocols you want filtered, and whether to allow broadcast packets to be bridged. By restricting certain types of Ethernet packets, you can save considerable bandwidth.

Many devices also offer an option that lets the bridge function as a router. And one of the most powerful options available on some of the bridges is the ability to do point-to-multipoint bridging. Moreover, Cisco, Lucent, and Pinnacle offer products that can accommodate more than one radio. By assigning each radio a different frequency, the capacity of the bridge and even the coverage area can be extended, or multiple buildings can be connected to form a metropolitan area network. But keep in mind that point-to-multipoint configurations can be difficult to engineer because radio interference, bandwidth considerations, and other issues must be taken into account.

How Far Can You Go

Because designers must take into account the RF effects of a system's Fresnel zone (the pattern of electromagnetic radiation that is created between a transmitting antenna to the receiving antenna), bridges should be elevated sufficiently off the ground to ensure reliable operation.

Producing a strong, reliable signal using bridging techniques can be challenging. According to industry sources, 2.4 GHz frequencies need to have clear line of sight, and the Fresnel zone must be at least 80% free of obstacles. There are also other major technical variables that affect range, e.g. radio output power, radio receiver sensitivity, antenna gain, and path loss.

An antenna transmits and receives radio waves. The focused strength of these waves is referred to as "radiated energy," which is measured in terms of gain in decibels (dB). Gain in wireless antennae is typically specified in dBi—the resulting decibel measurement in relation to a theoretic isotropic radiator, which is equal in all directions. The gain in the antenna focuses the transmitted signal towards the targeted area of coverage. It also focuses incoming energy on the receiving side.

Gain must be considered when selecting a bridging antenna. There must be gain on both the broadcasting and receiving side to establish stable links, but not so much as to exceed the legal radiated power limitations of 4 watts (+36 dBm) maximum effective radiated power (ERP). The ERP, the total amount of power actually transmitted through

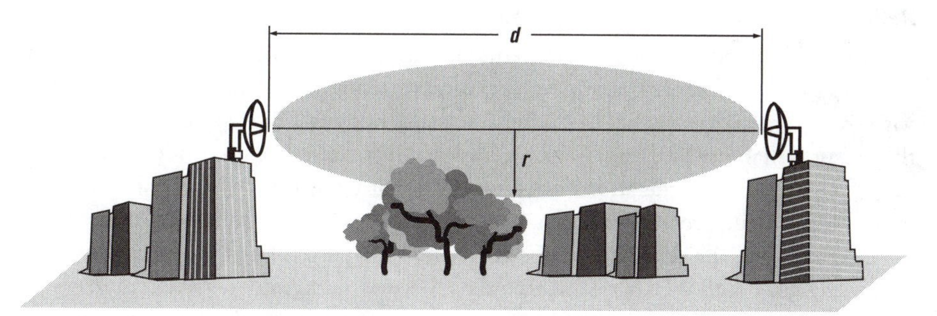

Figure 9.6 When designing an outdoor wireless link, the Fresnel zone, which is the elliptical area immediately surrounding the visual path as shown in this graphic, must be taken into account. Typically, 20% Fresnel Zone blockage introduces little signal loss to the link. Beyond 40% blockage, signal loss will become significant. The Freznel zone will vary depending on the length of the signal path and the frequency of the signal, although the radius of the Fresnel Zone at its widest point can be calculated by the above formula, where d is the link distance in miles, f is the frequency in GHz, and r is the radius off of the center line of the link in feet. Note that this calculation is based on a flat earth, i.e. it does not take the curvature of the earth into consideration. The effect of this is to budge the earth in the middle of the link. Thus for long links have a path analysis performed that takes the earth's curvature and the topography of the terrain into account.

the system's antenna, is the product of the transmitter's power output, the cable's power loss, and the antenna's gain capability.

Path loss is the attenuation that occurs when radio signals pass through air, and while generally easy to calculate, it varies with frequency (the higher the frequency, the higher the path loss per yard or meter). In over-simplified terms, at a given frequency, high radio output power combined with high antenna gain and good receiver sensitivity translates into improved range.

Most professional installers will add a *fade margin* to their overall range calculations. This margin will vary depending on the geographic and climatic conditions of different geographic areas. This may introduce atmospheric and multipath-related fading that must be added to the free-space path loss. It's not unusual to see engineered fade margins of 20 dBm for high-reliability applications. Your vendor or reseller should be able to help you with this; several of the vendors include technical information and range calculation utilities on their websites.

Of course, in the real world, you are subject to government regulations, which vary somewhat by jurisdiction. In the United States, the FCC imposes a number of restrictions on unlicensed radio operations, and vendors must gain FCC certification of compliance in order to sell their products. Key regulations involve acceptable waveforms, radio output power, and EIRP (effective isotropic radiated power), which is a combination of radio output power and antenna gain. (See Appendix III: Regulatory Specifics re Wi-Fi for more detailed information.)

All else being equal, receiver sensitivity (a radio's ability to separate a very weak signal from lots of noise) is the most important technical design element. This is where RF engineers earn their salaries.

Reliability

Because radio-based fixed-wireless systems (and that's what bridging products are) operate at relatively low frequencies, you won't experience many of the weather-related problems with them that are typical of the higher frequency home-satellite systems. You can further ensure reliability by turning to a professional installer, who can conduct a thorough path analysis prior to installation. By using directional antennae and building in an appropriate fade margin, professionals can guarantee 99.99 percent reliability in most areas.

Interference is the other side of the reliability coin, particularly for devices operating in the 2.4 GHz band. WLAN systems, cordless phones, garage door openers, and myriad other low-cost radio devices can affect a fixed-wireless system. Again, these problems can often be managed by using alternate radio channels or through antenna polarization.

Some of the reliability risk of point-to-point systems is associated with the systems' dependence on line of sight (LoS) communications–if LoS is broken, the link will go down. Syracuse University recently experienced just such an outage when a crane, participating in the construction of a new parking garage, parked directly between two buildings connected via Wi-Fi. Such situations can be dicey to manage, so as with any mission-critical system, you should put contingency plans in place.

Security

Security is always a top concern for wireless implementations, but it's not nearly as big an issue when signals are sent between buildings via a bridging set-up. The reason? First, most bridging products use proprietary radio-signaling schemes, and second, point-to-point systems employ highly directional antennae. Thus the most viable attack is literally a man-in-the-middle approach, though the man would need to be suspended in mid-air to intercept the signals. And if that's not enough, most vendors support some kind of encryption system.

However, there is always the risk of denial of service (DoS) attacks mounted using inexpensive, easy-to-conceal radios and antennae. In fairness, motivated criminals can mount DoS attacks on most information systems, so in that sense, wireless isn't unique. Nonetheless, this is just one more reason why an increasing number of organizations choose to implement their fixed-wireless systems in the 5 GHz bands, where more channels are available and the risk of interference is lower.

CASE STUDIES

North Okanagan-Shuswap School District

At a time of tight budgets in the educational sector, it's not hard to see why this Salmon Arm, British Columbia, School District chose wireless technology to provide the high-speed links it needed to run its operations. The decision by the North Okanagan Shuswap School District to install and implement WaveRider Communications Inc.'s NCL135 wireless bridge router currently saves the organization more than $700 in monthly leased line charges, and gives staff at the School District's technical center high-speed access to its central headquarters, located four kilometers (2.5 miles) away. The central headquarters currently houses all servers for the School District's intranet, as well as student management, financial services, and administration systems.

The Technical Center maintains a database of all electronic teaching and educational resources that are available to teachers. It also conducts research and development and testing on new software for the School District's operating systems and application programs. Robert Spraggs, network systems technician at the Technical Center, expects that it will take less than a year for the School District to see a recovery of its investment in the wireless bridge technology.

The WaveRider wireless bridge solution virtually eliminates any surprise expenses that would strain the School District's annual budget, says Spraggs. "When we leased lines from the telephone company, we would incur packet charges in some locations that we hadn't anticipated. With a wireless solution all costs are established when we sign the contract. There are no unexpected transmission charges at a later date."

Spraggs adds that another important advantage of choosing a wireless solution is that the School District is no longer vulnerable to network breakdowns because a line has been accidentally cut by a contractor or construction crew. "If a line is severed, our network goes down and our Intranet links are disrupted. With the wireless solution, we have more control over network reliability. I can troubleshoot network problems much more quickly. Equipment maintenance is also easier because we only have to call one number if we have any questions or need assistance."

Spraggs also says that the wireless routers enable the School District to be more productive. The high-speed links enable the Technical Center to quickly update the resources database at the School District's central headquarters and test new software programs. The biweekly resources update, which includes resource center catalogs, films, multimedia items, and electronic science resource kits, took up to two hours to perform prior to installation of the wireless router. The same task is now completed in ten minutes, giving teachers more timely access to new educational resources and teaching tools. Faster access to results of systems and software tests and research on new wide area network connectivity options also allow quicker decisions on new updates or software purchases.

Spraggs feels that in choosing wireless connectivity the North Okanagan-Shuswap School District is pioneering a growing trend, "The use of this technology to provide high-speed access will increase significantly in the coming months as organizations look for more cost effective and efficient ways for establishing high-speed connectivity."

Spraggs expects more educational institutions and businesses to implement high-speed wireless links. "Wireless technology is very stable and offers the throughput of fibre. Because you recover your costs so quickly, it is difficult to justify not choosing a wireless solution."

***Note:** The WaveRider NCL135 wireless bridge router is being used in other applications as well. Oakdale, a small city near Minneapolis-St. Paul, Minnesota has implemented new wireless link between the city hall complex and the public works building a half a mile away in order to give public works staff instant electronic access to the city's infrastructure.*

Clemson University

Clemson University was founded in 1889, a legacy of Thomas Green Clemson, who willed his Fort Hill, South Carolina, plantation home, its surrounding farmlands and forest, and other property to the state of South Carolina to establish a technical and scientific institution.

In 2001, Clemson decided to test out a wireless bridging application to link its Administrative Programming Group (APG) to the central campus. The APG is located about 1.5 miles (2.4 kilometers) off campus, and is situated across Lake Hartwell from the central campus.

Prior to implementing a wireless bridge, the University paid a monthly fee to lease telco circuits to connect the 30 programmers located at the APG facility to the campus network. With the wireless bridging solution in place, however, the University can now put the monthly circuit cost to better use and still provide the APG personnel with high-speed, 11 Mbps connectivity.

Another Clemson University location that is benefiting from wireless bridging is Muster Farms. This University research farm is located 5 miles (8 kilometers) from campus. Since only a few personnel actually need access to the University's network assets, the location was never outfitted with a leased line. Whenever anyone at the Farm needed access to the University's network, they either made a trip to the campus or accessed the network via a slow dial-up modem. Wireless bridging now enables the Muster Farm people to have fast connectivity to the Clemson Network.

The University also has, or is in the process of providing, network connectivity to a number of off-campus County Extension Research Centers. Some of these centers have several buildings, and wireless bridging has proven to be a very cost effective way to connect them to each other and to the University's central network.

The Rolling Stones

The venerable rock and roll band, the Rolling Stones, have gone Wi-Fi, albeit with the help of Clear Channel Entertainment, the marketing and management arm of the tour. The band members use their Wi-Fi network to keep in touch with their loved ones back home, but the band's entourage generally uses Wi-Fi to manage the 2002-2003 Forty Licks world tour as the band travels from venue to venue. Specifically, the band members use the WLAN to send and receive emails and Word documents, while roadies and others use the network to share CAD drawings for stage and seating arrangements for the tour's venues.

The Stones' Wi-Fi network is built upon 3Com 802.11b wireless LAN equipment, including 3Com Access Point 8000s and their 802.11b wireless LAN building-to-building bridging gear. These are then connected to a satellite uplink from Hughes and a downlink antenna system. Approximately 140 Wi-Fi-enabled laptops allow the production office, promotion office, band and crew to access this mobile Wi-Fi network. This enables the tour's management, and the band, to always be connected not only to their loved ones, but also to the home office. The latter allows for rapid re-working of extra performance bookings, changes to seating plans, last-minute promotions, business negotiations with venue operators, and ticket pricing decisions.

The set-up is completely mobile, thus it can be set up at each concert venue within about an hour after arrival at a new venue. The band's IT crew can tailor the system, based on each unique concert site around the world, to provide complete coverage despite concrete, steel, or other traditional barriers.

Fans also benefit from the Stones' Wi-Fi system. The flexibility of the network allows the Stones' tour promotion team to pass on much more information about each performance to the band's fans. This allows the worldwide Stones fan community to stay current with the tour via the www.rollingstones.com website. Fans who have full membership on the site (either by having bought a concert ticket, or by paying a membership fee), have a "virtual ticket" to follow the tour around the world.

The band is also interested in doing regular live webcasts of the tour and has, in fact, successfully tested the system several times. There are still lots of technical "bugs" to be worked out, though. While some of the obstacles are technical, others are rooted in copyright and performance contract issues. However, the tour did run live streaming audio and video tests in Atlanta, with a live webcast of the setup of the stage.

The tour's management has found that Wi-Fi does help the Forty Licks world tour go more smoothly. For instance, behind the scenes, the network speeds up the preparation of the stage for the band's performance. By using radio links instead of cables, the Stones' production company can set up a venue-wide computer network far faster than normal, because employees can easily share detailed floor plans of venues and the Stones' stage set. This helps them to fine-tune the stage setup for every stop on the tour.

Todd Griffith, the satellite Internet specialist for the Stones tour, built and runs the Stones' mobile IT setup. He says that the Wi-Fi system evolved over a period of a couple of years. Originally, Griffith says, the tour tried to encourage the local performance facilities to put a DSL or Cable Modem drop into each concert venue. However, the advantages of touring with a complete, permanent networking solution, rather than recreating it at each concert facility, proved to make more sense.

If a venue has no clear line of site to the satellite, Griffith then goes to a backup plan—DSL, which is then connected to the 3Com access points. According to Griffith, the quantity of data handled by the system when it's up and running is equivalent to a small to medium-sized office. There are usually anywhere from 50 to 120 end-users accessing the network at any one time. The majority of the data that traverses the Wi-Fi links are email and word processing documents, but there are some larger file transfers also, e.g. CAD drawings of the stage set or the concert seating.

Griffith also laughingly says, "Sometimes we have to find a Starbucks that has an access point service to do last-minute upload and download. Sometimes I get asked why I go to Starbucks so much. I say, it's work."

According to Griffith, the business case for creating the wireless data system for the Stones is as follows. First, the satellite Internet component of the system allows Stones fans to follow the tour via the www.rollingstones.com website. The Stones' website is updated on a daily basis by people within the tour group.

Second, Griffith explains that there are real morale benefits to having high-speed Internet connections wherever the tour goes. It makes the long times away from home more bearable for the band and crew.

However, Griffith does admit that deploying and managing the mobile Wi-Fi system has been a challenge. He says that he found there to be a steep learning curve.

CONCLUSION

Wireless bridges can stretch the meaning of the word "local" and give business users unexpected flexibility in extending corporate LANs and educational institutions' ability to extend their tech budget. Wireless bridges can have the biggest impact when deploying a temporary network, tying HotSpots together, and creating long distance links to wired networks.

Almost all wireless LAN vendors offer bridging products. We discussed in this chapter some of the leading enterprise-class bridging products, but there are also vendors such as LinkSys and Breezecom that offer inexpensive bridging products.

Chapter 10: Contrasting Deployments

Like any new technology, Wi-Fi can solve some problems, but create other challenges. Providing appropriate network coverage can be the most difficult issue faced when designing a wireless local area network (WLAN). This chapter, hopefully, will give the reader valuable insight and practical knowledge of different aspects of wireless networking technology, some of the general issues that each Wi-Fi flavor presents, and awareness of specific problems that may arise during implementation of the various Wi-Fi technologies. We will begin with a couple of practical deployment scenarios.

PRACTICAL DEPLOYMENTS

Let's explore practical deployment scenarios for both 802.11a (current and optimized) and 802.11b, including three-cell and eight-cell configurations.

***Note:** I want to thank CMP Media LLP and Jung Yee and Hossain Pezeshki-Esfahani, the authors of "Understanding Wireless LAN Performance Trade-Offs," published in the November 2002 issue of Communication Systems Design magazine, as the following text and metrics relies heavily upon that article.*

To cover areas greater than that provided by a single access point (AP), network designers may deploy multiple APs in a hexagonal cellular arrangement. Each cell has one AP at its center, so that end-users can access the WLAN from anywhere in a cell. To maintain performance levels and effective data rates, the bit error rate (BER) should not exceed 10e-5 anywhere in the network.

To make the problem tractable, assume at any given time that there is only one station, or wireless computing device, transmitting in each cell. The collision avoidance and back-off behavior of the 802.11 Media Access Controller (MAC) make this a good assumption in distributed-coordination function mode, while in the point-coordination function mode the assumption is nearly literally valid. Owing to the high degree of asymmetry of traffic flow, you should also assume that the only potentially jamming stations are the APs in other cells.

From the spectral masks and channel allocations described by the IEEE, you can infer that the fraction of spectral leakage (l) from one adjacent channel is 4.282853E-3 (-23.68 dB) for an 802.11a network, and 3.63273E-4 (-34.40 dB) for an 802.11b network.

As you should now understand, the 802.11b specification, in its current form, has three channels available for use. This number is important because it dictates how 802.11b access points are placed to cover wireless geography into areas called cells. 802.11a, on the other hand, has eight channels, and therefore while each specification's radio broadcast patterns are roughly the same, programming cells are decidedly different between the two standards. And it can get really complicated when planning a network with dual-channel or hybrid 802.11a/b access points.

***Note:** If you associate access points as cells, each cell has a broadcast pattern available for client hardware. The client hardware will associate with access points at a specific data rate that is a function of the quality and signal strength between the two devices.*

Three-cell Networks

Consider a network of three cells (Fig. 10.1). Since 802.11b affords only three frequency channels, one of the three cells (for example, Cell 3) experiences adjacent-channel interference (ACI) from the other two cells.

The greatest interference noise power (IN) is experienced by an end-user belonging to Cell 3 located at Point G in Fig. 10.1. Using Equation 6, the IN value at G is as shown in Equation 1.

$$IN = \frac{2x1xP_r}{L_o R^N}, \text{ where } L_o = ((4\pi f)/C)^o$$

Equation 1.

Similarly, the desired signal power is given by Equation 2.

$$P_{RX} = P_{rx} \times (L_o R^N)^{-1}$$

Equation 2.

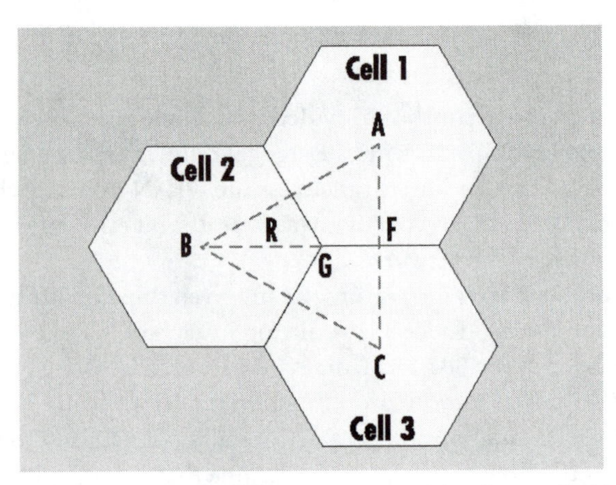

Figure 10.1 In an 802.11b network of three cells, one of the three cells (Cell 3) experiences adjacent-channel interference (ACI) from the other two cells because the 11b standard provides for only three frequency channels.

The signal-to-interference noise ratio is therefore equal to SINR = (2f)-1, which is 31.39 dB. By comparing this number with Fig. 7.4's 802.11b numbers, one sees that all 802.11b data rates can be supported. The size of the network in Fig. 10.1 can be calculated by finding the radius of each hexagon. (This is a function of the Tx power and the minimum Rx sensitivity.) As mentioned previously, it is necessary to use a minimum Rx sensitivity of -70 dBm due to ACI. Note that a +15-dBm solution can cover 8844 square yards (7398 square meters) at 11 Mbps, while a +19-dBm model will offer better range (137 feet/42 meters as opposed to 101 feet/31 meters), and therefore a larger total area of coverage (16,349 square yards or 13,670 square meters).

The 802.11a standard has more channel flexibility. If a 54 Mbps transfer rate is desired, the WLAN designer can choose, for instance, channels 1, 5 and 8 for Cells 1, 2 and 3 to eliminate adjacent-channel interference. In this case, the APs in Cell 1 and Cell 3 can radiate at +23.01-dBm EIRP, while the AP in Cell 2 can radiate at +16.02-dBm EIRP. Consequently, Cells 1 and 3 each cover 1370 square yards (1146 square meters), while Cell 2 covers 525 square yards (439 square meters), with a total coverage of 3265 square yards (2730 square meters). Although this total coverage area is less than the 802.11b examples, the WLAN has a transfer rate capacity of 54 Mbps (and a throughput of about 31 Mbps).

Having calculated coverage dictated by the IEEE 802.11a standard, what can real-world WLANs deliver? With an optimized 802.11a solution, designed to deliver 23 dBm, the coverage of a three-cell deployment with 54 Mbps will reach 3265 square yards (2730 square meters).

Alternatively, if a transfer rate of 12 Mbps is preferred (to reduce costs or the number of APs), then channels 5, 7 and 8 can be used for Cells 2, 3 and 1 respectively. This will result in ACI only between channels 7 and 8 and all APs can radiate +23.01 dBm. In this example, the interference noise power is as shown in Equation 3.

$$IN = 1 \times P_{rx} \times (L_o R^N)^{-1}$$

Equation 3.

So the least SINR is f-1, or 23.68 dB. As a result, all but the two highest Orthogonal Frequency Division Multiplexing (OFDM) rates are accessible. Maintaining at least 12 Mbps requires a cell radius of 47.59 meters (congruent hexagons), so that the total area covered is now about 21,109 square yards (17,650 square meters).

Eight-cell networks

An eight-cell 802.11a network is depicted in Fig. 7.4. If all of the cells are congruent, then all of the APs must transmit at the same power level. This requires the maximum power of 40 milliwatts (+16.02 dBm) and, in this case, a computing device belonging to Cell 4, located at Vertex G, experiences the greatest interference.

The sources of interference are the two adjacent-channel interferers, one at 2R and the other at root 7R. Therefore, the least SINR experienced by any station throughout the network is given by Equation 4, or about 31.49 dB.

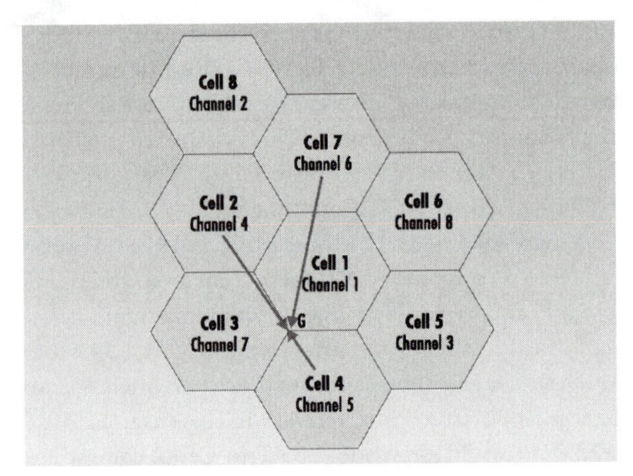

Figure 10.2 For an 8-cell 802.11a network, which allows for 54 Mbps operation, the greatest interference is experienced by a station belonging to Cell 4 located at Vertex G.

$$SINR = \frac{1}{l\left[\frac{1}{2^N} + \frac{1}{(\sqrt{7})^N}\right]}$$

Equation 4.

Comparing this number with the S/Ns for 802.11a in Fig. 7.4, we see that a data transfer rate of 54 Mbps is accessible throughout the WLAN. The total area covered in this configuration is 2490 square yards (2082 square meters).

An 802.11b WLAN can be used to cover a similar geometry as in the 802.11a example above. The configuration outlined in Fig. 10.2 was carefully designed for an 802.11a application to mitigate co-channel and adjacent-channel interference, with the most interference being experienced by a computing device belonging to Cell 2 located at Vertex G. In comparison, a similar 802.11b configuration (Fig. 10.3) leads to a co-channel at 2R, a co-channel at root 7R, two adjacent channels at 2R, two adjacent channels at R, and one adjacent channel at root 7R. Therefore, the least SINR experienced by any device in the network is as shown in Equation 5, or about 7.45133 dB.

$$SINR = \frac{1}{\frac{1}{2^N} + \frac{1}{(\sqrt{7})^N} + 21 + \frac{21}{2^N} + \frac{1}{(\sqrt{7})^N}}$$

Equation 5.

Comparing this with the S/Ns for 802.11b in Fig. 7.4, it is clear that a transfer rate of 11 Mbps is accessible throughout the 802.11b network. Note that a +15-dBm solution can cover 23,593 square yards (19,727 square meters) at 11 Mbps, while a +19-dBm model will offer coverage of 43,597 square yards (36,453 square meters).

As a side note, a concern for 802.11b/g systems is the proposed 22 Mbps Packet Binary Convolutional Coding (PBCC) extension (802.11g), which requires approximately 13.8 dB of S/N for a bit error rate less than 10e-5. In such a case, as Equation 6 shows,

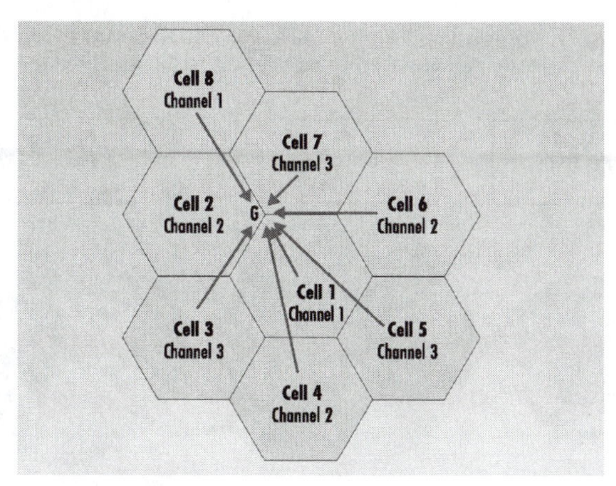

Figure 10.3 In an 8-cell 802.11b configuration, the least signal-to-noise ratio experienced by any station in the network is about 7.45133 dB, allowing for full 11 Mbps data transfer rate.

802.11b networks requiring eight or more APs, 22 Mbps would not be accessible throughout the entire network.

$$10\log_{10}(S_{RX}) = 10\log_{10}(S_{rx}) - 20\log_{10}\left(\frac{4\pi f}{C}\right) - 10N\log_{10} r \text{ for } r \geq 1$$

Equation 6.

Future 802.11a solutions that employ low-power, high-performance architectures may be able to transmit up to the maximum FCC limits. As a result, they will offer the range and data rate flexibility that will allow IT managers to tailor solutions to their specific applications and budgets, making an optimized 802.11a system the perfect option.

COMPARING 11a AND 11b IN AN OFFICE ENVIRONMENT

Let's now compare an 802.11a AP to an 820.11b AP in a typical cubicle office environment.

A 15 dBm 802.11b AP's transmission and reception will cover a circular area with a radius of approximately 100 meters (328 feet), while providing a maximum 11 Mbps data rate, thus the coverage area would be approximately 37,578 square yards or 31,420 square meters (p*100m2) at a top data transmission rate of 11 Mbps.

***Note:** This example assumes an open office environment and uses the A. Kamerman path loss model that can be found in the paper entitled, "Coexistence between Bluetooth and IEEE 802.11 CCK Solutions to Avoid Mutual Interference," by Lucent Technologies Bell Laboratories, January 1999, also available as IEEE 802.11-00/162, July 2000.*

Now, assuming the RF transmit power and E_s/N_0 are held constant (i.e. same link budget), an 802.11a AP's coverage equates to a circular area with a radius of roughly 164 feet (50 meters) for a comparable data rate (9 Mbps), equaling a coverage area of approximately 9388 square yards or 7850 square meters. (Note: there are various potential 5 GHz path loss models that show decreased range compared to 2.4 GHz solutions of 45%, 50% and 75%.)

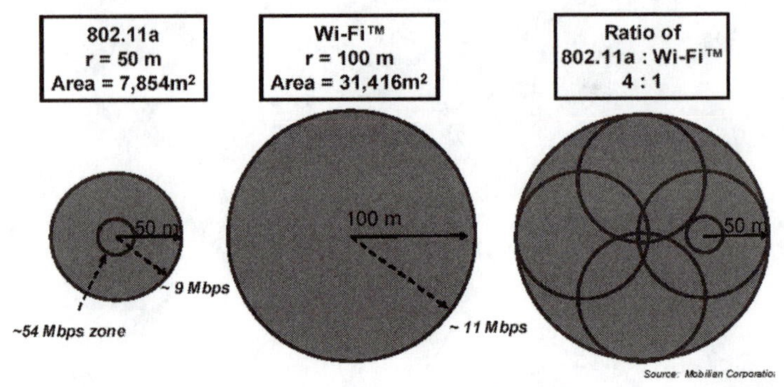

Figure 10.4 Coverage Comparison of 802.11b and 802.11a. There are significant differences between 11a and 11b in terms of each standard's achievable communication range between the access point and the computing device, and the corresponding service coverage area.

Because of the fundamental difference in signal wave propagation, to achieve similar coverage areas, greater numbers of 802.11a APs are needed to cover the same area that a lesser amount of 802.11b APs could cover. (See Fig. 10.4.) But with the increased number of 802.11a APs, the data rate is only comparable to 802.11b's data rate, not 802.11a's highest data rate of 54 Mbps. This is because (1) the E_S/N_0 threshold and corresponding bit error rate are more stringent for higher data rate transmit schemes, and (2) energy dissipates as a signal moves away from the transmitter. Thus the further the receiver is from the transmitter, the more difficult it becomes to decipher a message. 802.11a provides high data rate (36-54 Mbps) levels close to the AP (within about 33-49 feet/10-15 meters), making it attractive for dense user environments that also require high throughput, but its data rates are closer to 9-12 Mbps at ranges over 100-140 feet (30-40 meters).

Power Considerations, Relative to Data Rate and Signal Range. You should now understand that 802.11a systems are able to achieve about 50% of the range of 802.11b systems, holding operational variables constant (RF transmit power, receiver sensitivity, etc.), and requiring a data rate approximately equal to 802.11b's 11 Mbps. This limitation in range is caused by the more severe path loss of the 5 GHz spectrum, and the stringent E_S/N_0 requirements of 802.11a's higher data rate modulation techniques.

Yet, 802.11a systems can achieve data rates similar to 802.11b at ranges approaching those of 11b systems, although to do so requires an increase of approximately four times the RF transmit power of an equivalent 11b system, or from 40mW to approximately 200 mW11. This increase may have a negative impact on the mobile client's (i.e. an end-users' computing device) battery life. But some will argue that at very close ranges (less than 30-45 feet or 9-14 meters in an office environment), an 802.11a computing device will spend less time in transmit mode due to its high data rates, and therefore expend less energy than an 802.11b computing device burdened with similar data traffic amounts.

802.11a's more complex modulation techniques, however, require greater power to maintain a suitable E_S/N_0. Still, with multiple variables in the equation (environment, data rate, etc.), it is hard to know the breakeven point at which an 802.11a system expends less energy transmitting the same amount of data as an 802.11b system.

Another consideration is the efficiency of the RF power amplifier, which is generally significantly worse for 802.11a's OFDM relative to 802.11b's PSK12; however, efficiency numbers for OFDM are typically 1/3 of the PSK (phase shift keying) equivalent. As a result, the total power consumption when transmitting 9-12 Mbps OFDM at 5 GHz can be significantly higher than a 2.4 GHz equivalent, even at ranges of only 130-165 feet (40-50 meters).

The end results suggests that an 802.11a system is more power-efficient and may be better suited for transmitting high data rates over small, densely populated areas, while an 802.11b system is more efficient over greater distances. It is interesting to note, however, that OFDM technology in the 2.4 GHz band (i.e. 802.11g) will always surpass its 5 GHz counterpart (i.e. 802.11a) in power efficiency due to the path loss issue.

Unlicensed Available Spectrum and Spectral Efficiency. All three WLAN standards (802.11a, b and g) use network passbands that cannot overlap, thus limiting the number of simultaneously operating networks to the number of available channels within the unlicensed spectrum, divided by the width of the passband. As the number of WLAN users increases and the bandwidth requirements of applications grow, more WLAN infrastructure must be deployed to support the users, and available throughput must increase. Therefore, the amount of unlicensed spectrum available and the efficiency with which the standard uses the spectrum are important considerations.

The 802.11b specification allows for DSSS (Direct Sequence Spread Spectrum) to break the ISM band into 14 overlapping 22 MHz signal bandwidth channels. (Each operating network requires a single 22 MHz channel.) DSSS provides an 11 Mbps data rate to the network users. Thus the 2.4 GHz band's 83.5 MHz will support three non-overlapping, simultaneously operating 11b networks, with roughly 33 Mbps of data rate (11 Mbps x 3 networks) to be shared among common users across the coverage area. (See Fig. 10.5).

This configuration adequately distributes sufficient bandwidth to support the majority of data applications such as word processing, spreadsheets, email, server downloads, PowerPoint presentations, and FTP applications. DSSS cannot adequately support streaming media or large graphic files. However, as user density increases, AP transmit power can be decreased, which effectively serves to increase the number of networks within a given area, and therefore increases the total capacity of the WLAN.

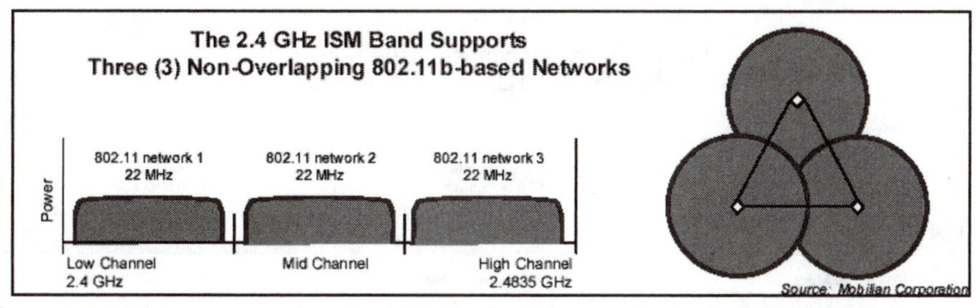

Figure 10.5 Three non-overlapping 802.11b-based networks in the 2.4 GHz band.

The 802.11a standard requires a 16.6 MHz signal bandwidth channel for one operating network. 11a's modulation technique, OFDM, is more efficient than the spread spectrum techniques that 11b uses (more bits/second/hertz), and provides up to 54 Mbps of data rate to network users. For the U.S.-based 802.11a standard, the 5 GHz unlicensed band covers 300 MHz of spectrum and therefore supports twelve non-overlapping, simultaneously operating networks (See Fig. 10.6).

Although you can find many sources that cite the figures "648 Mbps of total data rate to distribute across 12 networks (54 Mbps * 12 networks)," it is more likely that common network deployments will support only four simultaneously operating networks in a single area, and provide a 216 Mbps data rate (54 Mbps * 4 networks) within the relatively small range at which 54 Mbps is actually attainable. And, as you have already learned, the further the computing device is from the AP, the lower the attainable data rate.

Nevertheless, 216 Mbps of shared data rate is more than sufficient to support all the office productivity applications mentioned for 802.11b, *plus* highly bandwidth-intensive applications such as computer-aided design or manufacturing, large distributed databases, and streaming media. But remember, the maximum data rate of 54 Mbps data rate is within a relatively small coverage area. High bandwidth users outside that area would require either more APs for increased coverage, or those users will be subject to reduced throughput.

Thus, 5 GHz solutions are clearly dominant in terms of throughput, spectral efficiency, and available spectrum, but current network deployments and usage models indicate that 802.11b can adequately meet most user needs. When you consider the value of the 802.11g specifications, sticking with 802.11b for the short term may be a good choice. After all, 11g can provide the necessary data rate speed enhancement to allow current 802.11b users to deploy QoS and to access the same high-bandwidth applications that 802.11a users enjoy, while maintaining backward compatibility with legacy 11b networks. But 802.11a can provide greater overall capacity for smaller coverage areas that operate bandwidth intensive applications.

Therefore some WLAN deployments will require a mixed-standard environment in which both 802.11a and 11b-based technologies coexist. High-density, high-bandwidth common areas will be served by 802.11a, e.g. lunchrooms, conference rooms, team rooms, etc. Lower-density, greater coverage areas (e.g. warehouses, corporate office space, etc.) will be served by 802.11b/g-based technologies. Of course, such a mixed-standard environment implies that client devices should be able to roam between 802.11a and 802.11b/g networks. These dual-mode clients and APs are becoming prevalent.

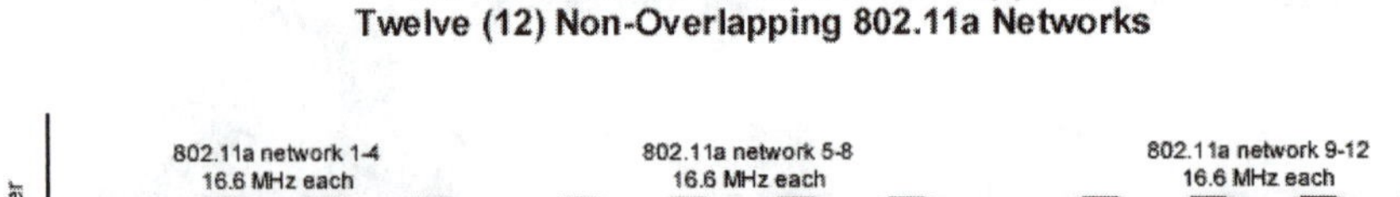

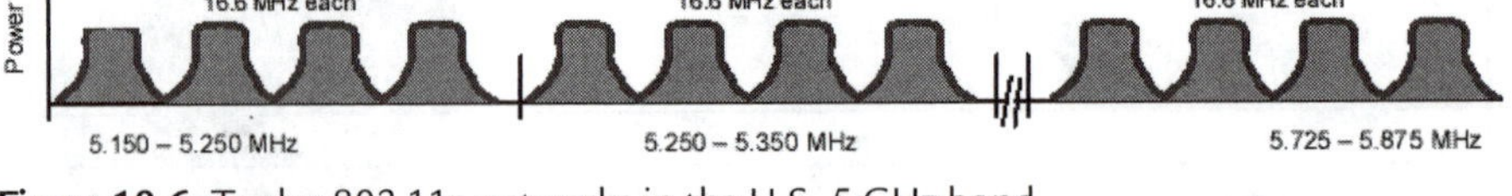

Figure 10.6 Twelve 802.11a networks in the U.S. 5 GHz band.

COMPARISON OF RADIO WAVE PROPAGATION AT 2.4 AND 5 GHZ FREQUENCIES

While the propagation characteristics of the 5GHz band have been studied less extensively than the cellular and PCS bands, there is an adequate body of analysis and measurements to draw from to create useful path loss models.

The physics principle of free space propagation found in Theodore S. Rappaport's Wireless Communications: Principles & Practice (Prentice Hall, 2000) states that as an electromagnetic wave's length grows shorter (i.e., higher frequency), the path loss of that wave increases according to a square law relationship. Also from electromagnetic wave propagation theory, the path loss increases proportionally with the square of the distance between transmitter and receiver in free space. Since 5 GHz radios operate at little over two times the frequency of 2.4 GHz radios, it can be shown that the same link budget can only be sustained over about half the distance as that of a comparable 2.4 GHz system. This is represented by the equation below, but also see Fig. 10.7.

lx = K(dN/lx2) [often expressed in decibels as lx (db) = KL + 10N log10(d) + 20 log10(fx)]

OR d = [(lx * l2)/K]1/N

Where:

lx = path loss

x = frequency band or standard ("a" for 802.11a or "b" for Wi-Fi and 802.1.1g)

K = constant

KL = constant for loss in dB = 32.44 when f is in MHz and d is in kilometers

d = distance

N = attenuation factor: N=2 for free space; N>2 in most practical environments

l = wavelength

f = frequency = c/l

c = speed of light = 3 x 108 meters/second (free space)

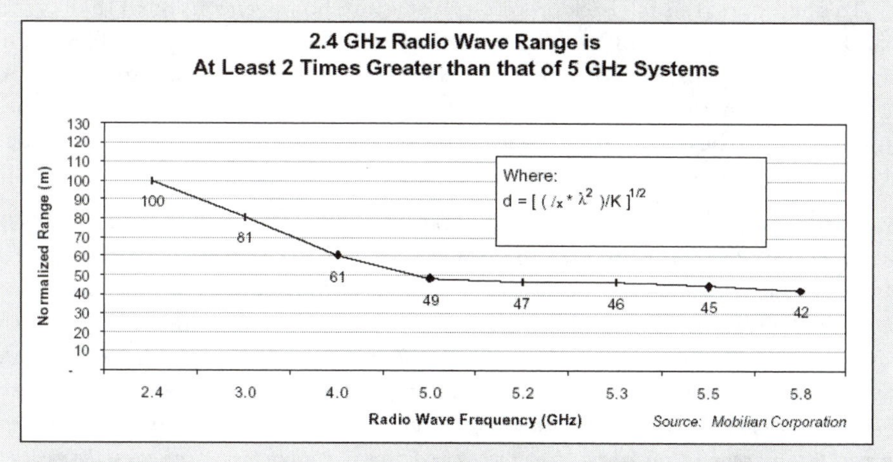

Figure 10.7 Comparative free space path loss between 2.4 GHz and 5 GHz systems.

***Note:** The Mobilian Corporation, a wireless systems company founded in 1999 to develop highly integrated, multi-standard chips, software, and reference designs supporting multiple wireless radio standards, issued an informative white paper entitled "2.4 GHz and 5 GHz WLAN: Competing or Complementary?," which the author relied upon extensively while writing this section. Additional information comes from the IEEE study of 5 GHz indoor path loss in urban homes, entitled "Measurements and Models for Radio Path Loss and Penetration Loss in and Around Homes and Trees at 5.85 GHz" (G. Durgin, T. S. Rappaport, H. Xu), IEEE Transactions on Communications, Vol. 46, No. 11, November 1998, pp. 1484-1496.*

THE "b" / "g" NETWORKING ENVIRONMENT

Adding 802.11g to an existing 802.11b network presents some challenges. As you know, the 802.11g standard brings much higher data rates to the 2.4 GHz band. But what you may not realize is that data *throughput* rates can drop significantly when 802.11g devices are introduced into a legacy 802.11b network (or vice versa). Some readers also might not understand that such drops in throughput are common to any wireless network where high-rate and low-rate devices share the same medium. Finally, you need to know that 802.11g provides protection mechanisms to mitigate such issues.

To gain accurate insight into the mechanics that underlie 802.11g WLANs, it is necessary to examine its operation under homogeneous (802.11g-only) as well as mixed-mode operation in which an 802.11g client shares an access point with a legacy 802.11b client.

***Note:** I want to thank CMP Media LLP and Menzo Wentink, Tim Godfrey and Jim Zyren, the authors of "Overcoming IEEE 802.11g's Interoperability Hurdles," published in the May 2003 issue of Communication Systems Design magazine, inasmuch as the following text and metrics relies heavily upon that article.*

The wireless networking is inherently more challenging than wired, especially since the effective range falls as data rate increases. For this reason, 802.11b computing devices (clients) located close to an AP can easily connect at 11 Mbps. However, these same devices will fall back to 5.5, then to 2, and finally to 1 Mbps data rates as the device moves further away from the AP, which causes the signal to weaken.

To understand this fall back process, consider two homogeneous 802.11b-only scenarios. In the first, scenario A_b of Fig. 10.8, an 802.11b AP is streaming large files downstream to two clients, each of which is proximal to the AP and operating at 11 Mbps. The second, scenario B_b, is similar to the first except one client/computing device has now roamed to the edge of the network and has fallen back to a data rate of one Mbps to preserve the network connection.

***Note:** An AP and all of the associated clients are collectively referred to as a basic service set (BSS). Within any BSS, data can flow downstream (AP to computing devices), upstream (computing devices to AP) or in a bidirectional manner (combination of upstream and downstream traffic). Although every data packet is immediately followed by an acknowledgement frame (ACK) in the opposite direction (unless the packet is received in error), the direction of data flow refers to the direction of the data packet. Thus, downstream data flow refers to a packet exchange consisting of a downstream data packet followed by an upstream ACK. When considering the case of downstream traffic flows, it should be kept in mind that the network throughput is determined by the speed at which the AP can access the medium.*

When two computing devices are queued traffic at the AP, packets are on average sent to each device on an alternating basis since all 802.11 networks use a carrier sense-multiple-access/collision avoidance (CSMA/CA) channel-sharing scheme. As a result, before the AP can transmit, it must contend for access to the medium. In Fig. 10.9, "a" depicts Scenario A_b in Fig 10.8, in which a series of 1500-byte packets is being transmitted downstream from the AP to two clients at 11 Mbps.

Note that each packet is followed by an ACK and a contention or "back-off" interval. Importantly, when the AP has queued traffic, data packets are sent to Client 1 and Client 2 on an alternating basis. Under the conditions depicted in Fig. 10.9 a, actual network throughput is about 7.2 Mbps. This does not include TCP/IP or Application Layer overhead. Therefore, if network throughput is measured by means of a large file transfer, results would be somewhat lower (more about this later).

Now consider Fig. 10. 8's Scenario B_b, which is an equally as plausible as that depicted in Fig. 10.9 b. Assume that Client 1 roams out to the edge of network coverage. In this condition, the AP would require a staggering 12,794 microseconds to transmit the same 1500-byte packet and receive the ensuing ACK from Client 1. The time required

a)

Scenario	Description	Total 802.11 throughput	Client throughput	
			Client 1	Client 2
A_b	Two clients operating @ 11 Mbps	7.2 Mbps	3.6 Mbps	3.6 Mbps
B_b	Client 1 @ 1 Mbits/s, Client 2 @ 11 Mbps	1.6 Mbps	800 Kbps	800 Kbps

b)

Scenario	Description	Total 802.11 throughput	Client throughput	
			Client 1	Client 2
A_a	Two clients operating @ 54 Mbps	30 Mbps	15 Mbps	15 Mbps
B_a	Client 1 @ 6 Mbits/s, Client 2 @ 54 Mbps	9.2 Mbps	4.6 Mbps	4.6 Mbps

Figure 10.8 The effect of a roaming client is similar for both the 802.11b a) scenario, and b) scenario. The AP will alternate transmissions between Client 1 and Client 2, and network throughput will drop between 70 and 77 percent.

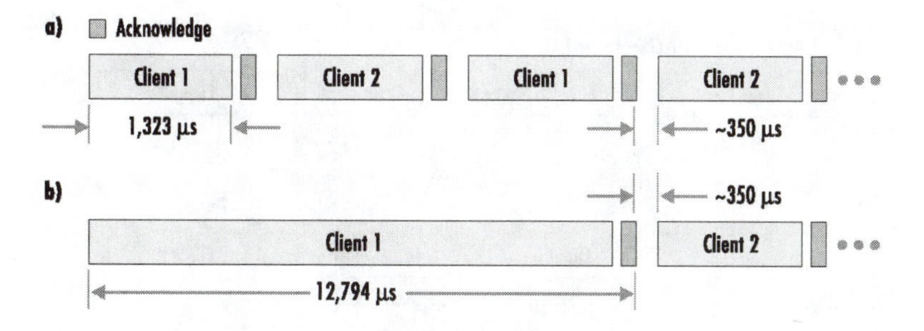

Figure 10.9 When two clients (computing devices) are close to an AP (a), the data rates are similar. But, as Client 1 roams to the network edge (b), its rate drops quickly, slowing considerably the time taken for the packet transmit and ACK.

for the same exchange between the AP and Client 2 would remain unchanged at 1323 microseconds.

It is important to remember that for downstream traffic the AP will alternate transmissions between Client 1 and Client 2. From b in Fig. 10.9 it should be clear that Client 1 dominates airtime due to operation at 1 Mbps. As a result, network throughput falls from 7.2 Mbps in scenario A_b to just 1.6 Mbps in scenario B_b (Fig. 10.8). This represents a drop in network throughput of 77 percent. In this situation, network throughput is shared equally among clients. On an individual basis, each client will realize 800 Kbps effective throughput. The drop in throughput is not a result of a network problem. Rather, it is a completely predictable consequence of network operation in a mixed-rate environment.

That scenario is analogous to the conditions under which most 802.11g network tests are being conducted (downstream transfer of large packets). It is worth emphasizing that the 77 percent drop in network throughput is not due to any interoperability issue, nor has Client 2 reduced its data rate.

IEEE 802.11a networks will exhibit similar behavior, giving a 70 percent drop in total throughput (see Fig. 10.8, scenario b).

Up to this point, the scenarios have been relatively simple, consisting of only two clients/computing devices receiving downstream traffic from the AP. But a WLAN's throughput depends on several factors, including the relative amount of time devoted to operating at peak data rates as compared with the amount of time spent operating at lower rates. For situations involving many devices, network throughput depends on the number of high-speed devices relative to the number of slower devices.

Due to the fact that the operation of an 802.11g WLAN involves devices using different waveforms, the situation is slightly more complicated. However, the same basic dynamics are applicable. Let's consider 802.11g network behavior under conditions analogous to those described above, again using two scenarios. In scenario A_g, two 802.11g clients are operating at 54 Mbps (Fig. 10.10 a). In scenario B_g, one 802.11g client is operating at 54 Mbps, while one legacy 802.11b client is operating at 11 Mbps (Fig. 10.10 b). In both scenarios all clients are performing large downstream file transfers.

For scenario A_g, note that for downstream traffic, packets are again shared equally among the clients. However, due to much higher data rates and a smaller con-

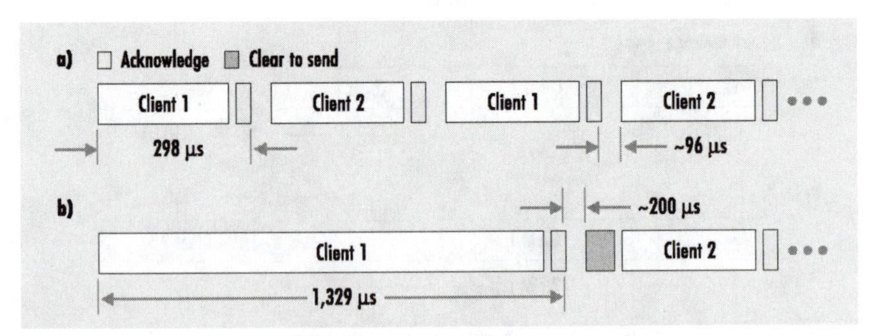

Figure 10.10 802.11g-only networks (a) can hit 30 Mbps throughput. But, when a legacy 802.11b client is introduced (b), protection mechanisms kick in. Still, the WLAN's throughput will drop to 9.3 Mbps (including TCP/IP).

tention period between packet transmissions, network throughput is much higher (30 Mbps).

Now consider what occurs if a legacy 802.11b device is substituted for one of the high-speed 802.11g clients. To ensure backward compatibility with legacy 802.11b equipment, a number of changes occur. The most obvious is the use of so-called protection mechanisms. 802.11g devices can recognize both legacy complementary code-keying (CCK) 802.11b packets as well as OFDM packets. However, legacy 802.11b devices are incapable of recognizing OFDM transmissions. Because CSMA/CA is a listen-before-talk algorithm, this would present a problem unless other measures were employed.

In mixed-mode operation, 802.11g OFDM packets are preceded by a short clear-to-send (CTS) packet transmitted using legacy 802.11b modulation (CCK). The CTS packet conveys the length of time required for the ensuing high-speed OFDM packet and ACK. Legacy clients receiving this information will remain idle for the specified period of time, thereby avoiding collisions with the OFDM packet exchange. Although the CTS protection mechanism results in a marginal increase in network overhead, this effect is dwarfed by the fact that even while operating at 11 Mbps, legacy 802.11b clients are much slower on a relative basis.

For scenario B_g depicted in Fig. 10.10 b, overall network throughput is reduced to 11.2 Mbps, representing a drop of about 64 percent. The resulting drop is significant, but is actually less severe than what already occurs in 802.11a and 802.11b networks under similar conditions, as described previously.

Because the AP alternates transmissions to clients when sending queued traffic downstream, each client receives the same effective throughput regardless of the fact that they are operating at much different data rates. This may explain why some early evaluations of 802.11g equipment indicated that data rates fell back to 802.11b levels while in mixed-mode operation.

Measuring an 802.11g WLAN's throughput via downstream file transfer is not an unrealistic approach, though it does represent only a single scenario out of the thousands that may be encountered in actual use. In practice, it is possible for dozens of computing devices to be associated with the AP at any point in time. The throughput realized depends on several factors.

One of the most important issues is the number of high-speed 802.11g computing devices relative to the number of 802.11b devices. A WLAN consisting of a majority of 802.11g devices will operate a larger fraction of time at higher rates, and a lower fraction of time at lower rates. The opposite is true if a majority of the devices are slower legacy 802.11b devices. The estimated 802.11g throughput, with various mixes of client types, is shown in Fig. 10.11. This includes TCP/IP overhead and a 15 percent correction factor for network efficiencies. Note that for downstream traffic, throughput is shared equally among clients.

The figures in Fig. 10.11 do not include Application Layer overhead, so throughput tests based on large downstream file transfers will render marginally lower results. Bear in mind that these figures represent total network throughput, which is shared among all active clients. The highlighted values along the diagonal axis all represent scenarios involving ten clients. Moving from upper left to lower right, the number of 802.11g

802.11b clients	0	1	2	3	4	5	6	7	8	9	10
10	**5.9**	6.2	6.5	6.8	7.0	7.2	7.4	7.6	7.8	8.0	8.2
9	5.9	**6.2**	6.5	6.8	7.1	7.4	7.6	7.8	8.0	8.2	8.3
8	5.9	6.3	**6.6**	6.9	7.2	7.5	7.7	8.0	8.2	8.4	8.5
7	5.9	6.3	6.7	**7.1**	7.4	7.7	7.9	8.2	8.4	8.6	8.8
6	5.9	6.4	6.8	7.2	**7.6**	7.9	8.2	8.4	8.7	8.9	9.1
5	5.9	6.5	7.0	7.4	7.8	**8.2**	8.5	8.7	9.0	9.2	9.4
4	5.9	6.6	7.2	7.7	8.2	8.5	**8.9**	9.2	9.4	9.6	9.8
3	5.9	6.8	7.6	8.2	8.7	9.1	9.4	**9.7**	9.9	10.2	10.4
2	5.9	7.2	8.2	8.9	9.4	9.8	10.2	10.4	**10.7**	10.9	11.1
1	5.9	8.2	9.4	10.2	10.7	11.1	11.3	11.6	11.7	**11.9**	12.0
0	0.0	22.1	22.1	22.1	22.1	22.1	22.1	22.1	22.1	22.1	**22.1**
	Number of 802.11g clients										

Figure 10.11 In a mixed 11g/b environment, the throughput (including TCP/IP overhead) depends on the number and type of clients associated with the access point. The figures represent total network throughput.

clients increases and the number of legacy 802.11b clients decreases. Note also that network throughput increases substantially as the ratio of 802.11g clients increases.

So far we've focused entirely on downstream transfer of large files. This is entirely appropriate given that to date, evaluations of 802.11g equipment have largely been conducted in this manner. However, there are some interesting points that come into play if one considers more generalized traffic flow in a mixed-mode 802.11g network.

As described previously, when data flow is exclusively in the downstream direction, the AP is the only node that may be contending for medium access. This is not true in general. Under typical operating conditions, multiple nodes may well be attempting to access the network at any given moment.

The issue of slower nodes consuming a disproportionate share of airtime was a point of active discussion within the IEEE's Task Group g as the standard was being developed. If every node has a statistically equal chance of accessing the medium, slower nodes will dominate traffic flow. As we have seen, this is precisely what occurs when traffic is exclusively downstream. Since the AP is the only node contending for medium access in this situation, every node having queued traffic will receive an equal number of packets.

Under more typical conditions, traffic flow in a WLAN is bi-directional and, in the case of upstream traffic flow, multiple nodes are contending for medium access at the same time. The CSMA/CA medium access mechanism is based on the fact that every node monitors the network prior to transmitting. Once the medium becomes idle, each node waits a specified time (50 microseconds for 802.11g mixed-mode networks), and then begins decrementing an internal timer, referred to as the back-off counter.

If another node begins transmitting before the back-off counter reaches zero, the decremented value is retained. When the medium becomes idle again and the client has waited 50 ?s, the back-off counter can begin counting down again. Once the back-off counter reaches zero, the client can begin transmission.

The back-off counter is initialized by selecting a random number of slot times from within a predetermined window. For 802.11g mixed-mode operation, a slot time is 20 microseconds in duration. Legacy 802.11b devices normally initialize the back-off counter by selecting a random variable from 0 to 31 corresponding to the number of slot times.

To combat the problem of slower nodes dominating airtime, faster 802.11g devices are given a statistical edge to help them access the network more frequently. For 802.11g devices, the normal selection window for the number of slot times is 0 to 15 slot times.

With this in mind, let's re-examine 802.11g mixed-mode operation with two computing devices in the BSS. This time, however, the direction of data flow is reversed. With data flowing upstream, both the legacy 802.11b device and the higher-speed 802.11g client will end up contending for the medium. However, because the back-off counter is statistically set to a lower number in the 802.11g computing device, it will get twice as many transmit opportunities (TXOPs) on average as the slower, legacy 802.11b device.

Referring to Fig. 10.12, it can be seen that with approximately a 2:1 advantage in terms of TXOPs, the network does a better overall job of balancing air time between 802.11g and legacy 802.11b computing devices. As a result, 802.11g mixed-mode throughput in this situation is 12.4 Mbps upstream, as compared with only 11.2 Mbps downstream (excluding TCP/IP overhead).

In mixed-mode operation, it is absolutely essential that a CTS packet transmitted using legacy 802.11b modulation precede each high-speed OFDM transmission. In the absence of the CTS packet, legacy 802.11b devices will transmit during an OFDM transmission, thereby colliding with 802.11g traffic. Use of the CTS packet in this manner is referred to as a protection mechanism. Responsible vendors will provide 802.11g products that properly support the use of protection mechanisms during mixed-mode operation.

However some vendors may not *properly* implement the protection mechanisms. One result of poorly implemented mechanisms would be that all 802.11g clients fall back to legacy 802.11b operation in a mixed environment. If all nodes fall back to 802.11b modulation and data rates, total network throughput will be the same regardless of the relative number of 802.11g devices in the network.

A simple test could be conducted by noting total throughput with an 802.11g AP that is streaming downstream traffic to a single 802.11b legacy client to establish a benchmark. If 802.11g devices are falling back to 802.11b rates in mixed-mode operation, overall network throughput will remain essentially constant, even as more 802.11g clients are added (recall that overall throughput is shared by each client).

On the other hand, if OFDM packets are transmitted without the use of protection mechanisms in mixed mode, the 802.11b devices will frequently begin transmission during 802.11g packet transmissions. This will almost always result in the loss of both packets. Recall that the 802.11g computing device still defers to the 802.11b transmission.

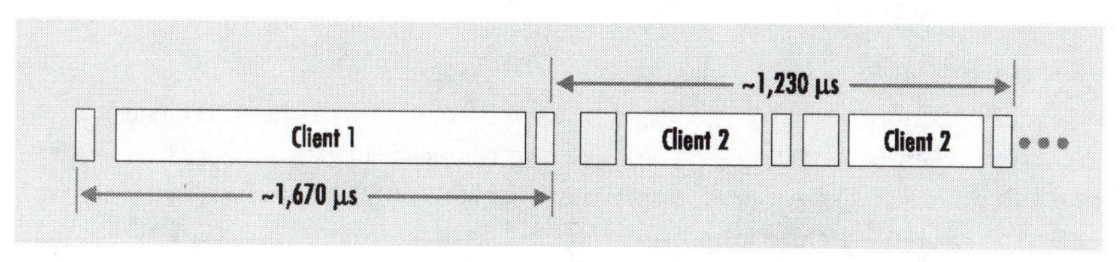

Figure 10.12 For mixed-mode upstream traffic, the 802.11g client will get twice as many transmit opportunities because its back-off counter is statistically set to a lower number. This doubles the upstream data rate.

Another test can determine whether the vendor implemented the protection mechanisms. The test involves upstream transmission of data with several legacy 802.11b devices and an equal number of 802.11g devices on the same AP. If protection mechanisms are not in use and the 802.11g devices continue to transmit at higher rates using the OFDM waveform, throughput for the 802.11b clients will actually be higher than for the 802.11g devices, in spite of the fact that the legacy equipment operates at much lower data rates. This is a strong indication that the product under test does not support protection mechanisms and will perform very poorly in a mixed environment.

PLANNING FOR A HIGH-SPEED DUAL-BAND FUTURE

While 802.11g and 11a both provide a 54-Mbps data rate, these two technologies will very likely be used in entirely different networking scenarios. Here's why.

IEEE 802.11g provides backward compatibility with 802.11b equipment, thus it can preserve an organization's investment in its existing 11b WLAN infrastructure. Since 11g builds on 11b technology, 11g is in a position to take advantage of the years of 802.11b silicon integration and the resulting reduction in power consumption, form factor size, and cost. But 802.11g is limited to the same three channels as 802.11b, thus scalability may become a factor if a WLAN's user density increases.

IEEE 802.11a, on the other hand, is not compatible with 802.11b devices. However, because it operates in the unlicensed 5-GHz band, it provides eight channels for enhanced scalability. It also is immune to interference from devices that operate in the 2.4-GHz band. But an 11a network means that you may be going into "unchartered territory," since there are very few 11a networks in existence at this writing.

To benefit from these new high-performance standards, some organizations will opt to design their WLANs using both bands. By using dual-band equipment that supports both 11a and 11b/g technology, companies can preserve their legacy client devices well into the future. 802.11b and 802.11g are ideal technologies when the premium is placed on power consumption, small size, and cost. And 802.11a provides the added benefit of eight additional channels with, perhaps, less interference.

Waiting for 802.11g. Should your WLAN plans wait until Wi-Fi Certified 802.11g devices are available and 802.11g has some experience under its belt? The answer is maybe. There are three or maybe four reasons that some might want to wait for 802.11g rather than to go for the immediate gratification of an 802.11b or an 802.11a installation:

- Lower power consumption.
- Longer range.
- Better penetration.
- Cost advantages over 802.11a devices. Lower-frequency devices are easier to manufacture. Eventually, however, the price gap between 11a/b/g devices will narrow. Rich Redelfs, CEO of Atheros, thinks that the price of 802.11a chips will be close to 802.11b/g chips "before long."

Another choice is to go with products that support both 802.11a and 802.11g (which by definition includes 802.11b). These products should be available in the near future.

A CAMPUS WLAN DESIGN

Many times, a facility will span more than one building or structure. LAN designers working in multi-building campuses have refined their approach to deploying networks in such situations. They now regularly use structured fiber/UTP wiring systems, a combination of Ethernet switches and routers, and a range of companion devices including firewalls, VPN concentrators, and traffic shapers. And now wireless LAN technology has been added to the mix.

Wireless LANs offer many benefits, but they also make the network designer's and manager's jobs significantly more complex. With conventional wired networks, the design process requires an understanding of how the Physical and Data Link Layers operate in a hybrid switched-routed environment. The designer transfers that understanding to the design of a Physical Layer infrastructure of UTP (unshielded twisted pair) and fiber. The design becomes a bit more complicated, though, when the medium turns invisible and unpredictable. Suddenly, amateur ham radio operators are getting more respect—they understand how radio works. Without that knowledge, it's tough to effectively design a campus WLAN.

If your task is to design a campus WLAN and you don't know much about radio frequency (RF), you need to get up to speed, quickly. Only with a sound understanding of RF can you design the "structured cabling system" of WLANs—an invisible collection of wireless ethers, over which 802.11b packets pass.

To put it more succinctly, when designing a campus WLAN, the designer's challenge is creating a stable cellular communication system with good data throughput campus-wide. This demanding task requires an understanding of how 802.11 radios work, the differences between vendor implementations, and the effect of varying building structure elements and sources of external interference.

You'll also need to think about core network services—including IP address management, authentication, encryption, access control, accounting, and maybe even quality of service that must be delivered to the wireless network's end-users.

Radios

Radio has been around for more than a century. A fluctuating current in a wire is transformed into radio waves and transmitted through the air, where it is received by other radios. With a WLAN, every device is a transceiver, capable of both transmitting and receiving radio signals. By employing any one of a variety of radio modulation schemes—essentially, playing around with the shape of the individual 2.4-GHz sine waves—radio waves can transmit digital information.

Unfortunately, predicting the behavior of a specific WLAN system in a specific environment is challenging. Using identical components, the effective system range may be well over 100 meters (328 feet) in one location and less than 50 meters (164 feet) in another.

A number of variables can affect transmission range, including building layout, construction materials, a warehouse's contents, and noise sources. Experienced WLAN designers can walk into a building, give it a once-over inspection, and make educated guesses about how the system should be designed. For the rest of us, it's trial and error. Fortunately, site survey tools, which are available from most enterprise-oriented ven-

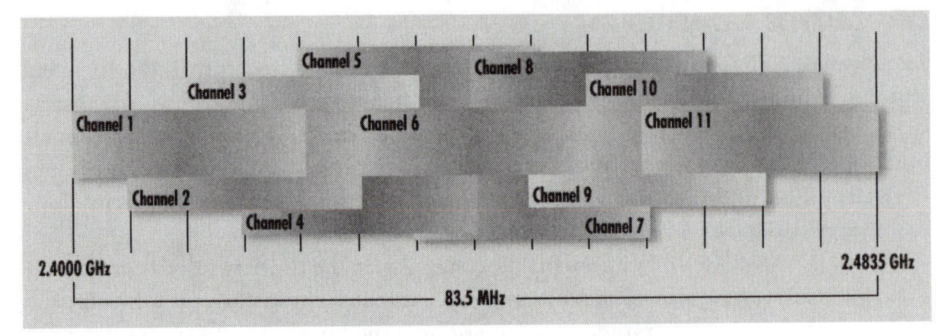

Figure 10.13 Approximate spectral placement of 802.11b channels.

dors, have improved significantly over the past several years. They are a reliable resource when it comes to customizing a WLAN system.

Range Limitations—Friend or Foe

Some people consider the range limitations of radio to be a big problem, but in fact it is the main ally of a wireless system designer. That's because range limitations let you reuse frequencies. As shown in Fig 10.14, designers usually work with channels 1, 6 and 11—three non-overlapping channels—to maximize bandwidth. In other words, a designer could theoretically install three access points (APs) in a room, each transmitting and receiving within a distinct range of frequencies, and with no interference to one another.

While in some rare circumstances, there might be a reason to install three APs in a single room to take advantage of the greater aggregate bandwidth, in most cases there's a different challenge. Assume a building requires 21 APs to deliver service to all users, and seven APs are installed on each of the three non-overlapping channels (1, 6 and 11). The designer needs to ensure not only that cells overlap (to avoid dead spots), but also that an AP on Channel 6 isn't interfering with another access point in the building that's also operating on channel 6. Fig. 10.14 shows a sample cell layout that ensures full coverage while avoiding interference.

Of course, providing full coverage while avoiding interference is much easier to do on paper than in real life. In the real world, the designer needs to think in three dimensions and factor in the possibility that a cell on the first floor could interfere with a cell on

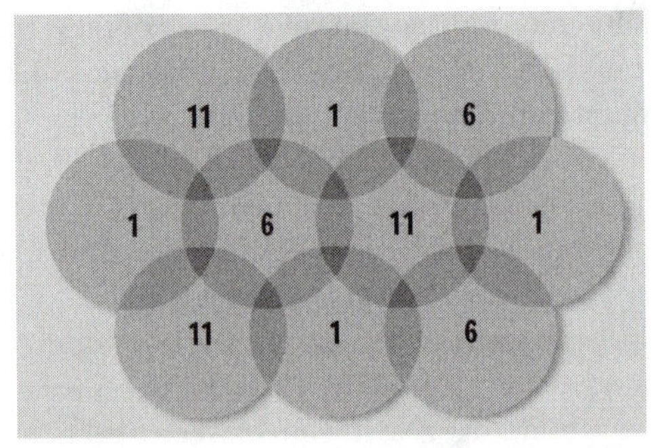

Figure 10.14 This figure shows an 802.11b overlapping cell arrangement using non-overlapping radio channels 1, 6 and 11.

the second floor. This limitation in the number of available channels at 2.4 GHz is one of the primary appeals of 802.11a, which offers eight non-overlapping channels at 5 GHz, though cell diameters usually are smaller.

Designing a Cell Plan

Laying out individual coverage cells takes time. Start with building plans, estimate coverage based on raw distances and workspace configuration, or the mobile employee's pattern of movement in their daily shift. You might, for example, work with 100 foot (30.48 meters) radii, each requiring an AP, and sketch out some locations.

Note that some APs and network interface cards (NICs) can be configured to reduce the output power of the radio, effectively shrinking the RF cell radius and reducing user contention in high-density environments. However, because you can't control the output power on all 802.11b products, a microcell design can get tricky. Unless you are in a position to strictly enforce the wireless devices used on the network, a single rogue device could wreak havoc. Fig. 10.15 shows a microcell design with all APs transmitting at 10 milliwatts. A single client device operating at 100 milliwatts can effectively interfere with multiple cells.

Now it's time to head into the field, equipped with appropriate tools. Most WLAN vendors offer site survey utilities that let you temporarily install APs and to measure signal levels at various locations. During this phase you are focusing exclusively on the RF design, so you do not need an active Ethernet connection to the AP but you should still use this phase to consider the feasibility of running Ethernet to the various possible AP locations, since it will be an eventual requirement. This is also a good time to consider selecting a product that supports "power over Ethernet" (PoE) to avoid the necessity of providing 110-volt power outlets for each AP.

***Note:** Because it's awkward to operate a laptop while moving around, consider using a handheld device, like a PDA when performing the site survey. This type of device makes a great survey tool. The author makes this suggestion even though she does realize that the power and flexibility of the underlying site survey applications for use with PDA operating systems, like Palm and Windows CE, may not be particularly mature. Some professional installers also carry gel-cell DC batteries and DC-to-AC power inverters with them so they can position access points in virtually any location, even if an AC outlet is not nearby.*

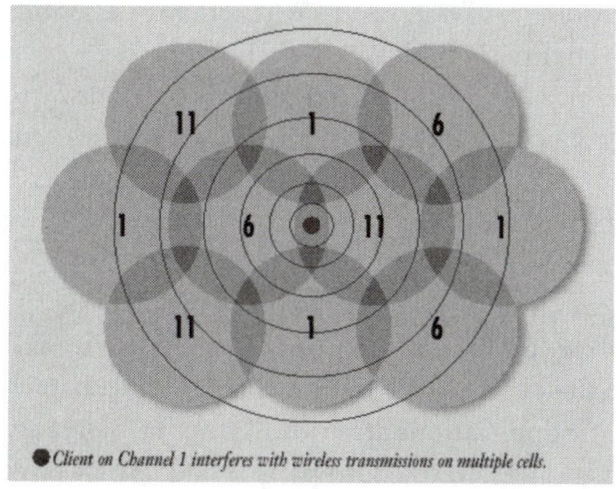

Figured 10.15 Microcell design with interface from a single high-power client.

Antennae

Another important variable to consider is the type of antenna. Antennae usually provide signal gain by radiating RF signals in a predictable beam pattern. For example, the antennae shipped on most APs are omni-directional, which means that the antenna will transmit a 360 degree beam width roughly in the shape of a doughnut, where the antenna pokes up through the hole in the doughnut. Thus, if you install an AP with an omni-directional antenna in the corner of a building, it will radiate along adjacent hallways as well as out to the parking lot. Also the alignment (polarization) of an omni-directional antenna can affect its transmission pattern, i.e. think about turning the doughnut on end.

Some vendors, including Cisco and Symbol Technologies, offer a variety of antennae for use with their APs. Many of these antennae can provide additional gain—thereby increasing range—by altering the direction and beam width of the radio signal. Patch antennae, for example, can radiate signals using an 80-degree beam width instead of the 360-degree beam width of an omni. Other antennae, like ceiling mounts, are not designed to provide additional gain but rather to blend into the physical environment, with the AP typically concealed above the ceiling.

In designing a campus WLAN, be aware it may not be legal to purchase APs from one company and configure them with third-party antennae. This is because when vendors submit their products for FCC certification, they include an antenna, and it is the combination antenna-AP or antenna-NIC the FCC certifies. So consider purchasing APs from a vendor that provides multiple antenna options.

Beyond RF

Some might argue that the site survey, though technically complex, is the easy part of designing a WLAN. The tougher challenges are assessing and meeting bandwidth requirements, ensuring security, and implementing an appropriate management infrastructure—the same issues LAN designers have wrestled with for years. Address each of these challenges in stages.

First, determine how much bandwidth is needed throughout the physical environment. Pay particular attention to the density of users and typical per-user bandwidth requirements. For example, in a warehouse where only a few users share a vast space, you want to have as large a cell size as possible. Think high-gain antennae. On the other hand, in conference rooms and classrooms where many users must contend for access using the same radio channel. Think smaller cell sizes.

Unfortunately, the number of concurrent users is only one factor driving bandwidth requirements. The other is the bandwidth intensity and relative "burstiness" of the applications. That's not only difficult to estimate at the outset, it's difficult to project. Although if you perform a user-needs analysis as set out in Chapter 14, it will be a bit easier to anticipate these requirements.

In most environments, a single 802.11b channel, which typically provides effective aggregate throughput of about 6 Mbps, can support 30 to 50 users, maybe more. But since in essence, we're back to the old days of shared-media Ethernet, bandwidth monitoring will be important. If specific applications are critical, you may select an AP that allows some level of traffic prioritization. However, third-party products may provide more flexible traffic shaping (though they can add significantly to the cost of the implementation).

Putting It Together

Once you've studied the RF characteristics of the campus, evaluated bandwidth requirements, and laid out your AP-cell design, the next step is to figure out how to integrate the WLAN with the existing wired network. This encompasses both technical and policy dimensions.

On the technical side, you need to develop a security plan and figure out how to tie the access points to the LAN's switching infrastructure, factoring in the management of IP addresses and application roaming requirements. The security strategy should consider authentication, privacy, access control, and accounting. Some WLANs are wide open; others need to meet high security standards. Most of the major vendors, including Cisco, Agere, Proxim and Symbol, offer their own security frameworks that, while based on open standards, may lock you into that specific vendor's APs and NICs. Consider third-party management and security products from vendors like Bluesocket, Columbitech, Ecutel, Funk, NetMotion, NetSeal, ReefEdge, Vernier and others to avoid vendor lock-in. Finally, many organizations use standards-based VPN gateways and VPN clients on all mobile devices to provide a security overlay on their WLANs. There is much more information concerning WLAN security options in Chapter 17.

How you tie APs into your existing network infrastructure will depend on its architecture and the capabilities of the existing Ethernet equipment. For example, with lots of bandwidth and fairly advanced Ethernet switches, you can establish wireless VLANs—maybe even a single wireless VLAN—to manage addresses more easily and to enforce security policies. The wireless VLAN can then be separated logically from the campus wired LAN, and policies can be developed that determine who can cross that boundary. (See the discussion on VLANs in Chapter 8.)

The downside to the campus-wide wireless VLAN design is the same as any flat network: performance may degrade as a result of excessive broadcast traffic. On the positive side, it addresses one of the most challenging aspects of campus WLAN design, how to facilitate roaming users.

With a flat network, users maintain a single IP address. However, when WLANs are associated with IP subnets, roaming is more challenging. If the WLAN's primary need is to provide *portability* (not mobility), i.e. enabling end-users to move between subnets, it might be reasonable for them to simply restart their machines (or renew their DHCP leases) and get valid IP addresses from their new location. However, if *mobility* (i.e. walking while remaining connected) is a key requirement for your employees, think about deploying a system that facilitates seamless roaming. NetMotion, for example, serves as a proxy server for all WLAN traffic, thus facilitating roaming. Other solutions include using Mobile IP or customized VPN capabilities to accomplish similar goals. For a detailed discussion on provisioning seamless roaming, see Chapter 7.

CONCLUSION

This chapter only examines the different characteristics of 11a, 11b and 11g. The particular networking environment issues relevant to making a decision as to the make-up of specific wireless network architectures are discussed in subsequent chapters.

Don't forget post-deployment issues. Many organizations excel in designing their WLANs but give little thought to ongoing maintenance and troubleshooting. If your goal is to provide four-nines reliability, you'll probably need to invest in additional hardware and software. Read Chapter 8's section entitled "Four-Nines Reliability" for more information on management, maintenance, and troubleshooting wireless networking systems.

Section IV: Building a Wi-Fi Business

Chapter 11: The WISP Industry

The Wireless Internet Service Provider (WISP) industry envisions a world where people routinely use high-speed wireless networks to access email, corporate networks, the Web, and more—whether they are traveling or going about their daily business in their local neighborhood. For the purpose of this book, the WISP industry encompasses all businesses (for-profit and non-profit) that operate under new, although varied, business models with a common theme: providing publicly available Internet access via Wi-Fi technology.

At the moment, the industry is struggling in its bid to find the right revenue model. As the reader will grow to understand, it will take time for the industry to determine the right revenue models for each segment of the industry. So while there is money to be made with Wi-Fi, there are still issues that need to be resolved. Until then, all within the WISP industry will run into difficulties when it comes to producing a consistent, positive revenue flow.

What exactly is a WISP? The term "wireless Internet service provider" or "WISP" is commonly used to refer to any entity that provides publicly available Internet access via a wireless local area network (WLAN), Wi-Fi or not. However, the term is too all encompassing in that it refers to all service providers in all segments of the WISP industry: individual HotSpots (they provide wireless access to the Internet only from a specific location), the HotSpot operators (these are the enablers—companies that provide the network, hardware, technical expertise, and back-end management for the individual HotSpots), and the aggregators (a group of virtual network wholesalers that help to tie the individual HotSpots and enabling networks into a larger, cohesive roaming network). Added to this mix are other business models including "brands," infrastructure providers, ISPs (Internet Service Providers), application service providers, and the telecommunication carriers (wired and wireless).

In an attempt to bring some kind of clarity, the author (and others) has divided the industry into layers of similar business models:

Venue Layer. This layer encompasses all of the individual entities that operate HotSpots out of a physical location. A venue operator has in place at least one access point and a high-speed Internet connection, which is made available to the public. The access point is used to provide connectivity between the venue customers' computing devices and the Internet. The access point also enables the end-user (the customer) the freedom to move about at will, as long as he or she stays within the access point's radius. This freedom of movement is referred to as "roaming," albeit, very limited roaming. Venues include business models such as hotels, cafés, airports, train stations, and convention centers.

HotSpot Operator Layer. This layer consists of the HotSpot providers. Companies in this layer maintain a network that venue owners can use when they contract to offer a specific HotSpot operator's service. In exchange, the HotSpot operator provides the necessary gear, support and network to provide Internet access to the venue owner's customers. Cafe.com, Wayport, and Surf and Sip, are just a few companies that reside in this layer.

Aggregator Layer. This layer is vital to the growth of HotSpots, because this is the layer where it is most practical to enable roaming between different HotSpots, and to provide the necessary back-office services, e.g. authentication, accounting, settlement and so forth. Because of the inherent fragmentation within the HotSpot marketplace, venue owners and HotSpot operators are beginning to see the benefit of partnering with aggregation layer companies such as Boingo, GRIC, and iPass.

Brand Layer. Companies within this layer typically partner with companies in the aggregator layer to offer Wi-Fi access to their already established customer base. Examples include T-Mobile, Sprint PCS, and AT&T Wireless.

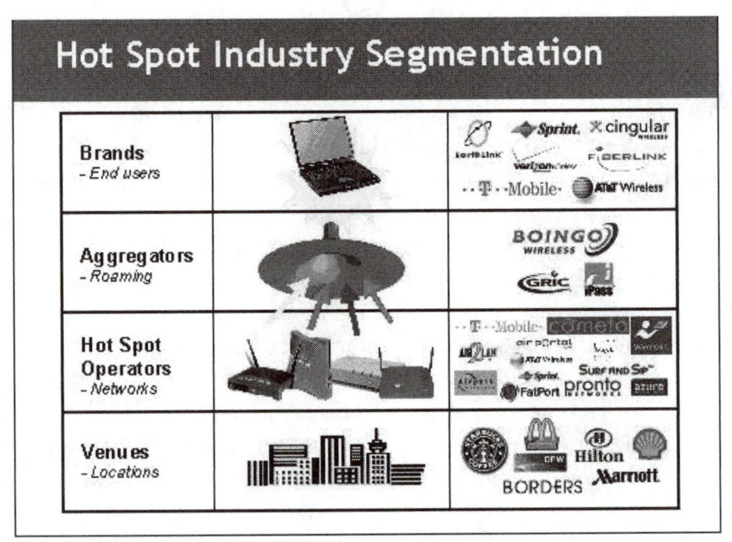

Figure 11.1 This graphic depicts the four layers within the WISP industry. As in the ISP space, the most successful companies will focus primarily on one industry layer, partner between the layers, and compete within their own layer. *Graphic courtesy of Boingo Wireless Inc.*

Other participants—hardware, software, application, and connectivity providers—work within and between the layers. There's also the telecommunication carrier group, which has yet to define its level of participation within the WISP industry.

Let's now examine each of these layers more closely.

THE VENUE LAYER—OPERATING A HOTSPOT

The main selling point for installing a HotSpot in a venue is a very low barrier to entry for the small venue owner: the necessary equipment costs only a few hundred dollars and the connection to the Internet can be as low as $50 per month for DSL or Cable Modem service. Of course, larger HotSpot operations require additional equipment, antennae, and faster connections (perhaps a T-1 line), but even for a large HotSpot, the cost is still orders of magnitude lower than what it would cost to set up a node to access a cellular network. (Wi-Fi uses unlicensed spectrum versus cellular's very costly licensed spectrum; the equipment necessary to set up a HotSpot costs hundreds of dollars versus hundreds of thousands or millions of dollars for cellular; and a Wi-Fi connection, even when using a low-end DSL connection, is significantly faster than the highly touted "NextGen" services provided by cellular carriers, which typically deliver throughput of between 40 and 60 Kbps). The actual speed experienced by HotSpot users is determined by the HotSpot's connection to the Internet, which can range from a low-end DSL (384 Kbps) connection to one or more T-1 (1.5 Mbps) lines; both offer speeds that are much faster than the data transfer rate of any other affordable technology.

Individual property owners and lessees offer wireless Internet access either by vertically integrating and operating their own HotSpots (the build-it-yourself model), or by licensing the right to deploy HotSpots within their venue or on their property to a HotSpot operator. In the latter model, the HotSpot operator usually will provide all of the necessary hardware, software, and back-office services (and perhaps even the backhaul connection). Some HotSpot operators will even send a crew to the venue to hook everything up.

***Note:** In this book the term "backhaul" typically refers to the high-speed link to the Internet. However, the term can, at times, refer to the physical location in each Local Access Transport Area (LATA) from which an interexchange carrier (e.g. AT&T) provides services to the local exchange carrier, and possibly directly to end-users. This Point of Presence (POP) enables a carrier to price a service or calculate mileage, but it may not actually contain the equipment providing the service. The carrier uses internal private lines to carry the information to the office containing the equipment. Also referred to as a "virtual POP."*

The average small venue needs only a simple network design, since most of the wireless network's functionality lies within the access point (AP), which is the basic building block of any wireless local area network (WLAN) infrastructure. Normally, these small networks are designed so that the AP is integrated with a broadband router in order to allow wirelessly networked computers to share a high-speed Internet connection. A simple network requires: an access point and router (or an all-in-one product such as Linksys EtherFast Wireless AP + Cable/DSL Router w/4-Port Switch); a high-speed connection to the Internet (DSL, Cable Modem); a computer with an operating Ethernet adapter, a wireless network adapter card and a CD-ROM drive, running

Microsoft Windows 98 or better and with at least Internet Explorer 5.0 or Netscape Navigator 5.0 or better installed; and finally, a network cable (for initial setup).

Note: *To clarify an access point's functions a bit more, an AP is somewhat analogous to an Ethernet hub or switch in that APs allow computing devices with wireless adapters to participate in a local wireless network (LAN). All device-to-device wireless communications go through the AP. However, an important distinction between an ordinary Ethernet hub or switch, and an AP, is that the AP can also connect a WLAN to a wired LAN. This allows both the wired and wireless LANs to communicate with each other.*

Although simple, this network design provides:

* Direct communication between the AP and any customer's computing device.
* The AP controls access to the venue's internal network and the Internet.
* Roaming within the radius of the AP.

Of course, this setup precludes back-office functionality that would enable the venue owner to bill its customers for network use. (More detailed information on building a HotSpot is provided later in this chapter, but the reader should also read Section III: Practical Deployments.)

If you build it yourself, you must pay for 100% of the cost of deploying and managing the HotSpot infrastructure, but you also receive 100% of the revenue (if you charge for the service).

If a venue owner/lessee partners with a HotSpot operator, expect that operator to offer a variety of pricing options that can be passed on to the customer. The typical pricing plan includes all-you-can-use monthly subscriptions and pay-per-use plans, such as by-the-minute or unlimited use in a 24-hour period. However, HotSpot access pricing plans are as varied as cell phone plans. For instance, rather than following the typical

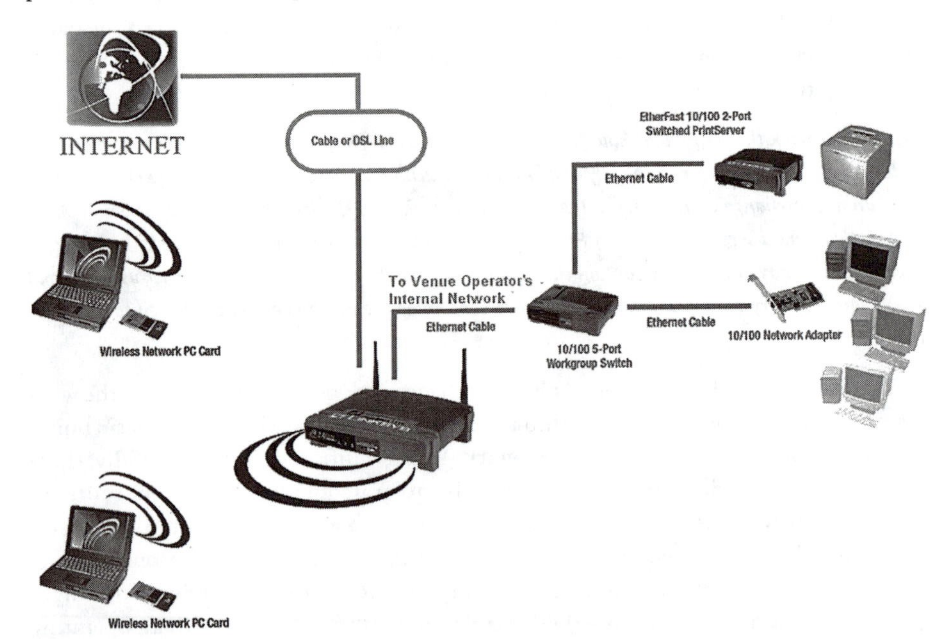

Figure 11.2 An example of a typical do-it-yourself HotSpot.

pricing plan, some HotSpot operators allow the venue to customize their pricing, letting them charge their customers whatever they like.

Here are some items to keep in mind when investigating the feasibility of partnering with a HotSpot operator: when a venue owner lets a HotSpot operator provide some or all of the hardware, software, and connectivity needed for an operational HotSpot, that venue owner must give the lion's share of the revenue it receives from the HotSpot operation to the provisioning HotSpot operator. However, the more a venue owner contributes to the up-front costs (e.g. the venue operator provides the pipe to the Internet), the higher its revenue participation. While relying on a HotSpot operator may sound like a poor business model, many venue owners are happy to let someone else handle the technology and back-office chores, since they are more interested in providing a HotSpot as an amenity and convenience to their customers than in earning direct revenue.

After choosing your method for implementing the HotSpot service, your next deci-

FOOD FOR THOUGHT

Consider these questions before deciding to host a HotSpot.

- Why are you becoming a HotSpot venue—to attract new customers, to keep up with competition, to offer additional amenity, or to add a new revenue stream?
- Will your customer base use the service?
- Will a HotSpot detract from your core business?
- Will uncertain quality of the HotSpot service and user experience negatively impact your business?

HAMPTON INN

Here is a good example to demonstrate many of the points made about venue layer HotSpots. Hampton Inn in Auburn Hills, Michigan, spent $33,000 to install Wi-Fi so it could offer free wireless Internet access to guests in all of its 124 rooms. The setup included a number of access points, which were connected via cable to the hotel's wired local area network. According to general manager Tom Keller, the response has been tremendous, with an average of between 20 and 40 guests logging on to its Wi-Fi network each day. Keller even provides wireless PC cards and a CD with the necessary drivers for guests who've never used Wi-Fi. "If I can get them to stay one extra night, it pays for itself," says Keller, who sees Wi-Fi as giving him an advantage over competing hotels. He and others in his industry feel that it won't be long before hotel guests expect free Wi-Fi, just as they expect hair dryers, irons, and TV. Hampton Inn, however, could have saved all or at least the majority of the $33,000 spent on the network build-out, if it had partnered with someone like Pronto Networks. Pronto provides its own wireless infrastructure solution for the larger venue owner via its Hotspot Networking System, which consists of a Hotspot Controller and a Hotspot Provisioning & Operations Support System. Of course, Pronto would extract a price from Hampton Inn for providing the gear and services.

sion is what kind of business model to employ. Many venues, especially those in the hospitality industry, offer free public Wi-Fi access. Others charge the user a fee to access their wireless network.

Costs and profits aren't the only concerns. Another concern is "who will own the end-user?" Some venues, such as airports or convention centers aren't that concerned over who owns the end-user relationship, although they might want to have their own brand as the first entry page the end-user sees when he or she launches their browser. Conversely, an airline or a hotel may be very concerned about who owns that relationship.

As the industry matures, revenue sharing plans are increasingly being negotiated. While a small hotel may be offered a small slice of the pie, a chain of hotels can negotiate for a larger slice, and if the property also regularly hosts conventions, the slice becomes even bigger. Moreover, high value premises such as airports and convention centers are gradually asserting their right not to grant exclusive Wi-Fi connectivity to only one HotSpot operator. Instead, they endeavor to accommodate a number of HotSpot operators. This leads to price discussions that can result in a different fee arrangement, e.g. a fee based on usage of the network in the premises. (An airport might, for example, want to benefit from growing variable revenues instead of a fixed fee.)

Whatever method you use to provide your customers with wireless Internet access, once you put up the signage to inform your customers that the venue is a HotSpot, you are in effect putting up a "Welcome Friend" sign for all wireless devices and their owners.

HOTSPOT OPERATOR LAYER

Numerous new businesses provide the technology, know-how, and back-office operations necessary for operating a HotSpot in a venue location. In this book, these businesses are referred to as "HotSpot operators," although they are also known as "resellers," "WISPs," or "microcarriers." Included in this layer are a multitude of small players, such as Deep Blue, Ikano, Surf and Sip, and Cafe.com, and a few large ones, e.g. T-Mobile (which has the largest footprint) and Wayport (with around 450 locations). This is the layer that holds the future of the WISP industry, since the marketing strategy of these companies is what's driving the current explosion in HotSpot availability.

Outside the U.S., the HotSpot phenomenon is just as extraordinary. Two major telcos in South Korea have installed around 10,000 HotSpots, and plan to ultimately deploy 25,000 in their efforts to saturate the country with Wi-Fi signals. In Australia, two telcos—Telstra and Optus—are rolling out HotSpots in retail outlets and other locations. In the U.K., British Telecom (BT) and several independent HotSpot operators are ramping up the HotSpot market. For instance, BT customers can buy DSL service together with a wireless modem and a subscription to BT's "Openzone" HotSpots. The Swedish cellular carrier, Telia, has launched its own HotSpot division, which it calls "HomeRun." HomeRun has already installed a few hundred HotSpots in locations throughout the Nordic peninsula. Telia recently entered into a roaming agreement with both BT and Finnish cellular operator Sonera, allowing either network's customers to roam freely into the other.

France's top three telecommunications providers have also joined the Wi-Fi hit parade. All three are aggressively deploying HotSpot networks. In addition, as of September 2003, France Telecom SA (through its mobile subsidiary Orange France), SFR, and Bouygues Telecom will allow a subscriber of one network to access a HotSpot owned and operated by either of the two other carrier's HotSpot network–a move touted by analysts as Europe's first high-profile domestic roaming agreement.

The start-up costs of a HotSpot operator business model are relatively low. Of course, there are the normal accouterments of any new business (office, utilities, etc.), but other than that, the software and gear to outfit a venue for wireless Internet access require relatively little overhead. Once you get past those needs, what follows is the execution of the business plan as it has been laid out.

The physical characteristics of Wi-Fi technology enable a HotSpot operator to start on a very small scale–covering one café, or one small apartment house–then enlarging its footprint from there (e.g. adding a neighborhood restaurant, gas station and laundromat, etc.). That is exactly how the southern California HotSpot operator, Cafe.com, got its start. The founders of Cafe.com, however, didn't just decide one day to become a HotSpot operator. They researched the market for this type of service in their geographical market, and learned all they could about the content and marketing side of the HotSpot operator business model. Once they had the confidence that their venture could succeed, they researched what gear and software they should use when setting up a HotSpot venue. With this information in hand, they decided to install HotSpots with access controllers from Tri-M, access points from GemTek, and NetNearU software for system monitoring, performing user authentication, and processing online payment.

Up to this point, the founders had used their own money to fund the venture, but the advance work outlined above positioned them to approach angel investors from whom they eventually raised $150,000. With everything in place, the Cafe.com network was born, one venue at a time.

Cafe.com's first venue client, the Novel Café in Santa Monica, initially operated in a controlled environment. The founders, Ronan Higgins and Chris Van Vleit, launched the HotSpot in "stealth mode," i.e. no fanfare, no press releases. They wanted to test the network under real network conditions. Even without any public announcement, the Novel Café HotSpot saw slow but steady growth in network usage. It wasn't long before both Higgins and Van Vleit were convinced that they had the right mix–technology,

WHAT IS AN ACCESS CONTROLLER?

An access controller is a hardware device that resides on the wired portion of the network between the access point(s) and the protected side of the network. (See Fig. 11.3.) This device provides the centralized intelligence behind the access point(s). It regulates traffic between the relatively open WLAN and the typically guarded internal network resources. Although access controllers can be used in a wide range of wireless LAN applications (e.g. a corporation might use an access controller in order to prevent an unauthorized user from getting entry to sensitive data and applications contained in its internal network), in the case of a HotSpot, an access controller regulates end-user access to the Internet by authenticating and authorizing those users based on a subscription plan. Many HotSpot-specific access controller products also provide access point capabilities, eliminating the need for a separate access point device, e.g. the Colubris Network CN3000 Wireless Access Controller.

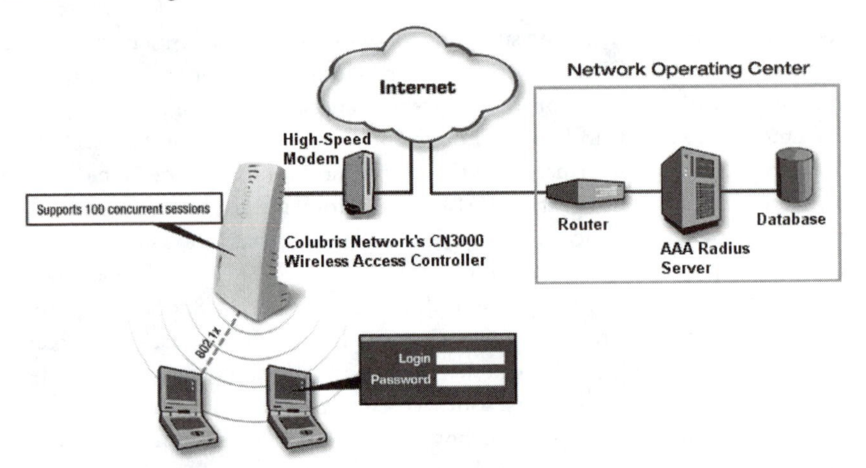

Figure 11.3 Example of a HotSpot deployed by a HotSpot operator using a wireless access controller. In this instance it is Colubris Networks' CN3000 Wireless Access Controller, which is designed for small to medium HotSpots, such as an Internet café or a small hotel. The functionality included in the CN3000 provides, among other things, access point capabilities, a full router, a customizable firewall, a RADIUS AAA (Authentication, Authorization, and Accounting) client, customizable login pages, per session per user access lists, an embedded VPN client for secure remote manageability, directed traffic to ensure the integrity of the HotSpot operator's back-office Network Operating Center (NOC), and more.

skillset and marketing—to expand their network. The HotSpot operator gradually added more cafés, and as of this writing, Cafe.com has HotSpots in nearly 20 venues and, according to founder Ronan Higgins, hopes to add at least ten more before the end of 2003.

Currently, the Cafe.com HotSpot provisioning model supplies the venue operator with all of the necessary hardware and software, which now consists of a Handlink wireless hotspot access controller, Belkin access point(s), cabling, and surge protection. If necessary, Cafe.com also will install the high-speed Internet service, which is provided by either DSL Extreme (Los Angeles) or Verizon. If necessary, Cafe.com personnel will go to the venue, install the access controller and access point, and do whatever is required to get the HotSpot up and running. Cafe.com bears all of the up-front setup costs. (The network layout would be similar to that depicted in Fig. 11.4.)

After the node is operational, Cafe.com provides all of the support needed to be a HotSpot, including taking care of customer authentication and billing, sending informational emails to customers, and providing the necessary marketing materials to help grow usage. When an end-user logs onto Cafe.com's network, the system knows the location used and the venue that signed up or "owns" that customer. This information is then used to send the venue owners their commission checks and usage statements. The idea is to make it simple for the venue owner, akin to how telephone companies provide service and share revenue for payphones that they locate in various venues.

To increase its network value, Cafe.com has entered into partnership with larger, wireless access providers. Its partnership with NetNearU provides a network operating center with a NetNearU management system that supplies Cafe.com with an established

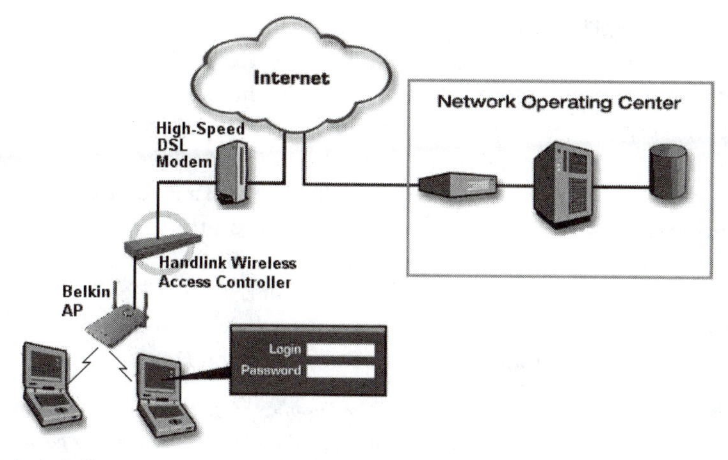

Figure 11.4 A Cafe.com venue owner's HotSpot network would look something like this.

remote management, credit card processing, and billing system, and also gives Cafe.com's network users access to the international GRIC network. Cafe.com also has entered into a two-pronged roaming partnership deal with Boingo Wireless, which provides (1) for Cafe.com to integrate its southern California-based Hotspots into the Boingo network, and (2) for Cafe.com to serve as a Boingo HotSpot reseller by transitioning additional HotSpots into the Boingo network using Boingo's installation solution.

Roaming partnerships, such as the ones into which Cafe.com has entered, are a win-win situation for everyone involved. The venue owner benefits through the HotSpot operator and the aggregator's marketing efforts. The HotSpot operator benefits not only by growth in its customer base, but also from a larger footprint for its network (e.g. Boingo Wireless's nationwide network and GRIC Communication's international network). For its part, the aggregator gains an infusion of new customers, and a larger footprint through the incorporation of Cafe.com's nodes into its network.

To counter the reluctance of some venue owners to contract for HotSpot placement, Ronan Higgins sometimes enters into a verbal agreement with the venue owner in order to develop a relationship and to prove the HotSpot's value to the venue. Once the HotSpot is up and running, the venue owner is usually willing to sign a formal HotSpot provision agreement. However, just because Higgins is willing to enter into temporary verbal agreements doesn't mean that he underestimates the importance of a fully executed location agreement. He knows that outside investors want to see signatures on paper binding the venue to the operator before they will even consider investing. He is also aware that success breeds competition—a successful venue will show up on the radar of other HotSpot operators who will be only too happy to steal a money-making venue from an incumbent HotSpot operator.

The HotSpot operators' market is still wide open and the individual HotSpot operator faces little competition. Currently, only a few thousand commercial HotSpots are in operation worldwide. The latest figures from the well-respected research firm Gartner Inc. estimates that by the end of 2003, there will be more than 75,000 HotSpots worldwide, up from a mere 6000 in 2002. (These numbers include FreeSpots—free Wi-Fi networks put up around libraries and other public spaces by local governments or by consumers who share their high-speed Net connections with neighbors.)

Ultimately, however as Gartner's data indicates, thousands of HotSpot operators, some of them very large, will flood the marketplace. And while this type of viral growth is healthy, it also causes fragmentation within the HotSpot layer. That's because anyone with a few hundred dollars to spend for a "HotSpot in a Box," access to a DSL line, and an "in" with the owner of a compelling HotSpot location, can become a HotSpot operator. And since even a large operator, such as T-Mobile or Cometa with their relatively unlimited bankrolls, won't be able to block out all of the small HotSpot operators, the fragmenting forces inherent in this layer's business model serve to ensure that no one single provider will ever own more than 10% or so of the total HotSpot footprint.

***Note:** A "HotSpot footprint" refers to the range or radius of coverage for a typical HotSpot, which with current technology can range up to as much as 500 feet/164 meters for each access point installed. As more access points are installed, the footprint enlarges.*

HIGH-SPEED INTERNET ACCESS VIA A TWO-WAY SATELLITE NETWORK

Mainstream Data and Airpath Wireless have created a joint solution to offer carrier-class satellite Internet connectivity for HotSpot providers in need of broadband connectivity in remote or cost-prohibitive locations. Mainstream's high quality two-way VSAT network, allows Airpath customers to order low-cost broadband access anywhere in North America.

Using the integrated solution, HotSpot providers who use Airpath solutions (e.g. the Airpath WiBOSS system, which provides back office processing including authentication, accounting, subscriber management, and roaming settlement) can provide wireless high-speed Internet access at locations such as remote marinas, truck-stops, campgrounds, resorts, and hotels located in rural areas—locales where it previously was impractical to offer high-speed Internet access due to a lack of Internet infrastructure or the high cost of telephone data circuits.

The combined solution seamlessly integrates the technologies and services of Airpath and Mainstream into a scalable and extremely cost effective package. "Mainstream's satellite offering is an ideal solution for HotSpot providers looking for carrier class service, and is a very cost effective alternative for our providers in locations where terrestrial offerings are non-existent or cost prohibitive," says Todd Myers, Chairman and CEO of Airpath. "Our providers can now expand their HotSpot build-outs throughout North America and beyond the reach of traditional telco circuits."

Scott Calder, President and CEO of Mainstream comments, "We believe that Wi-Fi is a major step forward in the quest for ubiquitous connected computing and are pleased to be able to play an important role in making that happen. We are excited to be working with Airpath to expand the number of locations where always-on broadband connections are available to those equipped with Wi-Fi."

The Road to Success

A successful HotSpot operator must operate its network at a low cost *and* maintain a high volume of traffic and revenue. A HotSpot operator's monthly per venue costs are mostly fixed: depreciation of equipment, network management and maintenance, HotSpot maintenance, and possibly the cost of the HotSpot's pipe to the Internet. Then, once traffic volume overcomes these fixed costs, any additional revenue is pure margin. Thus, for profits to grow, it is essential that the operator sign as many partners as possible.

Most HotSpot operators assume that they will expand their initial footprint. After all, subscribers who have invested in a wireless LAN card and a subscription will want to connect to the operator's network in many different geographical locations.

HotSpot operators can expand their networks in a number of ways:

The most costly is to deploy access points in as many venues as possible. This means a massive investment in both time and capital, not only to build out the network but also to negotiate with venue owners, ISP's, service vendors, and manufacturers. In addition, since each HotSpot must be connected to a local ISP, the HotSpot operator must work within a complex system architecture consisting of gateways, servers, customer service and so forth. Finally, the end-user must be directed to a HotSpot by some type of marketing campaign, and be educated on the authentication process, security, and billing procedures through a combination of venue owner education, documentation, and customer support.

An easier and less costly way to expand a HotSpot operator's footprint is for the operator to enter into partnership agreements, i.e. bilateral roaming agreements with other HotSpot operators. This method eliminates some of the complexity within the WISP value chain (i.e. fewer ISPs and venue owners to deal with), although it still requires that the HotSpot operator possesses the skills to keep everything working smoothly especially in the area of authentication, authorization, and accounting (AAA).

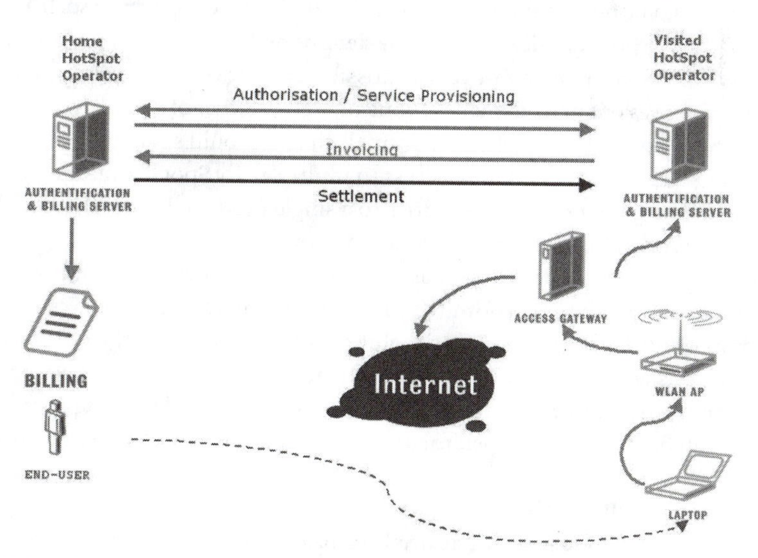

Figure 11.5 When a HotSpot operator enters into bilateral agreements, it must have the wherewithal to adequately handle all of the back-end tasks.

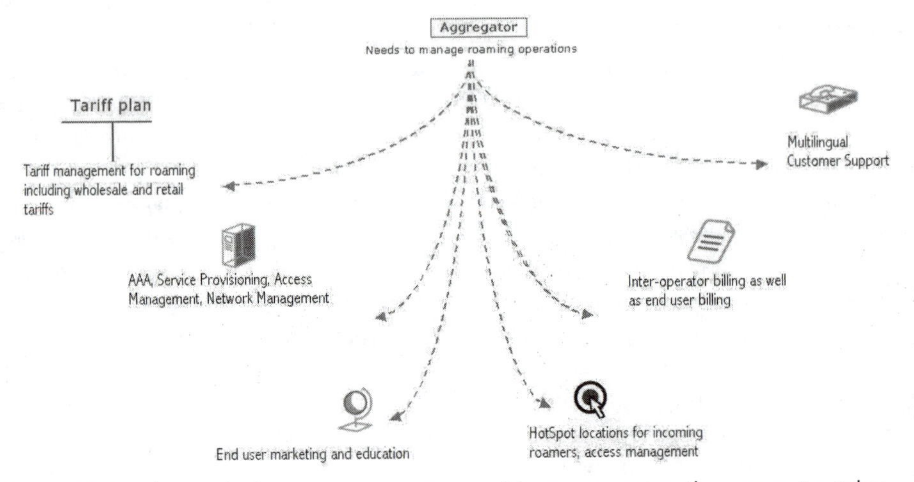

Figure 11.6 When a HotSpot operator partners with an aggregator, the aggregator takes on most of the back-office chores.

***Note:** A "value chain" refers to a group of organizations that work together to progressively add value in response to a market opportunity. The end result is that materials and/or services are transformed into something that can be purchased by the final consumer—the end-user.*

However, an even better approach may be to partner with an aggregator. Such a partnership can (1) take the headaches out of roaming agreements since the aggregator handles all of the negotiations concerning roaming–contracts, expansion of geographical coverage through the continuous addition of new roaming partners, etc., and (2), which is perhaps more important, spares the operator the back-office drudgery.

What if a HotSpot operator doesn't want to share its network, even for a fee? Well, in that case, it would be more difficult to recover the fixed costs of operating such a network. After all, HotSpot profitability is inevitably determined by traffic, which means opening the network to as many end-users as possible. Moreover, a non-sharing HotSpot operator that moves out of its domain of control (e.g. an owner of a small hotel chain selling its HotSpot operator services to other small chains) would soon find its venue owners becoming rebellious, demanding access to multiple HotSpot operators–they know the folly of holding their customers captive to a single product, be it coffee, hamburger, mystery books, or a HotSpot operator's closed network.

Hardware and software deployment as well as the management of the infrastructure to support an ever-expanding footprint require excellent co-ordination between many different components within the operator's value chain. This is especially true for the venue owners, since this group typically requires some handholding. Thus, unloading some of the back-office infrastructure, roaming agreement details, and other business processes to an aggregator can help a HotSpot operator and its venue owners to grow their businesses.

Minimizing Up Front Costs

Let's now look a little closer at what needs to be considered before jumping into the HotSpot market. Begin by investing some time in doing research. At minimum, your research should answer the following:

- Is there a market for wireless Internet access in your area?
- Will you set up your own network, or will you use a virtual network model (i.e. partner with an aggregator)?
- How many HotSpots can your operation handle over the next year, two years, or five years?

FRANCHISE REGULATION

This section is provided courtesy of the law firm of Wiley, Rein & Fielding LLP, a U.S.-based law firm specializing in wireless communications issues. Included in the firm's client list are the Wireless Information Networks Forum (WINForum), which pioneered the Unlicensed Personal Communications Services band at 2 GHz, and the Unlicensed National Information Infrastructure band at 5 GHz; UTAM, Inc., the joint industry frequency coordination body for 2 GHz LTCS spectrum; and the Wi-Fi Alliance (formerly the Wireless Ethernet Compatibility Alliance or WECA).

In July 2002, the firm issued a Wi-Fi primer entitled, "Wi-Fi—802.11: The Shape of Things to Come." That document includes sage advice, which with Wiley, Rein & Fielding LLP's permission, is set out below.

"Some participants in the Wi-Fi hotspot market fail to consider whether their business model subjects them to franchising regulations. Even if aggregators or hotspot providers do not think of themselves as selling franchises, they may be considered franchisors based on the setup of their business operations. If considered a franchise, a business is subject to additional regulation. In particular, in many states, franchisors are required to register with the state and to comply with disclosure procedures when marketing franchises. Additionally, franchisors are subject to additional substantive laws that prohibit them from engaging in certain practices and govern other elements of the franchise relationship such as requirements for termination.

"Generally, there are three elements that are necessary for a business model to be considered a franchise relationship. First, there is a trademark element that requires a franchisee to distribute products or services under the franchisor's service mark or trademark, Second, there is a control and assistance element that considers whether a franchisor exercises significant control or provides significant assistance to its franchisees in an ongoing relationship. Lastly, there is a required payment element that considers whether the franchisee is paying the franchisor either directly or indirectly for the privilege of operating under its service mark.

"Notably, as discussed above, the business models of many aggregators and individual hotspot providers feature many of the characteristics of a franchising model. Thus, before selecting a particular business model, a company should seek counsel as to whether their model may be considered a franchising arrangement in order to ensure proper compliance with existing laws and regulations. Moreover, companies should fully understand the advantages and disadvantages of adopting a franchising model.

"(Note that this discussion of franchising refers to franchise as the licensing of trademarks for distribution purposes, not as a special privilege granted by government such as in the case of cable franchises. Thus, aggregators and hotspot providers may engage in business relationships similar to those that McDonald's creates with its franchisees.)"

Want more information on franchise regulation? Contact Peter Klarfield at pklarfel@wrf.com or David Koch at dkoch@wrf.com.

- How do you envision the scale of the business—regional, national or international?
- How will you contain your operational costs while building a customer base?
- How much, and in what ways, do you want to segment services and users?
- Do you have a clear understanding of the relationship between your company and its venue owners?
- In what ways will your company support and benefit from inbound and outbound roaming? If you do decide on a roaming model, will uncertain quality of service and end-user experience have a negative effect on your HotSpot operation?
- What gear will you support and in what combination?
- What type of authorization, authentication, and billing methods will you provide?
- What will your network look like and how will it be managed in the beginning, then at the end of one year, two years, five years?
- What is your sales and marketing strategy? What will it be in two years or five years?
- What will be the total marketing costs needed to introduce your brand to the marketplace?
- Will the company court, or work with, content providers? If so, what will be the requirements for these providers?
- Will the company seek to obtain additional revenue through ads? If so, when? Immediately, two years down the road, five years?
- What type of customer relationship management requirements will be expected of your company? And what will be the customer service requirements? How will you provide these items?
- How will your company deal with problems such as fraud, repair, claims, etc.?
- What are your partnership strategies, especially in the area of compensation?
- What is your exit strategy? If it is acquisition or merger, how will you build out your network to minimize integration work?

Other related issues that you need to address will depend upon your specific business plan (e.g. self-funded, angel investors, or whatever), revenue model (subscription only, ads, etc.), specific circumstances (e.g. locale and existing competition). Furthermore, well thought-out business and marketing plans must be drawn up and adhered to for the project to have any chance of real success.

THE AGGREGATOR LEVEL

Aggregators can help to overcome fragmentation problems within the WISP industry by taking the fragmented layers and aggregating them into a single service. (The more HotSpot operators an aggregator can incorporate into its network, the more opportunity for that aggregator and the HotSpot operators.) Although companies operating in the aggregator layer may have different target markets and business strategies, they all have one thing in common: the hope that they can exploit Wi-Fi's potential by helping the end-user to obtain access to as many HotSpots as possible.

Many smaller HotSpot operators realize that to exponentially grow their network requires not only a huge investment in time and money, but also the skills to combine the varied components of the WISP value chain, and to enable all of those components to work in tandem. They also know that due to their size and budget, it wouldn't be long before they found themselves overwhelmed with running a network, generating sales, arranging roaming affiliations, and handling the necessary marketing and customer service chores. Is it any wonder that many of these operators (and venue owners) begin to see the advisability of investing in only the front side of the value chain–focusing on end-user acquisition and the resulting relationship–and letting others deploy the infrastructure, handle the branding and billing relationship, and the roaming technicalities?

Note: ***Other terms for the aggregator group are "virtual WISPs" as most do not really have an actual physical network; "brokers," since they broker deals between companies within the WISP Industry; and "managed service providers," since their business model is based upon managing Internet access service for others. The companies in the aggregator layer typically provide centralized authentication services in order to compute and validate the broadband traffic; fix the airtime prices in which they trade; operate as intermediaries between HotSpot operators through buying and selling HotSpot operators' airtime minutes; and fix tariffs for roaming between diverse HotSpot operator networks.***

Most national and regional aggregators' operations are similar to Boingo Wireless Inc. Boingo, a start-up whose log-on and authentication software provides a wrapper around existing HotSpot operators' networks that allows end-users access to a variety of HotSpots using a single account. Boingo strikes wholesale access agreements with HotSpot operators, and then consolidates the venue HotSpots into a single seamless network.

Note: ***Boingo also provides WLAN capabilities to brands, such as Cingular, General Electric, Pepsi, and Sprint, that might contract with Boingo to provide the whole WLAN set-up—software, technical support and back-office services.***

A different tack is being taken by Wireless Retail, Inc., a well-known member of the wireless industry–through its recently launched aggregator service, FootLoose Networks, Wireless Retail plans to facilitate the build-out of a carrier-neutral, unified national network of HotSpots. Wireless Retail hopes that its new aggregator service will become the catalyst for cooperation among the nation's wireless carriers, by providing a straightforward, cost-sharing model for the build-out and ongoing maintenance of a single aggregated national network. FootLoose will offer enterprise customers and consumers alike multiple service and payment options, including subscribing for service through their own wireless carrier's data plans, enabling the HotSpot charges to be incorporated into their current wireless phone bill.

Another approach is hereUare Communication Inc.'s business model. This aggregator neither owns nor operates public wireless networks, nor does it actively pursue end-users. Instead, it focuses on three ingredients that are necessary to the creation and growth of the WISP marketplace: (1) the extension of partner branding through an aggregated network of wireless locations, (2) the creation of a universal experience for wireless users on a global, roaming network of wireless locations, (3) enabling HotSpot operators and venue owners to deploy public for-pay wireless through its eCoinBox device, which performs router and Domain Host Control Protocol (DHCP) functions. Thus the modifications to any existing wired network are minimal, if any. The hereUare system consists of an "eCoinBox," and an access point. All the venue owner (or HotSpot operator) needs to do is to install both at the venue location. The venue owner (or HotSpot operator) is responsible for:

* A high-speed Internet link (T-1/DSL/Cable Modem).
* A wireless infrastructure, i.e. an 802.11b WLAN-enabled computer, a functioning Internet browser, and a null-modem serial cable.

All other components necessary for the operation of a fee-based HotSpot exist remotely and are hereUare's responsibility. (It's interesting to note that the hereUare eCoinBox technology is integrated into the Colubris Network's CN3000 wireless access controller depicted in Fig. 11.3.)

Of course, not all aggregators focus on the consumer market. According to Perry Lewis, GRIC Communications' manager of business development, GRIC only deals with large carriers. "We don't sign up the end-user." GRIC's business model differs from most Wi-Fi aggregators, in that it gives its client (carrier or enterprise) the ability to supply a subscriber with one user name, one password, and one monthly invoice. Then the end-users can use either a GRIC wireline access point, or a wireless HotSpot, anywhere in the world and get connectivity. GRIC provides all the billing and settlement functionality, but it's GRIC's client who bills the end-user. Lewis says that GRIC is "focused on a global footprint. The business user doesn't always travel in North America."

There are more than 800 HotSpots supported by GRIC's TierOne Alliance. John Rasmus, GRIC's vice president of marketing, adds, "We are in the business of putting RADIUS authentication servers on these networks so that little islands of Wi-Fi are tied into a true global wide-area network. The key issues are security and network management. And things are moving along quickly in the industry to address those concerns. Also, we offer a portfolio of security solutions that are embedded with our client and access servers."

iPass Inc. is another network aggregator with an international bent. It bypasses the consumer market and pulls together various types of networks so it can sell access to the enterprise market. It also facilitates roaming between various carriers by taking care of all the settlement issues and back-end needs.

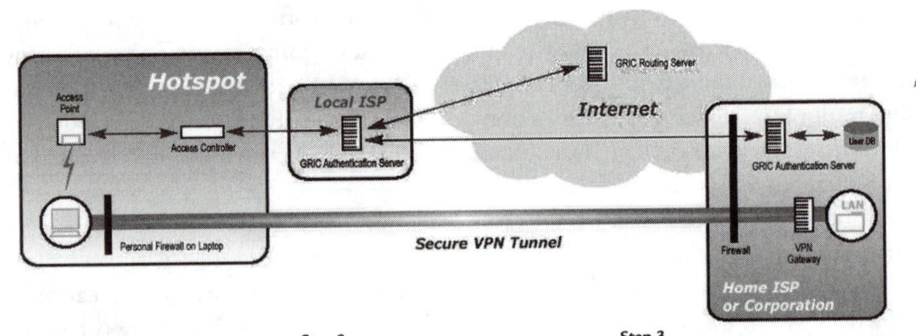

Step 1
Upon entering the public "hotspot" (airports, hotels, etc.) the user launches the GRIC Client from the laptop or PDA, selects "wireless" from the Access Method menu, and then chooses a hotspot location from the hotspot list.

Step 2
Clicking the "connect" button on the GRIC Client connects the user to the Wi-Fi network at the hotspot, and starts the authentication process through the authentication and routing servers located around the globe. The user is then authenticated and gains Internet access.

Step 3
The GRIC Client automatically launches the user's browser and VPN client in the background to establish a secure VPN tunnel back to the user's corporate network.

Figure 11.7 Anatomy of a HotSpot from GRIC Communication's point of view. *Graphic courtesy of GRIC Communications.*

Since both GRIC and iPass target the carrier and enterprise markets, they are quality conscious that if one of their client's end-users goes to a HotSpot and the network fails to operate properly, the client (which may be a large corporation or another aggregator) may drop their subscription. These aggregators make sure that their HotSpot partners meet a defined set of standards. GRIC's Lewis explains, "It isn't worth it to have customers complaining about service. It's a footprint versus quality issue."

Many within the industry believe that aggregators are the solution to making mass-market Wi-Fi services a reality. According to Analysys Research Ltd., more than 21 million Americans will use HotSpots by 2007; but that leap in public Wi-Fi use can be propelled only by a proliferation of HotSpots, which Analysys predicts will grow to more than 41,000 in 2007. In-Stat/MDR is another research firm that has taken a stab at estimating growth in the HotSpot market. This time the measurement is projected revenue. According to the In-Stat/MDR research, HotSpot revenue will grow to close to $225 million in 2007. "I think aggregators play a valid role in bringing these isolated locations [HotSpots] into a larger footprint," says Amy Cravens, an industry analyst with In-Stat/MDR.

The story is about the same in Europe. Gartner, Inc. says the number of HotSpots will increase in Europe, from a mere 70 at the end of 2001, to 15,000 by the end of 2003, and 43,000 by 2008. And research from Research and Markets indicates that 80 percent of European urban mobile data users will use HotSpots by 2005.

These positive projections have lit a fire under the aggregators. "We've seen incredible, overwhelming interest in Wi-Fi and what we are doing," says Sky Dayton, founder of Boingo Wireless Inc. As the industry matures, the aggregator sector will find itself with only a few *major* players (probably fewer than five).

There is immense complexity in negotiating with and aggregating thousands of individual HotSpot operators, providing systems for network management and authorization, authentication, and accounting. Aggregators can drive significant traffic to HotSpot operators, helping them cross the line of fixed costs into positive cash flow. Furthermore, it's expected that aggregators will serve as the vital link between the HotSpot operators and the major brands, and that the brands will drive the mass-market adoption of mobile Wi-Fi access.

Some aggregators may choose to act as clearinghouses for other HotSpot operators operating in different regions. In this case, the aggregator signs agreements with many local HotSpot operators, and then makes those HotSpots available not only to its own subscriber base, but also to roaming subscribers belonging to other HotSpot operators and aggregators.

BRANDS LAYER

Brands include cellular carriers, ISPs, PC manufacturers, and enterprise remote access providers seeking to offer a HotSpot service to their customers.

Unlike the cellular industry, which has a government- and cost-mandated limit on competition, the WISP industry's fragmentation makes it impractical to build out a single Wi-Fi network and then to sell those services exclusively to one's own end-users. For instance, suppose a cellular carrier were to build out its own HotSpots and place them in thirty airport venues. The carrier still would be operating under a problematic business model, since there is little overlap between the coverage that its HotSpots provide and the coverage that its customers will demand. Roaming agreements are essential. However, multilateral roaming agreements of the type found in the cellular industry are nonexistent in the WISP arena.

Even the brands that forge direct relationships by entering into bilateral roaming agreements with both large and regional HotSpot operators (e.g. Wayport, aXcess2go, and Airpath Wireless), in an effort to provide their customers the Wi-Fi Internet access they expect, may find they need an aggregator to tie the diverse HotSpots together into a single network.

BOINGO WIRELESS

Boingo, with its ever-growing nationwide network based upon partnerships with Earthlink (an ISP), Fiberlink (a provider of remote access for enterprises), and a variety of wireless networks, provides the end-user with a single, seamless Wi-Fi experience. At this writing, Boingo had aggregated the networks of more than 25 HotSpot operators (e.g. Air2Lan, Nomadix, Surf and Sip, Wayport, to name a few), representing over 1200 HotSpot venues. Although this number seems anemic, it is more than any other national aggregator has managed to cobble together. Furthermore, Boingo actively cultivates a pipeline of thousands of additional venues.

Boingo provides its partners (in particular the carriers and ISPs) with a single, turnkey aggregated roaming network that is uniformly branded. For example, when the user opens their wireless device in a Wayport hotel HotSpot or a SurfAndSip café HotSpot, the user is presented with a live signal branded, "[Brand] Wi-Fi." When the user selects the signal, the software prompts them for a username and password, which is passed through Boingo's back-end systems to the carrier or ISP's authentication systems, and then connects the user to the network. At no time is the user aware of who the underlying HotSpot operator is or what other carriers might also have roaming relationships with that location.

Boingo offers service level agreements (SLAs) with all of its commercial HotSpot operator partners, and its NOC (network operations center) actively monitors each of their HotSpots at a granular level, 24x7, to ensure a high level of reliability.

Moreover, through its software, network aggregation, support, and billing services, Boingo helps major branded communications providers (e.g. T-Mobile) add Wi-Fi capabilities for their customers.

Boingo also operates a call center dedicated to helping end-users quickly resolve driver issues, software conflicts, hardware problems, and other Wi-Fi-related technical issues. As an option, Boingo can provide private-label support services direct to end-users, or higher-tier support to an operator, carrier or ISP partner's own support personnel. Boingo also provides a single aggregated accounting stream to carrier and ISP partners in a wide variety of industry-standard formats. Optionally, Boingo's billing system can be used to bill end-users directly, or on a private-label basis. This service is made available to carriers or ISPs who are interested in getting an offering to market faster than they can prepare their own billing systems.

Boingo allows carriers and ISP partners to have the option of building their own HotSpots,

The Carrier Case

For the telecommunications carrier (wired or wireless) or ISP looking to offer a compelling Wi-Fi experience to its customers, there are two options. The first is to enter into roaming agreements with a large number of different HotSpot operators, and build systems to integrate and monitor them as a single network.

This approach crosses the segmentation lines, with the operator possibly occupying multiple layers at once. For example, experienced telecom providers deploy commercial HotSpots at the HotSpot operators layer, and launch branded Wi-Fi services at the brands layer (that's what T-Mobile in the U.S., Telia in Sweden, and BT in the U.K. are

or cutting direct deals with larger HotSpot operators, and still use Boingo's aggregation, network monitoring, and clearinghouse functionality to tie their networks together. Through Boingo's software and network aggregation services, partners get a single turnkey network, one NOC, and one stream of reporting data. This allows its partners to offer their customers a truly compelling Wi-Fi experience, without committing a significant amount of capital.

In addition to aggregating networks from HotSpot operators, Boingo fuels the infrastructure build-out at the grass roots level through its "Hot Spot in a Box." (Hot Spot in a Box 1.0, a $540 access point from Colubris, is available now. Hot Spot in a Box 2.0-based products from other equipment manufacturers should be available by mid-2003. The price of these products will be under $300.) The Hot Spot in a Box technology allows any Wi-Fi access point to operate as a commercial HotSpot. A Hot Spot in a Box-enabled access point is easily configured to communicate with Boingo's back-end systems for user sign-up, authentication, billing, and settlement.

Envision a future when every DSL line, cable access line, or T-1 line has a Wi-Fi hub on the end of it. Every one of these broadband endpoints has the ability to become a commercial HotSpot. You don't need to operate a café, retail store or hotel to consider the potential of this. If you live in the middle of a residential area or a small farming community, and have a DSL or cable broadband connection, you could become a HotSpot just by flipping a switch on a Boingo Hot Spot in a Box-enabled access point. At the end of the month, you would receive a check for all Boingo subscribers who connected to the Internet via your network

Note: *There may be commercial and regulatory issues to be addressed before sharing or reselling a residential high-speed connection. Generally speaking, DSL and Cable Modem providers and ISPs include clauses prohibiting bandwidth sharing if the service is intended for residential usage. A typical business-grade service package, however, will usually allow this kind of activity. Thus, since every service provider's terms of service agreement is different, before you begin sharing your high-speed Internet service with others, check with your provider.*

Boingo also offers a "Boingo Ready" certification program for higher-end access control devices designed for large HotSpot installations, such as those at convention centers, hotels and airports. This program certifies that hardware will plug seamlessly into Boingo's back-end systems for sign-up, authentication, network management, and support. Several vendors, including Nomadix, Colubris, and Vernier, ship Boingo Ready products.

doing). Since the carrier group has the capital and expertise to deploy networks, as well as established relationships with millions of customers to whom they can promote their HotSpot service, it's not too surprising to see these carriers entering the WISP marketplace through multiple WISP layers.

But even established carriers will need to maintain their focus, if for no other reason than to obtain economy of scale within a specific layer. If not, the carrier will have difficultly resolving questions, such as: if the company signs up new venue owners, what happens to the brand? What happens when roaming into other provider's networks? How is that handled? Who owns the customer as various operations travel through the layers?

Thus companies that decide to operate under a multilayer business model may need to run each layer as though it were an independent entity to be successful.

The other option is to work with an aggregator (e.g. Boingo, GRIC, iPass) that has already assembled a multitude of HotSpot operators into a single system, and that can provide the necessary management oversight and client software to uniformly brand the user experience.

Established telecommunications companies are welcomed into the HotSpot space, due to their deep pockets and customer clout. However, U.S. carriers are approaching the WISP industry tentatively. That's the exact opposite of their European counterparts, who are moving full-steam ahead. "In Europe, all of the 3G rollouts have been delayed, and so carriers are saying, 'If we don't it, others will,'" says Henning Klemp, marketing director at Birdstep Technology, a provider of IP connectivity software for WLAN systems. Klemp adds, "The established carriers not only have a large subscriber base, they are experienced with the ins and outs of roaming agreements."

Furthermore, Klemp suggests, "They also believe that if they can educate the people about high-speed access, then they will have a user mass that will more easily adapt to GPRS, CDMA, and UMTS, when that evolves. In the U.S., the operator market is too fragmented. That has enabled companies like Boingo to take a stronghold."

Another motivation driving carriers to the HotSpot market is that there's money to be made. "The dollars that wireless operators are able to tap from the subscribers is another reason [HotSpots] are rolling out," says Hans-Arne L'orange, Birdstep's CEO for the Americas.

The Roaming Conundrum

The next hurdle faced by the telecommunications industry is the end-users' inability to move between the different networks without losing their connections. The logistical challenges that this kind of roaming presents are tremendous. Still, many in the WISP industry are positioning themselves to meet these challenges, and some, such as Telia and BT, have already moved to address them. These two telecom giants are among the largest HotSpot operators in Europe, and they are doing roaming (but no money changes hands). Roaming is offered more as a footprint-expanding exercise rather than as a revenue builder. This is primarily because the nascent WISP industry doesn't generate enough roaming traffic to justify outsourcing settlement operations to the usual external clearinghouses, which have years of experience with telco-related billing.

Red-M, a leader in enterprise-level wireless management solutions is tackling the roaming dilemma. Steve Gallagher, director of business development at Red-M, contends, "Enabling a PDA user, for example, to go from WAN [wide area network] cellular to [Wi'Fi's] WLAN to enterprise Ethernet is a market requirement that needs to be resolved before any type of public hotspot takes off in large numbers. The carriers are driving this. A user subscribes to a carrier for voice and wide-area data service, GPRS, and now in addition to that, the carrier offers a hotspot service over WLAN while the user is in those areas. They wrap it up under one billing solution and it's much more compelling."

According to Gallagher, Red-M's Genos solution supports multiple wireless technologies through a platform that it has designed "to overcome device, user, and location management as well as integration of different wireless technologies." The Genos software integrates WLAN into an overall platform that could facilitate ubiquitous connectivity.

Birdstep Technology has also taken steps to address the roaming issue. It's IP Zone Server 1.5 software is specifically designed to facilitate a handoff between a WLAN and a WAN. The IPZone Server 1.5 provides cross-platform authentication so that any computing device with a wireless connection and a web browser can obtain access to the Internet. Moreover, the software achieves secure and continuous connectivity between HotSpots and GSM/GPRS networks through integrated support for Mobile IP client software, an application that allows users to connect and re-connect across different types of infrastructures without application downtime or user intervention.

Trond Lunde, sales engineer at Birdstep clarifies. "We take care of the IP connectivity, maintaining secure tones, true application state, and making sure that the transition is hidden from the user as he or she moves from one network to another. We do a handover by fooling the application state. So that if you're downloading a PowerPoint presentation, for example, and you move from a HotSpot to GPRS, the only thing you'll notice is that your connection speed changed."

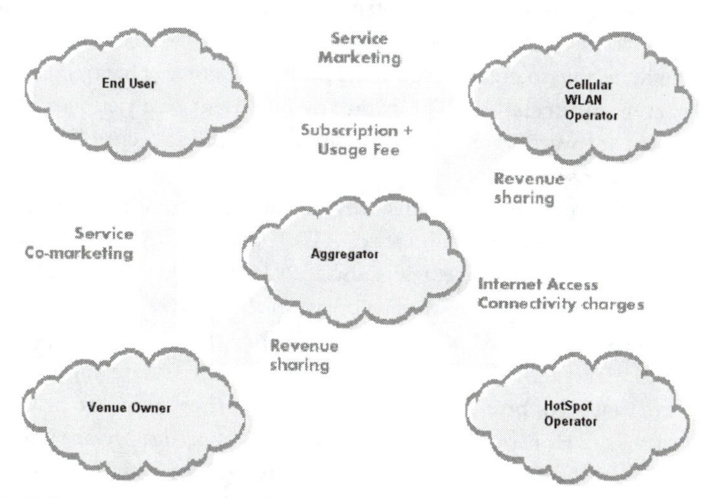

Figure 11.8 Cellular carriers can establish partnerships with members of the WISP industry to market their services. This graphic depicts a cellular HotSpot operator that chooses to not deploy its own access network. Rather, it buys capacity from networks deployed by others through the services of an aggregator, although it does manage its own customer relationships and billing (i.e. the end-users pay the cellular provider for HotSpot usage).

There are two different angles on the Birdstep mobile IP software. It can be designed for network operators or for use in a device.

Interoute, a Pan-European telecom provider, has opened what it calls an "802.11 roaming exchange." Nick McMenemy, Interoute marketing director, says, "A lot of companies are competing for the marquee sites but there are roaming conflicts. One may have an agreement with Costa Coffee, another with Starbucks. We're neutral and offer services based on our billing engine and presence in most European centers."

Interoute uses proxy authentication software to determine whether users have a right to roam in a location and then the provider takes care of the "how many cents per minute" calculations to apportion the roaming fees between the provisioning HotSpot (the one providing the wireless Internet access), and the operator with whom the end-user has a relationship.

Boingo Wireless, of course, also has its eye on the cellular provider market. In March 2003, it announced its Boingo Platform software, which, according to the company, melds the wide area networks of cellular carriers with the WLAN's of HotSpot operators. The company expects that, by June 1, 2003, anyone with a laptop equipped for 802.11b and cellular connection (i.e. equipped with a PC Card with GPRS, CDMA 1x or iDEN technology) will be able to roam between 802.11-based HotSpots and 2.5G cellular networks. The Boingo Platform is said also to enable carriers to offer subscribers both traditional cellular phone service and Wi-Fi data service, while maintaining brand identity and avoiding the necessity of building the appropriate back-end systems and installing the necessary roaming technology.

Ed Rerisi, director of research at Allied Business Intelligence, suggests that packaging Boingo's network of HotSpots under a private label is a smart move. He feels that Boingo's strategy is both offensive and defensive. By increasing its network's footprint, it can "thwart the big guys," For the Hotspot Operator layer, "coverage is the name of the game," says Rerisi. And through the sales of its Platform software to cellular operators, Boingo will be able to greatly expand its area of coverage. Through a Boingo statement, the aggregator declared: "The company has spent two years and $10 million developing client software, back office systems, and roaming technology that can now be leveraged by other service providers."

What's most interesting is that this move allows Boingo to compete directly with the likes of Cometa Network (an alliance of AT&T, IBM and Intel), which is itself hoping to sell back-office services to carriers and ISPs looking to start their own HotSpot operation. Rerise says, "Boingo wants to position themselves against Cometa." And, according to Rerisi, Boingo is "in a good place" since the company is positioning its combined cellular and Wi-Fi network as "a complete wireless data solution."

Boingo says that its re-branded software will give carriers access to private and public Wi-Fi networks, a HotSpot "sniffer," and a built-in virtual private network (VPN) that allows HotSpot operators to provide end-users with an entirely secure method of accessing the Internet or their corporate network. The aggregator also says that its software provides something new to the industry—the ability to filter out the signals from competitors, meaning users can only connect to their subscribed carrier or ISP.

***Note:** The term "sniffer" refers to a software program designed specifically to scan the airwaves for a Wi-Fi signal sent out by a Wi-Fi-enabled HotSpot.*

In fact, T-Mobile and Boingo announced that they had entered into an agreement to co-develop unique end-user facing software and services, making it easier for T-Mobile's customers to discover and access its HotSpots and its nationwide GPRS cellular network. The end product will be the same aforementioned Boingo Platform software, but with an added client-service solution designed specifically for T-Mobile. The end product will include an intuitive sign-up and log-on screen (a slick, single button sign-on); easy-to-use help files that will address the most frequently asked questions; a built-in, searchable location directory so end-users can find the nearest T-Mobile HotSpot; a streamlined setup process to enable end-users to connect easily and to move between private and public networks; the ability to integrate special site-specific content and promotions for T-Mobiles brand-name location partners (e.g. Borders Book and Music Stores, Starbucks, American, Delta and United Air Lines, Kinkos). Future versions of the product will allow customers to designate their network preference or to opt for the network connection with the best available speed. It is noted, however, that this agreement is silent on T-Mobile customers roaming into Boingo's network or vice versa.

Cellular carriers such as the oft-mentioned T-Mobile, along with Sprint PCS (a Boingo investor), AT&T Wireless (which provides HotSpots for the traveling business user), and Verizon (through pay telephone HotSpot locations), have all entered the WISP arena.

THE LAG IN INVESTOR CONFIDENCE

The WISP industry is still in its infancy, and there are plenty of opportunities available to the entrepreneur. However, so long as there are uncertainties about the economics and the characteristics of the HotSpot model, most large investors will remain on the sidelines. Let's examine some of the reasons for this lack of investor confidence.

Fragmentation

One poorly understood characteristic of the WISP marketplace is fragmentation, of which there are two types: one good, one bad.

Fragmentation *across elements of a network* of HotSpots (e.g. the venues, the infrastructure, aggregation, and branded services) is good because it helps to classify and type tasks so that the right tasks are handled by the right companies. For instance, billing tasks go to say, the aggregator, but infrastructure tasks are handled by the HotSpot operator.

Whereas, fragmentation that *crosses networks* is bad because it forces end-users to deal with many different settings and accounts. Here's an example. During the course of a day, a business traveler has his or her morning coffee with a colleague at a Starbucks, lunches at Novel Café, and stays overnight at a Hilton, all of which the end-user frequents because they offer Wi-Fi connectivity. However, each of these venues requires different authentication procedures and different log-on screens. That's bad fragmentation. As HotSpots proliferate, so will bad fragmentation. At least until there is unified roaming across all HotSpots. Until then, industry growth will be hampered.

In its early days, the cellular industry struggled with the same fragmentation issues that the WISP industry faces today. The cellular industry managed to overcome fragmentation within geographical regions (e.g. North America, Europe, and Asia), but there is still fragmentation within international regions (e.g. roaming is virtually impossible between North America and any other region and vise versa).

Only an industry-wide, unified roaming plan will negate the effects of fragmentation and allow the WISP industry to continue its monumental growth. If you remember, cellular adoption only took off after seamless roaming within large geographical regions became available. The same is true for ATMs—far fewer people used ATM cards when they were limited to use at only specific banking locations.

The Experience Quotient

Some end-users find the experience of locating and accessing a public wireless network difficult because Wi-Fi is invisible. Unlike a cell phone which, when turned on, will automatically take care of the connection sequence, there's no way for a potential user to know a HotSpot is present without noticeable signage or "sniffer" software. Furthermore, due to the limited battery life of most Wi-Fi-enabled computing devices, end-users are reluctant to leave their computer on. Even Boingo's vaunted sniffer software requires turning on the computing device before the software can do its job.

Also, the industry (as a whole) has done a poor job of advertising HotSpot availability. While this situation is changing, more needs to be done. Even if an end-user knows that a Wi-Fi signal is available, getting connected can be onerous. The end-user is expected to know the network's SSID (Service Set Identifier, a 32-character identifier for wireless LANS that acts both as a simple password and as an advertisement for a specific Wi-Fi network), how to turn on and off security settings, and how to program configuration software on their computing device.

Then there's the requirement that each time a user signs onto a new HotSpot operator's network, they must sign up for a new ISP, a new password, and a new billing arrangement instead of being able to port to their corporate virtual private network (VPN) connection or existing ISP. Many Wi-Fi users find the process burdensome and unnecessary.

ECONOMIC FUNDAMENTALS

As mentioned previously, a HotSpot operator can choose to go it alone, but there are risks: excessive customer acquisition costs, insufficient revenue streams, backhaul costs, and poor footprints. Thus, it's hard to find a viable economic model for an independent HotSpot business outside the venue layer. But venue owners and HotSpot operators who take the "community" approach may find that their model has sound economic fundamentals. In addition to the normal fees received from a local or "owned" end-user, the HotSpot operator finds that:

- The aggregator pays its partners (typically HotSpot operators) a fee for adding its access points to the aggregator's network footprint. The fee is usually a wholesale fee ranging between $1 and $2 per connect day (24 hours per user per location) for each connection generated through the aggregator's distribution channels, which might also include original equipment manufacturers, ISPs, and major telecommunication carriers.
- The aggregator takes on some of the more onerous tasks that can drag down its HotSpot partner, e.g. back-office chores and roaming agreements.

* The value chain is simplified.

Thus, it's possible to create an economic model on the basis of not only end-user fees, but also payments (based on some assumptions about traffic) from the aggregator based on roaming agreements. The principles of roaming revenues are that:

* End-users will pay more if roaming outside their home network. This is the same pricing model that cellular uses.
* It's the visited network providers that set the roaming charge that the end-user pays. For example, the retail roaming charge could be $1.50 per hour versus a normal $1.00 per hour usage fee.
* Every end-user using a HotSpot outside of his or her home network generate revenues for both the home and the visited HotSpot network chain (venue, operator, and aggregator).
* Hosting visiting end-users provides the network chain a guaranteed revenue stream since all payment risks are borne by the home network chain.
* Roaming charges are dynamic, i.e. they can change over time and may differ from partner to partner.
* There is a difference between what the end-user is charged and the whole inter-network fee.

For instance, a busy location with numerous visiting end-users (say, an airport lounge) can generate hundreds of roaming Wi-Fi connections per day equaling hundreds or even thousands of dollars in revenue per day. Even a small HotSpot may find that it hosts as many as ten or 20 roaming connections a day.

Small Can be Good

One of the ways "small is better" is that small venues may be more important to the overall health of the WISP industry than the larger venues because a viable national network most likely will be built upon the foundation of small HotSpots. (The number of small venues will far exceed the number of larger venues; the very propagation of HotSpots in small venue locations is what will eventually enable the aggregator layer to provide nationwide synchronized roaming.)

Another is low operating expense. As illustration: Surf and Sip provides high-speed wireless Internet access, mostly for small venues—cafés, hotels, gas stations, and other public establishments. But its 150 or so locations are set-up for a limited number of end-users and thus can get by with reasonably priced DSL connections. (Interestingly, most of Surf and Sip's Hotspots generate a positive cash flow.) And if the venue owner already had a high-speed pipe to the Internet (e.g. DSL or T-1) in place, then it can use that same service to provide the backhaul for its HotSpot service, resulting in the overall provisioning expense for the HotSpot being much lower than a venue that had to subscribe to a high-speed service specifically for wireless access to the Internet. Conversely, a higher volume location such as an airport, hotel or convention center generally requires more expensive equipment in addition to one or more T-1 lines (DSL usually won't do). This means that larger venues incur higher monthly fixed costs than the smaller venue. However since the traffic volume in these HotSpots is also usually higher, the costs may be offset.

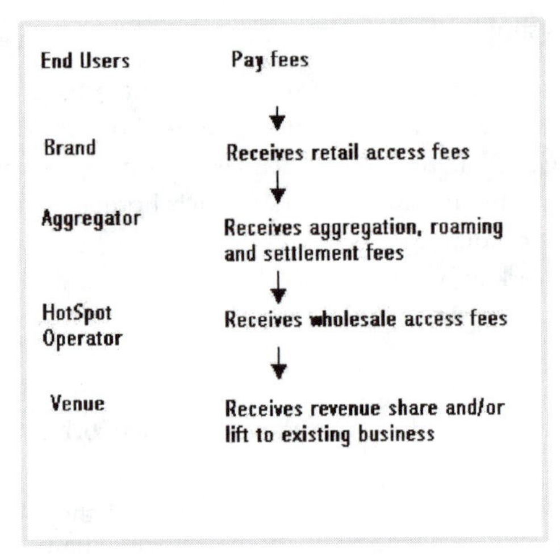

Figure 11.9 The brand charges its end-user and pays the aggregator fees for network aggregation, roaming, settlement, support, and software. The aggregator in turn settles with the HotSpot operators, paying a wholesale connect fee for each of the brand's user connections.

The small venue also gains the most from indirect benefits. Although a small HotSpot can receive revenue through roaming agreements within its value chain, these venue owners typically have a limited marketing budget, thus they reap benefit from venue listing on HotSpot operators' and aggregators' websites as an affiliated location, mention of its locale in value chain members' press releases, marketing efforts on its behalf by value chain members, and more customers drawn to the location because of the wireless Internet access. If the venue is near a travel hub, the benefit can be enormous–business travelers are increasingly seeking out Wi-Fi service.

Also, offering a Wi-Fi network as a carrot to attract business (often offering the capabilities as an amenity rather than charging a service fee) can foster loyalty, increase the amount of time customers spend in the venue, and create the opportunity to sell more goods and services. Moreover, the wireless network can provide a platform for future value-added services like location-based marketing ("We have a special on 2003 calendars and datebooks in aisle 3"); loyalty programs ("Welcome back, Ms. Wright, in honor of your fifth visit to our restaurant, we want to offer you a free order of fries with the purchase of our super value meal."); content distribution ("click here to download the latest video game to play until you reach your final destination.")

Thus, ignoring the Wi-Fi phenomenon can not only mean missed revenue, but inaction also holds the potential of incurring dissatisfied customers, lost business and lower profits.

An Evolution

The pricing structure in every layer of the WISP industry is evolving as the industry continues to search for a balance between the price that needs to be charged and the price the market will bear. To illustrate, as this book went to press:

T-Mobile, which previously charged for data transfers at the rate of $.25 per Megabyte above 500 Megabytes per month, now offers unlimited data transfer. Its service options include: a pay-as-you-go plan that charges $6.00 per hour, $.10 per minute thereafter, and a prepay option that costs $50 for 300 minutes (the prepaid minutes expire 120 days after purchase).

Wayport provides unlimited data transfer and its service operations include daily and monthly rates, which vary per venue. For example, at an airport the daily rate might be $7.00, and at a hotel it may run as high as $10.00. The "day" designation may also vary from time of purchase to midnight, a 24-hour period, or from time of purchase to check-out time the next day (in the case of a hotel). The monthly rate is $30 for a one-year commitment with a cancellation penalty or $50 for a month-to-month subscription.

Surf and Sip, on the other hand, offers prepaid cards in increments of 30 and 120 minutes (the hourly charge is $5.00); or one-day ($5.00), seven-day ($20.00) and 30-day ($40.00); the cards are activated at the time of purchase. This HotSpot operator also offers a monthly rate of $20 for a one-year commitment with a cancellation penalty or $30 for a month-to-month subscription.

The contractual area—the various agreements that are executed throughout the industry segments—is also still evolving. The standardization of these agreements, especially those entered into between the venue operator and HotSpot operator, is one of the keys to the success of a sustainable WISP industry. These standardized agreements must resolve such issues as: venue exclusivity, degree of cost and revenue sharing, ownership of infrastructure, ownership of the end-user, roaming criteria, and length of contract.

John Marston, a business development executive for the electronics maker Toshiba, advises the HotSpot industry to pay attention to the market and avoid being overly optimistic about subscriber numbers. His advice is not to spend too much money; the HotSpot operator must "think about every dollar they spend and how they expect to acquire locations." Still, with universal roaming, ease-of-use and enough in-venue marketing, HotSpot economics will become quite compelling in the near future.

THE POTENTIAL MARKET

Right now, the entrepreneurial companies drive the WISP market in the U.S., whereas in Europe and elsewhere, it's the established telecommunication operators. But no matter who's in the driver's seat, HotSpots will generate approximately $12 billion in revenues in 2007, according to an early 2003 Visant Strategies study entitled "3G and 3G Alternatives: 3G vs. Wi-Fi vs. 4G." This predicted growth becomes even more realistic when HotSpot operators team with cellular NextGen operators to provide high-speed wide area coverage coupled with even higher speeds at the HotSpot, as discussed in Chapter 13. (These and other partnering strategies will become more and more commonplace as both industries realize that they need each other for growth.)

It's the business professional that will be the prime marketing target, especially the traveling professional. Travel industry statistics indicate that business professionals on average take five business trips per year with the average business trip lasting three nights. For example, the World Bank (based in Washington, D.C.) sends its employees on more than 18,000 overseas trips every year. All of these professionals must keep apprised of what's happening on the home front—their email, managing their daily business tasks, overseeing revisions, updates to corporate documents, etc. The fact that Wi-Fi enables them easily to access their corporate files can be a godsend, especially during the ever-expanding wait time at airports, the three-hours-to-kill at a convention center waiting for the next lecture to begin, and when wide awake in a hotel room due to jet lag.

The website, Outlook4Mobility.com, recently conducted a "Wi-Fi Survey of Industry Professionals," which indicated a great deal of interest in the HotSpot model. For instance, 92 percent of those surveyed indicated an interest in making use of HotSpots, both among respondents who have previously made use of a HotSpot and those who have not. However, the survey also found that 70 percent of respondents feel there are not enough HotSpots yet deployed to make the service worthwhile, and 74 percent are not willing to contract with multiple providers in order to make use of the service. The survey also points out that the demand for wireless access is price-sensitive—most business travelers indicate that they want high-speed service and are willing to pay *something* for it. But most respondents questioned some of the pricing models being used (e.g. most were unwilling to pay a monthly fee when they only use the service a few times a month).

That "something" may be far lower than HotSpot providers expect. A mid-2003 survey conducted by ForceNine Consulting finds that while demand among consumers for Wi-Fi is quite strong, the sweet spot for pricing is no more than $1 per hour—much less than the $6 per hour being charged at Starbucks.

Despite the muddle pricing models, Gartner projects the worldwide number of HotSpot users to grow to 9.3 million users by the end of 2003, up from 2.5 million users in 2002. Of that 2003 total, North America will account for 4.7 million users, followed by Asia/Pacific with 2.7 million, and Europe with 1.7 million.

Global Branding

In early 2003, noting Wi-Fi's rapid advancement in the wireless service provider sector, the Wi-Fi Alliance (formerly the Wireless Ethernet Compatibility Alliance), expanded its mission to include other Wi-Fi-related programs. One such program is called the "Wi-Fi ZONE program." The Wi-Fi ZONE was initiated to create a global brand for easier recognition of public access HotSpots. The program sets a minimum standard of quality for HotSpots before they can label themselves a Wi-Fi ZONE or display the ZONE logo (as depicted in Fig. 1.2).

The Wi-Fi ZONE brand can be used in conjunction with whatever brands are also applicable to the venue provider, e.g. Starbucks Coffee shops can show the Wi-Fi ZONE brand along with the T-Mobile logo to indicate they offer *quality* wireless Internet access to the public.

That is just the first phase of the Wi-Fi Alliance's ZONE program. The second phase is slated to include use of what it calls "the Wi-Fi ZONE Finder tool," a searchable database of qualified HotSpots (similar to what's available at www.80211hotspots.com). For more information on the Wi-Fi ZONE program visit www.wi-fizone.org.

Other non-profit organizations also address the branding issue. Pass-One (founding members include Wayport Technology, US Open Point Networks, France's Wifi-com, Canada's FatPort, and Sweden's Tele2) addresses the industry's lack of a roaming standard and how to coordinate activities within the WISP marketplace. Pass-One's main task is to specify the rules for authentication (SIM and browser-based), authorization and profit transfer, access management (including service barring), accounting and billing, tariff planning, settlement, and overall service delivery. But, like Wi-Fi Alliance, Pass-One is also creating a service mark or logo that could indicate that a particular venue allows Wi-Fi access to all Wi-Fi subscribers. When an end-user sees the logo, he or she will know that (1) quality Internet access is assured, and (2) that no matter which HotSpot operator the end-user pays subscription fees to, he or she can use any HotSpot of choice with no fuss or muss. The logo is intended to signify the same type of service that credit cards offer, i.e. you see a VISA logo you know that whether your VISA card is a Bank One card, a CitiBank card, or a Capital One card, the card will be honored.

Recognizable signage identifying HotSpot locales should help improve the public's awareness of the availability of HotSpots. "Even at locations where service is available, people are not really aware of it," Phil Belanger, a Pass-One spokesperson says. "A service mark will give that awareness."

In the Asia-Pacific region, a group of telcos signed a memorandum of understanding to establish a regional roaming framework for Wi-Fi HotSpots. The founding members of the Wireless Broadband Alliance include China Netcom, Korea Telecom, Malaysia's Maxis, Singapore's StarHub, and Australia's Telstra. These telcos currently control 8600 HotSpots, but that number is projected to rise to 20,000 by the end of 2003.

The key objectives of the Wireless Broadband Alliance are to drive the take-up of public access WLAN services among business travelers, establish a HotSpot service mark to indicate pre-determined technical standards for quality of service and conformity in user authentication and access.

Marketing Pointers

Before deciding that a particular venue is suitable as a HotSpot location, check out the venue's customer traffic. How busy is the location? The most successful HotSpots will have a fair amount of trade prior to offering wireless Internet access. Also look at the customer demographics. Are they college students, employees from local office buildings, residents of a local retirement community, nannys with their charges in tow, medical staff, overflow from a nearby hotel or convention center? A HotSpot's ideal customer base tends not only to be professional and technically savvy, but people who use a HotSpot as an "office away from the office," e.g. business travelers, freelancers, sales people, independent consultants, journalists, writers, graphic designers and telecommuters.

Ideally, prior to launching the HotSpot, put your marketing techniques in place to drive traffic to the venue and to entice the customer to use your wireless service. For example:

- Literature extolling the virtues of Wi-Fi and instructions on how to use the HotSpot's features should be prominently displayed at the HotSpot's entryways and made available for customers to take with them when they leave.
- Signage displaying the SSID, company logo, and other pertinent information should be visible.
- HotSpot operators should encourage (via incentive programs) venue owners to promote and sell its Wi-Fi services.
- Offer regular Wi-Fi customers free use, on the stipulation that they actively promote the HotSpot, and, when available, aid new customers in setting up an account on their computing device.

- Develop a referral program whereby customers can receive free service when they refer a new customer.
- Offer sign-up bonuses, such as a free wireless networking card if a customer signs up for a 12-month contract.
- Notify the local technical community via direct mailing (email or regular) of HotSpot locations in their area.
- List your locations in online HotSpot directories, e.g. Broadband Wireless Exchange Magazine (www.bbwexchange.com), Ezgoal.com, Hotspot-Locations.com, 80211hotspots.com, WiFinder.com, and Swiss Hotspots.ch. Also register with Riverwalk Software. This company maintains a directory of HotSpots/FreeSpots which is available for download to computing devices in any of the following formats: in Microsoft Excel, Palm Address Book, Handmark's MobileDB, and comma delimited ASCII formats. HotSpot providers can add their sites to the database by emailing wifi@busdevcenter.com. Freeware providers can add the directory to their site by contacting busdev@makewireless.com.
- Issue press releases every time a new HotSpot goes online. Or when your company reaches a milestone, e.g. 100th HotSpot goes online, or one millionth end-user to sign up for your service, partnership agreements and so forth.
- Send out monthly or quarterly newsletters via email to everyone that's signed up to receive such mailings.
- List your website with all major search engines and develop linkage relationship with other websites.
- Advertise HotSpot locations in local newspapers and magazines.
- Develop promotional programs with venue owners, e.g. buy two donuts with a large coffee and get one hour of free access, or buy one hour of Internet access and get a free coffee.

ROAMING AGREEMENTS

Roaming will play a significant role in the development (or nondevelopment) of a nationwide network of HotSpots, providing end-users with seamless access in diverse geographical locations.

For the WISP industry the term "roaming" refers to the agreement between two or more HotSpot operators whereby every signee's subscribers can access the other signees' networks. (Whereas the ability to seamlessly travel from one coverage zone to another within the same HotSpot operator's network without losing the session is called a "handover.") Roaming agreements are the answer to the "islands of connectivity" syndrome that dogs the average HotSpot operator. The typical roaming agreement regulates and stipulates the technical and commercial conditions of such access and is a pre-requisite to initiating roaming services.

Phil Belanger, a Pass-One spokesperson says that, based on lessons learned from the cellular industry, enabling roaming between different HotSpot operators is a key prerequisite for the success of the WISP industry. "The classic model for distributing income is the cellular world, in particular GSM. In Europe, GSM is very successful in getting roaming across what used to be very separate domains."

To prosper, the HotSpot industry, as a whole, needs unified roaming since it is the best way for end-users to avoid the hassle of needing to sign up for multiple accounts in

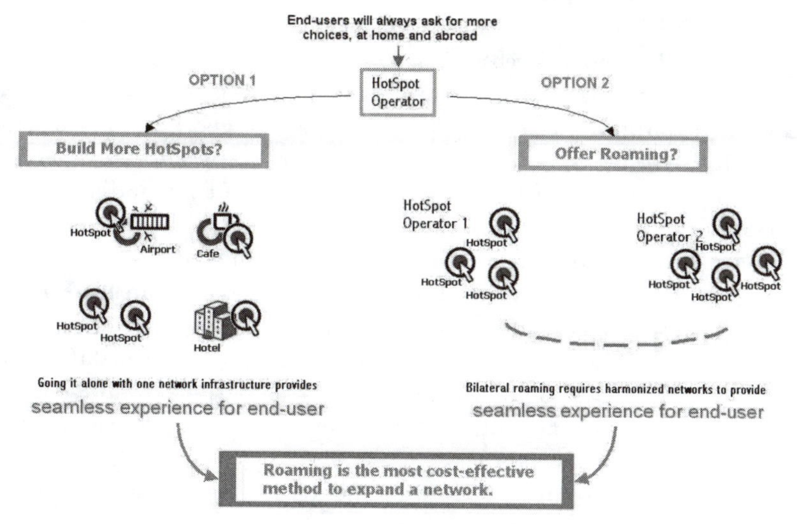

Figure 11.10 This shows the choices available to HotSpot operators who want to provide their end-users with an ever-wider coverage area, but are unwilling to go through an aggregator. Both options require the operator to build and maintain extensive back-office systems. But perhaps the bilateral roaming model presents the most challenges. Roaming end-users demand the same user experience no matter which network they are using. To provide such uniform experience, however, requires that the roaming partners harmonize their operations, e.g. AAA, assess management, service provisioning, content, billing, and end-user assistance.

order to obtain ubiquitous connectivity. Roaming also means more opportunities for recurring revenue throughout the WISP industry.

At the moment, HotSpot operators have a number of choices when it comes to expanding their area of coverage in an effort to provide network subscribers the unlimited coverage they demand: go it alone (similar to what T-Mobile is doing); grow their network via individual bilateral roaming agreements (this is how many currently operators provision roaming capability); partner with one or more aggregators and hand the roaming details over to them; or join Pass-One's roaming initiative.

HotSpot subscribers look for places to access the Internet, whether they are within their home network's territory or not. That's what roaming between non-related HotSpots (such as between WayPort's HotSpots and Cafe.com's HotSpots) provides. For the HotSpot operator, roaming offers the possibility to expand its footprint without investing in additional infrastructure.

***Note:** A "bilateral roaming agreement" is a roaming agreement signed between two service providers directly, whereas a "multilateral roaming agreement" is a roaming agreement signed between a HotSpot operator and a central legal entity representing a community of member HotSpots, such as an aggregator or Pass-One. Such an agreement allows for flexible tariffs with certain roaming partners, while the entire administration is centralized.*

Going it alone requires a huge bankroll. For the HotSpot operator with a vision of a nationwide chain of HotSpots, the network needs to have a robust back-office system and a diverse workforce. The necessary team includes a sales force to sign-up venues, staff to design and maintain a website, a skilled technical staff to manage and maintain the network, and a visionary executive group. A high-profile marketing campaign is also needed to drive traffic to the new HotSpots. Systems to support the growing HotSpot network will also be needed—AAA, asset management, service provisioning, content, billing (format content and methodology), customer support, and end-user assistance.

If a HotSpot operator brings bilateral roaming into the mix, however, the staff can concentrate on its strengths such as providing HotSpot capabilities to the hospitality industry (e.g. RoomLinX) or to travel venues (e.g. Wayport). By entering into bilateral roaming agreements, HotSpot operators allow authorized outside non-customers to use their network. However, this scenario requires HotSpot operators to take on different roles—a HotSpot operator is both a home network for its own customers and a visited network for the roaming customers of another network, i.e. they offer service to their own end-users as well as external end-users.

The home operator assures the payment of usage by its own customers, and guarantees the income from inbound roaming to the visited HotSpot operator. This reduces financial risks. Thus, a HotSpot operator offers its network capacity to incoming users because payment is pledged by a "guaranteed payer." Sound complicated? It is. This model requires the roaming agreement signatories not only to put in place extensive back-end systems, but also the roaming partners must harmonize their operations in order to provide their subscribers with a seamless roaming experience.

NetNearU, a provider of management and billing systems for HotSpot systems, has set up roaming agreements with several prominent HotSpot operators. According to Cody W. Catalena, the provider's executive vice president, "Operators that deploy a Net-

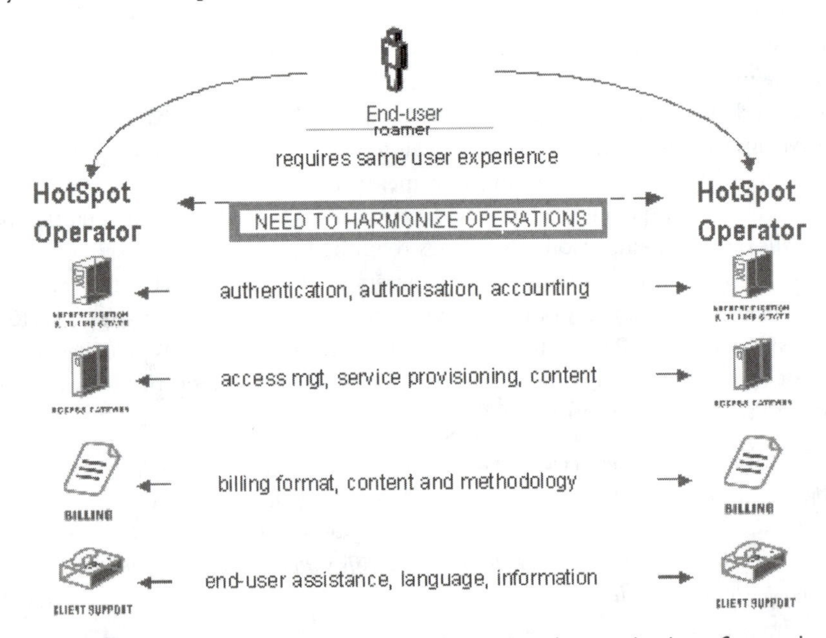

Figure 11.11 The challenges to offering bilateral roaming—harmonization of operations is essential to providing end-users with a successful roaming experience.

NearU system can provide seamless roaming for their user bases and others to roam into that network, and receive the roaming revenues. The revenue model is signing up your own user base, and then being able to accept users for other networks on your network and get a fair share of the revenue. Roaming will follow the same model as the cellular market, where first you were charged for roaming, but now can get the same rate for roaming in a national plan."

Another option is virtual roaming. This model requires an aggregator and multilateral roaming agreements. Most in the industry feel that for a national network of HotSpots to become a reality, virtual roaming must become the norm. And although aggregators are crafting partnerships with various HotSpot operators, especially in North America, there is still much to do before national and international roaming is a fact of life.

When a HotSpot operator becomes a partner to a *multilateral* roaming agreement, that operator turns most of the headaches (e.g. record keeping and housekeeping) over to the aggregator. No wonder many operators feel that entering into multilateral roaming agreements is the best way to offer their end-users more value. By signing such agreements, a HotSpot operator can offer a magnitude of HotSpot accessibility outside its own network without becoming burdened with more back-office chores. The financial advantage also is compelling since the payment flow is focused on the roaming partners and the inter-operator wholesale charge typically includes a mark-up fee.

Thus, both the bilateral and multilateral roaming models enable a HotSpot operator to reduce the need for additional investment in network infrastructure while optimizing the capacity of its existing HotSpots. Both also have the potential of increasing the operator's overall revenue stream.

There's one more model to investigate—Pass-One's approach to universal roaming. In September 2002, Pass-One members decided to open their networks to each other's end-users. To facilitate such roaming capability, the organization agreed upon an "inter-WISP roaming framework" that sets the roaming policies applicable to new members (e.g. the charging principles—wholesale and retail tariff, charging reference per time and timeslot, roaming relationships, invoicing and settlement rules, and a guaranteed minimum set of service levels that will warrant a global end-user experience). The result is that end-users of, for example, Wayport (U.S.), FatPort (Canada), Open Point Networks (U.S.), Wificom (France), and Tele2 (Sweden) are allowed to use the combined network of all HotSpots operated by each member.

There is also Wi-Fi Alliance's WISPr approach. While WISPr doesn't offer a roaming model, per se, it does create *recommendations that facilitate* inter-network and inter-operator roaming with Wi-Fi-based access equipment. WISPr's charter is to "describe the recommended operational practices, technical architecture, and Authentication, Authorization, and Accounting (AAA) framework needed to enable subscriber roaming among Wi-Fi-based Wireless Internet Service Providers (WISPs)." The resulting roaming framework is designed to allow end-users with Wi-Fi compliant devices to roam into Wi-Fi-enabled HotSpots for public access and services. Users can be authenticated and billed (if appropriate) for service by their home network provider.

That framework is laid out in a WISPr document released in February 2003, entitled "Best Current Practices for Wireless Internet Service Provider (WISP) Roaming." It is available for downloaded at www.wi-fi.org/opensection/downloads/wispr_v1.0.pdf.

The document, in short, recommends that the WISP industry adopt a browser-based Universal Access Method (UAM) for the public to access the Internet via Wi-Fi technology. The UAM allows a subscriber to access a HotSpot's network with only a Wi-Fi network interface and Internet browser on the end-user's device. It also recommends that RADIUS (Remote Authentication Dial In User Service) be the back-end AAA protocol that supports the access, authentication, and accounting requirements of WISP roaming. To that end, the document describes a minimum set of RADIUS attributes needed to support basic services, fault isolation, and session/transaction accounting.

Another organization taking a stab at addressing roaming issues is the Asia-based Wireless Broadband Alliance. It adopts a holistic approach by establishing three working groups to address the legal and administrative framework for roaming agreements, the development of a uniform technical approach, and the cross-promotion of HotSpot related services via a single international brand. Its inter-operator roaming platform is expected to be operational by June 2003 and the Alliance says that it is in discussions with other international telcos with regard to expanding its coverage worldwide. It is thought that the Wireless Broadband Alliance will seek to leverage existing telco roaming and clearinghouse facilities.

THE TECH SIDE OF CHOOSING A HOTSPOT

Whether you are a venue owner who wants to run your own HotSpot, a HotSpot operator, or a FreeSpot organization that wants to set up a HotSpot, the first consideration must be the location. Your venue must be in a desirable locale, be able to attract a viable number of customers interested in wireless high-speed Internet access, and be capable of receiving high-speed Internet connection, e.g. DSL, Cable Modem or T-1.

Once you've settled on the venue, obtain floor plans and/or layouts of all areas that the wireless network is expected to cover. These are the reference tools the installer needs for a feasibility study. For instance, such plans/layouts can help to determine, based on the venue's needs, where the access points should be located to accommodate the necessary wireless coverage. Mark those locations on the floor plans/layouts. Bear in mind, though, that the feasibility study only provides theoretical data. Only after performing a complete site survey can an installer determine which areas to include in the wireless "net" and which areas to avoid due to signal barriers (e.g. radio waves can pass through walls and glass but not metal), cable length restrictions, and so forth. From the beginning, take pains to ensure that everyone involved in the HotSpot venture is aware of such caveats. These same reference tools can also be used to coordinate installation of the HotSpot, and once installation is completed they can be invaluable when someone unfamiliar with the terrain performs routine maintenance, troubleshooting, etc.

After it is determined that a HotSpot is feasible, the next step is to obtain a signed location agreement authorizing the installation of the HotSpot network. If a HotSpot operator is involved in the process, that operator must make certain that it is given full rights to monitor remotely all activity that occurs on the network, and that the document includes a provision that authorizes the placement of marketing literature (posters, flyers, instructional leaflets, etc.) in prominent locations throughout the HotSpot.

With those items taken care of, the installer is in a position to conduct the site survey. A site survey allows the installer to understand any variables not shown on the floor plan that may affect wireless coverage in a specific area.

Some of the factors that will affect whether a wireless net can envelope a particular area include:

Signal Barriers. Reinforced metal in the structure of concrete walls can block a radio signal from traveling between rooms. Floors with steel girders and other metal material can block wireless signals from traveling between floors. Even in open areas, surrounding large metal walls can cause signal reflection that will cause a reduction in the data throughput rate.

Drop Points. It's advisable that the installer design a network so that each drop point (equipment installation areas) is easily accessible. This facilitates future maintenance and/or troubleshooting. Will there be ceiling drop points? If so, does the venue have on hand the ladder or scaffolding for the installer to reach that area? Another consideration is availability, i.e. when will the areas around the individual drop points be available for installation personnel?

Electrical Outlets. Each access point will need to plug into an electrical outlet (unless Power-over-Ethernet APs are used). Is there an outlet readily available where the AP(s) is to be placed? If not, one needs to be installed. Don't use extension cords.

Network Requirements. Although a HotSpot provides wireless Internet access, some cabling is still necessary. The cabling connects the Access Point(s) to the wired network where the backhaul is located. The typical cable used in such instances is CAT 5 cabling which has a length limitation of 328 ft. (100 meters) per cable run. Therefore, if the coverage area is more than 328 feet away from the data pipe to the Internet connection, one or more intermediate devices (hub, router, switch) along with the necessary power source will be needed to connect the backhaul point with the coverage area. Label the cabling (at each end of the run) as it is installed, describing each drop point location. Most venues won't want the cabling (or access points for that matter) to be visible to customers. The installer will need to run the individual drop points in such a manner that the cabling is hidden from view, such as above tiled ceilings, or covering the cable so that the run is aesthetically acceptable. As far as the access points, hubs / routers / switches are concerned, these devices should not be seen, nor should they be accessible to the general public. Once cable installation is completed, the installer should provide a written installation confirmation setting out that all cable segments have been tested thoroughly, and specify the length of each installed cable segments. Attached to the confirmation should be a cabling diagram indicating exactly where the cabling is located in reference to the floorplan/layout.

After completing the site survey, sit down and discuss the viable coverage areas and how it might impact the practicality of installing a HotSpot at that particular location.

BUSINESS PLANS AND LEGAL ISSUES

A good business plan helps a business obtain financing, arrange strategic alliances, attract key employees, and boost the entrepreneur's confidence. A business plan should sell the start-up to the world.

An AT&T study asked entrepreneurs to rate their overall success. Those that had written business plans (42%) rated themselves more successful than the entrepreneurs who said they had skipped the business plan stage (58%). Some of the ways in which a business plan can help a WISP related company to succeed include:

Obtaining loan approval: Since there are more businesses seeking loans than banks have money available, only those that make the best case will receive funds. A well-thought-out business plan can improve a company's chances of getting the loan because it carries an important message—the company's executives are serious enough to do formal planning. In the eyes of the loan officer, this makes their company a better risk and more deserving of a loan than applicants without a viable business plan.

Obtaining venture capital: A business plan is the first thing a potential investor asks to see. Investors use business plans as a screening device. If they are intrigued, they will ask the company's executives to come in for further discussion.

Arranging strategic partnerships: Many times a small start-up will align itself with a larger company to carry out joint research, marketing, and other activities. Such partnerships can mean gaining access to important financial, distribution, and other resources. But before a large company will even consider a strategic partnership, it will want to see their business plan.

Obtaining contracts: Before a start-up can convince another company to sign on the dotted line, they must convince the potential client that the company will be around long enough to fulfill its contractual obligations. A well-written business plan can go a long way toward reassuring a reluctant prospect.

In addition to the technical and commercial considerations discussed in this chapter, there are also legal concerns that members of the WISP industry must address. The most prominent may be liability arising from end-user abuse of the network.

Here's one way a HotSpot can run into problems. The Internet access that a HotSpot provides typically is made available via NAT (Network Address Translation), which allows administration of dynamic IP addresses by the serving ISP's DHCP (Dynamic Host Configuration Protocol) server. This means that the HotSpot's entire network is viewed as a single IP address (the HotSpot's). If an end-user performs any illegal or immoral activity via that access point, the first place law enforcement will visit is the HotSpot location. To alleviate risk, HotSpots should require end-users to login before they can gain Internet access. The login process means the HotSpot servers have captured the end-user's personal information.

Even if the venue owner provides free wireless Internet access, in order to minimize risk, it should require all end-users to register, and then authenticate the user every time they log onto the network. At a minimum, the venue owner should require its end-users to read a legal disclaimer and click on an "agree" button before they are granted access to the network.

Again, I lean on the expertise within the pre-eminent law firm of Wiley Rein & Fielding LLP (WRF) to address such issues as structuring a business model to avoid regulation. As WRF set out in its Wi-Fi primer entitled "Wi-Fi—802.11: The Shape of Things to Come":

"The Federal Communications Commission regulates, among other things, communications activities by common carriers. Services—like most ISP offerings—that are considered 'enhanced' are not, however, considered common carriage and therefore are not subject to direct FCC regulation at this time. Because there are a variety of significant tax, mandatory contribution (e.g., Universal Service Fund, Telecommunications Relay Service Fund), and other regulatory obligations (e.g., requirements to charge rates that are just and reasonable and not unjustly or unreasonably discriminatory, submission to the jurisdiction of the FCC formal and informal complaint process) attendant to common carrier status, many communications providers have sought to ensure their operations are classified as enhanced, and therefore unregulated by the FCC.

"While the issue of whether a particular provider is an enhanced service provider or a common carrier can be complex, and may warrant consultation with regulatory counsel, providers seeking to avoid common carrier obligations should, at a minimum, avoid selling service consisting of "raw" transmission capacity for a fee. Typically, ISP services considered enhanced include such a transmission component that is ancillary to the core offering, but coupled with the Internet access functionality (which typically includes enhanced attributes such as content, storage, email, and protocol conversion) that is the offering actually being sold. While most HotSpot providers and aggregators, for example, thus fall into the category of enhanced service providers, it is possible that coupling the hotspot offering with certain other types of services (e.g., a voice telephone service) could trigger FCC or state regulation."

The document can be downloaded from the WRF website at www.wrf.com/db30/cgi-bin/pubs/WiFi_Primer_Final.pdf. Want more information on common carriage and FCC obligations? Contact Eric DeSilva at edesilva @wrf.com.

Other legalities the WISP industry should be aware of include general end-user contracting guidelines, privacy issues, and liability issues, including those that might arise under the Communications Decency Act and the Digital Millennium Copyright Act. Wiley Rein & Felding LLP's aforementioned Wi-Fi primer also touches on all of these issues.

THE VALUE CHAIN

A value chain is a group of organizations working together to progressively add value in response to a market opportunity. As such, a value chain spans vertical and horizontal relationships within and across industries. While there are many different ideas as to what segments make up the WISP Industry's value chain, it basically comes down to brands, aggregators, HotSpot operators, venue operators, backhaul service providers, gear manufacturers and vendors, technology providers, and end-users.

Brand ownership, value chain management (where aggregators play a huge role), and role specialization (venues versus operators) are crucial elements in the WISP Industry's overall global marketplace. Big brands, such as cellular operators T-Mobile, Cingular and Verizon, are likely to become industry front-runners because their deep pockets can make the Wi-Fi transformation from a hobbyist fad to big business. But these big guns still must focus on their core competency, which for many isn't Wi-Fi. Many within this group will choose to partner with others in the industry, rather than using their capital to build out a Wi-Fi network on their own. The same is true for the technology sector (e.g. Cisco, Intersil, and Sony), which will probably steer clear of the Wi-Fi network.

Each company must determine how best to use their capital as a participant in the Wi-Fi value chain. That could mean a role as a brand owner, a technology provider, an infrastructure provider, a HotSpot operator, an aggregator, or some other meta-market yet to be determined. That's not to say that the various sectors won't overlap and won't share in ownership of a company that has a strict focus on Wi-Fi networks, e.g. IBM, AT&T, Intel and the Cometa Network, or the Toshiba/Accenture partnership.

While the nascent WISP industry has thrown out challenges to the communications industry's establishment, these telcos have been slow to assemble the right mix of participants in their value chain to meet the WISP industry's vision and needs. For the WISP industry to grow and prosper, its vision has to be backed up with robust delivery. That delivery requires speed, flexibility and scalability, which is why entrepreneurs like Boingo, FootLoose, and Surf and Sip are so important. And while some time in the future, the establishment may try to muscle the others out of the way, in the end it will probably be a mix of the new and old that gives the providers the impetus to speed Wi-Fi toward universal acceptance.

To understand what I mean, let's look at how Cometa Networks envisions the WISP Industry's value chain. Cometa's target market is the established telecommunications carrier, both wired and wireless. To convince this target market to get on-board, Cometa has drawn up an ambitious sales and marketing plan: to sell the target market on the promise that together they can build a national, enterprise-class Wi-Fi network, which the carriers can resell to their customers.

Once the Cometa carrier partners are onboard, they then must convince their customer base, especially their corporate customers, e.g. Coca-Cola, General Electric, Hewlett Packard and World Bank, that coverage will be reliable and so pervasive that the corporate community will change its business processes to accommodate this new form of data access.

But that's just the beginning. To attain its ambitious national footprint, Cometa must also sell the concept (and share revenue) to national brand retailers. After all, they are the group in the trenches—the group that will eventually host the projected tens of thousands of HotSpot locations. As any HotSpot operator can tell you, convincing these sometimes reluctant retailers of the benefits of hosting HotSpots in their venues can be difficult, even when promising to share revenue based on the percentage of total revenue their venues bring in.

Cometa executives admit it is a huge undertaking. The Cometa network includes a value chain that encompasses not only the carriers and venues, but also the HotSpot installs. While IBM is in charge of this portion of the value chain, it will need to work with others, e.g. vendors, engineers, integrators, and network aggregators such as Boingo, FootLoose, Gric, and iPass.

A NATIONAL WI-FI NETWORK

The companies attempting to build large-scale Wi-Fi networks take different approaches. One is the top-down network approach where the networks are built in the traditional way—by network operators who then charge a fee for access. The other is the bottom-up network approach whereby the network is built through a loose federation of HotSpot operators who offer access to all within the "federation." While both approaches have their problems, the top-down network model is dependent upon users willing to pay to use a fragmented service, whereby different venues are served by different networks with no single network providing wide coverage. This means that the end-user will need multiple accounts and passwords to stay connected.

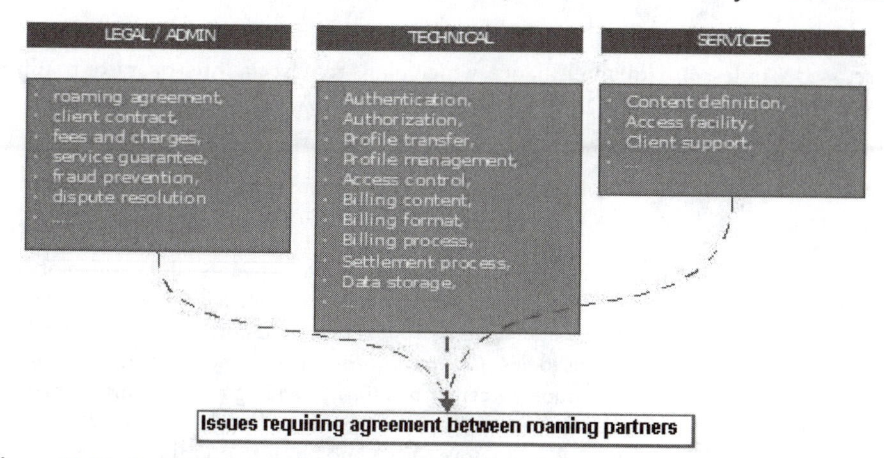

Figure 11.12 Aggregators are best suited to handle the legal, administrative, technical and services issues that arise during multilateral roaming partnerships.

The bottom-up approach presents a management nightmare unless there is a middleman to amass and manage the federation of HotSpots, such as an aggregator.

No matter which approach is used, the industry is coming to terms with the fact that roaming is essential and thus has begun moving in that direction—with the help of the aggregator crowd. Still, a concerted national effort is needed to create a national, virtual, single Wi-Fi network.

Some type of confederation is needed to put this virtual wireless network together. Although some aggregators might try, conventional wisdom is that neither a large aggregator (such as GRIC) nor a wireless network provider like Cometa can build, support and maintain a national network. The scope and scale of operation that is required for such an undertaking requires a massive organization. One such organization that has stepped to the plate is Pass-One.

One of the first tasks Pass-One took on was the crafting of a standardized multilateral roaming agreement. It hopes to have a final draft of this agreement completed sometime before the end of 2003. According to Phil Belanger, a Pass-One spokesperson, "It's similar to the ATM card model." But, Belanger warns, "It is only compelling if it's global."

When completed, the roaming agreement will require that all signing members, large or small, roam with every other member. Pass-One expects that within a year after its multilateral roaming agreement is finalized, any new member that joins the organization will automatically sign the agreement, allowing their end-users to roam with the other members of Pass-One. "The idea is the same as public access Internet. You might be a subscriber with a HotSpot in Nebraska, and when you get on a plane and go to Monte Carlo, if you see the Pass-One logo you won't have to give the local venue anything to use it. You won't have to verify. You will just get the service, and get one bill." Belanger also indicates that One-Pass would like to work with other groups, like Cometa.

It's Pass-One's hope that this will lead to a large international footprint. As more members come onboard, their end-users will enjoy the beginning of an international wireless network powered by Wi-Fi. But HotSpot operators must become a member of Pass-One to participate in this potentially international WLAN network.

Another organization working on a pervasive virtual wireless LAN is the previously discussed Wireless Broadband Alliance, which plans to have an inter-operator roaming platform in effect before mid-2003. Look for this group to move quickly to expand its coverage across national borders.

COMETA

AT&T, Intel, and IBM have formed the company Cometa Networks, along with global investors Apax Partners and 3i. The new company plans to build a large number of Wi-Fi HotSpots in 50 U.S. metropolitan areas by 2005. These HotSpots are slated to enable end-users to retain consistent logins, IDs, passwords, and payment methods as they roam. This one-stop Wi-Fi shopping center has IBM providing the back-office infrastructure, AT&T delivering the IP network and backhaul, and Intel using its integrated Banias mobile processing software to drive the technology into portable devices.

Cometa plans to sell its services to two distinct market segments. The first consists of large venue owners, such as regional or national convenience store chains, gas stations, and restaurants. The plan is to split the costs of installing the HotSpots. However, the company hasn't shared its revenue-sharing plan.

IBM will install the WLAN gear. AT&T Wireless will handle the initial authentication and backhaul connections by working with local carriers for DSL or T-1 connections.

The second consists of telecom carriers (both wired and wireless) that want to add a wireless LAN service to their data offerings.

Cometa's chief operating officer Joe Gensheimer suggests that a number of Cometa HotSpots, while bearing the logo of a particular carrier or retail chain, may end up working like today's ATMs, which handle a broad range of different brands of ATM cards and credit cards. "We're neutral. There might be five people connecting to one HotSpot. A Sprint customer will think it's a Sprint HotSpot, the iPass customer will think it's an iPass HotSpot, and an AT&T customer will think it belongs to AT&T."

Cometa's partners, whether a wireless or wired carrier, ISP, or some type of WISP, can then resell the services to their customers (venue owners or individual HotSpot operators), if they so desire.

The big question: "Does the emergence of Cometa spur or further confuse the WISP marketplace?" On the cellular/Wi-Fi partnership level, companies such as Cingular, which have hinted that they'd rather buy their way into the WISP marketplace, are delighted that a group of industry heavyweights have taken on the task of building a nationwide Wi-Fi network in which they can participate. On other levels, carriers such as T-Mobile, which already has significant investment in HotSpot operations, will look to Cometa to help fill any gaps in its coverage.

Cometa's co-founder Larry Brilliant believes that Cometa's projected ability to accommodate users in a seamless way provides a perfect business model. He says that the multiple competing market channels Cometa can tap into through wholesale agreements can achieve scale, although it will take some time for the new network to reach critical mass.

Cometa's business model, however, hinges on one critical factor: the willingness of Cometa partners to share the same network and to agree to use AT&T's backhaul services. For although the facility and backhaul cost savings look good on paper, Cometa may have a tendency to build HotSpots only in locales it finds most economically beneficial, for example, only where AT&T already has backhaul in place and where Cometa can strike favorable deals. These locales, however, may not be where HotSpot operators want to target key subscribers and differentiate service offerings. And, in fact, that issue has already come up in Cingular Wireless's negotiations with Cometa.

Still, although Cometa is poised to become the largest HotSpot provider in the U.S., it probably won't be the single-source solution. Not only for the reasons listed previously in this chapter, but also because the company doesn't plan on aggregating access to already established HotSpots. Brilliant and others believe that closing off roaming with other Wi-Fi providers is the only way to ensure quality of service and offer customers a simple way to sign on and register themselves on their network. Although the company did mitigate its stance somewhat by saying that if HotSpot operators want to meet Cometa's technical specifications and allow the company to migrate their HotSpots into its network management center, Cometa would be amenable to including them in their network.

Nonetheless, Cometa should come to terms with the fact that partnering with aggregators like Boingo, GRIC, and iPass is the only way for it to provide optimal nationwide coverage. That may happen. On March 4, 2003, Cometa Networks announced that it had signed iPass as its first partner in its ambitious plan to set up a network of 20,000 HotSpots in the next two years. The realization of such a plan would represent a giant leap forward for Cometa. On iPass's behalf, Cometa is just one more service partner, albeit on a larger scale than its other partners (e.g. Wayport). iPass uses its partners to provide the HotSpot locations, Internet access, and backhaul connections while iPass works behind the scenes handling the authentication, billing, and settlement for the HotSpot operators. John Russo, vice president of marketing for iPass, commenting on the partnership deal, said, "We'll take the Cometa [Wi-Fi] infrastructure and make it part of the iPass virtual network. It extends the number of useable hotspots, the user continues to see only the iPass user interface."

So, as you see, Cometa's business model is a bit different than what we've seen previously. But if it works, it could definitely accelerate the Wi-Fi explosion.

CASE STUDIES

Hotel Okura Amsterdam

Hotel Okura Amsterdam, is one of Amsterdam's leading business and convention hotels. As such it is committed to offering the best accommodations and service in the area by focusing on total guest satisfaction. Hotel Okura also wanted to be the first hotel in Amsterdam to have a wireless local area network (WLAN) and to offer wireless broadband Internet facilities in its lobby and lounge.

The hotel found itself fielding more and more inquiries from customers about the possibility of high performance broadband Internet facilities in their rooms. The hotel realized that by providing its guests with the ability to wirelessly access the Internet from anywhere within its facilities it could increase revenue by driving more business.

After researching the issue, Hotel Okura's management chose IBM and Cisco to help the hotel provide the Internet access its customers were increasingly demanding. The management's research indicated that those two technology providers had the ability to meet their total hardware, software and financing needs.

IBM Global Services - Integrated Technology Services (ITS) helped the hotel design and implement its new WLAN. ITS developed a plan for moving the current network to a WLAN and advised the hotel on technical issues such as network infrastructure and security. Then ITS moved forward with a project plan, which it carefully aligned with hotel activities, resources and timeline. Finally ITS deployed all of the WLAN components.

The new wireless infrastructure offers guests fast and safe access to the Internet and the hotel's intranet.

Hotel Okura management says that it anticipates that by offering high-speed Internet access for the guests, who are mainly business customers, it will be able to increase its occupancy rates from 65 percent to 85 percent. According to a hotel spokesperson, higher occupancy rates also will result in an increase in guest spending on other services (i.e. dining) by approximately 1.5 million per year.

The hotel also anticipates that by offering wireless broadband service, it will increase guest satisfaction and reinforce the hotel's image as an innovative facility.

Pechanga Resorts & Casino, Temecula, California

To be successful in the hotly contested hospitality industry requires that a venue operator offer its guests a distinctive array of services, including wireless Internet access. In 2002, while building its new hotel in Temecula, the resort's management, recognizing the competitive advantage of providing high-speed wireless Internet service to business guests, searched for a wireless solution that would be easily managed and monitored, and could work with its wired Ethernet network.

After consulting with a local value-added reseller (VAR), Technology Integration Group, and considering wireless products from both Cisco and 3Com, Pechanga management chose the 3Com solution, mainly for its ease of installation and use. For example, they were impressed with the 3Com 802.11b Wireless LAN AP 8000's web-based management, which enables administrators to easily adjust configuration parameters, run diagnostics, and monitor performance from any networked web browser.

The wireless LAN provides coverage in the convention center, ballrooms, meeting rooms and swimming pool of the new 522-room Pechanga Resorts & Casino in Temecula. The WLAN enables Pechanga guests with mobile computing devices (laptop or PDA) to stay connected, throughout the hotel property. Powered by the practical, high-value 3Com solution, the 2700-employee hotel hopes the WLAN will help to attract thousands of additional business customers per year to its facilities.

"The 3Com wireless solution allows us to provide our business guests with a unique first-class amenity," said Rod Luck, director of information technology, Pechanga Resorts & Casino. "Thanks to our 3Com wireless solution, our on-the-go guests easily connect to their corporate intranets, conduct videoconferences and do everything they would be able to do in the office—and we've gained a valuable competitive advantage."

For instance, 16 3Com APs were strategically installed so as to link end-users to the hotel's wired Gigabit Ethernet network and three 3Com Ethernet Client Bridges (ECBs). This wireless infrastructure enables Pechanga to offer its convention guests a variety of choices for Internet connectivity. Any guest whose computer lacks wireless capabilities can effortlessly connect to the Internet by plugging the device into one of the wallet-sized ECBs and they are off and running—able to quickly and securely conduct business with customers and clients.

The 3Com APs can support any Wi-Fi certified PC Card, allowing guests to easily connect to the Internet using their pre-existing system configurations. But if a guest wants, the hotel does provide 3Com Wireless LAN PC Cards to plug into their laptop. With features such as 128-bit encryption and Dynamic Security Link, the 3Com PC cards protect against data theft. The cards also interoperate seamlessly with guests' corporate virtual private networks (VPNs) and security solutions to ensure trouble-free operation. This, in turn, helps to eliminate end-user support burdens that might fall on Pechanga employees' shoulders.

Since each 3Com AP extends wireless coverage for up to 1000 feet and can support up to 256 simultaneous users, the resort complex can accommodate meetings and conventions of virtually any size. Luck adds, "3Com's high-value wireless solution enables us to provide a convenient, premium-quality service that greatly enhances business travelers' experience and satisfaction." He concludes, "As a result, our new hotel and convention center is better equipped to attract and retain them as customers, and we couldn't be more pleased."

The World's Largest HotSpot

Technology giant, Cisco Systems; consulting firm, Cap Gemini Ernst Young; and RATP, the agency that runs the Paris Metro, have formed a partnership called "Wixos" in order construct a system that can envelope Paris in a Wi-Fi "cloud." The partnership hopes to make Paris the world's largest HotSpot.

The consortium is installing two or three Cisco access points outside each of Paris' 372 Metro stations and then linking these APs to an existing fiber optics network in the subway tunnels. This allows RATP to provide the Internet backbone. The "cloud" also will extend to the street, thereby allowing anyone in the vicinity of an AP to access the Internet.

The initial pilot project is centered along Bus No. 38's route. Interested users can subscribe to the pilot, which initially is offered free of charge.

Jean-Paul Figer, Cap Gemini's chief technology officer, told the *International Herald Tribune* that he thought of wireless connectivity as similar to a television remote control: "We believe that giving this connectivity will develop a lot of new applications. Trucks, buses, cars, the same application has much better value if you get this kind of mobility. It's exactly like your TV remote control. It's only three meters, but it changes your life."

It is estimated that installation costs to blanket Paris will be less than $11.5 million. Although the Wixos partnership is building the system, the Internet connection is to be provided by other, separate commercial companies; for instance, Bouygues Telecom, Club Internet, Tele2, TLC Mobile, Wifi Spot, and Wifix.

CONCLUSION

The WISP industry is like the settlers of the American Old West. Its members are visionaries who see the future, create the future, and develop the future to ensure that HotSpots go forth and populate the land. Because of their cost-effective, sustainable, and scalable infrastructure, HotSpots and the entire WISP industry will be the basis for an IP communications infrastructure of the future.

There is a high level of interest and potential demand for Wi-Fi services, a fact that is well documented through numerous industry surveys. And yes, there are unresolved issues that have caused Wall Street to hesitate as others race to jump on the Wi-Fi bandwagon. Yet, given the high level of activity in the Wi-Fi arena, it shouldn't be long before many of the naysayers' concerns are settled to everyone's satisfaction.

Chapter 12: Building a HotSpot

As the reader should now understand, it is possible for established businesses to create additional revenue streams and for entrepreneurs to build a HotSpot without a large capital outlay for expensive, complex network hardware. The previous chapter discussed the "run of the mill" hardware necessary for the proper deployment of a small HotSpot. Turnkey solutions, however, provide an even easier method for HotSpot operators to deploy Wi-Fi service. That's because most HotSpot-related solutions provide an integrated group of functions and features, e.g. billing, security, and roaming and the provider offers them on a resale or revenue-sharing basis. Most offer the HotSpot operator (a/k/a "the reseller") a very attractive proposition—10 to 40 percent margins depending on the location and the reseller's sales and marketing efforts. Public access rates vary, but are beginning to settle to around $10 per day, depending on the venue's property type and geographic location.

HOTSPOT KITS

These solutions are dubbed "turnkey" or "HotSpot" kits and are increasingly available throughout the WISP marketplace. The kits provide a Wi-Fi access platform, as opposed to a Wi-Fi access point. These smart platforms deliver the fundamental services layer by providing standards-based services such as secure authentication, automated configuration, IP routing, firewalls, caching, remote management, usage tracking, and bandwidth shaping. Each kit requires only that the venue owner install the software, type in a little configuration information, and voila, a HotSpot is born.

Most of these kits offer the same basic benefits: a quick and inexpensive HotSpot setup; some type of revenue sharing plan; access to other network points (the kit provider's and perhaps its affiliates); software that, in most cases, solves many of the internal network security and management issues (but a closer look at the specific software is warranted); and back-end services such as AAA and customer support. But beware, some kits also may provide unexpected drawbacks, such as including the provider's brand on the end-user's start-up page rather than the brand of the operator or venue owner.

This section reviews some of the features and functionalities to look for when investigating the feasibility of installing a HotSpot using a turnkey kit. With most of the methods outlined herein, all that is needed is a broadband connection with a static IP address, a Wi-Fi-ready computer (PC or laptop), and the turnkey kit. A static address is a permanent IP address for which the ISP usually charges an additional fee—otherwise the ISP assigns a dynamic IP address that changes every time an end-user logs onto the service.

FatPort. This Canadian company has an interesting approach to deploying HotSpots. FatPort has named its turnkey solution, the "FatPoint," which is an all-in-one device that allows a new HotSpot operator to be up and running in a matter of minutes. The turnkey solution includes a FatPoint Server (an NS-GX which is i386-compatible 300MHz CPU with 128MB RAM and a 200mw Wi-Fi card with removable diversity antennae), marketing materials, and customer and technical support. FatPort utilizes the most secure encryption available to ensure end-users' personal data is protected. According to the company, the FatPoint solution also provides a network that is "as secure as online banking, so they can access their corporate networks from your business location, worry-free." The customer purchases time online or uses FatCodes (prepaid calling cards) sold at authorized dealers. FatCodes come in 15-minute, 60-minute, daily, and monthly increments. See Figs. 12.1 and 12.2 for more details.

FatPoint Intelligent Access Point

The FatPoint Intelligent Access Point is a true wireless-Internet-service-provider-in-a-box by its unparalleled ease of use. Anyone can offer FatPort wireless networking in their establishment just by taking a FatPoint out of the box and plugging it in. Three distinct service levels answer to every need.

	Price*	Features
FatPoint Complete	$199CDN/Month + an up-to 50/50 Revenue Split	Once you order the FatPoint Complete, we take care of absolutely everything - letting you focus on your core business. **Includes:** installed DSL Line & maintenance, FatPoint server & installation, staff training, marketing materials, customer & technical support, 1 free yearly account ***Only available in Canada***
FatPoint Express	$800 CDN + an up-to 50/50 Revenue Split	If you are located outside Canada, or already have your own high speed line, the FatPoint Express is an easy way of generating a new revenue stream hassle-free. **Includes:** FatPoint server, marketing materials, customer & technical support, 1 free yearly account.
FatPoint OEM	$800 CDN	FatPoint Server - Perfect for startup WISP's, community networks, developers and hobbyists, the FatPoint OEM is a stripped down version of our access point/authentication server. **Includes:** FatPoint Server

* Prices do not include local taxes or shipping - please contact fatpointsales@fatport.com for more details.

Figure 12.1 FatPoint allows a HotSpot operator to design a custom login screen that enables the venue or operator to brand the first screen subscribers see with news, special offers and up-and-coming events. Graphic Courtesy of FatPort.

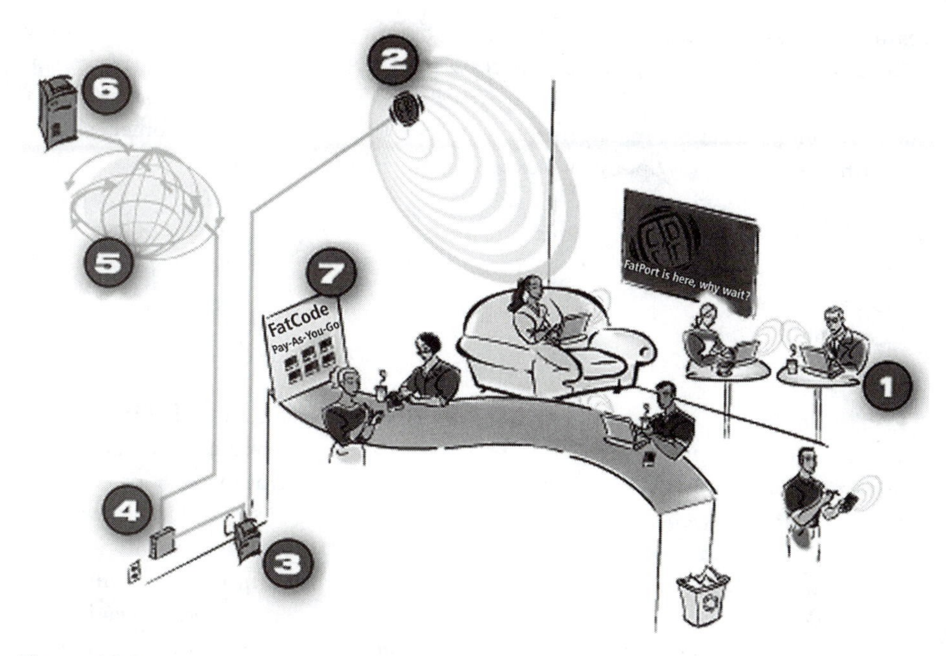

Figure. 12.2

1. The only piece of technology a customer needs to access a HotSpot is a computing device with a Wi-Fi 802.11b wireless card.
2. Just like in a cordless telephone, the Access Point sends and receives wireless Internet signals from the computing device. In FatPort's setup, the access point is contained within the FatPoint Authentication Server, a tiny box located in a cupboard or under the counter (at least in most locations).
3. FatPoint Authentication Server acts as the gatekeeper to the Internet and captures all requests for service from the customer's computing device.
4. The point where the high-speed connection to the Internet, typically a DSL line, enters the location. Using WWW Secure Tunnels, FatPoint sends the login requests and purchasing information to FatPort's FatCheck servers. Once connected, the end-user can access the Web, email, and even get to their company's internal network via a Virtual Private Network (VPN).
5. All transactions with FatPort are done through a secure "tunnel" via the World Wide Web. Data sent over this link is as secure as at a bank's website at 128-bit encryption.
6. FatCheck is the authentication server located at FatPort's Network Operations Control. It processes all login requests and purchases, then gives the venue's FatPoint service the green light to let the authenticated end-user surf freely.
7. Just in case an end-user isn't comfortable with purchasing time online, they can buy Fat-Code cards at a number of authorized dealers.

Graphic courtesy of FatPort.

Airpath Wireless. This provider offers what it calls an "Instant Hot Spot Provider" service. If a HotSpot operator is interested in a kit that allows it to "fly" its own brand, this service just may be the ticket. Airpath makes it very simple to provision a venue with wireless Internet access. The Airpath kit takes care of the hardware and software, letting the HotSpot operator focus on the front-end activities that are necessary for a successful business (e.g. adding new locations and building a subscriber base) without the capital outlay that normally would be necessary for servers, software, and other expensive network operations equipment. Airpath also takes care of the back-office details such as billing, subscriber management, network management, and customer care.

With the Instant Hot Spot Provider service, the HotSpot operator:

* Uses its own brand.
* Owns 100% of the revenue so it can create its own service pricing and policies.
* Can take advantage of free advertising links (ads, monthly specials, etc.).
* Owns the end-user.
* Controls its venues.
* Has immediate access to all of the benefits of the "AirPath Provider Alliance."

All the installer needs is the skill to use a simple web interface, since that is where the branding, service pricing, and policies are handled. This same web interface is also the management portal into all of the back-office processing, such as authentication, accounting, subscriber management, network management, roaming settlement, customer care, and more. The web interface, for example, provides the HotSpot operator with immediate access to AirPath's "Hot Spot Billing Suite," which is where the operator can choose its service pricing, and branding, and manage its accounts (both the venues and end-users).

Once everything is in place, new end-users can signup for access using the prices the HotSpot operator chooses (not what the service provider dictates), and later these same users will receive email branded account statements and usage summaries.

AirPath even provides real-time credit card processing using its own merchant account; downloadable template marketing materials that the HotSpot operator can customize with its own logo; dynamic reporting that gives the HotSpot operator immediate access to locations, roaming revenue, and more; and dynamic usage reporting that provides the operator with immediate access to end-user and location usage. Finally, the HotSpot operator can choose to support its end-users in-house, or optionally, AirPath will provide customer support.

Note, however, that AirPath's method isn't as simple as some others, such as FatPoint's or Boingo's, since the company doesn't offer a "one size fits all" approach for HotSpot deployment. This translates into "the potential operator must know enough technology to choose the HotSpot equipment that best meets his or her needs," although AirPath's staff will provide some advice. (The most popular HotSpot equipment such as gateways and bridges are listed on Airpath's website.) And since Airpath doesn't provide the equipment, its back-office systems are based on open standards, i.e. its systems work with all types of equipment and software.

Given that Airpath doesn't provide hardware, its annual services fees are minimal. AirPath charges a flat-rate per-location fee and a flat-rate per unique end-user login in a month. For example, unique end-user Janrey logs into the HotSpot's network two times for which Airpath charges the HotSpot provider a flat-rate fee of $5.00. But when another unique end-user, Texgal, logs on 20 times during that same month, the HotSpot still pays only a $5.00 fee to Airpath, even though the end-user accessed the network much more often. Furthermore, the company doesn't force revenue-shares and doesn't have minimums.

Pronto Networks. This developer of wireless infrastructure solutions provides three distinct ways for a HotSpot operator to set up individual HotSpots and to run its network.

One way is to use Pronto's "Hotspot Networking System." This solution allows HotSpot operators to establish and expand their network without expending a great amount of cash. That's because Pronto's Hotspot Networking System offers both the end-to-end infrastructure solution that enables the HotSpot itself, and the private-labeled, back-office operations for HotSpot operators that prefer to focus on front-end activities.

This system also provides HotSpot operators with all that is required to set up and manage a network of HotSpots (minus the pipe to the Internet). Pronto's complete plug-n-play system incorporates advanced authentication, authorization, billing, security, and roaming features, as well as several features that enable the generation of new revenue, such as location-based and personalized announcements and services.

The next method is what Pronto calls its "Networks Hotspot Managed Services." This solution provides complete infrastructure management services for HotSpot operators. When a HotSpot operator uses this solution, Pronto handles all deployment and provisioning of the venue location using the aforementioned Hotspot Networking System. Then Pronto's Networks Hotspot Managed Services provides all operational aspects of the HotSpot, including initial deployment and installation, network policy design and implementation, and ongoing network management. Pronto also manages the health and performance of the HotSpot operator's network, and provides extensive reporting capabilities. A HotSpot operator that goes with the Networks Hotspot Managed Services plan will incur little upfront capital expenditures since the plan requires only a minimal setup fee per venue, plus a minimal per user fee over time.

Furthermore, since Pronto's pricing model is largely based on the number of end-users, with the HotSpot operator paying in proportion to the revenue their network generates, the HotSpot operator can control its operating costs.

Pronto hasn't left out the "build-it-yourself" operator. For those who prefer to roll their own infrastructure, Pronto offers the Pronto Gateway and Server Software. Pronto's equipment provides an operator with the necessary infrastructure to build and support a network. The equipment solution can be configured to meet any client's specifications.

BOINGO AND PRONTO

Boingo Networks is not only targeting individual HotSpot venues, it is also doing its best to make it easy for a business to join the HotSpot operators club. Not only does it provide all of the traditional services of an aggregator (billing, customer service, marketing, etc.), but it also offers a product that it calls a "WISP in a Box." Of course, the product is in essence the Pronto "Hotspot Managed Services" re-labeled. Nonetheless, in an industry where scale is important, through this innovative approach of enabling HotSpot operators to have their own network using Boingo's underlying infrastructure, Boingo can more quickly grow its own network. And, of course, it provides an additional and compelling reason for a budding HotSpot operator to align itself with this innovative aggregator.

***Note:** In mid 2002, Pronto Networks and iPass announced two strategic agreements. One agreement allows for iPass users to seamlessly roam onto Pronto-operated WiFi networks, while the second agreement is a technology alliance whereby Pronto's Hotspot Networking System can be offered as a iPass-compliant solution.*

Toshiba/Accenture. The recently formed Toshiba/Accenture alliance hopes to capitalize on the surge in demand for fast on-the-go Internet access. Toshiba's Computer Systems Group technology services company and Accenture have joined forces. The two companies announced in March 2003 that they plan to have around 10,000 HotSpots up and running all across North America by the end of 2003.

The Toshiba/Accenture joint venture offers a turnkey solution for HotSpot operators, which includes hardware provided by Toshiba, and business and operational support from Accenture (e.g. network operations, billing and settlements processing, and help desk support). The kit costs about $200 and enables HotSpot operators to sell HotSpots to venue owners, and then set the product up for them.

As far as compensation is concerned, it has been left to the HotSpot operator to determine the cost of the service to its end-users. The resulting revenue will be split as follows: 50 percent of all earned revenue going to the Toshiba/Accenture venture, 30 percent to the HotSpot operator that installs the kit, and the remaining 20 percent to the venue owner. The new venture made no mention regarding who would bear the cost of the high-speed Internet connection.

GOING IT ALONE

Some HotSpot operators may, for any number of reasons, want to construct and set up their own network. For instance, a HotSpot operator may not perceive the benefit of being part of a larger network alliance (e.g. the HotSpot is off the beaten path and there are no other HotSpots nearby and few business travelers). Thus, getting the small group of end-users that frequent the HotSpot to sign up for, say, a Boingo account, is futile. Or the operator may have its own ideas about what services to provide to the venue owner and the end-user. Or perhaps the HotSpot operator wants to keep 100 percent of the revenue and to handle the billing functions internally. This makes sense if, the operator owns a group of small hotels, cafés, or convenience stores and has an infrastructure for billing and accounting support already in place.

Small coffee shops, restaurants and inns may not want to charge the end-user for wireless access and thus billing and accounting support aren't needed. This group, in theory, can simply create a HotSpot by plugging an access point into an existing wired network hub or router. Doing this, of course ignores security and billing issues. Most businesses will want a more professional, business-like HotSpot.

For those who want to "roll their own" HotSpot, all they must do is to follow a few simple steps:

First, determine the coverage area and bandwidth you'll need. Your HotSpot could cover a small coffee shop, a bed and breakfast, and/or an apartment building or it could encompass an entire neighborhood or park. You need a high bandwidth (a minimum of 386 Kbps) modem, or a PC equipped with a T-1 card (which can deliver 1.5 Mbps). Configuring such a nice fat digital pipe to the Internet presumes that you have enough

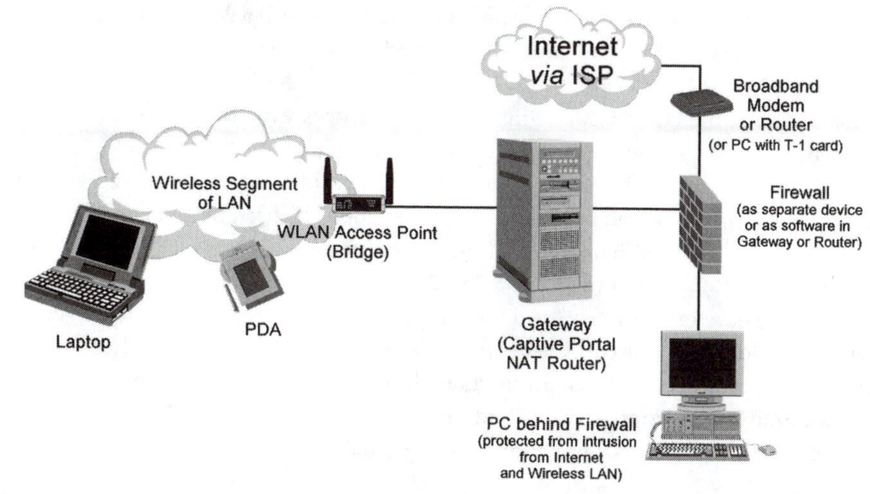

Figure 12.3 A "homegrown" Hotspot network layout. There are any number of ways a HotSpot can be designed and constructed. The one shown in this graphic is the minimum. Note that some vendors combine one or more of the standalone devices shown here into multipurpose units.

access points operating at a high enough bit-rate, so that your users can take advantage of this available bandwidth. (Short-haul access points can be small, commercial Linksys, D-Link or 3COM devices, or a higher-end dual-band Avaya or D-Link AirPro.)

Next, choose a different antenna if necessary. Larger-scale systems (e.g. apartment complex, hotel/convention center complex, or small housing development) may need something other than the antenna that accompanies the access point. An omnidirectional antenna mounted on a roof can cover a small neighborhood in a circular area around the antenna, while a more directional antenna (yagi, sector, panel, etc.) narrows the coverage area but increases the range of the signal. Typically, antennae can connect to an access point using a short cable with an "N" or "SMA" connector at each end.

Note:* *Don't use an amplifier to boost the signal over one watt because it will get you into trouble with the FCC. In fact, Wi-Fi providers must adhere to a number of regulatory standards. The most pertinent of these regulations are set out in Appendix III.

Finally, set up the equipment. To reduce signal loss caused by the cable, mount the access point as close to the antenna as possible. If the antenna is on a roof, then the access point should be housed in a commercial or homemade weatherproof enclosure. For example, the 802.11b airPointPro is available in three form factors: a standard indoor model, a ruggedized NEMA4-compliant model to withstand temperature extremes for outdoor configurations, and a "total" model with an integrated antenna for "one-box" convenience that can be placed anywhere. (See chapters 6 and 7 for more details on how to design and deploy a wireless local area network.)

Delivering power to an exterior-mounted access point is easy if it is compliant with the Power-over-Ethernet (PoE) standard, which allows power to be sent over unused wires in an Ethernet cable. In the case of the airPointPro, a "powerShot" PoE injector is included with every unit, which allows the airPointPro to be placed up to 100 meters

(327 feet) away from a power outlet. The unit even comes with a weatherproof RJ45 connector for both the "Outdoor" and "Total" models allowing quick installation of the Ethernet CAT-5 or Cat-5e cable.

Keep an eye on security issues. Small private networks, even wireless ones, often save money by sharing a single public IP address with all of the computers on the LAN. The router, computer, or other gateway device presents a public, registered IP address to the Internet and a private world of IP addresses (the "subnet") used by all of the devices on the LAN (or wireless LAN). The standard subnet for private networks includes IP addresses in the range of 192.168.0.0 to 192.168.255.255. In order for computers using these IP addresses to share one or a few "real" IP addresses on your publicly available gateway, your gateway device must run the proper software, such as Network Address Translation (NAT) type software. NAT products include Sygate's Home Network and Office Network products or Wind River's WindNET NAT. NAT is built into Windows ME and XP, where it's part of the "Connection Sharing" facility.

Implementing NAT can automatically establish a firewall between your internal network and outside networks, or between your internal network and the Internet. A firewall's purpose is to keep intruders out of your network, while still allowing people on the network to access the Internet. A firewall can be a separate, standalone device between a modem and a gateway, or it can be software running on the gateway.

Many basic access points act as NAT-capable routers instead of simply bridging to the network. In such a case, one could simply plug the access point directly into the Internet network. The gateway could also be configured as a "captive portal." Captive portals allow you to use a common web browser as a secure authentication device. In theory, they allow you to do everything on the network securely via the Secure Sockets Layer (SSL) and the IPSec tunneling protocol normally used with Virtual Private Networks. One type of captive portal software, NoCatAuth from the NoCat group, supports logins along with configurable restrictions to bandwidth and ports, based on whether the user is a trusted member or not. This comes in handy if someone at your HotSpot sends spam through your access point and your ISP threatens to shut down your account, if someone sends threatening or harassing email, or if someone hacks into your system and tinkers with government websites.

Several vendors offer business owners solutions for "going it alone."

Airpath Wireless. Many HotSpots will charge for their HotSpot service. This requires additional billing software. For many less experienced entrepreneurs, Airpath Wireless, Inc.'s turnkey package, "Hot-Spot-In-A-Box," includes everything needed to install a HotSpot—the network equipment, the end-user billing, even a branded portal. The Hot-Spot-In-A-Box solution features the Airpath Roaming Connector (ARC100C), which provides all the network equipment required to build a HotSpot bundled into a single unit. All you need is a high-speed Internet connection (and perhaps a special antenna and/or additional access points).

Hot-Spot-In-A-Box purchasers automatically become a part of the growing, worldwide provider-neutral Airpath Roaming Network. By being provider-neutral, Airpath Roaming remains transparent to end-users while providing a single-source of network integration for its members. Any telco, ISP or independent provider that has subscribers can share access to the same equipment or infrastructure by being a part of the Airpath Roaming Network. HotSpots with existing end-users retain control of their branding while offering the benefit of roaming anywhere within the network.

Airpath also offers custom designed solutions to accommodate all types of venues, including large hotels, convention centers, airports, marinas, and campuses.

IP3Networks. This company's of out-of-the-box products are helpful for business owners deploying their own HotSpots. Its "NetAccess for Wireless Hot Spots" is a 1U 19 inch, rack-mounted Linux device that comes with a 566 MHz Intel Celeron CPU and 256 MB of RAM, and provides plug-and-play access for end-users, as well as security and bandwidth management features, billing capability, complete logging, and extensive management tools. The HotSpot operator can even force page redirects if he or she wants to set up a website or portal. It's also compatible with the Boingo network.

This solution gives the HotSpot operator full control over duration, speed, and price; allows the operator to offer multiple classes of high-speed access; reserves different amounts of bandwidth for different areas (useful for small hotels that have conference rooms and guest rooms); and enables one-to-one static mapping that lets the HotSpot operator assign static IP addresses on a per-user basis. The operator could even create a special pricing plan that charges extra for a static IP address, which is important because some corporate VPNs won't allow entry into the corporate network without a static IP address.

The NetAccess product ships in two versions: the NA-25 (25-user license) has a suggested retail price of $1000, and the NA-50 (50-user license) has a suggested retail price of $1500.

Nomadix. This public networking provider offers what it calls the Universal Subscriber Gateway or USG. The USG is a stand-alone, 2U 19 inch, rack-mounted VxWorks device that can fully integrate with the Boingo network and billing system. Like the IP3Networks' device, the USG requires no client reconfiguration, even when static IPs or proxy settings are enabled. Standard features include 802.11 and web-based SSL authentication, RADIUS, AAA, and billing tools; a Home Page Redirect function that lets the HotSpot intercept browser requests and redirect users to a page of its choice; and bandwidth and billing options that are fully customizable.

CONCLUSION

The amount of activity in the HotSpot space is impressive. However, setting up a HotSpot requires some technical knowledge and most venue owners and even some HotSpot operators have no idea about rigorously architecting networks for quality of service and reliability.

Furthermore, many businesses and entrepreneurs that want to offer wireless high-speed Internet access to their customers have little experience with the technical side of networking and most cannot afford a to hire a consultant. For those businesses, one of the "turnkey kits" that are on the market may be just what the "tech doctor ordered." The technical expertise required to deploy one of these kits is minimal. In fact, most are designed with the small business owner in mind.

Now that you've reviewed some options for HotSpot/FreeSpot deployment, it's time to develop your rollout plans and put them into action.

Chapter 13: Wi-Fi and Cellular—A Dynamic Duo

The buying public definitely is attracted by all things wireless. When it comes to wireless data access, most of the buzz is about Wi-Fi. If you listen closely, however, you can still hear a slight murmur of excitement about cellular's "Third Generation" (3G) networks, but these 3G networks are being held back by high start-up costs, still uncertain technology and a shortage of supportive devices. Is it any wonder that many cellular operators ponder the question: can Wi-Fi serve the same purpose for less money?

CAN 3G AND WI-FI CO-EXIST IN THE SAME NETWORK?

Cellular technology revolutionized voice communications, and the industry has high hopes that—with the help of 3G technology—it will be the same with data communications. After all, cellular-enabled devices are in the hands of more than 1.3 billion subscribers, worldwide. Indeed for several years, industry experts considered cellular technology to be the most likely mode for wireless access to the Web. Cellular operators and equipment manufacturers invested billions of dollars on spectrum licenses and 3G related gear. Wi-Fi's emergence upset the established order of the wireless industry—forcing everyone involved to rethink their target markets, applications, partnerships, and strategies.

In an effort to protect their turf, some in the cellular industry are lobbying governments to limit WLANs' spectrum or even to make WLANs illegal. But others, such as the UMTS Forum (the industry group that represents 3G vendors and operators) take the position that wireless local area networks (WLANs) are "complementary," not competitors of cellular. UMTS's position is based on the fact that Wi-Fi has a short range but a powerful signal (i.e. fast) and cellular provides a long range but a comparatively weak signal (i.e. slow), and thus could be paired to create a dynamic duo.

To understand why each group feels as it does, let's consider the arguments put forth by each to support their position.

The proponents of a complementary relationship between Wi-Fi and cellular argue that:

- Wi-Fi is best at high-speed data transfer, with a potential for voice in the future after QoS concerns are addressed.
- 3G delivers superior voice and mobility and can fill-in wide-area gaps.

- There is a good business model for the cellular operator: hook users on wireless broadband with Wi-Fi then add 3G into the mix.
- Some in the manufacturing and vendor communities already integrate dual-mode (Wi-Fi/cellular) capabilities into their products, e.g. Qualcomm with its CDMA/802.11 chipset; Nokia and its GSM/GPRS/802.11 NIC Cards; Ericsson, which sells and deploys public Wi-Fi networks such as TeleDenmark in Europe).

Chris Land, director of strategy for Singtel Optus, a giant telco in the Asia Pacific region places Optus Mobile among the Wi-Fi detractors. He is adamant that Wi-Fi is no replacement for 3G. Instead, he says, "The difference between the two is the right to use a finite piece of spectrum for a period of time.... With 802.11, everyone can use the unlicensed spectrum. This will create problems in the long term, particularly as the space becomes congested."

The contingent that feels Wi-Fi will cripple 3G reasons as follows:

- Wi-Fi will render 3G technology redundant...any service 3G can provide, HotSpots can top it. Microsoft stated publicly that even with interference, Wi-Fi data transfer is still faster than any cellular technology, whether it's GSM/GPRS, EDGE, WCDMA or cdma2000. Thus HotSpots may cannibalize 3G operator's revenue. Even the UMTS Forum (which supports the dynamic duo scenario) suggests that 3G operators could potentially lose 12% to 64% of their Revenue to HotSpots by 2006.
- It costs a bundle to implement 3G. For instance, the provider must pay around hundred billion dollars for a 3G license versus Wi-Fi's free spectrum, not to mention the cost of building out a new 3G network. So why invite an upstart like Wi-Fi into their network mix?

The author takes the position that these two technologies could and should work as a team. Still, even among the supporters of such pairing, there isn't a consensus as to how such a partnership would work. Here's why:

First, the access points (the radio that creates the wireless field and manages network traffic) would need to be more complex than what's available today. In fact, these access points would be so complex that some of the heavy lifting would need to be done by a computer network.

Next, manufacturers of the popular diminutive cell phones will have to create a phone that uses a Wi-Fi network without draining its battery within a few minutes. (Motorola says it has solved this problem.)

There also is a dearth of hybrid gear, rendering most of today's network gear, access points, cell phones, laptops, and PDAs unable to make the transition between cellular and Wi-Fi. But this situation could be remedied quite soon—forthcoming plug-in cards for laptops will enable a user to access Wi-Fi, if it is available, and to fall back to a slower cellular connection, if it is not. Chip makers are in the midst of designing multi-mode chips to support both technologies; other vendors and manufacturers have their R&D departments working overtime to develop their own hybrids.

The normal business considerations, such as how HotSpot operators and cellular carriers would cooperate and how they would split revenue, would need to be untangled. This will be tough. The devil is in the details. For instance, items remaining to be worked out include how to provide a multilateral roaming platform, a common authentication and accounting system, and a shared billing standard. Bridgewater Systems, Intel, and others, however, are already addressing those issues.

Other questions to answer include:

* How does a network determine what pathway to assign a given user when that user signs onto the network—Wi-Fi or cellular?
* How do you make the transitions between the different networks seamless so that when a user switches from, say, an Instant Messaging (IM) or Short Text Messaging—better known as SMS (Short Message Service)— (which can be handled easily by cellular) to sending an email message or downloading a video clip (both would be better handled by a Wi-Fi connection), the user never knows he or she has left one connection method for another?

The answer may be in products by Ericsson, Lucent, Flarion Technologies and others that provide seamless handoffs between Wi-Fi and cellular networks.

Many in the industry are hard at work making the dynamic duo possible. For instance, Avaya, Proxim Corporation, and Motorola Inc. announced in early 2003 that they are collaborating on the creation and deployment of converged cellular, Wi-Fi, and Internet Protocol (IP) Telephony solutions. The solutions are enabled by an array of new products including a Wi-Fi/cellular dual-system phone from Motorola, Session Initiation Protocol (SIP)-enabled IP Telephony software from Avaya, and a voice-enabled WLAN infrastructure from Proxim. The jointly-developed, standards-based solutions support contiguous voice and data service to users across enterprise networks, public cellular networks, and HotSpots.

This chapter will discuss how the communications industry is tackling such issues. But first, the reader must understand how cellular technology works so the reader can more fully envision the many ways Wi-Fi can supplement and, in some instances, even replace cellular.

UNDERSTANDING CELLULAR

Millions of people use cellular telephonic devices—primarily cell (or mobile) phones, which are essentially small, sophisticated two-way radios with a built-in antenna. These diminutive devices, however, do much more than a two-way radio. Depending on the device, they can store contact information, provide calendaring functions such as to-do lists, perform simple calculations using a built-in calculator, send and receive email messages, obtain Internet-based information, provide simple gaming capabilities, perform basic computer-related duties (if a supportive operating system is installed), and integrate with other devices (e.g. cameras, PDAs, laptops, MP3 players, and GPS receivers).

Here are the basics of how cellular works:

When you place a cellular telephone call, the phone converts your voice to modulated radio frequency energy. The radio waves travel through the air until they reach a receiver at a nearby radio tower (i.e. cell site antenna) and base transceiver station (i.e. transmitters and receivers in a box or cabinet which is connected to the radio tower by feeders). If there are sufficient resources available (i.e. an open channel), the cell tower will forward the signal to a centrally located device called a "switch," "mobile switching office" (MSO), or "mobile telephone switching office" (MTSO) via special leased phone lines (e.g. a T-1 line) that connect the switch to many cell sites. This way the switch can hand all of the network's regional phone connections to the Public Switched Telephone Network (PSTN) and also control all of the base stations in that region. The switch then patches the signal to a channel on the PSTN, where it will continue on its way until it reaches the person you are calling. If the other person is using POTS (Plain Old Telephone Service), it will be delivered over the PSTN lines. If that person is using a cell phone, the call will be handed off to another cell and make its way via radio waves to the recipient's phone.

When you *receive* a call on your cell phone, the reverse takes place–the message travels through the telephone network until it reaches a base transceiver station (more commonly referred to as a base station) close to the phone's location. Then the base station sends out radio waves that are detected by a receiver located within the phone, where the signals are changed back into the sound of a voice.

A cellular call occupies a wireless channel as well as a PSTN channel, with both being held open until the call is completed. This means that neither channel can be used for any other calls until the cell phone call is discontinued. (There are exceptions.)

All cellular telephonic devices have special codes associated with them. These codes identify the device, its owner and the cellular service provider. For example, when a cell

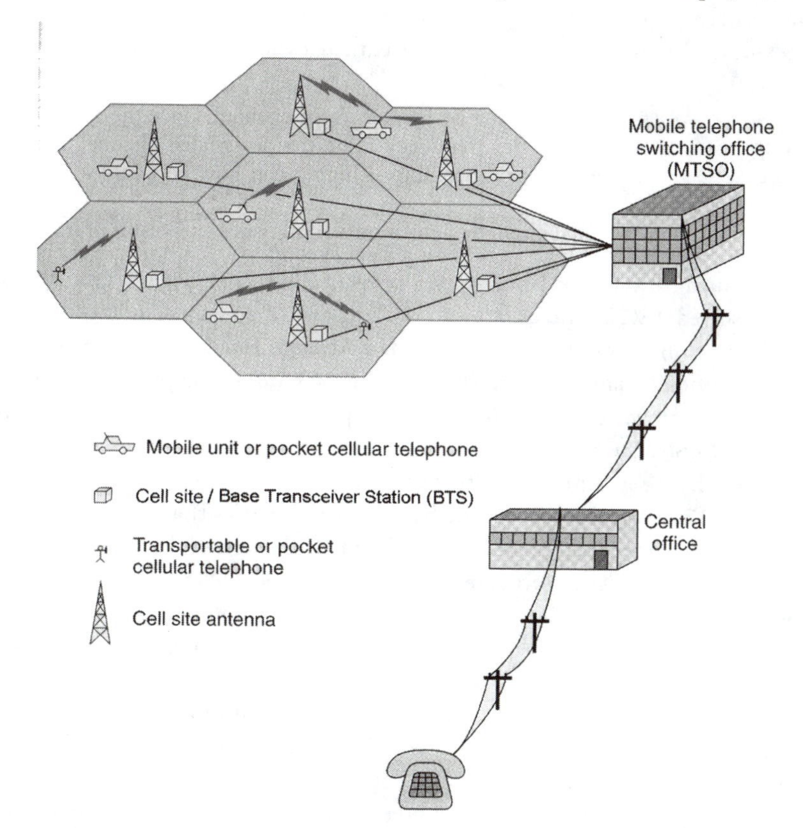

Figure 13.1 How a call traverses a cellular network. Each hexagon represents a "cell" which is served by an antenna (radio/cell tower) and base transceiver station.

phone is turned on, it listens on a designated frequency control channel for a System Identification Code (SID—a unique 5 digit number that is assigned to each carrier by the FCC). If no control channels are available, it displays a "no service" message. But if it obtains a channel and a SID, it matches the SID to the code programmed into the phone. If they match, the phone knows it is communicating with its home system. The phone also sends out its Mobile Identification Number (MIN—the phone number) and its Electronic Serial Number (ESN—a unique 32 number ID programmed into the communication device during manufacture) to the nearest cellular tower. This is then passed on to the MTSO, allowing the MTSO to track the phone's location. This way, when a call is received for that phone's MIN, the MTSO can forward the call to the right cell and pick the right frequency pair for the phone to use in that cell to take the call. Once that occurs, the MTSO communicates the frequency to the phone (using the control channel). After the phone and the tower switch to the right frequency pair, the call is connected.

If the cellular telephonic device is mobile and moves away from the cell with which it is communicating, the base transceiver station will note the diminishing signal strength and the adjacent cell will note the increasing signal strength. (All base stations listen and measure signal strength on all frequencies.) The two base stations, using the MTSO as a go-between, will coordinate so that at the right moment the phone will get a signal on the control channel telling it to change frequencies and then it will be handed off to the new cell.

Roaming

Roaming refers to the ability to use a cell phone anywhere, anytime throughout the world where compatible technology is available. Initially this was a very challenging issue for the cellular industry. The difficulty arose in trying to get various systems to communicate and pass routing and billing information to each other. (The same problems also will have to be solved if Wi-Fi and cellular networks are to share a subscriber base.)

As you now understand, when a cell phone is turned on, the cell phone listens for a SID on the control channel. If the SID does not match the SID programmed into the device, then it knows it is roaming. Since the phone also sends out its MIN (phone number) and ESN, the MTSO of the cell in which the phone is roaming contacts the MTSO of the home system (it does this based on the exchange, e.g. "305" if the phone number is "917-305-3336"). Once the home MTSO verifies that the SID is valid, the switch must then determine if roaming is possible, i.e. whether there is sufficient compatible equipment present. If roaming is possible, the local MTSO or "roaming switch" sets up a "Visitor Location Register" (VLR) registering the phone in its network and notifies the home switch of the phone's location so that it can route calls to the roaming switch as they come in. Outbound calls are handled through the roaming switch the same as they would be handled if the user were at home. Incoming calls are routed from the home switch to the roaming switch after sending a message to the roaming switch requesting a "Temporary Local Directory Number" (TLDN). This TLDN will be used to make a connection from the home switch to the roaming switch across the PSTN. (The process of registering the phone and notifying the home switch takes only a couple of seconds.)

Cellular's Evolutionary Path

Cellular technology evolved in three basic generations, with each successive generation being more reliable and flexible than the last. (See Fig. 13.2.)

Generation	Type	Time	Description
First	Analog	1980s	Voice centric, multiple standards (NMT, TACS etc.)
Second	Digital	1990s	Voice centric, multiple standards (GSM, CDMA, TDMA)
2.5	Higher Rate Data	Late 1990s	Introduction of new higher speed data services to bridge the gap between the second and Third Generation, including services such as General Packet Radio Service (GPRS) and Enhanced Data Rates for Global Evolution (EDGE)
Third	Digital Multimedia	2010s	Voice and data centric, single standard with multiple modes

Figure 13.2 GSM is the 2G technology of choice, but in North America it's more muddled with a mix of TDMA, CDMA and GSM being used for 2G cellular communications.

THE MARVEL OF THE CELL PHONE

The key to the cellular phone is the cell. By breaking a region into numerous cells and then spacing these cells fairly close to each other, cell phones can broadcast at very low power levels (typically 200mW - 1W). Because the cell phones can broadcast at such low power levels, they can be built with small transmitters and small batteries, enabling manufacturers to provide cellular devices that weigh less than three ounces.

Cell phones are remarkable. For an instrument with such a small "footprint" (i.e. it takes up very little space), it contains some amazing technology: antenna, liquid crystal display (LCD), keypad, microphone, speaker, battery and circuit board (the heart and brains of the phone). The typical cell phone's circuit board contains several computer chips, such as:

- The digital signal processor (DSP) performs signal manipulation calculations at high speed.
- The microprocessor manages all of the mundane tasks, such as controlling the keypad and LCD, handling the command and control signaling with the base station, and coordination of many of the functions on the circuit board.
- The analog-to-digital and digital-to-analog conversion chips translate the outgoing audio signal from analog to digital and the incoming signal from digital back to analog.
- The memory chips (ROM and Flash) provide storage for the phone's operating system and custom features, such as phone directory and calendar. Some phones also store the SID and MIN codes in Flash memory; others use external "Smart" cards for this purpose.
- The radio frequency (RF) and power sections manage the frequency channels and take care of power management and battery recharging.
- The RF amplifiers handle signals traveling to and from the antenna.

Cellular Networks and Data Transmission

Data transmission capabilities were added to most cellular technologies (GSM being the exception) much as an afterthought. For instance, Cellular Digital Packet Data (CDPD) is the technology used to add data capabilities to second generation TDMA (and FDMA) networks. This technology offers data transfer rates of up to 19.2 Kbps, quicker call set up, and good error correction.

CDMA networks, on the other hand, added data capabilities using one of two methods. A CDMA operator could use a circuit-switched approach (which involves setting up a data call in a manner similar to setting up a voice call except that digital data is transmitted rather than a voice signal). This allows data to be passed over the call circuit while the call is in operation. Or the operation could take a packet-switching approach, where the data is passed through the network without going through the typical call set up. This method requires that there be some form of arbitration to stop multiple users from sending data at the same time and interfering with each other. Because there is no need to set up a call, this method is sometimes referred to as "always-on." (Note: When the packet-switching approach is added to an existing 2G network, the resulting service is often referred to as "2.5G.")

2.5G

While the above data transmission techniques are still in use today in most 2G networks, efforts are underway to upgrade those 2G systems. These new hybrid "always on" networks (e.g. GSM/GPRS, EDGE, cdmaOne) are known as "2.5G." They can transfer data at speeds of up to 144 Kbps (faster than traditional 2G digital networks, but slower than *true* 3G networks). A phone with 2.5G services can alternate between using the Web, sending or receiving text messages, and making phone calls without losing its connection. Since it's these networks where Wi-Fi typically will be used to enhance their data capabilities, let's look a bit closer at some of their technologies.

General Pack Radio Service (GPRS). GPRS is a packet-switched data service that can overlay a GSM (Groupe Speciale Mobile or Global System for Mobile Communications) network and can be ported to TDMA (Time Division Multiple Access) networks. Theoretically, GPRS over GSM can reach 172.4 Kbps, but in practice, transmission rates range between 25 and 50 Kbps in the downstream direction (i.e. sending data to the consumer's cellular device) and 10 and 25 Kbps in the upstream direction. GPRS introduces packet data and requires Mobile IP (a protocol that builds on the Internet Protocol by making mobility transparent to applications and higher level protocols like TCP) with a real-world data rate of approximately 86 Kbps.

cdmaOne. This refers to a digital service that provides data service that operates at a theoretical maximum data rate of 64 Kbps. It supports packet-switched data on CDMA (Code Division Multiple Access) networks, operates in the 800 MHz or 1.9 GHz radio frequency bands, and uses a 1.25 MHz-wide carrier. Basically what makes cdmaOne a 2.5G technology is that it provides "bandwidth on demand," i.e. all users share a "bandwidth pool," which can be dipped into by each mobile phone for whatever purpose—voice, data, fax, etc. When one mobile phone is using less bandwidth, more is available for others.

Personal Communications Services (PCS). This is an all-digital service with a top data rate of 144 Kbps. PCS was specifically designed for U.S. operations and is available mainly in large metropolitan areas. The term "Personal Communications Services" was coined by the Federal Communications Commission (FCC) to describe a digital, two-way, wireless telecommunications system licensed to operate between 1850-1990 MHz (although the FCC's rules describe "PCS" as a broad family of wireless services without reference to spectrum band or technology). But true PCS also provides features like paging, caller ID and email functions. One main advantage of PCS technology is that it can increase a cellular network's call capacity—a PCS network using TDMA has 200 kHz channel spacing and eight time slots (instead of the typical 30 kHz channel spacing and three time slots found in earlier digital cellular systems). PCS networks can use CDMA, TDMA or GSM technologies.

Third Generation

The original goal of 3G was to create a truly global mobile telephone system. After all, POTS is compatible all around the globe, so why shouldn't the same compatibility apply to cellular service? The International Telecommunications Union (ITU) initially took the lead. It adopted the same kind of approach it took with ISDN (Integrated Services Digital Network), which could be described as "3G for wired networks" since ISDN is designed to carry not just voice, but also a whole range of data services at what was at the time considered to be a high transmission rate.

***Note:** The term, "third generation" or "3G" refers to pending improvements in cellular-based data and voice communications through any of a variety of proposed standards, with the immediate goal of providing data transmission speeds of 2 Mbps. However, any cellular network that supports packet-switching and promises data transmission speeds of at least 40 Kbps is oft-times mis-labeled "3G." But since technology to take cellular technology beyond 2G varies in its capabilities, perhaps a better vernacular for "beyond 2G" would be "next generation," "NextGen" or "nex-gen."*

The first attempt at formulating a 3G system by the ITU was the Future Public Land Mobile Telecommunication Systems (FPLMTS), which would eventually evolve into the International Mobile Telecommunications 2000 (IMT 2000) specification.

The ITU was slow on the uptake. So, in the interim, the European Union decided to create its own version of IMT 2000, which it dubbed "UMTS" (Universal Mobile Telecommunications System). Toward that end, the EU reached an agreement between existing GSM equipment suppliers—mostly European and Japanese manufacturers—for a single, common technology. The basis of this technology (which was originally referred to as "3G/UMTS") is today's W-CDMA (Wideband CDMA).

The introduction of W-CDMA put a monkey wrench in the ITU's plans for a "global cellular network," forcing the organization into a series of compromises. Ultimately, UMTS, as specified by the Third Generation Partnership Project (3GPP), was formally adopted by the ITU as a member of its family of IMT-2000 Third Generation Mobile Communication standards. The outcome is set forth in the latest version of the ITU document entitled "International Mobile Telecommunications-2000" (IMT 2000). UMTS is now therefore a part of the ITU's IMT-2000 "vision" of a global family of 3G mobile communications systems. However, because of the popularity of the term UMTS, the final 3G system goal is sometimes referred to as "IMT 2000/UMTS third generation personal and mobile services."

The IMT 2000 defines an "anywhere, anytime" standard for the future of cellular communications with a goal of making available all kinds of service—wide-area paging, voice, high-speed data, audio/visual, etc.—within a common system. The framers of the IMT 2000 also envisioned seamless service across the globe that merged (or at least made compatible) PSTN service, the various approaches to wireless (cellular, Wi-Fi and satellite), the Internet, and even cordless phone technologies. Moreover, they would like to see multi-function mobile terminals, worldwide roaming, and even a personal phone number or a Universal Personal Telecommunications (UPT) number to allow a user to receive any kind of communication, on any terminal, anywhere. Thus, IMT 2000 promises:

THE MIGRATION PATH

The cellular operators that go the 3G route will typically take one of two paths. The umbrella names for these two paths are: IMT-2000 MC for multicarrier CDMA (which includes 1X, 1xEV, and 3x) and IMT-2000 DS for direct spread W-CDMA. Let's take a closer look at each:

The IMT-2000 MC Path: Since CDMA already uses a wide (1.25MHz) carrier, the task becomes one of making the most efficient use out of what's already there. Providers currently using cdmaOne (the technology that provides services that are somewhat similar to W-CDMA) can rather easily migrate to one of the cdma2000 family of 3G technologies. This path is favored by cellular operators in the Americas and the Pacific Rim. It includes:

cdma2000 1X can double the voice capacity of cdmaOne networks and delivers peak packet data speeds of 307 Kbps in mobile environments.

cdma2000 1xEV includes cdma2000 1xEV-DO (Data Only), which delivers peak data speeds of 2.4 Mbps (although real-world speeds are more in the 300 Mbps range) and supports applications such as MP3 transfers and video conferencing.

cdma2000 1xEV-DV (Data Voice) can provide integrated voice and simultaneous high-speed packet data multimedia services at speeds that have tested as high as 3.09 Mbps.

1xEV-DO and 1xEV-DV are both backward compatible with cdma2000 1X and cdmaOne. Today, there are 26 cdma2000 1X and three 1xEV-DO commercial networks across the globe. Moreover, 22 1X and three 1xEV-DO networks are scheduled for deployment during 2003. For the current status of cdma2000 deployments, the reader can visit the CDG website: www.cdg.org/worldwide/index.asp and then click on the "View Entire Worldwide Database" link.

The IMT-2000 DS Path: GSM's evolution is from 9.6 Kbps circuit-switched (non-packet) data to the interim GPRS solution, then to Enhanced Data rates for Global Evolution (EDGE). EDGE is a further improvement to GSM/TDMA and it optimizes for GPRS. In fact, EDGE was developed specifically to meet the bandwidth needs of 3G-like services. Since EDGE works with current GSM networks and offers a top speed of 384 Kbps, it is the technology du jour in the GSM community. It's noted, however, that both GPRS and EDGE continue to use GSM's narrow 200 KHz carrier and TDMA technology.

* Increased bandwidth, up to 384 Kbps when a device is stationary or moving at pedestrian speed, 128 Kbps in a vehicle such as a car, bus, train or plane, and 2 Mbps in fixed applications.
* Circuit-switched and packet-switched services, such as Internet Protocol (IP) traffic and real-time video.
* Voice quality that's comparable to POTS.
* Greater capacity and improved spectrum efficiency than currently available in 2G networks.
* Ability to provide several simultaneous services to end-users and terminals, for multimedia services.
* The seamless incorporation of 2G cellular systems, in order not to have any discontinuity between 2G and 3G systems.

GSM's final evolutionary step to 3G is to switch over to W-CDMA, which is a huge leap—moving from its TDMA roots over to CDMA, and from a 200 KHz carrier to a 5 MHz carrier. New frequency allocations, infrastructure, and handsets will be required for this last step.

W-CDMA (Wideband CDMA), also known as CDMA Direct Spread or 3G/UMTS, is the 3G radio transmission technology that is expected to be favored by any European operator that build-outs a 3G network. W-CDMA can be built upon existing GSM networks and represents the obvious next step for current system operators. This 3G technology sends data in a digital format over a range of frequencies, which makes the data move faster, but it also uses more bandwidth than digital voice services. (W-CDMA has also found favor in the U.S. AT&T Wireless and a few other U.S telecos are slated to roll out W-CDMA networks in a few large U.S. metropolitan areas sometime prior to 2005.)

The only W-CDMA network that is fully operational at this writing is the aforementioned NTT DoCoMo network in Japan. The reason cdma2000 deployments are so far ahead of W-CDMA rollouts is because W-CDMA requires new spectrum allocations and cdma2000 operators can build their systems atop their existing CDMA infrastructure, allowing them to reuse much of the existing gear. Thus, cellular providers implementing W-CDMA networks must spend more time and money on their network build out, versus cdma2000 providers. For now, W-CDMA will do well only in areas in which regulators favor W-CDMA over cdma2000, such as Western Europe.

U.S. cellular providers AT&T Wireless, Cingular and T-Mobile are taking a torturous route to W-CDMA. AT&T Wireless's path toward 3G began with the installation of voice-centric GSM network on top of its existing analog and TDMA networks. It then installed a data-only GPRS system (which reduced the network's voice capacity). Next came an all-new radio access network (RAN) to provide higher-speed data using EDGE. Its most recent step is to install yet another all-new RAN (and potentially a new core network) for W-CDMA, slated for completion 2005.

Cingular, which initially straddled the fence about its technology choice, committed to the same 3G migration path as AT&T. T-Mobile (VoiceStream), already a GSM carrier, also committed to this same three-stage implementation path. But elsewhere in the world, the news isn't particularly good for cellular providers choosing this technology route—a host of W-CDMA carriers have announced delays in deployment.

* Global, i.e. international, roaming between different IMT-2000 operational environments.
* Economies of scale and an open global standard that meet the needs of the mass market.
* Common billing/user profiles, e.g. sharing of usage/rate information between service providers, standardized call detail recording, and standardized user profiles.
* Enhanced 911, which refers to the network's ability to determine geographic position of cellular devices and report it to both the network and the mobile terminal.
* Support of multimedia services/capabilities, e.g. fixed and variable rate bandwidth on demand, asymmetric data rates in the forward and reverse links, multimedia mail store and forward, broadband access up to 2 Mbps.

The ITU accepted five different standards as suitable for IMT-2000 (i.e. 3G) environments: Direct Spread, Multicarrier, Time Code, Single Carrier, and Frequency Time. Actually, only the first three standards are being adopted; the last two standards appear to be attracting very little interest. For further explanation of these five standards, see Fig. 13.3.

Adopting any of the IMT 2000 standards requires a lot for a new technology and new networking environments. The cost of building such a network will be exorbitant. Furthermore, it is highly unlikely that early 3G networks will be able to provide for all of the above requirements. This may cause many cellular providers to look at Wi-Fi in a different light.

Consider some of the limitations of 3G networks:

* Unlikely to provide end-users with a 2 Mbps data transfer speed, since to attain such speed would require that no other users be registered on a base station. Because spread spectrum systems (which all 3G standards are) transmit the same signal to multiple base stations simultaneously, finding a base station with no other users is highly unlikely.
* Cannot support both circuit- and packet-switched communications concurrently because of the complexity of implementation.
* The technology to seamlessly incorporate existing 2G networks does not exist. The only fully operational 3G network (at this writing) is Japan's NTT DoCoMo network and it doesn't support the Japanese 2G standard.

IMT Name	IMT Code	Other names	Regions of adoption	Specifying body
Direct Spread	IMT-DS	WCDMA, UTRA FDD	All GSM operators	Originally ETSI (part of UMTS specifications), now 3GPP
Multicarrier	IMT-MC	cdma2000 (3x)	cmdaOne operators	TIA (and 3GPP2)
Time Code	IMT-TC	TD-SCDMA, UTRA TDD	China	Originally ETSI (part of UMTS specifications), now 3GPP
Single Carrier	IMT-SC	UWC-136	Unclear, but a few TDMA operators	UWCC
Frequency Time	IMT-FT	DECT	Few, if any	ETSI

Figure 13.3 Official IMT 2000 recognized 3G transmission standards.

* Cannot provide international roaming between different IMT-2000 environments. But this could be remedied in the future–it was some time before GSM roaming became available.

The 3G Hyperbole

When the cellular industry realized the costly and technological challenges they faced in upgrading their current 2G infrastructure to 3G, some opted to use 2.5G technology as an interim step. But download speeds that 2.5G networks offer are, to put it bluntly, paltry when compared to Wi-Fi's speeds.

Compare the industry's marketing campaigns to reality. Many ads tout that these new 3G systems (actually 2.5G role-playing as 3G) are fast enough to blow the typical dial-up experience via ISPs, like America Online, out of the water. But the current crop of next-gen networks just barely manages to equal a typical America Online dial-up session–a fact that is documented by figures provided by the cellular carriers themselves.

Providing an explanation for the discrepancies, Keith Nowak, a representative for handset marker Nokia states, "We tried real hard not to hype that peak number [144 Kbps], but some other parties might have been more likely to hype it. In the long term, it's better to use the more realistic speeds."

Customer disappointment over a cellular network's data delivery doesn't bode well for an industry hoping to recoup the high cost of building these new networks by selling downloadable games or business applications needing speed to succeed. And, this gap between hype and reality has caused at least one analyst to state that the actual performance of the networks raises the question of why they were built at all. IDC analyst Keith Waryas said when commenting on the low end of a typical web session on new next-gen networks from AT&T Wireless, Cingular Wireless and T-Mobile, "You build this big network, and all you can offer is a 20 Kbps download? That's not much of an improvement over what the carriers already had."

In defense, Jim Gerace, a Verizon Wireless representative, said that his company has been insisting all along that it's more likely that a subscriber to its Express Network, which peaks at 144 Kbps, will experience data transfer speeds of between 40 Kbps and 60 Kbps on average. "Some will experience even better than that," Gerace said. "But 40 to 60 is the number we're sticking with."

AT&T Wireless's new mMode service (which uses the GPRS standard) replaces the carrier's PocketNet wireless web service (an 8-year-old effort that uses CDPD and operates at 19.2 Kbps with an average user experience of between 5 Kbps and 9 Kbps). A company spokesperson says that subscribers to its new mMode service can expect on crowded days to experience data transfer speeds of around 20 Kbps to 30 Kbps and the news gets even worse: speed can drop to between 10 Kbps and 20 Kbps if someone's trying to download, say, video using a heavily trafficked network. Those startling numbers come straight from AT&T Wireless's own data.

As justification for the new mMode service, AT&T Wireless Chief Technology Officer Rod Nelson says that its mMode technology enables "streaming media, which wouldn't even work at all over CDPD [the technology used by PocketNet]."

A T-Mobile representative reports that its average user experience on its highly promoted next-gen service is about 40 Kbps. T-Mobile launched its nationwide GPRS service as an extension of its GSM voice network in late 2001. The carrier is pushing its data network to the limit. Its "you had to be there" ad campaign introduces customers to the benefits of visual, wireless communications via T-Mobile's next-gen cellular network.

Sprint PCS's new telephone network uses the same technology as that of Verizon Wireless, and while its data transmission rates peaks at 144 Kbps, Sprint describes an average user's experience at between 40 Kbps and 60 Kbps (the same as Verizon).

Is it any wonder that some cellular providers are turning to Wi-Fi technology? These cellular operators need some way to meet customer's high expectations brought about by over-optimistic marketing campaigns. Wi-Fi can help these hapless operators to match the promotional campaigns that promise a nationwide network capable of speeds that make it easy to download (and send) music and video files via a cell phone. With Wi-Fi as part of their service package, cellular providers and their customers can use Wi-Fi technology to do the "heavy lifting." In other words, with Wi-Fi to augment the cellular operator's current network, a cell phone with a dual-mode Wi-Fi/cellular chip could, for example, use Wi-Fi to download a huge document directly to a PDA or laptop.

Also, since Wi-Fi networks can also be optimized to carry voice calls, these networks could be used in the future to improve cell phone reception in buildings, where cellular coverage is traditionally poor. According to IDC analyist Waryas, "there is some very real potential to offloading some of the voice calls onto Wi-Fi."

AN INTELLIGENT PAIRING

Today, more than ever, debt-laden cellular companies are looking for ways to bring down their per call cost and to attain even higher market penetration, especially in metropolitan areas. The pairing of Wi-Fi and these next-gen networks (whether 2.5G or 3G) could be an intelligent move, notwithstanding the fact that the technology for such pairing is still unperfected, and manufacturers are still wrestling with ways to address numerous technical problems (e.g. adding a Wi-Fi component to a cellular device's chip increases battery drain). However, once such issues are overcome, cellular and Wi-Fi technologies could actually fit together quite nicely:

- Next-gen networks transmit both data (albeit at a relatively slow speed) and voice traffic to mobile phones and PDAs. Wi-Fi only transmits data (at least for now) to laptop computers and PDAs.
- Next-gen networks provide regional coverage over huge geographical areas. Wi-Fi systems send signals over a very short range—generally less than 300 feet (91 meters).
- Wi-Fi offers the advantage of speed, an 11 to 54 Mbps data transmission rate (although the typical speed is probably more like 7 to 34 Mbps). Current next-gen networks top out at an optimistic 384 Kbps (in actual use, the typical speed ranges from 40 to 60 Kbps).
- It typically costs less than $1000 to deploy a Wi-Fi HotSpot that can accommodate ten users at once. The equivalent capacity on a cellular network costs about $10,000.

"Carriers are looking at 802.11 in a big way," says Dean Darwin, director of U.S. Business Development for RadioFrame Networks, a start-up that sells wireless LAN access points combined with picocell base stations. "Carrier X is afraid that carrier Y is going to out-execute them" and they will be left behind.

	3G	Wi-Fi
Data Rates	144kbps (outdoor) up to 2Mpbs (indoor). 2.5G can deliver theoretical data rates of about 171kbps with GPRS and about 144kbps with cdma1xRTT. However, the practical data rates will be about half the theoretical ones.	High data rates: 802.11b standard up to 11 Mbps and 802.11a up to 54 Mbps. The practical data rate is about half the theoretical data rate.
Security	Provides a more secure environment than WLAN	Currently security provided with WEP (Wired Equivalent Privacy) security mechanism using RC4 encryption which is very insecure. 802.11TGi will improve security using various mechanism like 802.1x, RADIUS, KDC, TKIP/ AES.
Handoff	Implemented as integral part of the specification.	Currently not implemented, however possible in the future with solutions like MobileIP
Cost of Network	Expensive to install: 3G base station costs about $250,000.	Cheap to install: access point costs anywhere from $1000 to $5,000
Coverage Area (per BS/ AP)	~5 mile radius	100-300 ft. radius
Spectrum	Licensed	Unlicensed (2.4GHz and 5GHz)
Supported Media	Voice and Data	Primarily data. VoIP in future
Connectivity	Seamless (anytime, anywhere)	Interrupted; only possible in limited locations

Figure 13.4 Comparing Wi-Fi and 3G.

"[Cellular] service providers have a very tough time providing enough capacity in high-density situations, such as conference centers, hotels, etc.," says In-Stat/MDR analyst Allen Nogee. He goes on, "If 2.5G and 3G should take-off, capacity demands in these types of situations will become even worse."

Next, factor in numbers cited by research companies such as Analysys. Its research results indicate that more than 21 million Americans will be using public WLANs by 2007. Moreover, the popularity of the service will generate over $3 billion dollars in service revenues by 2005. Analysys also claims that WLAN services will account for 25 percent of mobile data service revenues by 2007.

Consider the movement by Wi-Fi operators toward roaming agreements for carrying each other's data (e.g. Boingo and Wayport), and the birth of a process for providing a clearinghouse solution to enable roaming capabilities and a common billing platform between providers, all of which is discussed in Chapter 11. These elements provide a good business case for the "dynamic duo."

But Wi-Fi won't render 3G obsolete. Most industry experts agree that it's impractical to blanket a city with HotSpots, but they also argue that cellular operators would be foolish to ignore this new technology. Carriers "can roll out [Wi-Fi] services much quicker than 3G," says Shamir Amanullah, program leader at marketing consultancy Frost & Sullivan in Kuala Lumpur. They can also add Wi-Fi to their network at a reasonable cost versus the up to one billion dollars needed to set up a new 3G network.

Cellular Providers are Wary

Cellular providers sense that, like a shark, if they stop moving forward, they will die. So, despite the dire economic climate, cellular operators must keep their technology programs moving forward or risk losing it all. Although many in the cellular industry are investigating the potential of adding Wi-Fi to their service mix, they are also wary—they worry that no matter what they do Wi-Fi could encroach upon their market space. They are right to worry. For although current WLAN operators currently don't have an appealing customer value proposition nor a sufficient customer base to dominate the market, technology advancements and growing cooperation in the form of roaming agreements indicate that Wi-Fi operators could eventually capture the market, especially the attractive business customer segment. Consider these compelling examples:

Broomfield, Colorado. Recently the city and county of Broomfield, Colorado traded in their AT&T 9.5 CDPD service for a countywide Wi-Fi network. Broomfield installed enough Wi-Fi access points to provide total coverage of the county's 36 square miles. Greg Anderson, Broomfield's director of IT, called the return on investment on the $60,000 Wi-Fi LAN "astronomical," considering the $52 per month the local governments paid for each mobile unit using the CDPD service.

Currently, 23 patrol cars are equipped with Wi-Fi, and additional police cars and other municipal users can be added to the system at a hardware cost of approximately $800 per car. Anderson isn't even considering 3G cellular. "3G is slower than what I have now. Our system is much better [than 3G] and has zero dollars operating cost," he said.

Broomfield isn't alone. A growing number of U.S. local governments, including the California cities of Glendale and Oakland, and Orange and San Diego counties, have embraced Wi-Fi networks.

ITEC Entertainment Corp. The corporate world is looking at Wi-Fi in a new light—as something more than an indoor networking device. ITEC Entertainment Corp. is a good example of Wi-Fi's changing role in the corporate community. When ITEC Entertainment Corp. installed its wireless Transit Television Network it was looking for speed. Transit Television Network offers transit agencies a passenger communications solution that enables a transit company to provide its passengers with ADA-compliant, GPS-triggered automated stop announcements, real-time route information, and ride-enhancing news, weather and entertainment content.

"We push a tremendous amount of data back and forth, including MPEG video files," said Danial West, the company's vice president for strategic business development. Such transfers couldn't be supported by any of the cellular networks, whether 2G, 2.5G or 3G. That's why ITEC opted for Cisco Systems' mobile Wi-Fi technology based on 802.11b. In fact, West said that the company didn't even consider cellular. Wi-Fi is the technology that now supports ITEC's Transit Television Network system providing public transit information and entertainment content for transit companies operating in Milwaukee, Birmingham, and Orlando.

In Orlando, ITEC began with the installation of five access points to transmit information to what the company calls a "media engine" in 100 of the 250 buses operated by the Lynx Central Florida Regional Transportation Authority. The hardware setup includes a Wi-Fi receiver, high-capacity hard drives, and a Global Positioning System (GPS) receiver to provide route information, and play news and entertainment videos on monitors installed throughout the bus.

On top of the high data rate, the Wi-Fi network saves ITEC the $50 per month it would typically cost for access to a cellular data network plus the per-minute or per-packet charges that are usually tacked onto the average bill. West says, "That's a huge cost advantage for us."

A Packaged Deal

A hybrid network provides transparent continuous connectivity as well as the ability to utilize bandwidth-intensive applications. As such, this solution is being implemented across industries and markets. The vertical marketplace, especially public safety and transportation, has been using a hybrid network approach for several years.

Hybrid network solutions provide end-users with a combination of high-speed access and broad coverage. Case in point is Padcom, a Pennsylvania-based start-up company, which offers a wireless local area network (WLAN) solution it calls the "TotalRoam Ellipse." The solution uses a host/client configuration with a gateway attached to a user's LAN to integrate IP and non-IP networks. (A software component also resides on each of the mobile clients.) The public safety sector has embraced Padcom's solution. For example, police departments in Oakland and Baltimore are already using the software.

The Oakland Police Department installed Wi-Fi-based WLANs in different stations throughout the city. Motorola's wide area RD-Lap network is available if an officer can't obtain Wi-Fi connectivity. As an officer pulls into a station, the Padcom technology recognizes the unit is within range of a Wi-Fi connection and switches over. Back east (Baltimore, Maryland), the Wi-Fi set-up is similar—only in this instance, the Baltimore Police Department uses Verizon's cellular data network to take over when the officer is out of range of Wi-Fi connectivity.

In addition, Padcom is running a trial with the Sheriff's Department of Orange County, California. This trial uses the same technology that Oakland and Baltimore police departments use except that AT&T Wireless provides the cellular back-up.

Although the public safety sector is one of the first groups to latch onto the Padcom solution, Mark Ferguson, Padcom's director of marketing and strategic planning, believes the problems his company's technology solves are not just relevant to public safety concerns. Ferguson contends, "When dealing with wireless networks, the two primary problems are coverage and bandwidth. To get the kind of coverage you need, users are going to have to use more than one network."

Padcom, which already has existing relationships with both carriers and tranceiver manufacturers, offers a value add to the industry that Ferguson believes provides them the advantage. "Our technology is not tied to MobileIP, the standard for Wi-Fi roaming. Our system incorporates a prioritization scheme so even if you have wide area coverage when you enter a HotSpot's zone, the user's device is preconfigured to make Wi-Fi the priority network. MobileIP can't do prioritization," Ferguson said.

AT&T Wireless, a well-known U.S.-based cellular provider, is using Wi-Fi to supplement its indoor connectivity, while supporting 3G-like technology outdoors. Why? Because AT&T believes, along with many others in the cellular industry, that if their customers become accustomed to the capabilities that Wi-Fi offers, they might become more enthusiastic about trying 3G services once those networks are up and running.

The mother of all hybrid networks may be the one UPS is building. UPS is upgrading its existing network by using several wireless communication technologies. At this writing the wireless technologies being used in the UPS's pilot project include Wi-Fi, Bluetooth, infrared satellite-based Global Positioning System (GPS), and two cellular networks (cdma2000 1X and GSM/GPRS). To access the new hybrid network, the company has upgraded its current Delivery Information Acquisition Device (DIAD), a handheld device UPS developed in conjunction with Symbol Technologies. Once the company's drivers are equipped with the new DIAD IV, they will be able to connect to any of the six different wireless networks. While deployment of this network is still in the test stage, the express shipment giant does expect to begin full roll-out of its new hybrid system by the end of 2004, and it is expected that within just a few years, all UPS drivers will be equipped to communicate when, where and how they want.

The Synergy

The synergy between the cellular and Wi-Fi industries was marked recently when both Boingo Wireless founder Sky Dayton and John Stanton, CEO of T-Mobile, gave keynote addresses at CTIA Wireless 2002.

Dayton, whose company is the largest aggregator of public Wi-Fi access, told the gathering that the integration of Wi-Fi and cell networks is a case of "you got your chocolate in my peanut butter" and that while HotSpots are growing in airports, coffee shops and other public areas, for companies such as Boingo to flourish, they will need to partner with cellular carriers with extensive network capacity. (In December 2001, Sprint PCS invested $15 million in Boingo, saying it was very bullish on the Wi-Fi company.) Thus, there is little doubt that cellular operators will hedge their bets on next-generation networks by entering the Wi-Fi market. Given the relative strengths of each technology (cellular and Wi-Fi), it makes sense for the two to come together in some fashion. Upstart Wi-Fi operators see established mobile carriers as vital for expansion, while VoiceStream, Cingular, and other carriers see Wi-Fi as a way to reach users of lucrative data services. Products are already being announced and it appears likely that the market will see cellular-to-Wi-Fi roaming agreements soon, perhaps before the end of 2003.

The Big Guns

There was a flurry of Wi-Fi-related service provider news in 2002, and more action is on tap for upcoming years. Larger telco operators have existing services and the assets to figure out a Wi-Fi business plan and thus are well positioned to capitalize on Wi-Fi.

T-Mobile already has made its mark in the Wi-Fi world. The innovative carrier provides Wi-Fi-based Internet access to 2365 (and counting) HotSpot locations. T-Mobile claims this makes it the largest carrier-owned Wi-Fi network operator in the world.

T-Mobile, which has the advantage of both being one of the largest HotSpot operators and owning a nationwide GSM/GPRS network, is offering what it terms "an integrated GPRS/Enhance Data Rates for Global Evolution/802.11b service offering." The service allows T-Mobile customers to switch manually between Wi-Fi and cellular service. T-Mobile also has revamped its billing and subscription options so that each customer gets a single bill that lists all usage of T-Mobile services.

The carrier's future plans include developing an integrated WiFi/GPRS data card, giving customers seamless service between its cellular and Wi-Fi networks, and introducing dual-mode access devices (PDAs and phonesets).

Orange France, that country's leading cellular provider, and Air France have agreed to enter into an alliance where by Orange will install and operate HotSpots in all 54 airport lounges operated by Air France around the world.

It's likely that other telco operators soon will follow in the footsteps of T-Mobile and Orange France. Cellular providers understand that customers typically prefer a single provider that offers a variety of services to establishing relationships with separate providers. Thus a bundling of Wi-Fi with cellular can greatly improve customer satisfaction.

Here is a good illustration of how Wi-Fi is captivating the cellular industry's imagination. At the March 2003 Cellular Telecommunications & Internet Association meeting in New Orleans, the buzz was not about cellular technologies; rather Wi-Fi was on everyone's mind. The questions these cellular carriers, who want and need more data traffic, were asking included: "Is Wi-Fi a huge threat, or opportunity, or both? Where might it fit in with other present and planned data networks?"

However, if cellular providers do decide to offer Wi-Fi alongside their cellular services, they must allow their cellular customers to roam the airwaves for available Wi-Fi networks and then transition their signal from slower cellular connections to faster Wi-Fi connections whenever a Wi-Fi signal is available. This type of carrot and stick philosophy–enticing users to buy new phones that can take advantage of high-speed data access, and habituating users to the speed and convenience of such services will allow cellular operators to experiment with broadband services. They can then see if the billions they are proposing to spend on 3G is worth the time, money and effort.

Thankfully, it is becoming less and less difficult to combine Wi-Fi with current 2G or 2.5G networks, no matter which technology the cellular operator is using–TDMA, GSM, GPRS/GSM, EDGE, or CDMA. Consequently, it shouldn't be long before plenty of cellular carriers are racing to get a piece of the action. But the cellular industry will adopt Wi-Fi as it does everything–slowly and cautiously.

Business Opportunities Abound

Still, there are many cellular operators who feel that they must continue to roll out their next-gen networks because they have, after all, already invested billions of dollars in spectrum. Never mind that 3G technology is not quite ready for prime time, and that the customers aren't there either. Never mind that far more people use Wi-Fi than cellular to access the Internet.

Wide-ranging business opportunities exist for both the cellular and Wi-Fi industries. HotSpots present a perfect opportunity for cellular operators to familiarize their subscribers with the benefits of their upcoming high-speed wireless data access–3G. Charles Levine, CEO of Sprint PCS, raised that concept when he stated, "What we need to do is increase the number of people familiar with the concept of data-on-the-go." He's not the only one in the cellular industry who thinks that by pairing with Wi-Fi providers, the ensuing subscriber surfing experience will translate into increased interest in 3G.

While cellular operators are struggling to sell their data services, Wi-Fi is gaining momentum. Intel spent more than $300 million to advertise the launch of its new Centrino wireless-enabled laptop chips, much of which is focused on promoting public awareness of Wi-Fi. Cometa Networks, an industry partnership led by AT&T, IBM and Intel has been instrumental in providing customers from hotels to McDonald's fast food restaurants with turnkey networks. Toshiba and Accenture recently formed a partnership to exploit Wi-Fi technology. The new venture has already won a bid to set up HotSpots in hundreds of Circle K convenience stores and ConocoPhilips gas stations.

***Note:** McDonald's is interested not only in adding an amenity to attract more customers, but also in the savings to be had from a network where everything down to the milkshake machine's maintenance schedule can be accessed at a moment's notice. McDonald's already has Wi-Fi in its Australian, Japanese, Swedish, and Taiwanese restaurants.*

Sprint PCS, a direct investor in Boingo Wireless and a member of the Wi-Fi Alliance trade group, is stitching together a network of wireless HotSpots to cash in on the growing popularity of wireless networking, according to a Sprint PCS executive. Wesley Dittmer, the senior director for wireless LANs at Sprint PCS, says that the company has already signed agreements to let subscribers roam onto a number of different HotSpots, then be charged for wireless web access on their Sprint PCS bill.

Dittmer wasn't specific about when the service would be officially launched or how much access would cost. Nonetheless, he did tell executives gathered at the Wireless Airport Association meeting in Washington, D.C. in October 2002, that they would see Sprint get into the Wi-Fi market "soon."

Telstra Corp, Australia's premier cellular provider, is also hedging its bets. Telstra, which paid $302 million for a 3G license and committed numerous resources to implementing a 3G network, has taken steps to incorporate Wi-Fi into its service offerings. In August 2002, this provider and the Wi-Fi operator, SkyNetGlobal Limited, announced two agreements resulting in the Wi-Fi startup selling its network of about 50 hotel and airport HotSpots to Telstra, while beginning a relationship that will allow the Wi-Fi operator to sell services on top of Telstra's newly minted next-gen infrastructure. Commenting on the deal, David Thodey, Telstra Mobile Group managing director stated, "Telstra's mobile customers have an increasing need to experience seamless online services through broadband wireless access on various devices like laptop and palmtop computers and mobile phones, at convenience locations around Australia."

The Wireless Internet Service Provider (WISP) community is also playing an important role in the dynamic duo landscape. The aggregators, e.g. Boingo, GRIC, and iPass, are especially active. Amy Cravens of In-Stat/MDR, commenting on anticipated growth in the industry as a whole, says, "We expect wireless use to grow by nearly 300 percent this year and GRIC has taken a strong leadership position in the wireless marketplace. GRIC has already built the largest international wireless network using essentially the same model that has allowed them to achieve leadership in dial-up networking. No other network operator is adding international locations for business travelers as fast as GRIC."

Isaac Ro, an Aberdeen Group analyst, says, "People don't care how they access wireless broadband, so long as it works." Ro goes on to state that by creating its own Wi-Fi network, T-Mobile is hoping to "box-out" Wi-Fi providers currently attempting to establish a national foothold, e.g. Boingo. But that is a tough undertaking. If you recall, Sprint PCS has invested $15 million in Boingo, and says it is very bullish on the Wi-Fi company started by Earthlink founder Sky Dayton. The cellular carrier says it wants to serve mobile users wherever they are, and increasingly that includes Wi-Fi technology. Sprint PCS is also looking into developing a dual-mode card allowing its subscribers to move between Wi-Fi and its cellular networks.

Another carrier expressing interest in adding support for Wi-Fi is Verizon Wireless. Verizon has plans in the works to build Wi-Fi HotSpots and an extension of its DSL service in the New York area. It will use existing pay phones as the distribution vehicle. A spokesperson for Verizon says that the company will upgrade more than 200,000 pay phones in Manhattan to create Wi-Fi network nodes for its DSL subscribers.

European and Asian wireless carriers are also weighing their options for including Wi-Fi in their data service mix. In April 2003, British Telecommunications (BT) said it was in the process of launch its first Wi-Fi networks. And in early 2003, NTT DoCoMo began testing Wi-Fi in Tokyo by asking volunteers to check out Wi-Fi's performance when visiting certain websites or viewing live streaming video.

There is also Cometa, the new nationwide network established in 2002 by AT&T, Intel, and IBM, along with Apax Partners and 3i, to provide wholesale nationwide broadband wireless Internet access. Cometa is purported to cost a mere $30 million—chicken feed by 3G standards. When Cometa's U.S. network is completely operational, cellular providers may find that their customers will actually prefer to log onto the high-speed Wi-Fi network rather than their anemic 3G offering.

In addition, Cisco has reported heavy European cellular network operator demand for its WLAN infrastructure equipment. It's a logical move for the cellular industry. Wi-Fi services have a very real potential to cannibalize their 3G data revenues; operators can limit Wi-Fi's affect on their bottom-line by offering Wi-Fi services themselves.

It's clear that cellular operators' interest in hybrid cellular/WLAN networks is growing. This group sees Wi-Fi as an attractive and inexpensive way to quickly bring broadband service to subscribers waiting for 3G to expand beyond the initial pockets of coverage.

As carriers move into the Wi-Fi market, we're likely to see changes in the way HotSpots look and feel. For instance, InStat/MDR's Allen Nogee thinks "HotSpots would be owned by the service providers for use by their customers, and wouldn't be public."

"There is some very real potential to offloading some of [cellular's] voice calls onto Wi-Fi," said IDC's Waryas. "All the GSM guys are going to do it." Also, cellular providers "have a very tough time providing enough capacity in high-density situations, such as conference centers and hotels," according to Nogee.

Thus, as Alan Reiter, an analyst with consulting company Wireless Internet & Mobile Computing so aptly puts it, "these are all indications that the cellular operators don't see their cellular networks as the be-all and end-all of wireless data." Cellular providers who decide to support Wi-Fi—rather than avoid it—may find themselves making a very smart move. In the end, consumers don't care about the underlying technology—they just want wireless data access and they want it to be fast and reliable.

✪ DOES A DYNAMIC DUO MAKE GOOD BUSINESS SENSE?

Until recently, providing wireless data connectivity was brutally difficult, but thanks to Wi-Fi, now it's relatively simple. The technology exits to support the architecture for a dynamic duo relationship. But is there a good business model for such a pairing? Many within both industries are asking themselves just that question.

While there are many different business models for HotSpots and their integration with cellular networks, there are many issues that need to be settled. How will the cellular operators price Wi-Fi services? How will they integrate Wi-Fi connectivity into their marketing and telecom infrastructure? How should wireless operators and other vendors in the Wi-Fi "value chain" bundle Wi-Fi airtime with complementary hardware, applications, and other telecom services? Finally, how can Wi-Fi's with its limited range provide a pervasive communications network?

Here's how the Dynamic Duo might work. Comprehensive Wi-Fi coverage will be provided in densely populated areas, but in rural areas coverage will be more porous. Wi-Fi has a limited transmission range, so in less populated areas, establishing base stations every few hundred feet or meters is not feasible.

Wi-Fi will provide stationary "sit-down" broadband access while cellular networks will offer voice calls and narrowband data services. For example, a Sprint PCS subscriber could log onto the Boingo Wi-Fi network and see a Sprint PCS splash page and vice versa. John Stanton, CEO of Deutsche Telekom (DT) subsidiary T-Mobile, thinks that a dynamic duo makes perfect business since, "It's coverage where they want it and speed where they need it."

Perhaps the biggest challenge to a pairing of these two technologies is the establishment of a new value chain where all the partners are adequately compensated—a particularly tricky undertaking given the flat-rate and free nature of many Internet services. "There has been tremendous value erosion. One of the biggest challenges for operators is finding a financially viable model," at least according to Umesh Amin, director of new technologies and planning for AT&T Wireless. In the end, traditional cellular carriers will aim for an uneasy truce between the two industries with the hope that upstart Wi-Fi operators will rescue the future of their 3G networks.

South Korea, home of some of the most advanced and popular wireless data services, is already seeing the effect of the dynamic duo. In February 2002, Korea Telecom (KT), the nation's largest telecommunications company, started selling Wi-Fi access in addition to its regular cell phone service. The company has installed 1000 access points in major cities and plans to put in thousands more. In order to stay competitive, Hanaro Telecom installed numerous access points in places such as Burger King restaurants.

According to Hahn Won Sic, managing director of the fixed-mobile convergence business team at KT, Koreans expect high speed Internet in their homes and offices. He says, "They want to be able to get it everywhere." But Wi-Fi subscriptions haven't skyrocketed, although KT is forecasting 3.6 million Koreans will by using the networks by 2005.

✪ CONCLUSION

Rather than stealing customers away from cellular providers, Wi-Fi could actually foster a culture to drive their technology forward. So the question is not whether the cellular industry will offer Wi-Fi services, but to what extent. Whatever they decide, Wi-Fi certainly will not doom 2.5G and 3G systems, although it will most certainly change the cellular operator's business model. Carriers are going to have to face the fact that, at least in congested places, there are other ways of offering high-speed connections for data transfer than 3G. Also, for cash-strapped cellular operators, who have, for the time being, shelved their plans for a next-gen upgrade, Wi-Fi may be the saving grace, letting them compete with their more prosperous brethren by providing a way to deliver on the promise of always-on mobile broadband.

If Wi-Fi operators are left to go it alone and continue to spread their business model throughout the world, the Wi-Fi industry could capture up to one-third of the revenues cellular carriers had hoped to get from their corporate 3G users, alone. And if a cellular operator reads ARCchart's 2002 report, "3G by Stealth–802.11 Wireless LANs," he or she may just loose heart and throw in the towel. ARCchart estimates that by 2006, so many customers will be using Wi-Fi networks that there will be a 12 percent erosion of forecasted 3G data revenue. And that's the best-case scenario for 3G operators, the worst case predicts revenue erosion of up to 64 percent. Realistically, the cellular industry is almost forced to find a way to co-exist with Wi-Fi. It shouldn't be too difficult–the technologies are complimentary and they aren't truly competitive except in niche markets (e.g. the corporate community).

Soon the debate over whether the rising popularity of Wi-Fi will hurt or help cellular operators will be moot as these operators join forces with Wi-Fi firms to fill in spotty network coverage, reach indoor users and attract corporate subscribers.

But there is still a sticking point. At the moment, to take advantage of both strengths requires multiple devices. However, as the benefits of Wi-Fi are realized, the public will demand dual-mode Wi-Fi/cellular gear (especially cell phones, PDAs and laptops), creating an impetus toward gear that integrates both Wi-Fi and cellular functionality.

Yet, Wi-Fi and cellular face the chicken-and-egg scenario: users are not likely to buy devices with dual-mode Wi-Fi/cellular features until Wi-Fi is readily available, and the dynamic duo is unlikely to become widespread until there is a decent base of devices that can use both technologies seamlessly.

Section V: Wi-Fi for the Enterprise

Chapter 14: Wi-Fi in the Corporate World

The IEEE 802.11 family of standards is on a fast evolutionary track, with announcements made on a seemingly daily basis about improvements in Wi-Fi's ability to meet the business community's needs for performance, security, and manageability. Still, Wi-Fi is a relatively new technology. So despite the market pressures to adopt and deploy wireless capabilities, a pragmatic approach toward the implementation of a wireless networking plan may be best.

Some of the questions an IT professional might pose when considering a WLAN deployment include: How will the various 802.11specifications shake out? What will it cost to support this rapidly changing technology? How will a WLAN map into the organization's top priorities? How does it fit into the current network topology? How does it allow the company to reach its goals?

And don't forget the people factor. High-tech staffers who love to explore new technology will be among the Wi-Fi advocate's most ardent supporters; it will also be the group that will want to "push the envelope," so there may be a need to moderate their wireless initiatives. Other people-related issues that must be considered when planning a wireless network include resource management, training, workflow, and process issues such as troubleshooting, provisioning, and documentation. Thus, in the beginning, a WLAN might actually decrease productivity. But don't let such considerations detour a wireless networking strategy.

Virtual offices, virtual workspaces, and sending documents while on the go are all part of today's corporate environment. Employees have discovered that it's possible to access their workstation via a range of wireless devices—from their laptop computer to their PDA to their cell phone. While it's unlikely that employees will type long reports on their PDA or cell phone, sending and receiving email, and accessing contact information from such devices is already routine, and the technology exists to attach office documents and other files to emails sent by way of these diminutive devices.

A DISRUPTIVE TECHNOLOGY

Wi-Fi is a disruptive technology. Disruptive technologies are products or services that, when compared with the mainstream offerings of the dominant competitors, are simple, convenient to use, and inexpensive, and as such they create a new market. Examples of disruptive technologies include cellular phones when compared to wire-line phones; personal computers, which quickly managed to overtake the mainframe computer marketplace; and now wireless networks with their portable devices are taking the place of wired networks with their stationary computers.

Many times a disruptive technology will not initially perform as well as its mainstream competitor, yet because the new technology tends to have specific attributes that are missing from its more established competitors, it finds a following. A good example is the personal computer. The PC, like Wi-Fi, entered the marketplace as a disruptive technology. In the case of the PC, its more established competitors were the established mainframes and minicomputers. The performance of the early PCs offered by upstarts such as Apple and Compaq was limited, and they initially couldn't perform the computing applications that were in use at the time. Yet the cost and convenience of PCs were so compelling that *new* computing applications (WordPerfect, Microsoft Excel, etc.) quickly developed, to the detriment of established mainframe and minicomputer vendors like Data General, Digital Equipment, IBM, and Wang. Also, PCs' processing power improved so rapidly that they invaded the applications (e.g. database management and data processing) that IBM and others had dominated—and they did so with greater convenience and at a lower cost.

Sound familiar? Originally, considerable lack of enthusiasm existed for Wi-Fi's progenitor, 802.11 (mainly due to its bandwidth limitations), but when 802.11b came along with its relatively speedy 11 Mbps, a market for this disruptive technology quickly followed. Not only did the tech industry glom onto the technology, but so did the buying public. Go to a local Starbucks or McDonalds, and watch customers in their "virtual offices" writing emails, surfing the Web, holding impromptu meetings while they sip on coffee, and munch a snack at five or six bucks a pop.

Wireless networking is in the process of changing not only the way we work, but also how we define our place in the corporate environment. Observe the next meeting in a Wi-Fi-enabled organization—watch how the meeting is conducted, how the meeting is designed to be more interactive, and how information is more easily shared.

Just as PC makers grew by participating in improving technology so that they overtook the minicomputer and mainframe vendors, companies enabling Wi-Fi connectivity are growing at a breathtaking rate due to improved product capabilities. In doing so, these companies have overcome any barriers to this compelling technology.

Although Wi-Fi has been a disruptive technology for a while, the rapid technological improvements being made throughout the entire wireless networking industry are only now beginning to exert pressure on IT departments to adopt Wi-Fi. This pressure is likely to overwhelm companies that fail to move fast enough. Technically astute workers are probably already implementing unofficial wireless networks. You can either fight them or join them.

Whatever the choice, eventually, every business will have to acknowledge the existence of Wi-Fi and its applications. Once businesses accustom themselves to the idea that the disruption wrought by Wi-Fi actually may help them move forward to new levels of productivity and responsiveness, they will clamor for the mobility of wireless networking.

TAKE A PRAGMATIC APPROACH

Despite the groundswell toward wireless networking, in the big picture, Wi-Fi is just a blip on the LAN screen. Only a minority (albeit a growing minority) of organizations *fully* embrace WLANs, although more and more tolerate Wi-Fi within their networking ecosystem. Over the next few years, however, the rate of adoption will accelerate and in doing so, profoundly affect organizations and their IT departments.

When adopting Wi-Fi, however, the business community will need to face the problems of dealing with the changes that wireless networking not only requires, but creates. Despite the apparent simplicity of wireless networking, if not handled correctly, those problems can be significant.

First, there are the issues of security and quality of service. As has been discussed throughout this book, out-of-the-box wireless security leaves much to be desired. As noted in Chapter 17, a multi-layer security plan will need to be devised and implemented to meet the stringent security requirements of many organizations.

Managing and monitoring the network to maintain a certain level of Quality of Service also can be challenging. But unless your wireless network is expected to host bandwidth gobbling applications such as voice, streaming media, or CAD (computer assisted design), those tasks shouldn't be too onerous.

Next, most IT departments will need to either hire or train people who understand radio technology, since implementing a wireless infrastructure is fundamentally different from installing wires. As the networking infrastructure moves from wired to wireless devices that use radio frequencies, the IT department will find itself dealing with unfamiliar interference and propagation characteristics.

Furthermore, in most organizations, the WLAN will be an extension of an existing wired network. Since that may be quite complex, it's vital that prior to a WLAN deployment, an external audit of the wired network is performed to ensure the wired network can hold up its end of the new networking ecosystem. A WLAN project can fail if its wired counterpart doesn't provide a solid, stable, well-designed foundation.

Finally, integrating a WLAN into an organization's network ecosystem will require more infrastructure, i.e. more software, servers, and possibly staff. After all, the software that makes the aforementioned virtual offices and workspaces possible has to run somewhere, and someone has to manage it.

TAKE IT IN STAGES

To keep yourself and the deployment team from feeling that they are working in a maze, not knowing which direction they are headed–The deployment stage? The planning stage? Where does the check list fit in? What about technology? What comes first? What comes last?–you need a plan, and that plan should structure the WLAN project so that there are definable stages. Each of the stages should have clear-cut steps that need to be taken, such as defining security issues, performing an incremental survey, determining return on investment (ROI), and so forth. Here's an outline.

Stage One—the planning. This normally consists not only of performing a WLAN assessment and users' needs analysis, but also a site survey (which is covered in detail in Chapter 8), and any necessary IT staff training. Now is also the time to consider whether bringing on board an outside project manager may be necessary. One of the keys to a successful enterprise WLAN strategy is a project manager who pays equal attention to technology, process, and employee issues.

Many times Stage One will provide all the information necessary to determine whether or not a wireless LAN is in your organization's best interest. But often more will be needed before a final decision can be made. This takes us to Stage Two.

Stage Two—the pilot project. While some organizations will chose to skip this stage, others will insist on a pilot project to learn more about wireless networking to assess how it will affect the corporate culture, and to obtain real numbers for a required ROI assessment before they move on to Stage Three. And for others, the pilot project is part of the deployment stage since they will deploy their WLAN via a series of pilot projects.

Stage Three—the technology. Once the decision is made to adopt a wireless network, the next stage is to use the site survey results to determine which wireless standard (or standards, if opting for a dual-mode network) the network will be built around. Some technology testing may have taken place during the site survey, but there will be much more that must be considered. During this stage, the determination is made as to whether the gear chosen suits the organization's culture, the needs of the wireless network, the overall networking environment, and the budget allocated to the wireless project.

Stage Four—deployment. This is where the actual installation of the equipment and software, testing, and planning for operational support takes place. This is also the stage where user-training takes place.

Stage Five—ensuring service availability. Once deployment is completed, the wireless network is considered to be in production mode, fully capable of supporting user applications. People are utilizing the network's services. At this stage, mechanisms either should be in place, or being tested so that network service availability is assured. After the network is deemed operational, the network manager must constantly be aware of and monitor factors that could influence a decision to make a change to the existing infrastructure.

When done right, any organization that deploys a WLAN eventually will see a ROI—primarily because of productivity gain due to the employees' ability to obtain convenient access to critical information from any location, which, in turn, enables faster and better decision-making.

✪ ASSESS THE SITUATION

Evaluate the existing corporate culture and networking environment, as set forth in Chapter 8. If you are the Wi-Fi champion, take a leadership role in obtaining wireless technology buy-in from influential executives and managers. Achieve this by:

- Understanding the technological issues associated with wireless (e.g. security, interference, and the fact that the 802.11 series of standards is still relatively new technology).
- Understanding the costs and how your executive board feels about the issue.
- Creating a database that contains text written by journalists and analysts who cover

the wireless arena, and academics who conduct research into it. Also try to build relationships with some of those Wi-Fi supporters.

- Creating a list of wireless vendors and consultants using information from the aforementioned database.
- Coming up with a short list of trusted vendors to whom you would feel comfortable issuing an RFP (request for proposal) or RFI (request for information). It should be relatively simple to come up with a list by using the two aforementioned databases.
- Showcasing your findings to internal constituencies and external sources of expertise, e.g. consultants who can help with a WLAN deployment.
- Using the showcase to find a business unit manager who believes that the use of wireless would be beneficial for his or her staff.
- Forming an ad hoc group of interested parties to discuss wireless networking. The goal of such meetings should be to openly discuss wireless networking and the issues surrounding it, and to develop recommendations for implementing a wireless network within the organization's current network infrastructure.
- Implementing a *small* pilot WLAN project with the help of the aforementioned business unit manager. (A small pilot WLAN should take no more than a few weeks to deploy.)
- By using the data learned from the pilot project, it is easy to determine: what skills need to be developed to deploy and support a WLAN, a rough idea of the cost of an operational WLAN, the issues that may come up when dealing with wireless technology, and security and privacy concerns.

The astute WLAN enthusiast won't stop there. He or she will also look for patterns in the organization's purchasing records to discover which workers are already equipped with mobile computing devices.

Next, research and analyze the resulting data. Moving forward with a wireless strategy, without analyzing how wireless technology can help an organization, can bring unsound results. If you just buy into wireless networking without doing the necessary research and analysis, you might find that the WLAN project results in very little return on investment. For instance, say you've convinced a marketing manager to adopt wireless, but you didn't do your research. If you had, you might have discovered that a majority of the marketing staff couldn't use their PCs with any kind of proficiency. This would spell doom for any pilot program using that department's personnel.

However, through your research, you might have found out that the sales department had successfully introduced a computer-intensive sales force management solution within the last couple of years. Thus, the sales staff were (1) accustomed to using their computers as a sales tool, and (2) not adverse to new technology. That bodes well for any WLAN pilot program you might deploy within that department.

By following the above outline (and through the lessons learned during any pilot project), the Wi-Fi advocate will not only have created an opportunity to catalyze and manage conversations that can shape managers' and executives' decision-making, but also the IT staff will be better versed in what is needed for a viable WLAN project. This, in turn, will drive a wireless strategy and the subsequent WLAN deployment.

Users' Need Analysis

But more research is needed. The next step is to determine the actual wireless networking needs of the organization's employees. Weigh the results against what the organization is prepared to provide.

Normally, a wireless network is created because of some initial need or reason for the services that a WLAN can deliver. But a wireless strategy can also come about merely because of a desire to "stay with the times," or because "it's nice to have," as in the case of the H.J. Heinz Co. (see the "Case Study" section of this chapter).

Only through understanding the end-users' needs is it possible to match the wireless technology to the organization's business goals. While many times the analysis will indicate only a need to connect a mobile workforce to data residing on the corporate servers and the Internet, some user groups may need more. For example, the R&D department or manufacturing plant might need high bandwidth for their detailed schema. Or the survey may indicate that the executive suite desires wireless mobility, but only within specific areas, thus the network design might not need to *immediately* support Mobile IP. Or perhaps, after completing the employee interviews, it's determined that many within the organization need not rely on a WLAN for *all* their networking needs, which, in turn, means that fewer access points can serve more people.

End-user interviews are the only way to obtain the information for a User Needs Analysis. Unfortunately, in large organizations, it's usually impractical to try to interview every potential WLAN user. In such situations, it might be best to segment the potential end-users into user groups (i.e. secretarial, inside sales, outside sales, marketing, R&D, warehouse, executive suite, etc.), and then conduct personal interviews with only five or so employees within each group. Also interview departmental managers to learn their staff's tasks; how they perform them; how they interact with other departments, outside partners and customers; and how the managers feel wireless can help.

The interviewer must ask questions pertinent to the needs analysis report. Since each organization's reasons and needs for a wireless networking environment will differ, craft the questionnaire to fit your organization's specific situation. But there is a common ground to use as a guideline when drafting the questions. For instance:

- How many hours per day the employee spends using his or her computing device?
- How many of those hours are spent accessing the corporate servers, the Internet, or a partner's intranet?
- What type of information and/or data do the users access, e.g. text documents, CAD files, marketing materials from outside sources, graphic files such as slide shows, high-definition graphics, streaming media, schematics, blueprints?
- How do users utilize the data, e.g. review and analyze, draft or update, approve?
- How often will they use the WLAN to send and receive data?
- How often do they need access to corporate or Internet-related data when they are away from their workstation and how do they handle those situations? Why do they need access to information during these out-of-office visits—attending a meeting with co-workers, customers, outside partners?
- How often do they participate in videoconferencing?
- How often do they take work home with them?
- How often do they work through their lunch hour?

* How would they utilize a mobile computing device if the organization should provide such an item for their use?
* What is the graphical coverage area where employees will access the WLAN outside the normal workstation, e.g. conference rooms, cafeteria and perhaps, training room.
* What type of wireless computing device would the employee desire, e.g. how many will be happy with just a PDA and how many will require a laptop? When will they use these mobile computing devices—all the time, only when mobile, or only when, say, taking inventory or attending a meeting? (This will determine the battery requirements and power management requirements to look for when purchasing PDAs, tablet computers, and laptops.)
* The number of employees requiring only sporadic mobility, e.g. to attend a training session or a conference or trade show, or perhaps while located in a temporary office space.
* How large will be the typical information packet sent via a WLAN? What are the most bandwidth-intensive packets (streaming media, graphic files, blueprints or what), and how often will these types of files be sent and received via the wireless network?
* How many end-users will access the wireless network at any one time? How many in each user group (e.g. marketing, R&D, engineering, executive suite)?

The results of the users' need analysis will help to determine not only the flavor of Wi-Fi, but also the number and the positioning of access points.

WHICH WI-FI FLAVOR?

IT decision-makers, at least for now, are forced to deal with multiple WLAN standards—not only those that are available today, but also those that have yet to be commercialized. The immediate problem relates to incompatibilities between the IEEE 802.11 networking specifications, e.g. 802.11b, 802.11a and 802.11g, all of which have their advocates.

We dive into the minute details behind the entire 802.11 suite of specifications in various chapters (e.g. Chapters 6, 7, and 10), but for some readers, such particulars won't be necessary for their decision-making process. So, in this chapter, we take a more myopic view of the 802.11 standards, looking at them only in the context of a corporate networking environment.

For those that don't yearn for the fast lane, 802.11b, which operates at 2.4 GHz and offers a top data transfer rate of 11 Mbps at the access point, has obvious benefits: an extremely low cost, a large installed base, and a relatively good range. The most common types of wireless LANs in operation today support the 802.11b standard.

802.11a, however, can offer a fivefold increase in performance over 802.11b, it also operates on the less-crowded 5 GHz band (although you shouldn't put too much weight on this point), and has more channels (eight versus 11b's three). However, an 802.11a installation can be more costly than a network running 802.11b technology. And although 802.11a equipment has been available for some time, vendors of enterprise-class WLAN gear with sophisticated security and management options have only recently entered the market.

802.11g enthusiasts assert that this specification offers the best of both worlds–it runs like 802.11a (i.e. 54 Mbps), but provides full backward compatibility with 802.11b, and more importantly provides better signal characteristics than either "a" or "b." But, because of political and technical conflicts, the 802.11g specification (the ruling document to which 802.11g products must adhere) wasn't finalized until mid-June 2003. Thus there is no installed base, which means that like all new technologies, the kinks haven't been worked out yet. Another drawback is that the specification was only newly published, thus the Wi-Fi Alliance has only recently added 802.11g as part of its certification suite.

Despite these constraints, several manufacturers (e.g. Apple, Belkin, Buffalo, and Linksys) have shipped a variety of wireless networking gear based on chipsets that use a draft version of 802.11g. But beware, any products appearing before the final 802.11g specification is ratified can only say that such products *conform to a 802.11g draft specification* of an underdetermined draft version. The 11g products shipped in early 2003 were designed using a mix of specification drafts that emerged from the IEEE Task Group g meetings in September 2002, November 2002, and January 2003, respectively. There are significant differences between those drafts, which can affect both 802.11b interoperability and throughput performance of draft-802.11g products.

***Note:** The Wi-Fi Alliance is a nonprofit organization that tests equipment based on the 802.11 suite of specifications to ensure that the equipment is designed to work according to the specification and is interoperable with all other certified equipment. Only after the Wi-Fi Alliance's test labs give a piece of gear the "thumbs up" is the maker allowed to use the Wi-Fi logo. The great thing about the Wi-Fi certification is that interoperability of products built upon the same specification (e.g. a, b, and g) from different vendors is ensured if a product bears the "Wi-Fi" logo. Until 802.11g was finalized, however, the Wi-Fi Alliance had no way of guaranteeing that different 802.11g devices would work with one another, meaning that it will be take a bit of time after the IEEE's ratification before Wi-Fi certified equipment hits the marketplace.*

Implementing a WLAN requires that the deployment (or pilot project) team and IT department face many critical decisions, but the most important will be choosing the right wireless standard. The final decision will depend largely upon bandwidth needs and the current level of WLAN deployment within the organization.

"b" or "a"

Those struggling with the decision of whether to deploy 802.11b or 802.11a should concentrate on the value gained from the deployment, coupled with the bandwidth requirements of the applications that will run over the network. If the applications don't require bandwidth in excess of a shared 11 Mbps segment, and if it is predicted that there will be only a relatively small number of simultaneous WLAN users per access point (somewhere between 10 and 50 depending on usage patterns), 802.11b will do just fine. If more bandwidth is needed by a specific user group (e.g. marketing and its graphically intensive campaigns), the IT department can provide higher per-user throughput by reducing the number of users in each coverage area through the installation of more access points.

"b" or "a" Check List

Use this check list to help you in making a decision between an 802.11b or 802.11a:

* An environment that has many 2.4 GHz interference sources, e.g. Bluetooth devices, wireless phones, or microwave ovens. 802.11a, which uses the 5 GHz band, may be the better choice.
* A deployment that requires more channels. A larger number of channels allows more users to share the network and protects against interference from neighboring access points. The 802.11b's 2.4 GHz band is relatively narrow, providing for just three non-overlapping frequency channels, while 802.11a offers eight channels. However, if designing an 802.11a network to harness all eight non-overlapping channels, be aware that deploying external antenna for access points is a non-option, since FCC limitations require attached antennae, fixed transmit powers, and antenna gain limits on 802.11a access points. Still, the choice to forgo some freedom in favor of all eight non-overlapping channels can be advantageous in certain deployment situations.
* Deploying a WLAN in an organization that already has an installed base of Wi-Fi equipment. The more 802.11b clients installed, the greater the need to have access points that support 802.11b.
* An organization that has a large percentage of WLAN users who also must tap into wireless networks outside the organization where 802.11b is more likely to be predominant (e.g. HotSpots, a home wireless network, or wireless access outside the U.S.). An 802.11b or a dual-mode 802.11b/a network would be the best bet.
* 802.11b is better for transaction-intensive applications; 802.11a is better for bandwidth-hungry applications.
* Budget versus costs. Provisioning an 802.11a network usually costs 20 to 30 percent more than using 802.11b products. In addition, deploying an 802.11a network may incur additional costs due to different RF characteristics of the 5 GHz frequency.
* Data rates and signal range. Access points offer their clients multiple data rates. For 802.11b the range is from 1 to 11 Mbps in four increments, and for 802.11a the range is 6 to 54 Mbps in seven increments. A mobile computing device's wireless networking card will automatically switch to the fastest possible rate offered by the access point (although how different cards do it varies from vendor to vendor). Each data rate has its own unique cell of coverage (the higher the data rate, the smaller the cell), thus the minimum data rate for any given cell must be determined at the design stage. If the organization requires the WLAN to consistently provide only the highest data rate, the deployment will require a greater number of access points to cover a given area. Most network designs, however, will be a compromise between required aggregate data rate and overall system cost.

In the final analysis, the main reasons to consider deploying a WLAN based on 802.11b standards are:

* 802.11b products were the first to market, thus it is more likely to provide platform stability.
* The standard has been widely adopted by vendors and organizations alike so there is more experience with its deployment and maintenance.

* Interoperability between many of the 11b products on the market is ensured through the Wi-Fi Alliance's certification program.
* 11b's relatively high speed is more than adequate for many organizations' networking needs.
* If one of the WLAN requirements is to support a wide variety of devices from different vendors, 802.11b is probably the best choice.
* An 802.11b WLAN can be easily upgraded, usually through firmware updates, to 802.11g when advisable.

802.11a products are relatively new to the marketplace, so careful consideration should be given to their platform stability and performance. However, there are some instances when 802.11a will win out over an 802.11b installation:

* When deploying a WLAN in a multi-story corporate environment, 11a may be the best choice because it uses smaller cells and may encounter less interference from other installations within the building.
* When higher per-user throughput is needed. 802.11a offers eight non-overlapping usable channels per access point (versus three for 11b), so when an 11a network is properly provisioned, it can provide a shared medium that offers higher per-user throughput.

The downside is that since 11a operates at a higher frequency, it has a shorter range and requires more access points (APs) for a given coverage area than 11b. Thus, a network based on the 802.11a specification may be more expensive and may be unnecessary, especially for applications such as warehouse data entry and the typical corporate office-related applications that don't need the extra bandwidth.

"g"

Let's now look at the impact of the new 802.11g standard on the WLAN scene.

The advantage 802.11g affords is a top data rate of 54 Mbps, while maintaining full backward compatibility with 802.11b via complementary code keying (CCK) modulation. However, since 11g uses the same transmission type and modulation technique as 802.11a (i.e. OFDM) while remaining in the same 2.4 GHz spectrum as 11b, the 802.11g standard carries some range and performance characteristics from both 11b and 11a. Some of these features can limit 11g's effectiveness. Here is an example:

While 11g, like 11a, boasts a top data rate of 54 Mbps, it operates in the unlicensed portion of the 2.4 GHz spectrum, which means that the 802.11g transmissions are limited to the same three channels and crowded 2.4 GHz band as 802.11b, creating possible scalability and interference issues. Thus, although 802.11g is capable of a theoretical 54 Mbps throughput, it probably will not reach the same speeds as 11a, because of 2.4 GHz interference problems and the longer resend intervals required for backward compatibility with 11b. But, while 802.11g offers less range than 11b, it does provide greater range than 11a because of wave characteristics at the 2.4 GHz spectrum.

Furthermore, 802.11g circumvents some of 11a's FCC restrictions on auxiliary antennae and transmit power. Thus, there is more leeway in the use of creative antenna technology and deployment. Also, since 11g is compatible with 11b, the same antenna equipment from current 802.11b deployments may be used.

Organizations with a significant investment in 802.11b, but needing a performance boost without going the 802.11a route, may want to consider 802.11g technology as long as it bears a Wi-Fi certification label. However, even though 11g's increased throughput and backward compatibility with 11b may be enticing, don't ignore the five additional non-overlapping channels you get with 11a.

As WLANs become an increasingly integral adjunct to the corporate wired network, the higher speed standards—802.11a and 802.11g—will become the logical choice because of the projected needs of future bandwidth-intensive applications (e.g. wireless videoconferencing, voice-over-wireless IP, and the like). Yet, even when 802.11b is the standard used to implement a new WLAN, you may still want to give consideration to designing the WLAN around dual-band a/b or a/g access points. By installing dual-band access points, you can not only protect existing wireless investments, but also provide a migration path to the dual-band wireless world of the future.

When taking the dual-band approach, however, a comprehensive site survey must be conducted for *each* specification to guarantee adequate network coverage. This is due to 802.11b's and 802.11a's different signal strength, and requirements to adjust for interference and reflection characteristics.

HARDWARE

Selecting the right networking gear is far from cut and dried. A corporate WLAN, in particular, requires a thoughtful hardware selection process. While anyone can buy WLAN gear at most electronics stores, many of those products won't have the sophisticated security and management features that a corporate IT professional demands.

The most important piece of advice the author can give to a WLAN team regarding product selection is to buy wireless networking gear from established vendors that can offer enterprise-class support. A detailed examination of all of the components that might be found in an enterprise wireless network is provided in Section VII: The Hardware.

Selection Process

Follow the advice set out in "The Vendor Selection Process" section of Chapter 8. But don't be shy. As with any other technology purchase, pin the vendors down on precisely what they can deliver and what they can't. For instance, if they promise site survey tools—find out exactly what they mean. The same holds for network management and maintenance tools.

As with any project, calculate the overall costs of the WLAN. The results can be used to compare a wireless LAN with a wired alternative, perform a feasibility study for a specific application (e.g. deploying a WLAN to provide connectivity within a conference area), or provide the basis for the budget you're proposing to upper management. In all of these cases, take into account the hardware, software, and vendor support needed to install and maintain the new wireless networking environment.

PILOT PROJECT

Wireless networks often top an IT manager's pilot project list. Not only is the allure of enhanced mobility extremely strong, technologists are also innately fascinated by the technology itself. When it works, wireless networking is nothing short of amazing, a testimony to 21st century engineering ingenuity.

Like all new technologies, however, wireless networking solves some problems and creates others. Enthusiasm for wireless networking is often tempered by worries about security and privacy, authentication, interference, over-population of access points, and the needs of high bandwidth applications. A pilot project can address those concerns and pave the way for a full-scale WLAN deployment by confirming technology choices and establishing productivity metrics. A well-executed pilot project is also the means by which it's possible to overcome any design limitations that a full-scale WLAN deployment may encounter.

But getting from pilot project to production system requires compelling ROI. The costs and benefits of many mobile and wireless systems initiatives, however, are difficult to measure. Sure, you can slap together a budget for a new WLAN, but determining the TCO (total cost of ownership) that reflects the upgrades as the technology evolves is another issue. Just how much of a productivity increase can you realize by giving people untethered access to the network? Of course, you may have a situation where a WLAN can save you the cost of wiring, but in most environments where performance is a priority, you'll probably need to have a wired infrastructure in addition to wireless. Then it comes down to demonstrating ROI for two networks.

Here's the good news. Most executives who are in a position to sign off on big technology initiatives will respond favorably to a well-conceived wireless project, many times due to their own experiences with the technology.

Furthermore, the WLAN pilot should pay for itself, not only in productivity gains, but also in the timely data it provides. A WLAN pilot project provides a testing platform for the technology, and is a tool that can be used to collect useful data, to identify users most likely to benefit from wireless networking, to help build the in-house skills needed to deploy and sustain a wireless networking environment, and to show how a WLAN can deliver a credible ROI.

In many organizations, employees are already experimenting with wireless networking and "outlaw" wireless networks are probably already in operation within most large organizations' facilities. While the existence of unauthorized wireless networks may trouble some, it can lessen the efforts needed to document the business needs for wireless networking. Since early adopters are already Wi-Fi champions, they should be the first ones recruited in your effort to make a business case for a WLAN. These experienced end-users are important not only because they can help to influence the final decision-making process, but also because they can help spur adoption in their individual departments.

Form a Team

The first step in initiating a pilot project is to form a "pilot project team" to address both a wireless project's technical and end-user implications. This team should include representatives not only from the IT department, but also from various departmental constituencies. During the project team formation, look for members that can address:

- The broad technical requirements (e.g. technical design, scope of the pilot).
- End-user considerations (e.g. what the pilot should accomplish from the employee's view point).

* The assessment of technical considerations (e.g. security, encryption, etc.).
* A customer support model.
* Systems management considerations.
* Financial management issues.

If the pilot WLAN is successful, later on, the pilot project team's constituents can form the basis of a WLAN deployment team.

Have a Plan

After the pilot project team is in place, the next step is to plan the project. This means defining the goals of the project, its scope, the participants, the areas the pilot WLAN will cover, and the necessary training of participants and IT staff to help ensure the success of the pilot project.

Goals

Every pilot project will have its own set of goals, but the goals of every test WLAN should include:

* Demonstration of the feasibility of wireless data networking.
* A determination of engineering constraints.
* An estimate of the total cost of ownership (TCO).
* An estimate of the ROI achieved through wireless networking.
* A means to build interest in wireless networking.
* The development of a funding model for a full-scale WLAN deployment.
* In-house IT staff developing expertise in wireless deployment and operation.
* Understanding and resolution of security issues associated with wireless networking.
* The development of a support model for end-users.

Scope

The scope of the pilot project is dependent upon the size of the organization, the planned scope of the proposed full-scale WLAN, the size of the facility, the user base, and the pilot's budget. But the project must be based upon a good sampling from each location in which wireless connectivity will be available. If possible, include end-users with various job classifications, in different employee segments, and departments. In other words, try to populate the test program with a user group that will derive different values from wireless networking. Include opponents of wireless networking as well as proponents, and encourage executives or senior management to join the effort. Executive management buy-in is vital to the success of both a pilot program and any subsequent WLAN deployment.

For example, if a large enterprise wants to test the model for a large-scale deployment of a WLAN to serve, say, more than one thousand users, the pilot project team should conduct a pilot program over a period of a few months for up to 250 employees. Those users should be able to access the pilot WLAN from many representative areas of the facilities, e.g. different floors, different buildings, cafeteria, conference rooms. In a smaller organization that is considering the deployment of a WLAN to serve a staff of fewer than 500, the pilot program should include, for example, three or four top-level executives, the sales department, and perhaps some of the IT staff.

A successful pilot project will apply a usage model that makes sense in the organization's specific environment. For example, in a corporate environment, a popular site for a wireless LAN project could include executive offices, sales offices, training center, and/or a new office complex.

Plan the pilot around different kinds of wireless usage, e.g. large graphic and text files, multimedia files, and even videoconferencing. If the wired LAN commonly supports videoconferencing, implement at least one test videoconference session during the pilot program. It's important that during the design process, the team knows when and where these varying kinds of network data will be transmitted and received.

Coverage Area

The pilot WLAN's coverage area should provide a breadth of experience. Map the wireless zones that participants will use (conference rooms, common areas, cafeterias, offices, operational areas, and so forth), and determine the existing LAN wiring in those areas—access point hardware is hard-wired to the LAN. Work closely with the organization's

LAYING THE GROUNDWORK

Make every effort to launch the pilot project on the right footing. That is the only way a WLAN pilot can achieve the desired results. There are certain steps that should be taken before you actually execute the pilot program.

When deciding whom to use to test the pilot network, other than early adopters, target high-value users such as sales staff, design engineers, mid-level executives, and managers—users who would benefit from untethered connectivity. Target existing mobile PC users rather than first-timers. Include a number of technically savvy departmental heads, managers, and executives, as well—they can be instrumental in obtaining the necessary funding for a full-scale WLAN deployment.

However, don't limit the pilot project to just high-value end-users. If you don't use the pilot to study the impact of WLANs on support staff and other worker segments that would find mobility an asset, the pilot won't be able to provide you with a complete picture of the benefits that can be derived from a WLAN, or the problems that might be encountered when a full-scale WLAN is deployed.

Finally, sprinkle a few people from the WLAN opposition group into the mix. If you can win over that group, it may prove to be quite advantageous when you approach the stakeholders for approval for a full-scale WLAN.

Know what to measure before measuring it. While the timesaving factor is important in calculating productivity gains, it's difficult to measure; furthermore, timesavings can be less meaningful when it is applied to a diverse workforce with different job functions. So the best way to handle timesaving measurements is to track exactly how a WLAN end-user saves time when performing specific tasks, e.g. sending/receiving emails, accessing data while on the go, transferring files, attending online meetings, and connecting to the network.

Classify the pilot participants according to their job functions. This is because some end-users will derive more benefit from a WLAN than others, based on their job func-

facilities team to incorporate wireless technology in ways that will minimize reworking as the WLAN expands. If the organization's facilities span several floors of a building, or even extend over two or more buildings, plan the pilot around a coverage area diverse enough so that it is possible to test the mobility value of a wireless initiative.

As the coverage area is defined, the team also needs to ensure that the IT department can support the size and scope of the test network. While a large (rather than small), representative pilot project is a better test of usability and productivity, the project will be doomed to fail if it doesn't have the resources to support it.

Site Survey

Once you know the goals, scope and coverage area, you can plan access point placement to provide adequate wireless coverage. Review the Site Survey section in Chapter 8 to ensure the best coverage results are obtained.

Always keep in mind the following:

tions. Also, when determining the ROI of the pilot group's productivity gains, the value of their time (for example, salaries) will play an important role.

Here's an example on how a pilot's user base could be segmented:

- Engineering/product management
- Manufacturing
- Sales
- Marketing
- Support

After determining what to measure, take a survey to find out how the potential pilot participants do their jobs. Have them give you an outline of their daily routine, and ask them to identify the difficulties they encounter without wireless connectivity. Once you have that information in hand, you will have a good foundation for determining whether wireless connectivity can improve the workplace and can deliver an appropriate ROI.

Taking the segmentation approach will also help an IT department plan for deployment after the pilot phase, because it can begin the roll out of WLAN connectivity to first serve the workers who will benefit the most from wireless networking.

Inform all pilot participants as to what you expect from them before they enter the program. For example, you may want them to keep a log, participate in surveys, and undergo training. Also educate them on what service level they should expect from the pilot WLAN, and how to get support when something goes wrong.

Beware: If the participants are not prepared for what they must do, the pilot experience could become very messy. On the other hand, supplying them with extraordinary support during the pilot period may provide an unrealistic end-user experience, resulting in feedback that is too positive and that may not be reflective of an actual deployment.

The number of simultaneous users that an access point can support depends mostly on the amount of data traffic at the time. One user downloading or uploading a 30-minute streaming media presentation will have a more detrimental effect on the network's speed than 20 users simultaneous uploading a newly released corporate manifesto.

WLANs, like wired networks, provide bandwidth that is shared among its end-users, so a WLAN's performance, as gauged by the number of simultaneous users, hinges on the combined computing activity of those users. Scaling a WLAN, whether for a small, medium, or enterprise-size organization, depends on many factors including number of buildings, size of the campus, number of users, and the types of work or activities being done throughout the day. For example, each 802.11b access point has a maximum data *throughput* of approximately 7 Mbps, which is adequate for:

- 50 nominal users who are mostly idle and check an occasional text based email.
- 25 mainstream users who use a lot of email and download or upload moderately sized files.
- 10 to 20 power users who are constantly on the network and maintain large files.

To increase capacity, more access points can be added to give users more opportunity to enter the wireless network. Scaling a WLAN, relative to the number of employees and the type of network usage, is the best rule of thumb for estimating capacity. But use a careful site analysis to provide an accurate picture of the equipment requirements and configurations for your organization.

Wireless networks are optimized when the access points are set to different channels. For instance, an organization may place three 802.11b access points (with a range of up to 325 feet or 100 meters each) in three adjacent offices, with each unit set to a different channel. This, in theory, would enable an 802.11b network serving the number and type of users referenced in the above bullet list to share the maximum of 33 Mbps (11 Mbps x 3 APs) of total capacity (although no single user would ever have throughput faster than 7 Mbps). But in reality, wireless computing devices associate with the access point with which they share the strongest signal, so the bandwidth usually is not dispersed evenly among users.

When considering AP placement, usually higher is better (although there are limits), and open areas work better than smaller spaces. Look for physical limitations that might interfere with wireless radio signals: buildings, objects, and concentrations of people can affect a WLAN's coverage. Having multiple access points will extend the number of users the WLAN can support.

Training

Establish a training program. The success of a pilot often hinges on how well participants and internal Help Desk personnel are trained, and whether this support group can respond quickly to requests for help.

Now is also the time to start documenting how pilot project users will get started, how to use the WLAN, and how to obtain support.

Prepare trainers and draft a set of FAQs. Train the IT and technical support teams thoroughly in the new technology. Plan to have roving trainers during the first two weeks of the pilot, so they can help users through tricky spots. To guarantee 100 percent training participation, consider a deployment process in which pilot participants cannot receive their wireless hardware unless they get usage instruction at the same time.

First impressions can make or break the success of the pilot.

Computing Devices

Take an inventory to determine the current number of mobile computing devices. Next, determine which will need wireless network interface cards (e.g. PC cards) to communicate with the WLAN's access points. While a wireless pilot will be less expensive, and will yield higher initial value if you work with employees who already use mobile computing devices extensively, a test program also presents a good opportunity to equip participants with devices that have integrated wireless network support.

Metrics

Determine what the pilot project should measure. Measuring the outcome of a wireless pilot is vital to moving ahead to a broader network. Success metrics might include productivity hours reclaimed, adoption rates, increases or decreases in revenue and costs, and how well people understand and use the WLAN. To reach the end goal of a full-scale WLAN deployment, the pilot project team needs far more than a handful of supporters extolling the wireless network's virtues.

A pilot project is the perfect time to determine the expected ROI from your anticipated WLAN program. But, if after the project is complete, it is clear that its users would complain loudly if they were suddenly cut-off from wireless connectivity, the approval of a full-scale WLAN is not only likely, it has a great chance of success.

Security

Plan to test the strength of any security procedures put in place during the pilot program. Plan for technical support teams to assault your WLAN. Ask them to attempt to hack, wardrive, and eavesdrop into the wireless system, then get feedback.

Feedback

Determine how you will obtain end-user feedback. Ask the participants how they foresee using wireless day to day before the pilot begins, then create metrics and surveys that will reveal how they actually used and benefited from wireless. Intangible benefits should also be measured, such as a user's sense of convenience, job satisfaction, increase in perceived productivity, and better life/work balance.

To increase the amount of user feedback, consider providing incentives for participation and reporting. Keep in mind that early adopters are often very enthusiastic, so numbers may be slightly inflated (another good reason to include nay-sayers in the pilot program).

Deploy the Pilot

Once you're satisfied that everything is ready, deploy the pilot WLAN. Be sure you have staff on hand with the necessary experience and expertise to adjust and support the live system.

Initiate end-user training classes. By following the training program that the pilot project team set up earlier, it will be easy for users to train on their new systems (preferably at the same time they receive their Wi-Fi-enabled mobile computing device). This is also a good time to set appropriate end-user expectations for support, and to formalize how they should request it.

Take the system live, check and cross-check everything as you go. Collect intermediate metrics for ROI reporting. Address technical issues as needed during the pilot. Maintain constant security vigilance by closely monitoring access to the WLAN. Be prepared to react quickly to seal security leaks or to handle other problems.

Early on, ask the best technical support staff to try to hack into the system or interfere with its functioning. Document the results and, if necessary, immediately act to re-secure the network.

Obtain feedback from users during deployment. It will help to ensure the pilot's success and aid future planning.

Learn

Collect user feedback from the very first day of live deployment until the project ends. (This can be even more important than technical feedback.) Consider implementing an

INTEL'S WLAN ROI MODEL

Intel's worldwide information technology division, Intel IT, with the help of Intel Finance (a strategically focused organization that operates within the Intel corporate structure) succeeded in the difficult task of linking return on investment (ROI) to productivity gains from wireless LANs. They built an ROI model that became the backbone of the Intel IT's business case for deploying WLANS. The model demonstrates how WLANs can deliver business value to an entire organization by measuring "soft" benefits (greater employee satisfaction, increased productivity and accuracy, etc.), which although difficult to measure, can add tremendous value to the business case for WLAN deployment.

The ROI model uses the formula

Productivity benefits - Start-up costs - Sustaining costs = ROI

It focuses on productivity data derived from various WLAN pilot programs that Intel IT put into place. Intel Finance's analysis shows that an average of 11 extra minutes of productivity per week will pay for a WLAN, but that the productivity gain by most pilot project end-users was much more. (Of course, the actual calculations require that such items as tax consequences, depreciation, time value of money, and percentage of users within each user group are also factored into the equation.)

This method demonstrates tremendous ROI that doesn't take into account "hard" benefits associated with lower network costs, or elusive "soft" benefits, such as faster decision-making. Instead, the Intel method uses a very logical way to demonstrate ROI—through quantitative productivity benefits, although even with this ROI model other soft benefits can be used to bolster the business case (even if no dollar figure is attached).

After the productivity benefit figures are in hand, it's time to "put the icing on the cake."

informal system in which IT personnel roam the pilot project coverage areas for on-the-spot input from users during the first week of live deployment.

Take surveys, but understand that although an IT department can learn a lot about its pilot program from surveys, it must not rely on surveys alone. Surveys do not always tell the whole story. Throughout the pilot, use varying tactics (as outlined in this chapter) to collect data from users to measure the impact of the WLAN.

Activity logs are good for capturing an end-user's experience as it happens. (Surveys require users to remember experiences there were in the past.) Ask end-users to keep a record of their experiences, including the time they save using the WLAN. If using activity logs, prepare the users for the effort that requires.

Set up a lab to test the end-users' skills in connecting and using the WLAN. Run a test session at the beginning of the pilot project, and another toward the end of the project, to see if there was any improvement. (If so, factor that improvement into the ROI time-saving calculations.)

Conduct formal interviews with each participant somewhere in the middle of the program and again at the end of the project. These interviews can be used to help to measure the productivity benefits of the WLAN pilot, but they can also point towards problematic areas so that solutions can be applied. Those interviews, as well as infor-

In other words, it's time to add in the tangible benefits including IT departmental savings and lower network costs, and to put a figure on the more difficult to quantify intangible benefits (e.g. employee satisfaction). By laying out a projected ROI in this manner, there is more credibility in the final numbers.

Intel took the annual per user productivity gains and factored them into a much larger equation. The calculations accounted for the already established start-up costs, sustaining costs for WLAN support and service, time value of money, tax consequences, depreciation, percentages of users within each user segment, and more. The final figures show that over a three-year period, Intel would see sizable returns on their WLAN investments.

Here's an example. If an organization were to provide 32 users with WLAN connectivity, the total cost of ownership would be $20,000 over three years and the WLAN would deliver a benefit of $300,000 over the same three-year period. And if you up the ante to 150 employees equipped with WLAN capabilities, the total cost of ownership is only $60,000 for the three years, but by providing those employees with wireless networking capabilities, the organization could see a benefit of $1,000,000 over that same period. An enterprise that offers WLAN connectivity to 800 end-users would incur a total cost of ownership of $400,000 over a three-year period, and the WLAN would deliver a benefit of $5,000,000 over the same time frame.

Intel IT's pilot programs demonstrated a net present value (NPV) ROI of over $4.6 million in its large-building scenario. In the medium-building scenario, the NPV ROI over three years was just under $1 million. In the small-building scenario, the three-year NPV ROI was slightly more than $250,000.

While not all organizations are the same, and thus their deployments and ROI may not be the same as Intel's, these figures are still compelling.

mal interviews and observations of users actually using the WLAN technology, can add to the IT department's knowledge and serve as a useful reality check when looking at the data gathered.

Evaluate

Evaluate the pilot project's metrics. Review ways to improve systems to meet needs that the pilot addressed, e.g. smoother setup or support for users, and needs it didn't address such as additional access point placement.

Evaluate the impact of wireless networking. If the Intel IT program is any guide (see text box), employees who already had a mobile computing device will actually use that device more when they become wireless-enabled. For instance, the Intel IT pilot project found that the participants actually emphasized the fact that they liked being always on-call and responsive to customer and co-worker needs.

The Intel IT pilot program also found that:

* 68 percent of the participants used the WLAN continuously, or most of the time, during working hours.
* 62 percent of the participants, when given a choice between a wireless and wired connection, preferred to use a wireless connection whenever possible.
* The average participant saved a significant amount of time in meetings because the wireless connection made the user more productive.

Start-up costs for deploying WLANS[1]

		Large Building (800 users)		Medium Building (150 users)		Small Building (32 users)	
	Unit Cost	Units	Extended Costs	Units	Extended Costs	Units	Extended Costs
Capital Items							
VPN Box	$5-10,000	1	$10,000	1	$7,000	1	$5,000
DHCP Server	$3,000	1	$3,000	1	$3,000	1[2]	$1000
Sniffer	$5,000	1	$5,000	0	$0	0	$0
Switch	$2-3,000	8	$24,000	1	$3,000	1	$2,000
Power over Ethernet (PoE)	$1,500	8	$12,000	1	$2,500		
Spares/Backups			$3,000		$500		$500
	Subtotal		$57,000		$15,000		$8,500
Initial Expense Items							
Access Points (APs)	$449	96	$43,000	12	$5,400	2	$900
Cabling/installing APs[3]	$1,000	96	$96,000	12	$12,000	2	$2,000
Client NICs	$90	800	$72,000	150	$13,500	32	$2,880
Installing/configuring NICs	$175	800	$140,000	150	$26,250	32	$5,600
	Subtotal		$351,000		$57,000		$11,300
Cost Per User			$510		480		$620
Total Deployment Cost			$408,000		$72,000		$19,800

[1] Example costs only. Does not include sustaining costs. Costs and needs will vary depending on the company. All numbers are rounded.
[2] Cost for a gateway.
[3] Includes project management.

Figure 14.1 This graphic gives an indication of the infrastructure start-up costs for three potential WLAN deployments: large (800 users), medium (150 users), and small (32 users). Intel IT's calculations include the initial hardware, software, and labor expenses to build the WLAN, but they do not include sustaining costs for supporting a WLAN over time. (But those costs must be accounted for when calculating ROI.) From Intel Information Technology's May 2001 white paper entitled "Wireless LANs." *Graphic courtesy of Intel.*

The pilot project can provide definitive proof that WLANs save end-users significant amounts of time. And, although many positive results arose from Intel IT's WLAN pilot, timesavings proved the most important in terms of obtaining quantitative metrics—the same can be true for your pilot WLAN.

Evaluate the costs. Another benefit of instituting a pilot project program is that your IT department can get a firm grasp on the costs of deploying and sustaining the wireless network. For example, a pilot will show that the total cost of deploying a WLAN goes up when more users are added to the wireless network, *but* the *cost per user drops;* the actual figure is dependent upon the size of the deployment.

The pilot may also provide proof that certain costs will no longer exist by the time a WLAN pilot deploys. For example, mobile computing devices (notebooks, PDAs) that were purchased prior to 2003 will probably require the purchase of a wireless network interface card (NIC) before the WLAN can be accessed. Computing devices purchased after that date will most likely come with an embedded wireless NIC.

Let's study Fig. 14.1. When preparing these estimates, Intel IT assumed several things: the deployment of 24 wireless access points per 100,000 square feet of office space; 12-15 simultaneous users per access point; "large building" calculations based on four floors of 100,000 square feet (9290 square meters) each, for a total of 400,000 square feet (37,160 square meters); "medium building" of 50,000 square feet (4645 square meters) with one or two floors; and "small building" of 8500 square feet (769.5 square meters) with one floor. Intel IT also assumed that the WLAN would be built using the 802.11b standard.

After determining start-up costs, factor in the sustaining costs, e.g. the burden rates associated with support personnel. (If WLAN support isn't assigned to one group, but shared across the board, estimate how the WLAN support time will be proportioned.) Then do your calculations on the resulting figure. This will allow the pilot project team to get the most accurate picture of sustaining costs for the WLAN. For example, Intel IT estimated the time required to support a given installation, and then multiplied it by the burden rate of the support personnel.

Report

Report the findings, including ROI values, to stakeholders to get approval for a broader WLAN deployment. Provide the outcome to all your pilot users as well. After all, they have a vested interest in the project and will want to know what became of their efforts.

So far, most organizations have kept their WLAN deployments small. And while security issues may be one reason for keeping a WLAN on the small side, another reason large organizations shy away from large-scale WLANs is lack of understanding. When only considering the hard benefits (e.g. less wiring), they can't see how wireless is less expensive to deploy than its wired counterpart. Despite the fact that the cost of wireless hardware is in a freefall, it is still a significant expense, and what is saved on cabling can be gobbled up by site survey costs, network management, and security software. So for many, a WLAN is only a convenience—to enable authorized visitors and road warriors to connect when necessary, or to provide connectivity in a hard-to-wire or temporary location.

Thus in many executive boards, the jury is still out on "going wireless." It's up to the pilot project team to convince them that wireless in many instances is better than wired. So any time an organization announces that it plans to open a new office, talk to the decision-makers about going wireless—the average WLAN deployment for a new office of between 50 and 100 people costs $20,000, vs. $250,000 to get a wired LAN up and running.

Try to get some of the organization's top executives to go wireless, even if just when they are traveling. Many will quickly become advocates; it never hurts to have a few allies in the executive boardroom.

RETURN ON INVESTMENT

WLANs are clearly a way in which IT can add value to the overall company. But management typically will not fund technology adoption without a business case supporting a tangible return on investment. This means that WLAN enthusiasts must quantify the ROI associated with the adoption of a WLAN environment.

As you read the following, keep in mind that there are many ways you can quantify the costs and benefits of deploying a WLAN, and that the WLAN technology can vary from organization to organization. For example, one organization may occupy a large fairly open retail space in a mall complex, another may be located in an 1880 ten story office building, still another in a campus environment.

The first step in defining the total costs and benefits of a WLAN is to understand the value of adopting the technology. However, you will find that while some of the benefits derived from a WLAN have dollar figures attached, many do not. Therefore, measuring the ROI of a WLAN can be challenging. For instance, you must find a way to assign "soft benefits" such as productivity gains to a dollar figure. While quantifying these soft, intangible benefits is difficult, it can be done. Once you've put a value on these subtle benefits, you can add their total to the "hard" benefits total and you have a credible ROI.

While it may not be difficult to *explain* the hard and soft benefits of a WLAN, *quantifying* them to a skeptical audience of executives is another story. Some benefits, such as those that arise out of the time and money saved moving employees to a different workspace, or accommodating a temporary workforce when a WLAN is the data communications environment, are tangible and easy to calculate. But others, such as higher employee satisfaction and quicker informed decision-making, are intangible and thus are not easily measured in dollars and cents.

To use the Intel method (see text box), start by determining the organization's employee demographics. Then segment those employees into groups based on the potential value that could be gained from increased mobility. For example, a sales force that conducts much of its business outside the corporate walls would belong in one group. Another group might be composed of executives and managers who must attend a lot of meetings, but also need to stay in touch with the office. For an organization with a warehouse or a distribution center, there are, in all likelihood, many employees who might benefit from untethered connectivity. Another group might be populated by employees whose sole benefit from wireless connectivity is convenience. This group might include receptionists, secretaries, customer service employees, data processors, and other back-office employees.

The best way to determine the ROI for a WLAN deployment is to base the valuations on real data. You can get that data from a pilot project. That is if along with the pilot project, the team also performs incremental surveys of end-users, IT staffers, and management, and keeps impeccable records. When the pilot project is complete, submit the material to a financial specialist for rigorous analysis.

The following are some basic guidelines to help you in determining your WLAN project's ROI. Factors within your organization will necessitate some refinement of the following guidelines.

1. Incremental surveys: These should document soft benefits such as:

- Increased flexibility—WLAN users can work how, when, and where they want to work. This flexibility includes mobility within the office, at home, or at public HotSpot locations.
- Faster decision-making—anytime, anywhere access to information leading to quicker, informed decision-making.
- Higher employee satisfaction—employees should appreciate increased flexibility and access to the latest technology.
- Greater accuracy—when an employee can instantly access and transmit data from wherever they happen to be, they are more likely to avail themselves of the available information rather than guessing.
- Increased productivity—thanks to the ability to work anytime, anywhere.

2. Productivity measurements: Ask the pilot project participants in each user segment to estimate how much time the WLAN saved them, each day—overall, and while performing certain tasks. Then segment the responses by user group and calculate the user-perceived timesavings for each group. End-users will almost always over-estimate their timesavings. Therefore, to establish conservative estimates for your ROI calculations, you need to make two adjustments to these figures.

First, cut the user-perceived time estimates by 50 percent (for a conservative estimate) or between 25 and 30 percent for a more liberal estimate. Take the results and divide it in half again to recognize that not all timesavings necessarily contribute directly to higher productivity and increased ROI. For instance, taking the conservative line, if an outside sales person claimed that using the WLAN enabled one hour of additional productivity each day, the figures would show that he or she would achieve 15 minutes in actual productivity gains per day.

Don't worry about what will seem like paltry daily timesavings. As Fig 14.2 indicates, the end result will still show that a WLAN provides a huge ROI.

As an organization expands its WLAN to incorporate more end-users, the average cost per user drops dramatically and the ROI increases. But conversely, the average benefit per user also drops, because the WLAN was first deployed to the employees who would benefit the most. Thus when additional end-users for whom wireless connectivity isn't as vital to their daily work process are incorporated into a wireless network, the overall per user benefit will level off.

By determining the value of productivity gains, and then subtracting start-up and sustaining costs for the WLAN, you can find the actual ROI of the WLAN project. But to help build credibility in your ROI model, consider all indirect costs (that's what Intel IT did), including, for example, extensive employee Help Desk support.

	User-perceived daily time savings (in hours)	Adjusted for human judgment (50%)	Adjusted to reflect actual productivity gains (50%)	Fully adjusted daily time savings for productivity gains (in hours)
Engineering/ product management	1.49	0.75	0.37	0.37
Manufacturing	1.33	0.67	0.33	0.33
Sales	0.67	0.34	0.17	0.17
Marketing	1.80	0.90	0.45	0.45
Support	1.47	0.74	0.37	0.37

	Adjusted daily timesavings for productivity gains (in hours)	Hourly Burden Rate[1]	Productivity benefit per year/ per user
Engineering/product management	0.37	$60	$5,253
Manufacturing	0.33	$40	$3,126
Sales	0.17	$55	$2,165
Marketing	0.45	$55	$5,816
Support	0.37	$45	$3,886

[1] Burden rates are for example purposes only and will vary from company to company.

Figure 14.2 Timesavings equal productivity. These tables represent the Intel IT WLAN pilot project's calculations. The top table shows the user-perceived timesavings and the adjustments thereto. The bottom table shows the resulting translation of the timesavings into annual WLAN productivity gains. This was done by taking the daily timesavings for productivity gains (last column of the first table), calculating the value of each end-user's productivity gains, then multiplying each end-user group's average hourly burden rate-salary plus benefits-by the number of workdays per year (235). *Graphic courtesy of Intel.*

Even when an organization approves a corporate-wide WLAN, wireless connectivity won't be the "be all to end all." When there is wireless connectivity everywhere, it can create cultural problems. Thus before implementing a corporate-wide WLAN, institute polices that enable department managers and executives to find the line between the intrusiveness of wireless connectivity (e.g. employees sending and reading email during a boring meeting) and connectivity as a powerful, productive tool (employees researching a specific problem to help move a meeting to conclusion).

Access to a WLAN is not necessary for every employee group. Deskbound employees (e.g. secretaries, data processors, customer service staff) usually have no compelling need for access to a wireless LAN.

***Note:** The Wi-Fi Alliance offers a ROI Calculator that can help you to support your argument for deploying a wireless network. The ROI Calculator, which uses Microsoft Excel, determines ROI based on the productivity gains associated with increased access to an organization's networked assets. The premise is that making employees more connected increases productivity. The results are presented in terms of dollars saved, or payback in months, relative to the up-front costs for installation, hardware, and the recurring costs for IT support for the WLAN. The tool is offered as a free download from the Wi-Fi Alliance website, www.wi-fi.org/OpenSection/WLAN_Calculator.asp.*

ENLARGING THE WLAN'S SCOPE

Assume that the pilot project was a success and the stakeholders were impressed with the business case and ROI. The next step is to enlarge the scope of the pilot WLAN. But don't try to deploy a full-scale WLAN in one fell swoop. Instead, broaden the existing WLAN by expanding it selectively to serve the needs of the next level of mobile-users.

It will be easy—just repeat the process outlined for the pilot project. Take time to reassess the questions and decisions you made for the pilot in light of scaling the WLAN to increase its scope. Also be sure to consider the cost and ROI of equipping additional users with mobile computing devices.

Security

Despite all of the hullabaloo to the contrary, a wireless network can be secured effectively when the right steps are taken, and proper attention and diligence are exercised. The key is not to rely on just one form of WLAN security—apply multiple layers of security and institute a monitoring program along with authentication and encryption. If you follow the advice set out in Chapter 17, you should have a wireless networking environment that is at least as secure as your wired Ethernet network.

CASE STUDIES

Dow Corning

This $2.7 billion maker of silicon sealant and personal care products installed its first wireless network in April 2000, with five wireless access points and just over 200 wireless clients running on Windows 2000. Dow Corning's venture into the Wi-Fi world began when a vendor brought along an 802.11b product during a routine visit to Jim Marshall, who, at the time, was the company's telecom manager. Marshall was impressed with the product's simplicity and affordability. So much so that to help out Dow Corning's engineers, who complained that during too many meetings they lost a lot of time trying to access the company's wired network for important documents, Marshall set up an Enterasys RoamAbout 802.11b wireless LAN to extend the company's Ethernet to 185 IT workers at company headquarters in Midland, Michigan. Dow Corning's employees responded enthusiastically.

Laptops now give workers access to email, the Internet and project documents. Employees can work at their desk, in a conference room or in a colleague's office. Marshall says, "Staying connected from any location lets the users communicate more effectively and participate in different workgroups throughout the day."

Marshall goes on to say, "Our people embraced this technology so quickly; you see people stop wherever they happen to meet—in hallways and some of the strangest places—and collaborate, because they're always connected to the network no matter where they are in the building. If I tried to take out our wireless capability now, they'd kill me, because the mobility and time savings it provides is incredible."

After the network switched over to Microsoft's Windows XP operating system, Marshall was struck by how well the operating system handled the transition from one network to another. When a user suspends his system in one network environment and brings it up in another, Windows XP understands that it needs a new IP address and is dealing with new default gateways and network parameters—all transparent to the end-user.

"The stability of Windows XP in the face of a changing network environment is what really stands out," Marshall says. "When you move from a wired network to a wireless network to an infrared network, it's pretty much a seamless operation now. If you want to add hardware or drivers to a device, you can do it while the device is in use. You can update the firmware on the network card while it's in the machine, and it doesn't appear to ever lose the network connection. These improvements are worth their weight in gold."

Senior managers liked the pilot so much that Dow Corning decided to go with 802.11b as the network backbone for a new corporate office in Belgium. Then came another wireless LAN for connecting a warehouse and office building, and yet another for an engineering group in Kentucky. The company also plans to deploy wireless LANs in other buildings on the corporate campus.

"One weekend we put up an access point and started handing out wireless cards to people. Now it's prevalent throughout the organization," says Marshall.

Of course, it wasn't quite that easy. Between the ceilings and floors of Dow Corning's three-story campus building is corrugated metal with poured concrete, which effectively isolates signals on each floor. Marshall installed several Enterasys RoamAbout R2 access points on each floor and put them on the same IP subnet, so client addresses don't have to change as users move through the building.

Dow Corning also found that going wireless has some downsides. According to Marshall, there are rare times when bandwidth can max out, such as when a meeting is called and all IT developers are in a conference room logging on to the network from the same access point. But bandwidth bottlenecks are negligible, in the company's IT building, as well as on the IP network backbone in Belgium, which feeds 85 users that are heavy bandwidth consumers.

As these networks grow and wireless technology matures, Dow Corning will likely consider faster wireless technology. Among the more promising is 802.11a, the IEEE standard that brings access speeds up to five times that of 802.11b, to 54 Mbps. A downside is that 802.11a will require new access point radio cards and client NICs because it operates on the 5-GHz frequency range.

Going wireless saved the Dow Coring a ton of money. Including the cable, electronics and labor to pull the cable. According to Marshall, Dow spends about $300 for a wired port, compared with $180 for a wireless port.

Dorsey & Whitney, Attorneys at Law

Wireless LAN technology is also used at Dorsey & Whitney, a law firm with 22 offices worldwide. It installed a wireless LAN in its new high-rise building in Minneapolis. The WLAN enables visiting clients to access the Internet and their corporate intranets from the firm's conference center. It also allows lawyers visiting from other Dorsey & Whitney offices to have wireless access to the firm's corporate LAN.

Dorsey & Whitney spent $5000 for the wireless LAN access points, says Mike Ter-Steeg, the firm's legal technology consultant. The firm purchased four Cisco Aironet 350 access points that plug into Cisco routers to cover concentrated areas in the conference center. Such access points, which include transmission cards and antennae for transmitting data as far as several hundred feet, range from $200 to $1000, half of what they once cost. The adapters that provide a wireless interface to each user's notebook computer cost less than $100. Those adapters are still more expensive than high-speed wired Ethernet NICs, but TerSteeg feels that, when compared with what his company would pay in labor to get the wires behind the wall, they're well worth it.

Federal Express

At Federal Express, wireless networking is an important part of the overall networking infrastructure. This company has deployed wireless throughout two campuses at its headquarters in Memphis, and is treating the WLAN as a full-fledged member of its corporate network.

Ken Pasley, Federal Express's director of wireless development, says that FedEx's employees have come to expect the convenience of wireless access. If a group gathers for a meeting in a location that doesn't have wireless coverage, they'll move until they find access. However, although a WLAN connectivity speed as low as 5 Mbps can support a huge range of business applications, including enterprise resource planning and PowerPoint, Pasley does have to keep an eye on large file downloads, such as CAD/CAM drawings, since they can cause network slowdown.

FedEx was an early adopter of wireless networking. It started using proprietary wireless LAN technology in the late 1990s, albeit mainly in package sorting and aircraft maintenance areas. Gradually the company shifted its wireless assets from those early proprietary LANs to 802.11b to gain higher bandwidth. FedEx discovered that once 802.11b was in place, the company saw a 30% jump in productivity at its package sorting centers, says Pasley.

H.J. Heinz Co.

In 2001, the H.J. Heinz Co. deployed a WLAN in its Pittsburgh, Pennsylvania corporate headquarters. The reason? Well, according to Kurt Kleinschmidt, a senior network analyst at Heinz, "It was just nice to have. There was no real business issue that WLAN addresses."

To add mobility to their corporate offices, the IT department first researched the situation. It bought and tested access points, network interface cards, and client adapters from several different vendors. Then the staff analyzed the results. Cisco Aironet 802.11b series won the day, mainly because of the Aironet series' early adoption of the Extensible Authentication Protocol (EAP).

The next step was for the in-house IT staff to install and configure 20 Aironet 350 APs, set up wireless-enabled laptops for about 100 employees (mainly senior business managers and IT professionals), and teach them how to use the equipment to access the WLAN.

Heinz is happy with its decision. And although the company has no plans to provide enterprise-wide WLAN access anytime in the near future, its IT department will install APs elsewhere when an office requests it.

General Motors

At GM, the march toward wireless was jump-started by Ralph Szygenda, GM's vice president and CIO. Under Szygenda's leadership, GM spent the latter part of the 20th century investing $1.7 billion in new Internet applications, while reducing its IT costs by more than $400 million.

Over the last few years, Szygenda and GM's chief technology officer, Tony Scott, noted a growing interest among GM business managers in wireless capabilities. Of course, GM personnel were already geared toward wireless because of (1) the company's OnStar division, which delivers wireless applications to automobile consoles and (2) its manufacturing plant managers had installed wireless networks (usually proprietary in nature) early on, and thus were well versed in radio frequency and spread spectrum technologies. The latter group was clamoring for newer wireless technologies. Furthermore, like so many other organizations, the wireless action was already taking place internally at the workgroup level—end-users were playing with wireless hardware and implementing their own impromptu wireless networks.

In April 2001, Szygenda decided it was time for GM to develop a wireless strategy—one that would enable wireless networks to deploy globally. Toward that end he formed a wireless strategy task force, consisting of 30 people from various divisions of the company, and put Scott in charge. The group's initial assignment was to write a white paper outlining where wireless was currently being used at GM, and where GM could take advantage of wireless opportunities.

Sixty days later, the task force presented Szygenda with a strategy for untethering GM on all levels, from dealer sales floor, factory floor, and distribution warehouses to construction sites, corporate offices, and manufacturing plants.

First, the group recommended a framework for prioritizing potential wireless projects. While the company grew excited about the potential of wireless technologies and made it a corporate priority, it also realized that every wireless project would need to save the company money, and the projects achieving that goal the quickest would be the first to be implemented.

Next, the task force recommended that since the wireless projects would need to be able to deploy globally, they would need to be based on universal standards.

Then the group determined where the first wireless networking capabilities would be implemented: construction sites, office environments, and in manufacturing and material handling processes. Finally the group decided, after much debate, the individual projects where the company could see the fastest return.

The focus on quick ROI meant that, of the three broad areas of focus, GM concentrated first on deploying WLANs to quickly connect the trailers that surround the automaker's construction sites. Those temporary shelters are typically hardwired, adding time and expense to the process of wiring and cabling. The 802.11b infrastructure allows users to communicate over the WLAN from their laptops or handheld devices. When the project is over, the WLAN hardware is easily redeployed at another construction project. Scott says that savings from such deployments have been significant.

The next projects were aimed toward mobilizing the corporate office environment. According to Scott, this was to enable office workers to collaborate via email, PowerPoint presentations, and other content without having to find a fixed LAN connection. He also commented, "We're doing more collaborative work in conference rooms, also—so it's beneficial to be able to have people come in with their PCs without having to run wires."

Next came the task of untethering manufacturing and material handling processes in order to improve quality and reduce inventory costs. Like many enterprises, GM wanted its loading dock workers, for example, to be able to enter inventory information into ERP (enterprise resource planning) systems as soon as goods arrive from suppliers, which in turn, would speed its arrival to the manufacturing shop floor.

A new manufacturing plant near Lansing, Michigan, met the latter two requirements and became one of the first GM sites to deploy a WLAN linking administrative, security, human resources, and other departments.

GM also expects to save by eliminating or reassigning technicians who currently handle moves, adds, and changes for wired workstations.

"Our criteria for how we spend remains the same regardless of the economy," Scott said. "We're making a business decision for what enables business the best. We're looking for return on investment while managing a broad portfolio of activity."

A Change Of Thinking

Scott describes GM's wireless project as a significant "change of thinking" in how GM's application architects design offerings. He clarifies by saying that although wireless technology can enable new applications, developers must be aware of the bandwidth, data display, and security limitations of wireless transmissions and systems.

GM doesn't have a master plan for which applications will be wirelessly enabled, but Scott lists applications ranging from ERM, CRM, and inventory management, to the remote monitoring of forklift trucks on the factory floor (allowing forklift operators to enter information without leaving the vehicle), as clear candidates for wireless. According to Scott, "We justify these on a project-by-project basis."

Lessons Learned

There are lessons to be learned from GM's wireless strategy. First, if your organization is not currently developing a WLAN strategy, you're late to the game—especially since employees have probably already brought their own wireless devices into the facilities, to ease their workload with the help of ad hoc WLANs. Industry experts say that now is the time to join the Wi-Fi bandwagon, particularly for large, global, multifaceted enterprises like GM.

The Wi-Fi phenomenon, along with affordability advancements, convinced GM that it was time to officially embrace wireless, Scott said. "I compare it to 20 years ago, when PCs were first coming out and there was a lot of reluctance about whether or not PCs were viable. It took many years for us to change, and I'm not going to let that same thing happen in the wireless space." During the last few years, General Motors spent millions of dollars to install wireless LANs throughout its plants and offices.

Like in many large organizations, the catalyst for GM to adopt Wi-Fi was that a number of departments and manufacturing divisions had already started implementing different handheld devices and different wireless networks. Szygenda, Scott and others knew that they had to try to put order into this potential chaos.

The task force initiative helped GM identify and prioritize potential wireless networking projects, leading to a series of major wireless deployments. It also helped the company zero in on the 802.11b standard as a core enterprise technology for GM. The 11 Mbps data speed that 802.11b provides is more than sufficient to handle the applications GM expects its WLANs to support. In addition, it led GM to standardize its global wireless networking platform around Cisco Systems' Aironet suite of products.

Devices from that suite are integrated with Cisco's 10-Mbps LAN infrastructure. (GM also has a 100 Mbps and Gigabit wireline LAN infrastructure.)

Eventually, the initiative will lead to wireless projects that will touch practically every aspect of the company's operations, from marketing and sales to the factory floor.

Despite economic concerns, GM's eventual goal is to give its employees faster access, through wireless networking technology, to all kinds of business information as they roam GM's vast facilities. And its overall wireless objective is to gain experience in terms of extending this technology into as many areas as possible—distribution centers, dealerships, factories, transportation centers, corporate offices, and more. In order to do that, the task force knows that it can't rest on its laurels; it must continue to develop GM's wireless strategy as the technology unfolds.

Even as GM forges ahead with wireless projects, however, Scott and his team continue to exercise caution. While there are departmental pilots under way using Palm handheld computers in sales, service, and marketing areas, he said there are no plans to provide wireless access to mission-critical systems enterprise-wide, just yet.

Scott also anticipates a few problems. While the company expects that it will eventually see great savings from the wireless enabling of many of its applications, bringing wireless networking into the manufacturing environment may introduce safety concerns. For instance, the company is studying to what extent WLANs could interfere with the wireless communication spectrum that is currently used by manufacturing-floor machines.

And, despite being a Wi-Fi advocate, Scott believes that the technology still has a long way to go. "Right now ... the security model is something that has been a bit difficult from an administrative standpoint." The security concern surrounding 802.11b, for example, is what helped to persuade the company to standardize around products from a single vendor. It was felt that by going with a platform built around all-Cisco products, the company could avoid security gaps that might result from attempting to knit together products from different vendors.

Nonetheless, Scott believes the potential of wireless networking will eventually outweigh all the administrative headaches and current technical shortcomings. According to Scott, while GM is "not telling our guys that work in the wired space to quit their jobs because, certainly, wired has some pretty strong advantages, [GM's] belief is that wireless will change everything about the way we run our business."

CONCLUSION

The challenge for any organization deploying a wireless network is to make the wireless network secure, manageable, and easy to use. To meet this challenge, pay attention to detail. Document employee application use and bandwidth requirements and then choose the appropriate technology and products to accommodate both. Conduct a comprehensive site survey. Finally start small, with a pilot project or a limited deployment. To properly design and deploy a WLAN that can meet and even exceed everyone's expectations is a job requiring much more upfront planning and different ongoing diligence than the traditional wired networks.

For corporations, money is the bottom line. Of course, how additional employee productivity translates into dollars is difficult to quantify. Yet without a doubt, a more connected employee is a more productive employee. And high productivity is good economics.

Chapter 15: Vertical WLANs

Vertical markets or industries refer to the segmentation of industries by type and/or business processes. Manufacturing, aerospace, construction, healthcare, and education are each an individual vertical industry. There are also large corporations and conglomerates that span many vertical markets, such as General Electric.

It was among the vertical industries that wireless networking first achieved popularity. The mobility provided by wireless networks caused proprietary wireless solutions to be implemented into vertical markets such as warehouses, distribution centers, and manufacturing, long before the IEEE 802.11 standards hit the scene. But the standardization provided by the 802.11 series and declining price points have helped to make 802.11-based WLANs the cornerstone of wireless networking everywhere, trampling many proprietary standards out of existence.

New vertical markets—education, hospitality, retail and wholesale, transportation and logistics, healthcare, and others—are stampeding toward the Wi-Fi camp. Today, WLANs, based on the 802.11 series of standards, are used for a multitude of vertical applications to increase the effectiveness and efficiency of their daily processes.

Many verticals find that the critical impetus for their adoption of wireless networking is that Wi-Fi provides an effective means for their workers to connect wirelessly to the organization's central data bank. Some vertical markets where mobile workers need access to real-time information, and the ability to process that information, are as follows:

- The warehousing industry finds WLANs indispensable. These innovative networks can decrease the number of inventory specialists required, making warehousing systems more efficient, less costly, and more accurate.
- Manufacturers use WLANs in a number of ways. For example, a WLAN could enable a quality engineer to proactively investigate a defective production line, using a WLAN to record and analyze information on the fault, and fix the problem as quickly as possible.
- Retail and wholesale organizations use wireless networks to maintain accurate and timely manufacturing and inventory control processes. Inventory control is a high overhead activity, and errors such as shipping the wrong items, restocking charges and mollifying dissatisfied customers can be costly. But the combination of barcode

scanning and real-time wireless database updates can help an organization tighten its inventory management processes.

* In the transportation and logistics industry, companies like Federal Express and UPS have found that a WLAN can beneficially affect their bottom line. FedEx, for example, uses its WLANs to automatically move package data from its drivers' portable computers to the corporate database at the end of the day or when the van returns to its home base.
* Healthcare facilities use WLANs to enable their caregivers to access patient information instantly. With WLAN access, a healthcare professional can create new or refill existing prescriptions, access laboratory test results, dictate patient notes, or capture photos of patient's conditions and email them to a colleague for a second opinion.
* The hospitality industry is finding that wireless check-in and restaurant/poolside ordering is useful, especially in large facilities. Large hotel and casino complexes find that wireless connectivity allows managers to get "out from behind the desk" to personally interact with hotel guests while retaining timely access to back-end systems, enabling them to make real-time decisions that positively affect the property's financial performance. Restaurants use Wi-Fi to improve order accuracy, customer service, and more.
* Educational institutions recognize the advantages of untethering their students and staff. They quickly grasped the fact that a Wi-Fi network could satisfy the demands of thousands of Net-hungry students without dragging miles of category 5 cabling through large campuses and century-old facilities.

There are as many potential wireless applications as there are organizational categories. WLANs are used to keep tabs on valuable assets—from laboratory equipment to cargo—that need to be tracked at all times (in part to prevent loss, theft, or damage). Wireless sensors, tags, or transceivers can identify the location of these assets—and even monitor their condition remotely.

Wireless networking is quickly becoming a favorite tool at trade shows, conventions, fairs, and other events. Reliable network environments can be established rapidly, and then removed just as quickly when the event is over. For example, WLANs can manage temporary check-in / check-out facilities at events. Or enable reporters and journalists to access their centralized data (e.g. reports and statistics) from any location within an event venue such as the Winter Olympics.

Major metropolitan communities install access points on rooftops. (Many downtown areas are not much different from a college campus when it comes to size and geography.) The financial industry, too, has found WLANs useful, for example, in enabling securities traders to conduct transactions as they roam the stock exchange floor.

According to Gartner, more than 50 percent of all WLAN implementations through 2004 will take place within the vertical marketplace. Ken Dulaney, vice president and research director for Gartner, says that the applications for WLANs in vertical markets such as retail, transportation, and construction are endless. "Quantifiable results can easily be measured by baselining nonwireless productivity and costs, and measuring them against wireless-based communications systems."

The benefits of wireless networking are tangible and quantifiable. For example, New Orleans' Oschner Clinic and Hospital deployed a WLAN within its facilities and quickly found that time in its emergency room was down 20 to 25 minutes per patient, because medical personnel use the network to research treatments and register patients at their bedside.

TAKE IT IN STAGES

Data networks are growing ever more complex, due to pressure from business applications, e-commerce, greater connectivity between the corporate LAN and the Internet, and an increasingly mobile workforce. With 802.11 technology on-board, an organization can enable wireless mobility throughout a campus, or connect LANs together for a fraction of the cost of traditional Wide Area Network (WAN) technologies. Wireless LANs also make it much easier to add or move workstations, and to provide connectivity in areas where it is difficult to lay cable. An additional benefit is that the entire wireless network can be managed from one location, anywhere in the world, and there are enhanced security/access controls available to thwart hackers or intruders.

As discussed in previous chapters, ensuring a robust and stable wireless network requires that a deployment team take the time to study the organization's networking environment and its users' requirements. This includes determining the organization's overall business strategy, preparing a needs analysis, and assessing the current network infrastructure to determine whether a Wi-Fi network is feasible.

***Note:** This chapter is written with the understanding that the reader will read this chapter in conjunction with the other chapters not only in this Section V: Wi-Fi for the Enterprise, but also Section III: Practical Deployments.*

Next is intensive preparation. This is the stage where comprehensive site surveys are performed to ensure there is acceptable data throughput available for the type of usage *and* network coverage area; the WLAN gear options are studied; pilot projects are implemented and the results studied to determine if the project's WLAN technology matches the organization's business and security goals; and the WLAN business case, with projected return on investment (ROI) and total cost of ownership (TCO), is presented to the stakeholders.

Once a WLAN project is approved, a sound deployment plan and an impeccable security strategy based upon all of the previous work must be put in place. This is the stage where the deployment team configures, installs, and tests all of the WLAN's components for data integrity, signal strength, and coverage. It also ensures that appropriate security measures are taken and security policies are implemented; the proper network management and maintenance systems are fully operational; and the end-users are properly trained.

But deploying a WLAN to provide connectivity for a corporate headquarters involve different requirements than those needed for a WLAN that serves a vertical inudstry. Let's visit some of the areas that might cause special concern.

VERTICAL HARDWARE CONSIDERATIONS

Most of the networking hardware is common to every type of application, but there are exceptions. For example, specific vertical situations may require special antennae and/or client computing devices. In these situations, hardware selection may require a bit more thought before a final decision is made.

Not Your Every-Day Computing Devices

The typical wireless computing devices—laptops, PDAs, and tablet computers—oft-times aren't suitable for use in vertical industries. A large number of enterprise workers, including shop floor personnel, and warehouse and field service personnel, use mobile computing devices as they go about their daily tasks. The rough-and-tumble work environment and wide variety of applications that these workers face requires more rugged and diverse mobile computing and communications technology than the typical personal digital assistant (PDA) and laptop provides. Thus, ruggedized notebook computers, vehicle mount computers, pen tablet computers, handheld computers, and wearable computers are among the computing devices that may be used to access a vertical WLAN.

Many vertical industries, especially schools and hospitals, use wireless carts because of the hands-free mobility that they offer. All wireless carts come with a built-in wireless network card, but other than that, their features vary widely. For example, a wireless cart for use in an educational environment might be equipped with a number of laptop computers, a printer, an access point, a powerstrip, and Ethernet cable. Whereas a cart equipped for use in a healthcare facility might come with a flat-screen-monitor computer and a battery pack, so there is no need to look for an electrical outlet to use the cart's equipment. A specific situation may require mixing and matching components, for example combining a medical cart from EMS Technologies with a ruggedized computer from Amrel. Or carts like the ones Computerstation, Inc. offers, that provide a "clean room" environment, i.e. computer enclosure so the wireless computing device can be used confidently in a dusty, grimy, or damp work area.

Look for *computing devices* that can take the best advantage of a wireless network, but try to not limit the organization to a single vendor's device(s), even for a vertical application. Instead, select best of breed devices. Technology has advanced to the point where a well-designed wireless infrastructure can support any vendor's standard PCI bus radio boards and PCMCIA radio network interface cards. Also, you should be cognizant that it is possible to incorporate a variety of components into a wireless networking solution, for example, barcode scanners, time and attendance systems, scales, monitors and printers. Do your homework:

* Ask if the vendor offers a wide range of terminals and radios that will fill a variety of communications needs.
* Consider computing devices with proven technology. Be sure that the chosen device has been tested in a production environment, especially for mission-critical applications.
* If the organization is trying a new application, such as a voice-over-Wi-Fi system, keep your old system operational until you are certain the new application and its supporting devices work as expected.
* If the application requires barcode scanning, does the new system easily support such usage? If so, how?

* Does the device offer flexible data capture features? For example, is the handheld capable of operating as a "batch device," i.e. does it have ample memory to store large "look-up files" in secure, non-volatile memory for on-the-spot processing without the need for wirelessly accessing a server?
* Can the device decode the most popular barcode symbologies including, if needed, two dimensional (2D) symbologies such as PDF417 and microPDF?
* Can the handheld device support a variety of barcode scan engine options, including no-scan, standard, high-density, high-visibility, and long-range scans capable of reading bar codes to 15 feet (4.6 meters)?
* What type of data output does the device support? Text only, text and graphical, color, black and white, etc.
* Will the device's battery hold its charge for an entire day's shift, with some to spare? Test the device before buying. Some might hold a charge only when used moderately, but when used during, e.g., the annual inventory process, it might not hold up to the task.
* Is the device designed to perform in harsh environments and to withstand rough handling? Does it meet the IP65 standards for moisture and particle resistance?

Note: *One dimensional (1D) barcodes are an array of parallel, rectangular bars and spaces arranged according to the encodation rules of a particular symbology. They enable human readable data in machine readable form, whereas 2D barcodes can carry up to 1.1 kilobytes of information, including photographs and graphics. 2D barcodes are often used on driver's licenses, on passports, and to provide customs officers with information on goods that must pass through control.*

Barcode Scanners

Barcode scanners read various types of barcode symbology, either through laser scanning or imaging technology. These devices scan and decode barcodes, helping to automate the data collection and transmission process, and thus improve productivity and reliability by limiting human intervention and reducing unreliable paperwork.

Organizations can choose between pure barcode readers, which are typically shaped like a gun (and include a trigger), or more multifaceted handhelds or PDAs with barcode reading capabilities. Scanners come in several design forms with varying degrees of functionality. For jobs that require a simple scan, an untethered barcode scanner is the appropriate choice. For situations that call for barcode scanning as well as data input, PDAs or handhelds with barcode reading capabilities may be more appropriate.

AN ALL-IN-ONE DEVICE

Symbol Technologies just recently began shipping a new ruggedized Pocket PC that sports support for voice-over-Wi-Fi as well as a built-in barcode scanner. The device, called the Symbol PDT 8146 Imaging Computer, is designed for mobile workers in industries such as transportation and logistics, public safety, and retail. According to Symbol, it is the first handheld to merge Wi-Fi voice over IP with imaging and data capture capabilities, enabling workers to speak to each other over 802.11b networks. VoWi-Fi is quickly becoming the hot IT item for the mobile worker segment.

Barcode Printing

Barcode printing products typically create 1D or 2D barcodes on label stock. These devices allow output from information technology systems, such as a WLAN to be displayed quickly and easily to better manage the supply chain. When considering the purchase of a barcode printing product, look for a device:

* That is fast and reliable, since barcodes often are rapidly applied to a large volume of items, such as on an assembly line.
* Durable enough to operate in tough environments, and flexible enough to print multiple printing formats, including text and graphics.
* Offers interoperability features. Although most printers operate on a proprietary software language, which has historically made implementation difficult and interoperability between different brands of printers impractical, there are vendor or third-party software solutions (i.e. middleware), which ease system implementation and allow for better interoperability.
* That can be managed from a central location so management can receive remote error notification from a malfunctioning printer, and that can diagnose and repair the error on a remote basis.

Portable Printers

Some vertical operations (e.g. retail, healthcare) require workers to tote around a handheld printer in addition to a portable computing device or barcode scanner. If this is the case, look for a device that is small and light enough to be used comfortably in a handheld operation, and that can print barcodes and text with high-quality print resolution. Another option to consider is a printer that can "cradle" its companion handheld device.

TRAINING CONSIDERATIONS

Vertical industries find that training workers to use the wireless system and the supporting devices can cost more and take longer than expected. But the proper training must be undertaken if the WLAN project is to be successful. For example, a handheld can bring with it a whole new set of problems that even those familiar with a PC or a laptop may have trouble comprehending. For example, automatic layouts and forms can cause no end of peculiar little problems. And if there are workers who are using mobile computing for the first time to replace paper and handwritten systems, end-user training may need to be comprehensive.

To illustrate, when Old Dominion Freight Line rolled out its WLAN, it found that it was the first exposure to any kind of computer for many of its drivers. To help ensure success for its WLAN initiative, the motor transport company hired trainers, who physically visited every service center to train the drivers on how to use the equipment.

While some drivers caught on quickly, it took more time for others to learn the ropes. Typically, the Old Dominion's trainers spent a week at each service center teaching 20 or so drivers. Then, after a group training session, some instructors would re-enforce the training by riding along with any drivers struggling with the new process. The trainers found that going one-on-one with the drivers allowed the workers to ask questions as problems arose during their workday. Finding answers, rather than frustration, allowed the drivers to become comfortable more quickly with the whole wireless process.

WAREHOUSES AND DISTRIBUTION CENTERS

Warehouse and distribution centers traditionally require reams of paperwork. A WLAN and a staff, equipped with handheld computing devices with barcode scanning capabilities and portable printers, can eliminate much of the paperwork and at the same time reduce errors and decrease the time needed to move items in and out of the facility.

Warehouse and distribution centers (along with other vertical industries) use barcoding technology because it follows international standards in structuring the content of information held in the barcode, thereby simplifying worldwide supply chain automation. Thus, WLANs and barcoding are almost ubiquitous across most forms of product delivery.

For example, when pallets of goods are received on the loading dock, a worker can scan the boxes to determine in which area of the facilities the products should be stored. The WLAN then accesses the appropriate servers to determine the storage location, and sends the information to the portable printer, which prints out a "put-away" label. A forklift operator can then move the item to the storage location and document the procedure using a computing device mounted on the forklift. Once the pallet is inside the facility, other staff can scan each item's barcode and, if necessary, enter other information from the handheld's keypad or touchscreen (e.g. the product's estimated ship date). The WLAN allows the inventory system residing on remote servers to keep track of all transactions.

Then, as products are needed to fulfill an order, the inventory system spits out the appropriate list of items, their storage location, and the order in which the products should be picked. Warehouse/distribution center employees than take that list, which is displayed on their handheld device, and pick the listed items for shipment. As workers remove the products from their respective storage bins, he or she scans the product's barcode and either scans or inputs the bin number, allowing the inventory system to be updated automatically.

In terms of a WLAN's operational effectiveness within a warehouse or distribution center, the improvement is measured by a marked reduction:

- In errors as a result of barcode scanning.
- In the need for safety stocks.
- In the need for stock counts.
- In physical inventory.
- In missed shipments, short shipments and back orders, which in turn, result in fewer lost sales.
- In time spent reconciling with suppliers and customers.
- In inventory write-offs.

Many times unconventional antenna usage may be utilized to provide the best wireless coverage in a vertical deployment situation. For example, in a typical warehouse or distribution center, there are usually large parallel racks of goods perpendicular to either the front or side walls of the building. The typical installation of several access points throughout the facility would not provide good coverage. But, by using a combination of directional panel and omni-direction antennae, full facility coverage could be provided. The APs with direction panel antennae would be mounted along the one wall, providing coverage down the rack rows all the way to the opposite wall, and an additional access point with an omni-directional antenna would be mounted in the area where the facility's office is situated.

Another way a WLAN deployment in a warehouse or distribution center differs from other types of wireless networking projects is that there is almost always a quick and impressive return on investment. That's because a WLAN enables faster inventory turnover, better equipment and facility utilization, reduction in labor overhead as a result of the elimination of paper-based task assignments, and the optimization of system-driven tasks.

Case Studies

Ca' del Bosco

Although wine making is an art with ancient roots, winemakers haven't overlooked the advantages offered by technology. Ca' del Bosco, an Italian wine producing firm in the heart of the famous Franciacorta area at the gateway to Brescia, produces 13 types of wine, which form the basis of around 200 different products. The firm also exports close to eight million bottles of wine every year. To support that kind of volume cost-effectively, the firm introduced an automation island within its production lines, using LXE Inc.'s 802.11b technology to speed up the handling of the grapes at the pressing operations.

The most hectic period of the year is harvest time when tractors arrive at the acceptance zone of the plant towing their loads of grapes, each coming from a different area of the vineyard. Each tractor's load is weighed on an automatic weighing machine, where a barcode label is issued and placed on the relevant pallet of grapes. On this label is encoded the type of grapes, area of origin in the vineyard, and day of harvest.

The pallets are then turned over to the "helping hands" of the forklift truck operator. Each forklift is equipped with a wireless LXE vehicle-mounted wireless computer with an integrated keyboard and a long-range barcode scanner. These devices are of rugged design and are water- and dustproof, so they can be easily migrated from outside to inside use without condensation problems.

The forklift operator scans the barcode labels, and an appropriate "thermal storage room" destination is displayed on the vehicle's wireless computer. (The grapes have to be placed in thermal storage to bring the temperature of the grapes to "dew point," which is the ideal temperature for pressing because it prevents secondary fermentation and it safeguards the aroma of the wines.) In total there are seven thermal storage rooms, spread over several loading floors.

To ensure that no mistakes are made, each storage room is marked with its own barcode label. Before entering the storage room, the driver must scan this label. He will then receive authorization or be refused entry, based on data in the barcode and data in the system's database.

Once the grapes reach the required dew point in the thermal storage room, they are ready for pressing. When that event occurs, a message is sent to a forklift operator via his wireless computer, telling him to pick up a specific pallet of grapes in a specific storage room and convey the pallet of grapes to one of 17 pressing areas. (There are different pressing areas for different types of wines.)

Ca' del Bosco's 802.11b WLAN operates within a wired Fast Ethernet (100 Mbps) environment on a Windows platform, with the wireless computers operating in a client/server mode. The WLAN not only takes care of the storage handling and management operations, but also monitors the entire operational flow.

According to a company spokesperson, positive advantages arise not only out of the wireless system's ability to provide real time monitoring of the lines, but also from the on-line information collected.

PepsiAmericas, Inc.

PepsiAmericas, Inc. is the second largest Pepsi-Cola anchor bottler, with operations in nine countries. The company manufactures, distributes, and markets Pepsi-Cola core brands, along with Cadbury beverages and other national and regional brands.

In every market served, the company's mission is "to make, sell and deliver beverages." To build strong customer relationships this bottling company has made its everyday goal to deliver first-rate service and superior products. The effort to meet that goal is seen in its day-to-day operations and was the catalyst for the recent overhaul of its sales and delivery teams' business processes via Wi-Fi technology. With Wi-Fi at the helm, PepsiAmericas has morphed many of its business processes into a more customer-centric pre-sell model.

With the company's previous system, its sales and delivery teams' job was to service a defined number of customer accounts each day and to take new orders, most of which were completed manually as the staff went about their daily delivery route. This method was ripe for errors, and those errors could impact the entire supply chain. This led PepsiAmericas to re-assess its sales and delivery business processes and to explore new technology-based solutions.

In analyzing the overall cost of doing business, the bottler found that its inventory management system needed overall, especially in light of the fact that the company's inventory had grown from 55 SKUs a decade ago to the current level of 300 SKUs. PepsiAmericas' existing technology also offered limited access into customer demand, and trucks often returned to the distribution center with as much as one-third of the load still on board. With more than 100 U.S. distribution centers and 50 or so trucks operating out of each warehouse location, it's easy to multiply the additional costs of doing business caused by such an inefficient system. As a result, PepsiAmericas made improving the movement of its inventory its top priority.

In order to streamline processes, increase efficiencies, and reduce errors, PepsiAmericas deployed a Symbol Technologies wireless mobile computing system—a solution broad enough to balance the different needs of both mobile teams (sales and delivery) with the needs of the distribution centers.

For the pre-sales process, the company's account sales managers use Symbol PDT 8000 wireless computers to take customer orders on-site. These devices help to ensure that delivery trucks are stocked efficiently and accurately for the day's routes (on-site orders are remotely uploaded to the central order and routing system using cell phone connectivity).

The rugged PDT 8000 wireless handhelds also empower the sales and delivery staff with access to the information they need to strategically upsell to customers.

For direct store delivery operations, the handhelds provide each driver with an automated tool to track inventory, record deliveries and print invoices to an RP 2000 portable printer. Using the Spectrum24 High Rate wireless local area network (which is based on 802.11b technology) along with their PDT 8000 handheld, drivers can download information including customer list, orders, pricing, and the day's route before leaving the distribution center each morning.

Each handheld is also Direct eXchange/Uniform Communications Standard-enabled for instant data synchronization, facilitating the electronic transfer of receivables and new orders directly into the customer's database.

With the addition of Symbol's RP 2000 portable printer, drivers easily generate invoices at the customer's location. Upon the driver's return to the distribution center, route information is uploaded quickly and efficiently to the host computer via the WLAN.

Symbol's strategically placed wireless access points guarantee the broadest coverage in the distribution centers. As the account sales managers and delivery drivers return to the distribution center at the end of each day and upload their data, it is relayed by the WLAN to back-end systems for route settlement.

Fast access to current information makes it easier to locate and load products on trucks for improved inventory management. Additionally, marketing, sales, or operations management may also use this current information to analyze delivery progress, track returns, manage inventory, or to provide more effective marketing offers.

To complete the wireless mobility solution, Symbol supports management of the wireless network through a remote monitoring service using SymbolCare services 24x7 support, Symbol Wireless Network Management System (WNMS) software, and Symbol AirBEAM. This software is designed to streamline the entire wireless mobile computing device management process, including updating files, applications, and operating systems wirelessly from remote locations.

For PepsiAmericas, the wireless solution provides a more streamlined process, increased efficiency and accuracy, and improved productivity, which, in turn, enables the company to better serve its customers—and that is the key to competing more effectively. And when a mobile worker saves enough time to make an extra stop to another account or sells more products to an existing account, PepsiAmericas gains a strategic advantage over its competitors. Just one additional stop per day for each driver or account sales manager adds up to several hundred additional sales opportunities per year from each worker in the field.

Extended Technologies Corporation (Xtek) is a company that designs and implements software solutions to facilitate communications for all mobile sales and service applications. This includes order management, data retrieval, merchandising, field services, and remote/decentralized distribution. Xtek provided PepsiAmericas with the software applications, including Route XPress NP for delivery (pre-sells and conventional) and full service vending routes; Xtek's Xtended Gateway suite for the integration of RouteXpress NP with PepsiAmericas' host system; and Xnet Communicator for data distribution, mobile application development and mobile device management.

RETAIL

WLANs have long had a place in the retail industry through back-office applications such as receiving, mark-downs, price verification, and inventory management. According to the twelfth annual RIS News/Gartner study, more than 40 percent of retail respondents have adopted wireless and an additional 29 percent plan to adopt it by the end of 2003.

If a facility already has wireless LANs in place, adding wireless access points and devices in the public area of the store is simple. Symbol Technologies, which has the majority of the wireless scanner market (at least two thirds), sees strong movement to wireless deployment in a few key areas:

* Better customer service. Mobile registers for better customer service during seasonal or sales-related traffic changes within a store's departments and sectors.
* Mobile customer service. WLANs can help provide mobile customer service for applications such as line-busting, assisted shopping, or checking inventory availability at another store.
* Self-service. WLANs provide kiosks connectivity, which are used to provide customer self-service options, such as bridal and gift registries, custom configuration, and can even be used for customer price checks, if scanners are attached.
* POS applications. Wireless registers can be employed for customer checkout. Of course, if the WLAN is to support POS applications, such as processing customer credit cards, robust security measures must also be in place.

Frank Riso, director of business development for retail at Symbol Technologies, estimates that about a quarter of midsize and large retailers have at least one pilot program in operation to test front-of-store wireless applications. For example, Macy's department store in New York's Herald Square places Wi-Fi-enabled registers on wheels, so it can add cashiers as needed during sales and other customer peak times. It also uses wireless kiosks for customer self-service price checks.

Riso says that supermarkets and discount stores use wireless register set-ups for sidewalk sales, and to equip springtime garden and December Christmas tree "departments" that are seasonally deployed in front of stores or in the parking lots. "The cost is small, about $300 to $500 to add the client bridge hardware to the register," Riso says.

"More and more, they're not building a store without building a wireless infrastructure," says Riso. In the U.S., Wal-Mart, Best Buy, and Home Depot all use wireless LANs in various aspects of their day-to-day operations. According to *WLANA Magazine*, the ROI for retail-based WLANs averages 9.7 months.

Case Studies

Home Depot

Home Depot, an early adopter of wireless LANs, uses proprietary two Mbps wireless LANs to access inventory applications in most of its 1400 or so stores. But Home Depot is installing 802.11b products made by Symbol Technologies in new stores, and has begun the process of upgrading other stores to 802.11b. Home Depot pays about half of what it paid five years ago for its proprietary wireless network. The fact that the new technology is much more efficient provides the incentive to upgrade. But the real takeaway for the home improvement retailer is that it's now much easier to scan products and then label merchandise as it sits on the shelf, rather than taking merchandise to a fixed terminal to do the same.

A notebook computer with a scanner connected to a wireless LAN enables employees to price and label products more efficiently, thus saving on labor costs. "It really did change our labor standards and the amount of labor hours set aside to do certain activities," says Dave Ellis, vice president of IS operations and networking at Home Depot.

But that is not the only way Home Depot utilizes its wireless infrastructure. Thanks to lower costs, new open standards, and a wealth of new applications, more and more retailers use wireless applications to improve customer services, e.g. to provide faster customer check out, to find relevant promotions, to locate products, and to print out coupons.

By capitalizing on its wireless technology, Home Depot keeps its customers happy, which of course, translates into more profits. The compelling reason for a sales associate to stay with a customer during a high-ticket sales event is higher profits. A PDA that can wirelessly access the company's inventory databases allows the sales associate to query inventory and product information without disconnecting from the customer.

Home Depot also extends its WLAN in other ways. How many times have you spent the better part of an afternoon wandering through a cavernous warehouse-style retail store, filling up your shopping cart—or worse, taking a quick dash through the aisles for a couple of quick items—only to find the lines so long at the check-out area that you can't even see the cash register. Not a scenario that results in a happy shopper.

Home Depot's enterprise mobility platform also underpins a rapid register program, known as "unleashed"—a wireless system empowering store associates to expedite the checkout process. Home Depot has deployed this "unleashed" application in all of its stores to enable Home Depot associates, using wireless handheld devices, to scan a customer's purchase while they're in line.

It works like this: when lines get too long, the associates go to the customers for checkout rather than vice versa. Store associates use wireless handheld personal digital assistants (PDAs), which are equipped with barcode scanners, magnetic strip readers, and printers for generating receipts. They scan a waiting customer's items while they wait in line, print out a receipt, and hand it to the customer. When the customer reaches the cashier, their purchase record is retrieved and transacted.

This process is called "line busting," and it's one of several wireless solutions that are debuting on the floors of major retailers around the world. The company's CIO, Ron Griffin, notes that the increases in staff efficiency that resulted from the "line busting" process allowed the system to pay for itself within the first year.

The unleashed application, developed by 360Commerce, was built upon Home Depot's existing wireless infrastructure and point-of-sale technology, demonstrating the power of modular architecture. While its primary role is to speed up the transaction process, 360Commerce's Sr. VP of marketing, Christine Lowry, notes other benefits that the unleased program yields "It is also convenient for customers who purchase products from outdoor areas, such as garden centers, and a great way for stores to stay in contact with a customer through a multi-item, big-ticket purchase, such as a kitchen remodeling project."

Longs Drug Stores

Each of Longs Drug Stores' nearly 450 western U.S. retail locations stock, on average, more than 100,000 items. Thus a lot of support staff, paperwork, and computing power is required to keep on top of inventory and pricing changes. All the necessary data including point-of-sale (POS) information must then be sent over high-speed data lines to the company's central inventory database in Walnut Creek, California, to keep the company's records current.

With the retail system far from the store floor, keeping up with changing prices meant that individual store managers had to fill a shopping cart with every product sold, wheel it into a back room, scan everything, then return the products back to the shelves. Then the system printed out problem lists, and the store employees addressed those issues, in the aisles.

For instance, any change in pricing (e.g. manufacturer price increases, temporary price reductions, promotions) required new labeling on the shelves, and sometimes even on the items themselves. Regular changes in stock (holidays, seasonal, and back-to-school promotions) meant a constant cycling of products on and off the shelves. "It's a very dynamic environment," says Carl Britto, Longs' director of store technology planning. "That's what's so exciting about retail."

Longs Drug Stores knew it had a problem and sought ways to eliminate some of the costly and time-consuming steps needed to keep abreast of inventory and pricing. The answer was a wireless LAN and a "mobile manager 1000," a pushcart with a built-in wireless laptop from Symbol Technologies. With the mobile manager, store employees could take the cart to the product, and scan and transmit the information to the store's retail information system. According to Britto, once the system was designed and tested, all of Longs more than 450 locations were up and running within seven months. He adds, "The rollout was one of the fastest we've ever done."

Britto pointed out several critical steps that the retailer used to help quicken the pace. The first was a five store pilot project. This gave the company data on what needed to be done for a smooth deployment, as well as a means to calculate an estimated return on investment (ROI). Next, the company developed a complete rollout plan. So, as Britto explained it, "you don't have to double back" and repeat any steps that might have been missed. Then, the company rolled out the new technology and application before the wireless hardware was up and running, giving local managers time to learn the software.

As for the ROI, according to the pilot figures, it's estimated that a WLAN will reduce labor costs by at least 10 to 15 hours per week, per store. Thus the WLAN and its supporting gear are estimated to provide complete payback within 12 to 18 months. And Britto brags the company is "pretty close" to being on track to delivering on that commitment.

Longs is planning for the future. According to Britto, the next step is to use smaller, handheld devices so that store personnel can process customer returns and perform store audits quickly. Furthermore, the company is experimenting with connecting other store equipment (e.g. photo lab processing and video rental machines) to the WLAN, and perhaps even converting some cash registers to wireless to enable stores to hold sidewalk sales more easily.

HEALTHCARE

The healthcare industry is another vertical undergoing a transformation of its information management processes; the industry is moving to a mobile and wireless environment. In fact, the healthcare industry is perfectly suited to the use of wireless technology. Workers in this industry are very mobile; patient data is needed fast and on demand, and the wires for electrical power, monitor hookups, and other equipment leave little convenient space to plug in additional computer equipment. Is it any wonder that an ever-increasing number of hospitals are integrating wireless access into their network infrastructure?

A WLAN, paired with portable computing devices, allows the average healthcare provider to input, view, and update patient data from anywhere within a medical facility, which, in turn, increases the accuracy and speed of each patient's care.

Reduced errors, increased efficiency, and improved patient satisfaction, as well as a growing shortage in healthcare workers, are driving the healthcare industry's adoption of wireless networking technology. IDC, one of the leading global providers of technology intelligence, estimates the use of handhelds by healthcare workers will grow by nearly 40 percent annually through 2005.

The motivation that is fueling this industry's interest in wireless networking technology includes:

* FDA rules on medication barcoding.
* Improved accuracy in all levels of patient treatment.
* Improved patient care through the ability of caregivers to obtain real-time, point-of-care charts, records, and test results.
* Reduction in time-consuming and costly paperwork.
* The ability to track medication, testing, and specimen samples through the chain of care.
* Ensuring that the right medication, in the right dose, is given to the right person, in the right way, at the right time.
* Voice-over-Wi-Fi capabilities that enable nurse-call communication with doctors.
* An impressive ROI—*WLANA Magazine* found that for a healthcare WLAN to produce a positive ROI takes on average just 11.4 months.

While there are many benefits that can be derived from enabling wireless networking at the point of care, there are also challenges associated with operating RF devices in a medical environment. Proper network design, however, can help mitigate such difficulties.

Address 2.4 GHz Interference

Before installing a wireless networking system in a healthcare environment, several issues must be taken into consideration. For instance, because 802.11b and g systems emit and receive radio frequencies in the 2.4 GHz range, some interference with monitoring and lifesaving devices may occur if the network isn't properly designed. Although most interference is relatively benign, causing no serious effects or producing only low-grade interference such as snow on a monitor, there may be more severe interference, which might cause a medical device to produce erroneous readings, to reset, or to jam its communications.

The first step in preparing for a WLAN deployment should be to determine whether there is any potential interference with medical equipment. Some medical devices are not properly hardened for certain radio frequency levels, including frequencies common to wireless networking. This is particularly true of legacy devices. Check the standards and certification labels on all medical equipment. Then design the WLAN around the associated limitations.

A thorough site survey, along with a non-mission-critical test of all components in the system, will go a long way in minimizing the possibility of harmful radio interference. Ask the biomedical engineering department to participate in the test. This department is responsible for testing devices, and for defining the policies and procedures relating to such devices and their usage. Most biomedical engineering departments have a standard test set that is based on known industry issues and the electromagnetic devices that the health system has already installed. Knowing the location of problematic devices is extremely important, since those devices must be brought under the umbrella of the site survey process.

In a medical environment, where mission-critical decisions are being made all the time, avoiding interference problems is worth the cost of hiring outside professionals, especially for the site survey portion of the WLAN deployment. But also consider hiring a project manager who has extensive expertise in WLAN deployments within the healthcare industry. His or her expertise will prove to be invaluable throughout the deployment process.

Certification and Standards Compliance

Wireless LANs are subject to equipment certification and operating requirements established by regional regulatory bodies. So, as part of the WLAN preparation process, you must check to determine if the WLAN's equipment meets all of the local regulatory requirements. For example, if you read the "compliance" label on a Cisco 1200 Series Wireless LAN access point, you would find that it complies with:

- Standards: UL 1950; CSA 22.2 No. 950-95; IEC 60950; EN 60950.
- Radio approvals: FCC Part 15.247; Canada RSS-139-1 & RSS-210; Japan Telec 33B; Europe EN-330.328; FCC Bulletin OET-65C; and Industry RSS-102.
- EMI and susceptibility: FCC Part 15.107 and 15.109 Class B; ICES-003 Class B (Canada); CISPR 22 Class B AS/NZS 3548 Class B; VCCI Class B; and EN 301.489-1 and -17.
- Other: IEEE 802.11 and 802.11b Microsoft WHQL.
- Operates license free under FCC Part 15 and complies as a Class B device.
- DOC regulations.
- UL 2043.
- EMI and susceptibility: FCC Part 15.107 and 15.109 Class B; ICES-003 Class B (Canada); CISPR 22 Class B AS/NZS 3548 Class B; VCCI Class B; and EN 301.489-1 and -17.

Medical Electromagnetic Compatibility Standards

As another part of the WLAN preparation process, you must check to determine if the WLAN's equipment needs to meet the Electromagnetic compatibility (EMC) and safety requirements, which are required of medical devices used to provide direct patient care or peripheral support, e.g. telemetry patient-monitoring services, heart monitors, and defibrillators. Electromagnetic compatibility means that any equipment used in proximity to such devices will not cause harmful interference, but can accept harmful inteference, including that which disrupts service.

In the U.S., the Federal Drug Administration (FDA) is the agency that is charged with regulating EMC requirements for medical devices. In that agency's "Guide To Inspections of Electromagnetic Compatibility Aspects of Medical Device Quality Systems," it is stated:

> EMC generally needs to be designed into the product. Therefore, the following questions regarding EMC are included in the Design Control Inspectional Strategy (DCIS), dated March 1997:
>
> 820.30(c) Design input
>
> For an electrically powered device, where electromagnetic compatibility (EMC) should have been considered in the design, determine the following:
>
> How has EMC been addressed with regard to the device use environment? For example, the interface with other medical devices or the interference from other consumer products?
>
> If complaint or failure data for similar devices distributed by the manufacturer indicated EMCproblems, did the manufacturer use this information in establishing the design requirements for the new device?
>
> Identify any relevant EMC standard(s) used as a part of the design input process.

> That same document further states:
>
> AT THIS TIME FDA DOES NOT REQUIRE CONFORMANCE TO ANY EMC STANDARDS. HOWEVER EMC SHOULD BE ADDRESSED DURING THE DESIGN OF NEW DEVICES OR REDESIGN OF EXISTING DEVICES, ESPECIALLY IF EMC INFORMATION HAS BEEN REQUIRED FOR PREMARKET CLEARANCE.

In the EU countries, medical devices manufactured over the last few years must be designed to meet levels of immunity that are defined either in the specific device's standards or in the general standards for EMC. Any device that meets the standards' requirements bears a "CE-mark." (The CE designation, which is French for "Conformité Européene," indicates that the marked product conforms to the relevant EU directives.) Otherwise, manufacturers should be able to provide data about the maximum field strength in the wireless LAN band for safe operation.

Yet, the Medical EMC issue is muddled. Note this quote from the summary of the Medical Devices Agency publication DB9702 entitled, "Electromagnetic Compatibility of Medical Devices with Mobile Communications":

> Results from a large study based on 178 different models of medical device using a wide range of radio handsets are available in MDA Device Bulletin DB9702: 'Electromagnetic Compatibility of Medical Devices with Mobile Communications.'" The publication found that overall, "in 23% of tests, medical devices suffered elec-

tromagnetic interference (EMI) from handsets. 43% of these interference incidents would have had a direct impact on patient care, and were rated as serious.

The DB9702 publication, however, did state, "No significant levels of interference were detected from cordless handsets, local area networks or cellular base stations." But it was noted that the type of radio handset made a large difference to the likelihood of interference. At a distance of 1 meter (3.28 feet): 41% of medical devices suffered interference from emergency services handsets, and 35% suffered interference from security/porters handsets, but only 4% from cell phones.

You should know, however, that the test didn't specifically test Wi-Fi devices; it included all types of radio components.

A recent set of EMC tests conducted at the Walter Reed Army Medical Center to ensure EMC compliance should set your mind at ease. Biomedical Maintenance conducted an independent validation and verification study in 2002, using different vendors' 802.11b WLAN products: Symbol Technologies' Spectrum 24 and Symbol 2800 handhelds, and Cisco Systems' Aironet 340 and HP iPaq 3670 handhelds.

It initiated the tests by deploying a WLAN in the biomedical lab complex that had the access points set to maximum power output (100 mW). Then the testers used medical equipment—patient monitoring unit, vital signs unit, pediatric defibrillator—with documented EM susceptibility, and assessed interference at varying distances. The tests demonstrated no electromagnetic or radio frequency interference. Thus Walter Reed concluded that, based on preliminary findings, the tested biomedical devices are not compromised through the use of Wi-Fi technology. (It is also noted, however, that comprehensive testing is pending with the FDA to generate a more thorough report.)

Nonetheless, before committing to a wireless LAN installation, a healthcare organization should undertake field strength measurements with the proposed WLAN equip-

SPECIFIC MEDICAL CONCERNS

Many people worry about Wi-Fi interfering with hearing aids or pacemakers. The possibility is remote. The tests that have shown interference could occur have been conducted with cordless and cellular phones held up to the ear, not the typical access point and mobile computing device, which are generally nowhere near a person's hearing aid. As far as the pacemaker is concerned, while it is noted that older pacemakers are subject to interference from microwave ovens, the newer models are not. Furthermore the interference came from systems operating in the 900 MHz or lower bands, not the 2.4 GHz or 5 GHz bands. Also, changes and improvements in pacemaker design have helped to eliminate many interference problems.

To further ease concerns about possible EMC interference, it is noted that in 1996 tests were conducted by Greater Chicago's Ingalls Memorial Hospital to verify that no problems would occur between 2.4 GHz radios and pacemakers. Also, the Ohio State University Medial Center tested Wi-Fi components in its MRI facility and found no interference to the radios or the MRI unit.

However, to ensure safe operation of your WLAN system, obtain the services of a WLAN installation specialist to verify noninterference.

ment to ensure they are operating within safe limits. But, since Wireless LAN technology radiates at about 5% the power of a mobile phone, no serious problems should be encountered with a device/transceiver separation greater than 1 meter (3.28 feet).

HIPAA

In July 1996, President Clinton signed the Health Insurance Portability and Accountability Act (HIPAA) into law. One of the objectives of that law was to protect patients' health information against unauthorized access.

Failure to secure identifiable health information that traverses your WLAN can have costly consequences. Organizations can be penalized both for failure to adhere to HIPAA standards and for wrongful disclosure of health information. Thus HIPAA has great influence over how a healthcare organization designs and deploys its WLAN. Insurance companies must be confident that identifiable health information is protected from outside interests, and there must be reliable and appropriate access to health information stored in information systems.

To ensure that identified health information is protected, the proper security, privacy and confidentiality measures must be put into place. A healthcare facility's WLAN must use encryption and advanced security solutions, such as virtual private networking. It is also imperative that the site survey ensures that the WLAN's signals do not "leak" out of the intended coverage area, either inside or outside the facility.

Use as a guideline the advice set forth in Chapter 17 to help you develop a security program. Also, to ensure that patient confidentiality is maintained, it must be decided exactly who should have access to the WLAN, from what device(s), and what information can be obtained via the WLAN.

WLANs are Flexible

Because wireless networking is both flexible and scalable, a healthcare facility can take a modular approach to deployment rather than commiting to location-wide coverage immediately. For instance, a WLAN could be deployed in a single department, such as a trauma center or the facility's emergency room. As network use increases, you can grow

PROTECTING THE MOBILE COMPUTING DEVICE

Mobile computing devices are more susceptible to theft than their stationary counterparts. Thus precautions must be taken.

First, all patient information stored on these devices must be encrypted. Next, the devices themselves should be password protected. Then, end-users must understand both their obligations to protect the information stored on their device, and what the device is capable of accessing.

Finally, wireless security policies and procedures must be developed. Although it is recognized that the policies and procedures must be specific to the healthcare environment, use the "Wireless Security Policy" section in Chapter 17 as a starting point. Then ensure that all end-users agree to comply with whatever security policies and procedures that are put into place.

the system to cover a ward, a wing, or even entire buildings. Eventually you could even extend the WLAN across every square foot of a medical campus.

All Wi-Fi certified wireless solutions should seamlessly integrate with existing wired solutions, and work smoothly with all wireless equipment built upon the same 802.11 specification. However, if you initially implement wireless gear that provides added functionality such as Quality of Service and/or security, you may be locked into using only the initial vendor's equipment. That is because the vendor, in all likelihood, used proprietary solutions to provide the additional functionality.

Building Bridges

If your healthcare facilities include buildings that are not on the main campus, such as emergency medical centers and/or outpatient treatment locations, you can seamlessly connect them to the core network, using wireless bridge technology as discussed in Chapter 9. Wireless bridges save enormous time and cost incurred for dedicated lines or leased lines. For example, the Cisco Aironet 350 Series wireless bridges provide high-speed, long-range, building-to-building wireless connections.

Case Studies

Memorial Medical Center, Springfield, Illinois

Memorial Medical Center is the teaching hospital for Southern Illinois University School of Medicine. About 3000 of the hospital staff need daily access to patients' medical records. In September 2002, in an effort to improve patient care by providing fast access to standardized information, the hospital began work on a program to introduce electronic records that could be accessed wirelessly by physicians, nurses, and other hospital staff.

The hospital administrators understood that by providing the staff access to patient information as needed, no matter the location within the hospital grounds, it could eliminate error-prone paper documents, prevent unintended drug interactions, eliminate illegible handwriting through the use of electronic forms, and ensure that procedures were instantly documented and that they followed hospital, insurance, and government regulations.

At the outset, Memorial Medical planned to offer 802.11b wireless access just in locations that were hard to wire or where the staff was very mobile, such as in patient wards. But CIO O.J. Wolanyk argued that it made more sense to make the entire hospital wireless–including the parking structures and cafeterias, so doctors and nurses and other personnel could access information from any location within the facility. He also wanted the full coverage so the hospital could later implement Voice-over-Wi-Fi phones.

The hospital administrators gave their approval to Wolanyk's $900,000 wireless-everywhere project. When fully implemented, the WLAN will serve not only medical wards, but also other areas including accounting, billing, and human resources.

"Memorial's need for a wireless LAN could not be more mission-critical," said Wolanyk. "We were running out of places for PCs and had to reduce the $770,000 we spent annually on scanning medical records. The wireless network is the predominant means by which we access medical records. We are not experimenting here; we had to be confident that the solution would meet all our expectations."

Initially, Wolanyk thought there would be a need for around 300 access points, based on the square footage of the hospital. But after an outside firm performed an extensive site survey, it was determined that, with optimal access point placement, the facilities would need fewer than 150 access points.

After evaluating a number of vendors, Wolanyk and his staff decided that Cisco wireless access points, Cushman antennae, and ReefEdge Connect System (a server with Edge Controllers), had the most to offer. It was also determined that the wireless network would have three levels of security to ensure that patient data is protected, and that only authorized users can log onto the network. Those three levels consist of Windows NT Active Directory authentication, policy-based firewall access (to ensure that users get only access to the information they are allowed to see), and 128-bit Dynamic IPSec 3-DES data encryption.

The infrastructure was designed to provide a secure system with session persistence across subnets to enable seamless roaming between access points. The ReefEdge Connect System also ensures concurrent access to medical records by staff, and allows administrators to remotely manage the wireless LAN. Furthermore, the WLAN meets federal regulations set by the Health Insurance Portability and Accountability Act (HIPAA), according to Ajei Gopal, ReefEdge's CEO.

Everyone is pleased with the new WLAN set-up. The fully operational WLAN can directly support all tertiary hospital services including cardiology, cancer, burn, orthopedics, pharmacy, radiology, medical laboratory, and trauma.

Even though the hospital does not expect the system to greatly affect its per-patient care costs, Wolanyk is enthusiastic about the project. Having better, safer operations will be payback enough, he says. Wolanyk also "wanted to experience this myself." Toward that end, he uses a laptop computer as his primary computing device so he can work throughout the hospital. Since he has no office, he connects wirelessly or via Ethernet jacks just like the hospital's 550 on-staff physicians and 2400 employees. Those caregivers access the wireless system on a regular basis via laptops, PC tablets, and handheld computers as they move about the hospital's facilities. Even PC workstations are equipped to access the wireless network.

The ReefEdge Connect Server and Edge Controllers work together to ensure high availability, even in the event of network connectivity failures. To provide redundancy Memorial deployed standby Edge Controllers (EC100 and EC200) to monitor the primary ones. In the unlikely event of a failure, a standby Edge Controller can automatically reroute wireless traffic through a redundant path to the network in a matter of seconds, with virtually no disruption of user sessions.

That's not the end of Memorial's foray into wireless connectivity. Recognizing that doctors are not technology adopters, and realizing that it's critical that the care givers find the system easy to use, and that it actually helps to make patient care easier to deliver, Wolanyk asked doctors to help design the applications and work methods.

Wolanyk also wants doctors and hospital staff to recharge their devices easily without having to find an open power jack. He is planning on installing custom-built power racks throughout the hospital, so multiple devices can be recharged from one power outlet, instead of every recharger needing its own power jack.

Physicians typically affiliate with more than one hospital in order to better serve their patients. So Wolanyk also works with the other major hospital in the area in an effort to standardize medical information systems, including the wireless access. If and when the project is completed, physicians can treat patients wherever they are located and not worry about changing procedures or technologies.

Still to be decided, at this writing, is how the hospital will connect to the outpatient care facilities across the street. Although the staff that works in the main hospital doesn't often visit the outpatient building, and vice versa, Wolanyk would like to have both on the same network. The holdup is that the hospital must work with other tenants in the outpatient building, since several have wireless networks that overlap the outpatient unit's space.

St. Luke's Episcopal Hospital, Houston Texas

St Luke's in Houston, Texas, is located in the heart of the world's largest medical center. The hospital's wireless network originally consisted of non-802.11b Proxim access points (APs) and Cisco Aironet 802.11b APs. (The hospital is phasing out the Proxim APs in favor of an all-Cisco 802.11b WLAN.) Users connect to the network through Dell Latitude mini-laptops and NEC MobilePro handheld devices. The laptops run on Windows 2000 and NT, and the handhelds use Windows CE. All are equipped with Cisco Aironet 802.11b wireless NICs

The APs are distributed throughout several hospital buildings, and are used to access a variety of network applications in many different areas. The wireless computing devices are employed to obtain medical data as the hospital's staff make their rounds.

According to Gene Gretzer, senior analyst and project leader for the wireless rollout at St. Luke's, the WLAN was originally deployed in January 1998 in a single building using Proxim access points. Yet, while the WLAN was supposed to improve the efficiency of care at the hospital, it sometimes had the opposite effect. For instance, once the staff started using the wireless network, the staff noticed that the wireless signals didn't always maintain a constant connection.

The reason for signal fluctuation varied from ongoing construction within the building, to medical equipment, to roaming between floors and from building to building. Each time a connection was dropped, the user would have to get back within the range of a wireless access point, reconnect to the network, go through the log-on and authentication process, start the application(s) again, and re-enter whatever data was lost in the process. This wasn't what the IT department or the hospital administrators had in mind when they decided to go wireless. In fact, with the WLAN's then infrastructure, there were disruptions in many hospital-related activities.

Time and attention to detail are everything, and staying connected is absolutely essential. Is it any wonder that when the hospital surveyed the staff on their WLAN experience, the staff identified disconnected wireless sessions as one of their primary user complaints?

"People using the handheld devices had trouble maintaining their sessions walking around the floor," says Gretzer. "[The operating system] had a tendency to drop the connection." Naturally, this situation frustrated busy doctors and nurses, who resented both the waste of time and the distraction from patient care.

In addition, the hospital's physical construction itself caused problems with wireless signals. The walls in the original building were built using chicken wire, which interferes with radio waves. Some patients' rooms were also located in pockets that had weak radio signals. Nurses and doctors working in these rooms sometimes had to step out into the hallway to reconnect. Microwave ovens in the kitchenettes on each floor also occasionally caused interference.

To remedy the problem, St. Luke's turned to NetMotion Wireless Inc. and its Mobility Server solution. Mobility is a software solution that provides persistent application connection for wireless devices. Gretzer deployed Mobility software to 20 Microsoft CE client devices in March 2000.

The Mobility software maintains the state of an application even if a wireless device moves out of range, experiences interference, or switches to standby mode. Once users come back into range or switch into active mode, they can resume the application where they left off, without having to login or reenter data.

The product consists of a Mobility server and Mobility agents for the individual client computing devices. The Mobility server, which runs on Windows NT/2000, acts as a proxy between the clients and the network server that runs the applications. The server solution manages each application session to keep a continuous connection between the wireless device and the application server on the wired network, and provides several security features, including encryption and basic firewall functions. These features help to protect sensitive patient data.

The Mobility clients are then able to communicate with the Mobility server via UDP (although the Mobility software supports both UDP and TCP, and can use either protocol to communicate with applications). "We encapsulate application traffic using UDP to avoid many of the commonly understood issues surrounding the use of TCP

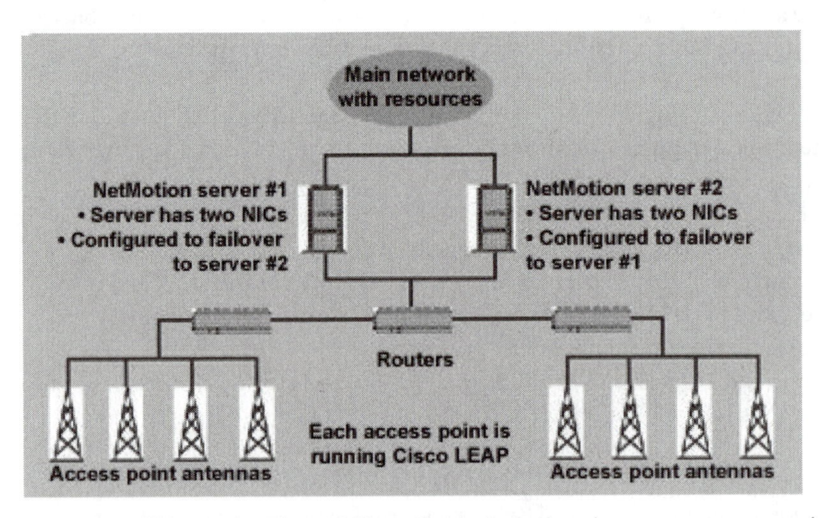

Figure 15.1 St. Luke's WLAN. The mobility software is deployed on a proxy server and individual mobile computing devices. The proxy server brokers IP addresses for the client computing devices, maintains network and applications sessions, and acts as a firewall between the wired and wireless networks. *Graphic courtesy of NetMotion Wireless Inc.*

in a wireless environment," says Emil Sturniolo, chief scientist at NetMotion Wireless. Those issues include:

- TCP will drop its virtual circuit when a wireless client goes out of range of a radio signal, forcing the user to login, reconnect to the network, and restart whatever applications were open.
- TCP attributes packet loss to network congestion or packet collision, even when the cause is a device going out of range. Under this assumption, TCP waits a specified length of time before resending the packets, which NetMotion says can degrade performance.
- When retrying to send lost packets, TCP may send data that the client has already received, thus eating up bandwidth unnecessarily. This may not be an issue on a 100 Mbps wired network, but bandwidth is at a premium on more constrained WLANs.

Sturniolo says that the Mobility software employs a guaranteed delivery system much like TCP, so that if a frame is lost, it gets retransmitted, shielding the application from network errors. Just how NetMotion does this is confidential.

In March 2002, the wireless project expanded to all three of St. Luke's buildings, but with the expansion came a host of new problems. "We changed from having the wireless on a single Virtual LAN—a single IP segment—to a multi-facility network," says Gretzer. The result was that when a user roamed onto a new segment of the WLAN, the client device would be assigned a new IP address, causing it to lose its current session. For undisclosed reasons, the hospital didn't deploy access points in the hallways that connected the three buildings. That served to increase the incidents where a loss of signal incurred as users moved to a new building.

Gretzer once again turned to NetMotion to help alleviate its mobility problems. Besides maintaining application persistence, NetMotion has a system for managing the IP addresses of mobile devices.

Each Mobility client has two IP addresses—a Point of Presence (POP) address and a virtual access. The POP address is the address assigned to the client by whatever WLAN segment it happens to be in at the time. If a user roams onto a new coverage area, it gets a new POP address. It is typical for roaming users to change POP addresses numerous times while on a shift.

Each client device also has a virtual address on the wired network. The Mobility proxy server tracks both the virtual address and POP address for each client, and acts as a broker between the wired and wireless networks. When the Mobility server gets data from an application server on the wired side, it forwards those packets to the client's POP address, and reverses the process to send packets from the client back to the application server.

Without this feature, clients who roamed onto a new WLAN segment would have to reconnect to the network every time the POP address changed. With NetMotion at the helm, transitions are invisible to the end-user, since the network and application connections are maintained at all times. NetMotion clients obtain their IP addresses from standard DHCP servers.

NetMotion is also assisting St. Luke's as the hospital swaps the Proxim APs for access points from Cisco. The reason? Well, according to Gretzer, "Proxim's pipe is only 1.6

Mbps and Cisco's is 11 Mbps—nearly ten times faster," and the hospital's wired network infrastructure is Cisco-based.

In the meantime, NetMotion plays a vital role in the interoperability of the two vendors' products. Support for both wireless systems once meant that nurses had to switch between devices as they moved about. On floors with Proxim access points, they used devices installed with Proxim radio cards; on Cisco floors, they could only use devices with Cisco cards. In moving from floor to floor, staff members would have to log out of one device, then log back in on a new one.

"With NetMotion, you just need two radio cards in one laptop," says Gretzer. "The Mobility system lets nurses and doctors transition seamlessly between floors that have the old Proxim and new Cisco." Thus staffers can continue to use the same machine, regardless of the technology in the APs.

Let's not forget security. If the thought of sensitive medical information floating blithely through the ether causes your heart to palpitate, you're not alone.

"Of course security is an issue," says Gretzer. "The Cisco network we put in uses Lightweight Extensible Authentication Protocol (LEAP), which is basically just a rotating Wired Equivalent Privacy (WEP) key, and WEP has been broken, so we wanted some extra layers of security."

NetMotion's Mobility software provides two of those extra layers. First, all the traffic between the Mobility clients and the Mobility server is encrypted using the Advanced Encryption Standard (AES) at 128-bits. Second, the Mobility server also functions as a basic firewall, separating the WLAN from the wired network.

"We treat the wireless side of the network as hostile," says Gretzer. "The mobile devices do not have access to the corporate network. They have to go through the Mobility server." The server itself will only talk to clients that it recognizes.

Gretzer says that he also employs other methods to secure wireless traffic. While he wouldn't give details, he did say that a security company, which performs penetration tests among other measures, regularly audits the hospital's network.

"We also do our own wardriving to see what our vulnerabilities are," says Gretzer. "You can receive a signal in the parking lot, but with the multiple layers of security we have, you can't get any usable information. In fact," he adds with a laugh, "we end up finding out everybody else's vulnerabilities. We're the only ones implementing higher-level security."

However, one prominent feature of the NetMotion system—its ability to maintain application persistence—can up the system's overall vulnerability quotient. Here's an example: suppose a nurse entering patient data steps away from the device in the middle of a session—anyone who picks up that device would have authenticated access to the application and the records stored there. (This applies to wired PCs deployed in the hospital as well.) And while Gretzer realizes that this is a risk, he says that the risk can be minimized through careful management. First, to enter or change data, users have to enter their user name and password again. Then, sensitive applications are set to time out after a specified period, if they're left open but unused. The staff has been trained not to leave sessions up and running, and to question anyone who picks up a laptop or handheld, if they don't recognize that person.

This kind of risk is familiar to every administrator. "Users want access, access, access and don't care about security, but we have to consider security and then functionality," says Gretzer. "There's a fine balance in that relationship."

The hospital has 21 floors dedicated to patient care, but not all of them have been wireless-enabled. Floors without the WLAN are still using paper charts. Gretzer laments that the transition to a paperless hospital is taking longer than expected. What was supposed to be a three-year project is stretching into five. One reason for the delay is simply a matter of convincing staff members to accept a more computerized working environment, and then training them to use the devices properly.

"We anticipated a faster rollout because we didn't see the barriers to the staff accepting the devices," says Gretzer. "A lot of nurses and doctors are set in their ways. They sometimes have difficulties adjusting to the new system."

Regarding performance, Gretzer says that the NetMotion software functions perform as advertised. "We haven't found any applications that don't work with NetMotion. We're getting the performance we need." And although proxy servers sometimes have a reputation for adding latency to a session, Gretzer says that any overhead that Mobility might be adding is negligible. This is particularly noteworthy considering that the proxy server encrypts and decrypts all of the wireless traffic, in addition to maintaining application and connection states for its clients.

In one case, nurses using the Mobility software complained of poor performance. "We turned off AES encryption because we assumed that may have been the problem," says Gretzer. In fact, the problem was with the radio cards. "A card was trying to hold onto a signal from a transmitter two floors above," he explains. After tuning the card to only talk to transmitters on its own floor, the problem went away. "We turned AES back on and the users didn't even notice a difference," says Gretzer.

There was also some problem with intermittent functionality of the Mobility proxy server, which necessitated a visit from one of NetMotion's senior developers. However, like any good administrator, Gretzer has a backup server standing by. While NetMotion supports failover, users still have to log back into their applications because the server's address changes. "It's a nuisance," says Gretzer, "but computers break down. The client computing devices know that if they can't talk to the main server, they switch to the backup."

Gretzer's has a couple of gripes associated with wireless networking. One is that if a staff member says his or her mobile computing device is running slow, it is sometimes difficult to determine if it is the computing device, the network connection, or the connection from the access point to the server. Normally, he would send a trace packet from the device to the server to see where the problem was, but his network management tool, Lucent's VitalAgent, doesn't work well with WEP or other data encryption methods. According to NetMotion, because WEP encrypts traffic from Layer 2 and up, traffic sniffers have difficulty tracking packets based on IP addresses because they're encrypted. Gretzer is looking for a monitoring tool that addresses that problem.

The other complaint is that his IT people must physically touch each laptop and handheld in order to deploy and upgrade the Mobility agents. He is evaluating different applications for wireless device management.

When asked about return on investment, Gretzer was hard-pressed to come up with an amount. "When you consider what the hospital staff does, it's hard to get a dollar figure." Rather than save money, he says that the point of the system is to make life easier for the staff. An internal report, however, estimates a 15 to 20 percent increase in efficiency with the wireless system. For instance, before the wireless network was installed, staff members would spend 30 minutes uploading new patient records to a laptop in the morning, and another 30 minutes downloading those records at the end of the evening. Now that the staff updates records in real-time, that hour can be spent on more important tasks.

According to Gretzer, the NetMotion system adds to this efficiency because staff members don't continually reattach to the network or start over when an application session is interrupted.

In terms of hard costs, he says that the NetMotion system has saved St. Luke's from having to add access points to transitional areas, such as the corridors that connect the hospital's three buildings. "It's nice to be able to close your laptop's lid, walk to the next building, and open it up and you're still where you left off," says Gretzer.

In the hurried environment of a hospital floor, time is at a premium for busy nurses and doctors. Technical glitches that interfere with their ability to provide quality patient care are unacceptable. NetMotion ensures that St. Luke's WLAN maintains application sessions and network connections, even as users move out of range or place devices on standby. This application and connection persistence improves productivity and allows medical professionals to focus on the most important task at hand: caring for the patients.

Gretzer is so pleased with the results that he's prescribing a Mobility client for every wireless device that accesses patient records. "As long as the applications work as needed, my users are happy," he says. For a network administrator, that's the best medicine.

HOSPITALITY

The need for mobile high-speed data access is here to stay. According to IDC, 45 million mobile business professionals need to access the Internet or their corporate networks daily and that number is growing.

Providing high-speed wireless Internet access as a service to customers (i.e. HotSpots) drove the hospitality industry's adoption of wireless networking technology, but operational uses have also emerged. Restaurants are the early adopters of wireless technology, although chains are quickly following in their footsteps.

Vendors offer a variety of wireless applications to help the hospitality industry manage overhead, to deliver faster and better customer service, and to run more efficient and safer kitchens. All of the applications can help a restaurant or café gain a vital competitive advantage.

But new technology can be a hard sell in the restaurant business. Whether they specialize in fine dining or fast food, restaurateurs know their success depends more on keeping their customers well fed than on keeping up with the latest high-tech trends. Still, there is a growing cadre of restaurant/café managers and owners who are using wireless technology for table management, point of sale (POS), kitchen management, and more.

"Wireless technology is still in the early adopter stage for many restaurants," says Lance Gallardo, a commercial solutions development manager for Cisco Systems. "But it's rapidly moving into the mainstream." New wireless applications can play a role in almost every aspect of a restaurant's daily operations. For instance, a WLAN enables "wait staff" equipped with PDAs to place customer orders directly from the table, allowing them to spend more time on the floor tending to customers' needs, rather than moving back and forth between the kitchen, the bar, and a fixed POS terminal.

"One thing wireless brings to any industry is the ability to minimize delays and increase employee productivity," Gallardo says. "If a server takes a customer's order over a wireless PDA, they eliminate one of the greatest sources of delay and frustration for the customer, while allowing the restaurant to turn over tables more quickly."

In addition, wireless PDAs simplify employee training by placing menu information and daily specials at their fingertips, instead of requiring them to memorize the menu and its daily changes. Restaurants can also use wireless systems to remind employees to suggest drinks and desserts, and to promote more expensive items, which can improve per-customer revenue.

Most restaurants already have a solid foundation upon which to build a wireless networking solution. According to technology integrator Ameranth Wireless, more than 80% of all restaurants in the U. S. use computerized POS systems and the vast majority of vendors specializing in hospitality-related technology report that their systems support wireless networking.

As in other industries, the main reason given for implementing a wireless network is that the cost for the wireless components (access points, network cards, routers, PDAs, etc.) is relatively low. And as a growing number of proven applications enter the market, even the most conservative restaurant/cafe owners will be prepared to embrace wireless technology. Gallardo says, "We're really still at the beginning of a long process. The more restaurants discover just how much this technology can do for them, the more excited they get about it."

Even fast food outlets are discovering that wireless networking can be a way to cost-effectively perform line busting, both inside the establishment and at their drive-through window. While the restaurant sector primarily uses wireless in pre-sales applications, some are introducing payment via wireless LANs.

Hotels, too, have joined the Wi-Fi parade. They are not only offering their guests wireless high-speed Internet access, they are also experimenting with wireless networking, in particular for check-in, restaurant/poolside ordering, and keeping track of guests' needs such as room service orders and laundry requests.

But it's the HotSpot model that has garnered most of this market's attention. After a day of travel, meetings, sales calls, or meetings, corporate nomads often return to their hotel rooms and log onto their home office network in order to respond to workplace or customer demands via e-mail, submit orders, file reports, review critical documents, or prepare for the next day's work. Traditional dial-up networking is too slow and unreliable to meet the needs of these road warriors. Hotels that can deliver secure, hassle-free high-speed connectivity have a powerful advantage in winning the loyalty and repeat business of these most valued customers.

Case Studies

Colours By The Bay, Singapore

Colours F&B introduces a novel concept in dining experience. It brings together a number of different restaurants under one roof to offer diners a choice of specialized cuisines ranging from Chinese to Italian, Japanese, Thai, Western, and fusion. There is also an international bar called Embassy, which offers live music and entertainment from around the world. The man behind these creative ideas is Andrew Tan. He is also the founder of the combined eatery concept, which he calls, "Colours By The Bay."

The restaurants making up Colours By The Bay are established names in Singapore: Shima Aji, ThaiExpress, The Garlic Restaurant (which serves Western and fusion food), Al Dente, Tien Yuan Kitchen (which serves Cantonese cuisine), and Local Bite. Each restaurant occupies about 2000 sq. ft. of space, although the Al Dente is spread over two floors, as is The Embassy (it is also the largest at 12,000 sq ft). Together all the restaurants take up over 26,000 sq ft, including a 4000 sq. ft. outdoor area.

The eateries offer their diners the choice of ordering items from any or all of the restaurants, regardless of where they are seated. Diners are presented with two types of menu: a specialized menu which is extensive and specific to each restaurant, and a "Colours" menu which is a compilation of several items from all the restaurants allowing the diners to savor his or her choice of cuisine.

Food orders are placed through a wireless cross-ordering system (the biggest system of its kind in the world) that was specially developed for Colours. Orders are entered via the latest model of the HP iPaq, which acts as a point of sale (POS) device.

The system also can, if a customer wishes, split a table's bill to reflect the different restaurants from which the table ordered. Or a group of customers can have their bill split as to what each person ordered, right down to an item such as half a bottle of wine, if two people wish to share its cost.

Now to be able to cross-order, cross-bill, and bill in a number of others ways is a tall order for a systems integrator. But that is what Panacea, a small firm that specializes in hospitality systems was asked to do by Tan.

Tan also found an unexpected added benefit from his WLAN—employing staff. "The biggest problem a restaurant faces today is that you cannot get staff," say Tan, who places his business in the small- and medium-sized enterprise (SME) category, as each restaurant only has around four wait staff. But he says that a restaurant of the same size with a wireless network on-site only needs two to three servers, which amounts to a saving of about 40 percent on staff.

Tan explains that without a wireless network, waiters are always running into the kitchen to place orders, spending a lot of time there, especially when they follow up on a late order. With a wireless network in place, however, all the waiter has to do is to key in the food items on his or her PDA, confirm the order with the customer, and then click the "send" button. The order reaches the kitchen over the wireless network.

Chris Tay, chief executive officer of Panacea, says that to provide seamless wireless coverage for Colours' combined 26,000 sq. ft. area, eight Intel 802.11b APs were used with each AP configured to act as a routing device. A minimum of 30 HP iPAQs and 40 Epson printers are used for all the restaurants, primarily for the kitchen staff to print out order slips once they receive orders over the wireless network.

Each restaurant is equipped with a Pentium IV 866 MHz PC with 256 RAM and 20 GB disk drive, running Windows 98 SE and printers to print bills for customers. The point-of-sale software is Compuware Elite 32. Also, a simple customer relationship management system is in place, so that individual restaurant managers can determine the demographics of their customers or calculate which are the best-selling dishes. Tan is also linking the ordering system with an inventory system to connect to suppliers to procure materials online. Moreover, he is connecting the restaurant system with local hotels so that guests can make a restaurant reservation through the hotel and charge food bills to their hotel account.

Each menu is compressed to about 250 KB of data, according to Tay. The iPaq H3850, equipped with 64 MB RAM and 32 MB ROM, has sufficient storage and memory to host the menu, including double byte characters for a Chinese menu. Each PDA stores a restaurant's entire menu, plus 10 items from each of the other six restaurants. The restaurant operator can synchronize all the PDAs being used by that eatery to update the menu, including any daily specials.

At the back-end is a Compaq Proliant dual CPU server running a Microsoft SQL database. There is also an uninterruptible power supply (UPS) and hot-swappable Raid 5 (redundant array of independent disks) to ensure that the data server is backed up and is available to the individual restaurant operators. Each order logged and confirmed on the iPaq is sent to the Compaq server located next to the kitchen. This is to enable each restaurant operator to check on sales on an hourly, weekly, or monthly basis, says Tan.

Tropicana Hotel & Casino, Atlantic City, New Jersey

The Tropicana in Atlantic City spans about four city blocks. Greg Dyer is the property's PC network manager. He oversees the IT staff and a network system serving approximately 1200 end-users.

When asked about the hotel's networking infrastructure prior to deploying a WLAN, Dyer said, "We have currently a combo of about 13 fiber optic closets throughout the property and 12 Windows NT servers, numerous POS systems, and slot rating systems. The slot rating systems show how long a customer has been playing the slot machines and also profiles the customers. IBM AS400 servers run the hotel purchasing, marketing, financials and sales systems. On the client side, we have over 800 networked PCs and appliances."

Then Dyer discussed the hotel's move to wireless networking. He explained that Tropicana decided to go wireless in the casino because it offers an alternative to traditional wiring. "My job is to evaluate new technologies that will help the hotel and casino streamline efficiency. I chose the Intel Wireless networking solution because it is a solid performing product," he said.

The WLAN was originally deployed using six Intel PRO/Wireless LAN access points (APs), which were placed in "special event" locations throughout the building, e.g. the atrium, casino, reception, and theatre areas. Those APs allow the property's marketing staff when they are within an AP's range to use mobile computing devices, which are equipped with Intel PRO/Wireless LAN PC cards, to access the wireless LAN for marketing activities, special events, and promos.

Just prior to implementing the wireless LAN, the marketing department had an event in a room in which they had a wired computer network. The Fire Department came in and shut down the event because the wires violated building code. Now, with a wireless network, the hotel and casino promoters can set up an event, and move it from one area of the hotel/casino to another, and have no worry about a visit from the Fire Marshal.

Dyer clarfies: "We move around so much with special events and promotional give-aways. Although Tropicana is the biggest hotel in New Jersey, it's difficult to find areas to hold special events. If you are a customer who meets a certain profile, you'll receive certain prizes. We give away anything from t-shirts to fine china.

"Every time we've had an event within the hotel, we've had to cable or network that area of the hotel–whether it's on the casino floor, in the lounge, in the pool deck area. We don't want to pull people into a back hallway to give them their prize. Some of the large convention areas are sometimes the only areas available, but it's not necessarily the best area to hold these events. We want our customers to feel at ease, so now that we have the Intel PRO/Wireless LAN network, we are not limited to a particular room. The staff uses their wireless networking laptops and they can easily monitor the prize giveaways and marketing event from their networked laptop.

"Sometimes we use the wireless network like a registration desk. At bingo, for example, the hotel staff uses their wireless networked laptops to register the clients."

Thus its wireless networking solution allows Tropicana to creatively manage special events, stay in line with the fire code, and offer new ways to monitor the casino floor.

Dyer said that the reason Tropicana chose Intel Wireless networking products to deploy throughout its hotel and casino was because "there is a reliable name (Intel) behind the architecture. We had looked at a couple other brands that we felt weren't as strong, but we felt comfortable with the Intel brand name. So far, the experience working on the Intel wireless network has given us excellent results: flexibility, no more cabling issues, and it's opened up a whole new creative solution for our special events group. And, we can move wherever the fire department wants us to hold the event!"

"We are considering opening up our conventions department to host computer/tech events where we can lease out wireless Internet access for our clients if they want to hold seminars or tradeshows," explains Mr. Dyer. "We are also exploring point-to-point bridging opportunities to link multiple hotel towers, thus keeping T-1 leased line costs down."

Dryer adds, "We are also considering running Intel PRO/Wireless LAN products on handhelds or PDAs on the casino floor. We'd like to be able to rate players right at the tables instead of running back and forth to the casino pits."

He went on to say that Tropicana chose to implement a wireless solution because it offers them more flexibility. "We wanted to get our feet wet, and test wireless LANs. The next step is wireless point of sale systems and remote cashier stations in the parking lots or beach areas."

FIELD SERVICE

Field service applications are oriented to external customer support and service professionals who go to the customer's or product's physical location. And because field service is by definition done in the field, any automation of such service activities will result in a high return on investment.

There are many different types of field service, some involve outside personnel, and some don't. For example, field service personnel install, maintain and inspect systems. But with a WLAN at the helm, some of those tasks can either be optimized or performed remotely. We will examine how, through the following case studies.

Case Studies

Barwon Water

Barwon Water is a large regional water authority in Australia responsible for 8100 square kilometers (or a little over 3126 square miles) southwest of Melbourne. Barwon's 106-person field staff includes service maintenance, water supply, sewerage, survey, construction auditing, and customer service specialists. This staff continually travels long distances to manage the company's $832 million asset base. The same staff must also service over 250,000 customers through a network that includes more than 5000 kilometers (3107 miles) of in-ground pipes, ten reservoirs, ten water treatment plants, and nine sewerage treatment facilities.

Although Barwon Water had invested heavily in computer technology, when field personnel were on the road, they had no access to the company's electronic information systems. Whenever they needed information such as schematics of a particular asset, detailed work instructions, or maps of an area, they had to drive back to one of five regional depots, locate the physical documents they needed, and bring them along into the field. They then had to return to the depot every night in order to return those documents and file forms detailing the work they did during the day.

"There's a lot of paperwork in the system," said Grant Green, executive manager, Customer Services. "For example, if a maintenance employee has repaired an asset, details of that repair have to be keyed into the system after the event. This opens the opportunity for translational errors."

Because of the remote geography, the field staff was often out of radio contact. Even when they were within radio range, communications were often compromised. When workers entered areas of poor or no radio coverage, dispatchers had little idea of exactly where the field workers were or what they were doing.

Recognizing that the ability to access crucial information in the field would significantly improve its field staff's efficiency, Barwon Water deployed a new service management system, integrating it with its call center and other key business systems. At the same time, Barwon also deployed FOCUS, an enterprise-wide mobility service request and dispatching application from Melbourne-based developer e-Wise. FOCUS is built on Microsoft's .NET Compact Framework, and allows Intel Centrino mobile technology-powered computing devices to access multiple, centrally-hosted databases that manage details of all work orders carried out by field staff. Call center personnel enter work orders into FOCUS as they field calls from customers. The orders are then allocated to workers by dispatchers, who can monitor workers' movements and status.

"Our core business drivers are business efficiency and customer service," said Green. "One of the drivers for our FOCUS system is to have a more efficient field workforce, and at the same time improve our levels of service to our customers. These improvements will come by ensuring field staff have the information they need, when they need it," he adds.

The full solution enables field personnel to use Intel Centrino mobile technology-powered PCs running FOCUS, which integrates and makes accessible to end-users Barwon's Geographical Information System (GIS) and other documentation. Use of both broadband and narrowband wireless networks are important because the right network can be used for different types of communications. Workers can log into the system from home to retrieve their allocated work orders, while supporting information, such as GIS data file updates, will be automatically retrieved and downloaded to the mobile PC via an 802.11b wireless network located at each depot.

"This is not just about dispatching a message for a job," Sheiman said. "It's about total bi-directional communications between corporate back-end systems and a mobile field worker. For this, it's no good to be running in cut-down mode; workers need strong enough computing power to be able to trawl through all the data they need to take with them. Standard mobile technology would have been inefficient, but Intel Centrino mobile technology [and the 802.11b network], opens up a whole lot of new opportunities to provide rich clients in the field."

Once workers are in the field, work orders, messages, and other information continue to be transferred to and from their PCs throughout the course of the day via a PC Card. The card accesses a low-speed State Mobile Radio (SMR) network providing coverage even in remote areas, thanks to a statewide network of repeaters. Although the SMR network can transmit small packets of information, it's unsuitable for large files used by GIS and other applications. Barwon Water's mobile solution combines the slow, far-reaching SMR with faster, but relatively short-range, 802.11b wireless LAN technology to move data as efficiently as possible.

Still the SMR provides the remote workers with any necessary updates when they are in the field, allowing the information to travel to the workers that need it, rather than *vice versa*. According to Green, wireless PCs enable employees to start and finish their working days from home; and by providing a rich mobile PC environment, Barwon Water is able to better provide its field service staff with all the information necessary to complete their tasks. He goes on to say, "Significant duplication of effort and cost is avoided, and collaboration between field crews, the customer call center, and job dispatcher is significantly more effective."

Also, by giving field service personnel instant access to the data they need, Barwon Water is able to streamline its customer service process and, in turn, see a rapid return on investment. And, by using 802.11b wireless connectivity at depots, and low-bandwidth radio network connectivity in the field, dispatchers can better track the location and status of each field employee.

U.S. Fleet Services, Inc.

As one of the country's largest on-site, commercial fleet re-fueling companies, U.S. Fleet Services, Inc., must provide accurate and timely information to its customers 24/7 while being responsive to their needs. But to provide such service requires a lot of time, which translates into a smaller overall profit. U.S. Fleet found a way to improve accuracy, shorten its billing cycle, keep its customers happy, and improve its profit ratio—an Intermec wireless solution, in the form of a new 802.11b wireless network. To access the wireless network, U.S. Fleet employees use Symbol Pocket PC-based 710 mobile computers with integrated wireless communications and barcode scanning. Then the company bought Symbol 782 portable printers to enable their employees to print invoices and statements.

The company undertook a huge wireless project—there were over 50 branches in 23 states that had to be integrated. So U.S. Fleet selected an outside project manager to assist with the new systems design and implementation. The company turned to Intermec and its professional services division to implement a new wireless data collection system and to provide a project manager to help U.S. Fleet personnel develop a deployment process.

The completed network connects each branch office with headquarters. From each office, branch managers set up drivers' routes by entering start and end times for customers' re-fueling needs. Then the system organizes the best routes and schedules. The driver synchronizes his or her wireless mobile computer with the network using the 802.11b standard and access points that are strategically placed within the offices, and then downloads the day's work.

During customer off-hours, drivers scan barcodes on each gas tank of a customer's truck, using the mobile computer's integrated laser scanner. Once the vehicle is identified, the fuel is pumped. The driver uses the computing device's integrated 802.11b radio to communicate wirelessly with the U.S. Fleet truck's electronic fuel meter, specifying the exact type and quantity of fuel, to the tenth of a gallon, consumed by the vehicle. All the data is maintained in memory on the CompactFlash card in the mobile computer.

U.S. Fleet reports that productivity has measurably increased—its drivers are no longer required to write down transactions, and that time saver alone enables most drivers to add another stop to their route each day. The timesavings also have enabled the company to steer the drivers to a path in which they can provide better customer service. The company expects a ROI payoff in less than a year.

After a driver completes his or her route and returns to home base, the next step is to wirelessly upload the fuel delivery data to the company's database. The data is then immediately available for invoicing and is published to the U.S. Fleet extranet, allowing customers access to detailed reports.

With U.S. Fleet's new wireless mobile system, customers now have up-to-the-minute access to account information, accurate invoices, detailed reports, route schedules, and complicated tax information on time, anytime, from anywhere.

MANUFACTURING

In the fast-paced manufacturing industry, companies seek wireless LANs and durable scanning technology for a variety of shop-wide applications. Large manufacturers like Alcoa, Boeing, Ford, and others have already rolled out wireless LANs in their manufacturing facilities.

But in many cases, manufacturers know that the staff of a certain area or division isn't performing as well as they might. While they know they have a problem, they just don't know the solution. "Customers don't necessarily come to us and say they need wireless," said Dave Worton, CEO of Made2Manage Systems, a company that ships its own mobile, wireless access solution, Mobile Manager. Instead, what they say is that they are "frustrated in getting access to production and inventory schedules."

In the past, the shop floor lacked good mobile computing devices—rugged and robust devices with Windows CE or Palm operating system (OS) capabilities. The products offered by vendors didn't excite anyone. They were big, clunky devices that, although they worked well enough, were not always suitable for factory shop employees to lug around all day. There were also issues concerning standards, awareness, and a fear of change. All served to cause a lack of interest in wireless LAN technology in the factory.

That's all changed. The excitement and media attention over all things Wi-Fi serves to drive interest in wireless technology, even in the manufacturing sector. Vendors like Symbol and LXE offer some pretty nifty, rugged mobile computing devices, designed specifically for use in the rough and tumble manufacturing world, and the 802.11 series of standards have addressed many other concerns that the industry had about wireless.

Another driver for the adoption of wireless LANs is good quality software. Vendors such as Oracle and SAP now cater to manufacturing applications on a mobile basis, creating applications that once could run only on stationary workstations.

However, any manufacturer, large or small, who is considering going wireless, should know that there are a few important issues to consider first. One is signal blockage. Another is interference. These and any other special circumstances within your specific plant facilities should be considered very carefully when planning a WLAN deployment.

Pay strict attention to the planning process, especially concerning placement of the access points, because as John Williams, data product specialist for Avaya Canada so aptly puts it, "the wireless spectrum doesn't go through stuff very well." And there's a lot of "stuff" in manufacturing facilities: ceilings, floors, walls, raw materials, products, and machinery.

Williams gives as an example a rolled paper plant in Nova Scotia that faced a difficult challenge implementing wireless. A big roll of paper is one of the worst interference agents because there's water in it and it's solid wood. One big roll of newsprint can stop RF signals dead in their tracks. And since these paper rolls were constantly being moved around the plant, interference was dynamic, not static. Thus, the company's WLAN had to be designed so that no matter where the rolls were dropped, the radio signals still could get through.

Whether a wireless LAN is right for your plant will depend on a number of factors, and each facility will have its own peculiarities. A consultant, integrator or wireless vendor can help sort through everything to ensure that the WLAN deployment and subsequent operation goes smoothly.

In the meantime, the software and hardware are improving and the opportunities to cut costs continue to mount. The benefits aren't going away any time soon.

Die castings manufacturer Meridian-Jutras operates out of a 50,000-square-foot facility. Every time Robert Bolton, the company's automation supervisor, wanted to hook up a programmable logic controller (PLC) to the company's wired network, or move machinery (which was happening regularly), Bolton was forced to figure out a way of relocating the cable. He also had to call in contractors to run Category 5 cabling, make sure it was properly shielded, and possibly put it in conduit. That was until the manufacturer went wireless.

Here's some sage advice from Bolton concerning the deployment of a wireless solution. "Like everything in manufacturing, it should be driven by cost, but I just knew it was technology I wanted to get in [the plant]. I saw a use for it in all departments. Don't let the upfront costs sway you against moving forward."

Bolton also says that although it's tough to cost everything out up front when payback might be a year down the road, after his wireless network was up and running, he was very pleased with the choice.

Case Studies

Dingley Press

Located in south-central Maine, Dingley Press is a large printer of mail-order catalogs. The company, which operates around the clock, mails around one million ink-jet-addressed catalogs every day, making it the second-largest "post office" in Maine.

The assembly of the catalogs is a complex operation. Pages, which are printed in groups called "signatures," are loaded onto pallets to await assembly. This means that literally hundreds of these pallets are stacked four-deep throughout the warehouse. The catalogs, which range from 32 to 128 pages, are assembled from the center outward, starting with the order form, which is stapled in the middle.

Customers frequently request multiple versions of a catalog, which complicate matters even more. A clothing catalog customer, for example, might want its winter catalog to sport two covers, one for mailing to people in the warmer climates and another for people in the northern, colder locales. The catalog destined for southern regions might also have extra pages of clothing for wearing in more temperate weather conditions. The company is oft-times required to assemble as many as two dozen different "signatures" for the same catalog, and all versions must be printed and bound for simultaneous mailing to get the best postal rates.

Accuracy and completeness of such complex inventory is critical. Tracking every pallet within the 350 x 450-foot building—as well as those located in off-site warehouses—and locating the right pallet, especially in an emergency (and there are emergencies), can be a nightmare. With as many as six companies' catalogs in production at any one time, "losing" a pallet or bringing the wrong one to the bindery can be expensive. So expensive that Dingley assigned two people to the full-time task of managing the stock.

Working together, the company's Information Systems and Warehouse departments devised a better solution—one based on their own in-house application and Cisco Aironet wireless technology. The Cisco and Dingley team originally connected Hitachi handheld Microsoft Windows CE touch-screen tablets to barcode readers that could decipher the tags identifying each printed pallet load. But, after several months of operation, Dingley replaced the handheld Hitachi units with forklift-mounted Intermec mobile terminals, because Intermec's rugged keyboard was better suited to the rigors of an industrial environment than the touch-screen tablets.

When a forklift driver takes a pallet to its warehouse location, he or she uses a scanner to identify the load, and then keys the section and row in which the pallet is placed onto the mobile terminal. The terminals are connected wirelessly via a Cisco Aironet 340 Series PC card client adapter to the server. Thus, when the driver inputs the location information onto the terminal and presses "transmit," the information is sent to the server via Cisco Aironet 340 Series Access Points mounted throughout the facility.

"If it's a new load, it's added to the inventory. If it's just being repositioned, the system updates the location. If the pallet is being taken to the stitcher for assembly, it's recorded as consumed. So we always have a running account of produced, consumed, and in-stock inventory," said Mike Martell, systems manager and principal developer of the customized application. "Also, we can strategically position specific pallets for quick access, which means there's no time lost searching for them because the wireless system has enabled us to maintain an accurate, readily available location of inventory."

Having accurate locations helps solve printing problems, too. If a color or alignment problem is spotted at the start of a printing run, it is likely that the problem actually began near the end of the previous load. Dingley personnel can quickly check and find how far back the error went, so they can quickly determine how many extras they need to print to catch up.

But an up-to-the-minute inventory of every pallet's location and status is just one benefit derived from the wireless system. Martell says, "customers sometimes call and say they need 25,000 of a particular book to carry to a trade show. Do we have sufficient stock? We can look up the combination of signatures and determine the highest number on hand of least-available components, which allows us to calculate how many full catalogs we can provide them. If we didn't have real-time inventories, we couldn't provide this service."

Five forklifts are equipped with Cisco Aironet wireless technology. Although the original forklift-mounted units were fitted with Cisco Aironet 340 Series PC Card client adapters, providing 11 Mbps of bandwidth, the new Intermec units use internal 802.11b adapters to connect with the WLAN. Four Cisco Aironet 4800 access points, each with a 5 dBi antenna, are mounted in the warehouse. "We could have managed with just two access points, but four ensure that no posts or machinery or high towers of stock will interrupt transmission, and they give us a measure of redundancy," said Martell. (Dingley is considering a similar setup for an external warehouse.)

Dingley selected Cisco Aironet products for numerous reasons. "Recognition of Aironet's support, capacity, and reputation for reliability was key. Aironet had a reputation for working properly," says Martell. Cisco Aironet also offered drivers for Windows CE (Dingley's primary platform), while many other companies did not.

"To put it succinctly, no one else had a wireless data collection system that provided ease of use, the ability to be mounted on forklifts, and customizability to suit our needs. The Cisco Aironet package solved a significant business problem for us," Martell explains.

In the near future, the wireless network will reach into the printer's manufacturing arena. If a pallet is not used completely, an operator can identify how many pieces are left over and create a return-to-stock tag. Martin says, "Rather than dropping Ethernet connections all across the manufacturing floor to reach the machine used to print these tags, we will use Cisco Aironet wireless connections, and they can create return-to-stock tags free of the restrictions of a cabled environment."

The Dingley tech support team uses handheld computers and the wireless network as they move around the campus on maintenance checks, accessing web-based controls for print servers, for the wireless access points, or any other remotely managed device. "If

I have a printer not working properly somewhere, I can access the web-based management system to test the print server," Martell says.

A number of laptops are also equipped for wireless operation. The vice president of finance and the comptroller use their 802.11b-enabled laptops in board meetings and in strategy discussions, where they can always access key figures instantly. To maintain the complex parts inventory, the Purchasing Department takes a wireless-equipped laptop to the parts cages to conduct inventory reconciliation online. They have found this faster and more accurate than writing everything down and returning to their desks to transcribe the data.

For sales representatives, consultants, and others who require network or Internet access, Martell can deploy a laptop with a Cisco Aironet PC card client adapter. This capability is also available for customers who visit Dingley to conduct quality checks on catalogs.

While the inventory system remains under development to improve its reliability, accuracy, and applicability to the companies' needs, the wireless component has become a solid, reliable part of the processing. According to Martin, by helping Dingley to prevent mistakes, to locate stock in substantially less time than by manual searching, and to maintain accurate inventory, the Cisco Aironet system has already paid for itself several times over. "It's reliable, flexible and secure."

Libbey-Owens-Ford

Libbey-Owens-Ford (LOF), with headquarters in Toledo, Ohio, is the second largest North American glass manufacturer. It supplies glass products for the transportation, architectural, and specialty markets. It has over 200 sites and approximately 7000 employees.

Over the past decade, LOF has sought to reduce cost and inventory levels, and increase productivity—every corporate executive's dream. The company decided the way to increase productivity was by giving more detailed information on products and inventory levels to floor personnel, so that they could make better business decisions.

The manufacturer's executives decided wireless networking might be the solution. After doing a fair amount of research, the company settled on 802.11b wireless networking, using Symbol's Spectrum24 wireless Local Area Network and LRT 3840 handheld computers.

Before going forward with the wireless project, Libbey-Owens-Ford faced myriad ongoing challenges. Inventory reporting was always behind by 24 hours, with tickets entered manually and handed in the next morning. "Inventory can only be as accurate as paperwork turned in," says Jeff Hobbs, the company's lead programmer/analyst. "It is only as accurate as the person entering the data."

In addition, performance reporting also ran at least a day behind. "We need to know how we are doing in real time," says Hobbs. On the plant floor, company personnel needed to know inventory levels. Scheduling needed correct inventories. Schedulers were delayed until reports were printed, which increased the possibility of missed shipments. LOF lacked personnel to enter data. It looked for cost reduction and elimination of unnecessary steps in the inventory process. "Inventory levels were kept at summary levels only," says Hobbs. "Turning of inventory was hard to control."

The manufacturer wanted the WLAN to standardize all of its plant processes, and to print customer-specific tickets at shipment time.

Enter Symbol Technologies, which first partnered with LOF ten years previously to supply the manufacturer with barcoding equipment and its original wireless LAN system, the Symbol Spectrum One network. "Symbol was there with the right equipment at the right time," says Hobbs. "The equipment has very good uptime even with rough handling. We have also had very good support from Symbol. The products have saved us time and money in tracking our inventory with more accuracy."

Once everything was laid out and everyone knew what was expected from the new WLAN system, LOF installed Symbol's Spectrum24 with LRT 3840 computers in a client/server Unix environment at two sites. The Spectrum 24 system complies with the 802.11b standard, provides wireless connection to an Ethernet LAN through multiple, SNMP-compatible access points. The system features one-piece PCMCIA and ISA bus cards, which can be installed quickly in terminals, laptops, and other devices for easy connection to the wireless LAN. Spectrum24 also runs seamlessly with the company's IBM RS/6000 system and IBM MVS mainframe. "Spectrum24 offers us open connectivity, fast communications, and fast installation, since Ethernet is already installed at most of our locations," notes Hobbs.

Over the years, the benefits of using Symbol's products in a manufacturing environment have become clear to LOF. These include reduced data entry errors, lower inventories by more control, increased performance on manufacturing lines, timely reporting, inventories kept in real-time, detailed inventory ticket by ticket, inventory turn around on a first-in-first-out basis, and a better understanding of the manufacturing flow process by personnel. "Personnel have a direct impact on our bottom line, including loss reporting; and downtime reporting," states Hobbs.

The Spectrum24 network is used for a variety of manufacturing applications, including LOF's physical and cycle inventories. The company has 20 Symbol LRT 3840 devices, 35 LL500 wedges, and LS 4000 scanners. In fact, the manufacturer uses Symbol products in a number of applications, including shop floor reporting systems, ticket verification during shipping, physical and cycle inventories, production reporting, and rack tracking. Eventually, the company plans on rolling out the Spectrum 24 system to all of its manufacturing plants.

Currently, LOF inventory, raw through finished goods, is stored with a unique ticket on each rack, box, or container. When the physical inventory is in progress, every ticket in the plant is scanned wall-to-wall, using Symbol series 3840 scanners. If a ticket is not found, the operator is asked to scan the part number and verify the data that is returned. If the operator agrees with the data, then the ticket is flagged, "to be added to inventory." Reports are generated that list tickets to be removed from or added to the inventory. "These reports are itemized in dollar amounts, so that the impact is readily apparent," says Hobbs. "All of our Symbol investments have paid back in a year or less." The big payback is more accurate inventory levels as well as less time spent doing physical inventories.

While LOF's executives are pleased with the Spectrum 24 system, there have been a few bumps in the road. "Training is our biggest area of challenge," admits Hobbs. "You are training people from all walks of life to walk into the StarWars arena. The best part is when someone is scared of the technology and ends up loving it. . . What else could we ask for?"

MW Manufacturers

This $200 million maker of windows saved $5000 on cabling costs alone by implementing a wireless LAN in one of its manufacturing plants, asserts Eric Martin, the company's corporate director of IS and business systems. And, according to Martin, the WLAN enables MW to save money every time the production floor changes, which happens an average of three times a year, because the network doesn't have to be rewired.

"We used to have wiring closets set up all over the shop floor so we could run a wire to a PC," he says. "Then a few weeks later, we'd do it again."

MW Manufacturers deployed a WLAN using Enterasys' RoamAbout technology to meet the changing demands of its manufacturing facilities. It also availed itself of the expertise of Classic Networking, a Pennsylvania-based network solutions provider for on-site installation, at both MW Manufacturers sites in Rocky Mount, Virginia, and Hammonton, New Jersey.

Cultivating a culture of continuous improvement, MW Manufacturers frequently implements new procedures. These improvements often result in the movement of heavy machinery and equipment. Wireless networking, however, allows MW Manufacturers to reduce costs and provide flexible, secure, high-availability access to information in an environment of continuous change.

"For Information Services, changes to the facility layout and configuration were a nightmare, as well as costly. While other groups in the company could change manufacturing processes on a daily basis if needed, IS was seen as a roadblock," says Martin. "We needed a technology that would allow our employees to make changes in their various departments at will, without worrying about their network connectivity—Enterasys' RoamAbout allows us to do this."

Utilizing RoamAbout wireless LANs at its various facilities provides MW Manufacturers with seamless connectivity to support a dynamic environment. In Rocky Mount, the RoamAbout outdoor solution provides building-to-building wireless connectivity throughout their campus, delivering "anytime, anywhere" access to shipping and receiving information. In Hammonton, the RoamAbout network covers 365,000 square feet and connects "end-of-line" desktop computers, barcode scanners, and file servers to follow the product from the production line to shipping. Remote time clocks and desktop computers in the shipping department are also connected via wireless.

In addition to the RoamAbout wireless products, MW Manufacturers is also implementing Enterasys' Matrix E5 modular switches and Vertical Horizon standalone switches to provide high-density, 10/100 Fast Ethernet to desktop users. The company is deploying Matrix E6 switches for high-performance Gigabit Ethernet across the network backbone. The entire network is managed and monitored by Enterasys' NetSight Atlas network management software, providing an easy-to-use, graphical user interface to complement the advanced network management application.

"Enterasys is dedicated to developing networking solutions that truly serve the business needs of our clients," says Mads Lillelund, executive vice president of worldwide sales and marketing at Enterasys Networks. "Coupling RoamAbout wireless connectivity with Enterasys' switching technologies and enterprise network management software enables a communications infrastructure capable of delivering the performance, availability and flexibility MW Manufacturers requires."

Despite clear advantages, wireless LAN transmissions can be troublesome. Building construction obstructs signals, and 802.11b transmissions pick up interference from competing devices on the same public frequency. Major culprits are electromagnetic interference from production equipment and any steel structure that can insulate the signal. Martin used signal monitoring software provided by Enterasys to determine that ten access points could provide complete wireless coverage for the company's 360,000-square-foot factory. He situated the APs near high-traffic areas because, Martin explains, the further a user gets from an access point, the slower the access speed.

EDUCATION

Educational settings from elementary schools to college campuses are turning to WLANs to provide network and Internet access. In fact, some of the largest active Wi-Fi networks in operation today are installed at campuses such as Howard, MIT, and Stanford. A late 2001 NOP World Technology survey conducted on behalf of Cisco Systems discovered that 35% of educational institutions in the U.S. have already implemented wireless networking in their organization. (For full copy go to newsroom.cisco.com/dlls/tln/pdf/WLAN_study.pdf.) This is far higher than in any other vertical industry. The reason for such growth is simple. The use of wireless LANs helps financially constrained institutions achieve more while staying within their budget. It allows them to introduce, at a very reasonable cost, leading applications, like e-learning, that enhance the learning experience.

The survey also discovered that the benefits of Wi-Fi in the educational sector were based around mobility, convenience, and flexibility. The ability to quickly deliver networking services anytime, anywhere without the need for cabling, is Wi-Fi's most distinguishing feature. (See Fig. 15.2). When translated into an ROI calculation across all industry sectors, the survey found that this represented a $550 savings per user, or an average $164,000 annual cost savings–not including ROI from productivity gains.

However, Wi-Fi's momentum in the educational arena was initially brought about because many institutions have buildings that are difficult to reach or expensive to cable. Most of such locations don't have connection to the school's network, because there is

Mobility within building or campus	78%
Convenience (no cabling)	74%
Flexibility (anytime, anywhere access)	65%
Easier to set up temporary spaces	68%
Lower cabling costs	63%
Easier adds, moves and changes	55%

Figure 15.2 Percentage of educational IT managers who perceived benefits from wireless LAN technology. *Graphic courtesy of Cisco Systems.*

THE UNIVERSITY OF TWENTE

This Netherlands university has built a Wi-Fi network, with the help of IBM and Cisco Systems, that extends over the school's entire 346-acre campus. The WLAN supports 6000 students and 2500 staff members. The network uses 650 Cisco Aironet 1200 series access points, and while most of the network uses 802.11b, there are sections run on 802.11a.

With a WLAN as part of the networking mix, the wired LAN infrastructure is minimized; it is only needed to support the necessary virtual LAN connections and APs in the classrooms and elsewhere.

no physical link between the central network and the networks serving these outposts. Or, if linked, it's via an expensive leased line.

Today's WLANs allow a scalable connection with any wired network infrastructure. Technology is also available that can enable students and faculty to roam about the campus while maintaining a stable network connection. Above all, retrofitting older buildings with networking using wireless technologies lowers the overall cost of ownership and facilitates faster installation. And, except for connecting the WLAN's access points, physical wires are not needed.

Wireless Bridges

As discussed in detail in Chapter 9, wireless technology can bridge buildings so that a single network can cover an entire campus. A wireless bridge not only provides connectivity, it also reduces server hardware and licensing costs, as well as bringing about a reduction in the management activity needed to build, monitor, and maintain multiple networks.

A WLAN also offers greater flexibility than its wired counterpart. For instance, there is a national shortage of classrooms, which is often addressed through the use of trailers configured as mobile or temporary class space. Although these temporary structures have heat and light, they often don't have access to the school's network or the Web. Schools that do make the investment to provide access often do so via a fiber-optic network link, which can be expensive (often costing at least $5000). Furthermore, the fiber-optic link has to be run each time a trailer is added or moved. Since trailers are modular, they are often moved, resulting in recurring recabling costs.

Wi-Fi, however, provides a better method. A single access point usually is sufficient for a number of classrooms, or an access point could be used to connect temporary classrooms, remote from the main building housing the central network. Thus, simply placing an access point and antenna near a window of the main building allows connectivity across, for example, a playing field to outlying buildings.

As discussed at length in Chapter 9, buildings at a greater distance can be connected using a bridging solution. Even if you can't see the other building from your window, a mast from the roof may provide the line-of-sight required for this low cost solution.

And if you need your network to reach outside your campus, Wi-Fi can also connect facilities within a specific geographical region. There are many wireless bridge products that can link networks and provide fast, cost-efficient integration of remote sites and end-users. Wireless bridges provide line-of-sight bridging at distances of up to 25 miles (40.2 kilometers) in FCC regulated countries, or 6.5 miles (10.5 kilometers) in the EU countries.

A wireless point-to-point or point-to-multipoint bridge can connect remote campuses, research field sites, and even community facilities to provide community-wide information, learning, or research networks. The technology also enables multiple buildings to share a single high-speed connection to the Internet, without cabling or dedicated lines. In addition, wireless bridges lead to the complete removal of recurring leased lines expenses, delivering tremendous financial benefits.

Mobile Carts

If space is limited and setting up computer labs is impossible, you may find that a mobile cart is the right solution for your institution. Wi-Fi allows any classroom to become a computer lab, as needed, through the use of computer carts equipped with laptops and perhaps a mobile printer. The only wired component for a mobile lab is a wire "drop" for the AP on the mobile cart to be attached. Of course, if the entire campus has been "Wi-Fied," there is no need for the drop.

This kind of wireless solution is reusable and thus allows Wi-Fi to quickly pay for itself.

Provisioning Multimedia

An educational setting typically also means bandwidth-hungry applications such as multimedia presentations. So unlike the other vertical applications discussed in this chapter, most of which have limited bandwidth needs, building a WLAN around 802.11a or 802.11g should be considered when deploying a WLAN in an educational environment. But, of course, the bandwidth capacity offered by 802.11g must be balanced against the fact that it is a new technology. As with any new technology, there are in all likelihood "bugs" that will need to be worked out. However, 802.11g offers a greater reach than 802.11a. In college and university settings where the WLAN will need to extend its reach to dorms as well as providing connectivity around the campus, that is an important consideration when determining which technology to use.

Addressing Wireless Concerns

While wireless campuses offer many benefits, they also pose new challenges for network managers who need to ensure that only authorized users access the network, and that wireless traffic isn't intercepted or tampered with. The management and maintenance of a wireless network infrastructure also presents different challenges than those of a wired infrastructure.

The educational sector's main concerns when it comes to wireless networking include security of personal data, Internet content filtering (for lower-level schools), and interference. Since a school's facilities are often located near businesses and homes, countering interference from other networks may be a challenge.

In many educational environments, it's the student or the facility member who makes the wireless NIC selection, not the school's IT department. Furthermore, there is a grow-

COAST-TO-COAST, EDUCATIONAL INSTITUTIONS ARE EMBRACING WI-FI.

- American University is getting rid of its POTS service and going totally wireless on its 84-acre campus serving 10,000 students.
- University of Southern Mississippi estimates that it costs $75,000 to wire a building versus $9000 to go wireless. So far this budget conscious university has installed 300 access points.
- Carnegie Mellon's Wireless Andrew is one of the largest Wi-Fi educational networks. It connects approximately 35 academic buildings, more than 30 campus residence halls, and key outdoor areas.
- The top (un)wired campuses in the United States include; Carnegie Mellon, Stanford, Georgia Tech, Dartmouth, MIT, Drexel, Indiana University, University of Delaware, University of Virginia, New Jesery Institute of Technology and SUNY Buffalo. (The Wireless Education List Serve has more information. It can be found at www.educause.edu/netatedu/groups/wireless/.)

ing number of mobile computing devices that are equipped with embedded Wi-Fi capabilities. With a variety of wireless networking equipment to choose from today, an educational institution must educate its user community, in order to create awareness of the type of Wi-Fi network that is deployed in its environment and to avoid confusion.

The primary reason Wi-Fi is making such inroads into the educational sector is because its most compelling advantages are cross-vendor interoperability and, if designed correctly, practical interference-free communication over reasonable distances. The main disadvantage—security, or the lack thereof—is problematic, but can be overcome.

Admittedly, provisioning security in a multi-vendor environment can be challenging. And although truly manageable and scalable, multi-vendor security solutions are not pervasive today, the industry is working on architectures that will provide security solutions independent of the wireless client adapter and operating system. Until then follow the advice given in Chapter 17.

Wireless Networking Check List

Here is a check list to help you determine your school's wireless networking needs:

- Do any of the buildings have historic status, asbestos, or any other features that make cabling difficult?
- How many locations/buildings are to be covered by the wireless network?
- Are any of the buildings more than one story? If so, is the entire building to be wirelessly enabled?
- What type of network is currently in place?
- How many users currently use the network? How many end-users is the network projected to support in one year, three years, five years?
- Are there any open hubs on the existing network?
- How does the school currently access the Internet? Does that access method cover the entire school? If not, how is it supplemented (by dial-up service)?

* Does the school currently have network printers?
* Have you determined the area the wireless network should cover?
* Have you performed a site survey?
* How many laptop computers will be needed? How many will need to be Wi-Fi-enabled?
* How many desktop computers will be needed? How many will need to be Wi-Fi-enabled?
* How many wireless network interface cards will be needed? What type?
* Does the school want to support wireless carts? If so, how many?
* How many students are expected to use the wireless carts?
* What percentage of spare parts will you want on hand?
* Will the end-users require help desk support?
* Will the IT staff require training on the new technology?
* Do you require professional assistance with the site survey?
* Do you require outside assistance with the deployment process?
* Will you require that specific applications be installed on new computers prior to receiving them?

Whatever the educational environment, be it primary or secondary grades, private or public schools, a wireless LAN solution can enable your institution to deliver technology-based instruction resulting in information on demand. It can free space currently used by computer labs and, at the same time, optimize the time of students using the computers because Wi-Fi enables access to information whenever and wherever the learning takes place.

Case Studies

Howard University

Howard University, a world-renowned university with 91 buildings and four campuses in the Washington, D.C. area, has installed a high-speed voice, data, and video network that serves all of its more than 25,500 students, faculty, and staff, whether they're in a dorm, classroom, library, lab, or roaming the campus.

Students can learn better using new technologies, says Charles Moore, who, at the time was Howard's interim vice provost and CIO. "We're connecting our students with a web of technology that makes it easier for them to achieve their education. It's really become almost a recruiting tool," he says.

Howard installed the $10.4 million network to meet its objective of providing integrated communications to each student. The network made a huge difference in how students view the school and how much more easily they get their work done, according to Pamela Cohen, a Howard graduate with a bachelor of science degree in information systems.

In many respects, Howard is catching up with what other schools offer, Cohen says; most other schools she's familiar with have cable-TV and modem lines in their rooms. But Howard's installation of high-speed Ethernet ports and wireless networking capabilities in students' rooms puts it ahead

The network (wired and wireless) is one of a series of ongoing projects to make advanced computing and networking technology available to Howard's students and faculty. "We've immersed our students and also our faculty in technology," Moore says.

Believing that technology is fundamental to education in the 21st Century, Howard University President H. Patrick Swygert pledged to provide access to the Internet for every student as of the 2000-2001 academic year. This led not only to an extensive, best-of-breed wired network, but also the installation of a pervasive wireless network from Cisco Systems.

The university constructed its network in two stages. As part of the first stage, the university completed construction of a $4.5 million computer "super lab" with 200 computer stations including PCs, iMacs, Sun, and Microsoft workstations (and Dell Computer servers) within Howard's Technology Center. iLab, as it's called, is open around the clock.

The university has three other technical labs on campus, one with a PictureTel Corp. videoconferencing system, one with iMacs equipped with flat-panel cinema-style screens and software for media and graphics projects, and a third with powerful servers and workstations for math and engineering students.

At the same time it built iLab, Howard also constructed two new libraries, each with hundreds of Ethernet ports, wireless LAN access, and extra desktop computers and notebooks for student and faculty use. For example, the $27 million law library has 690 Ethernet ports, and the $27 million health-sciences library has 700 Ethernet ports.

As far as the wireless network is concerned, Howard University's chief information officer, A. Burl Henderson, and his staff researched different options, and chose Cisco's Aironet wireless solution because, according to Henderson, "it was competitive with other products and also due to our long use of Cisco networking equipment." But Henderson also said the Cisco Aironet 340 series of products were chosen because of the features they offered, the ease of installation, and cost-effectiveness they provided, and their ability to integrate with the wired network infrastructure. Installation included Cisco Aironet access points throughout the residence halls, and network interface cards in user computers.

The University began its "Wi-Fiacation" with a pilot project in 2000, in which 3000 students participated. During the pilot program, the school offered free wireless interface cards to the students and promoted the new wireless access within residence halls. The pilot was a resounding success. Network administrators quickly drew up plans to add wireless connectivity to other areas of the campus, including libraries, common areas, and classrooms.

Hundreds of Cisco Aironet 340 series APs were installed, covering a total of 1,993,208 square feet (185,175 square meters) in the hallways of 12 residences. They integrate easily into the University's wired network via a single autosensing RJ-45 port. Each Cisco Aironet 340 series AP acts as a bridge, forwarding at media speed between the Ethernet and wireless protocols. The 250 APs are connected to the campus network via copper wire that is connected to a fiber optic infrastructure in each building.

"We engineered the network so that there would be no more than 20 students per access point," says Henderson.

Students bringing their own PCs to school are offered the loan of Cisco Aironet 340 series network interface cards (NICs), enabling them to receive wireless Internet access within 400 feet of an access point. Each credit card-sized NIC has a radio transmitter/receiver that establishes the connection with the access point, much like cellular telephones.

Students without laptop or notebook computers at Howard can use PCs and Macintoshes at the computing center in each residence hall. Each center is also wireless enabled.

Network administrators at Howard University saw an enormous increase in network activity after the Cisco Aironet wireless LAN was installed. "We routinely have over a thousand simultaneous users on the network at peak use hours," says Henderson. "I believe that this is a clear indication of the success of the implementation. Clearly we expect these numbers to increase, as our faculty become more accustomed to using the Internet as a research and teaching tool."

The popularity of the wireless network with residence hall students quickly caused plans to be drawn up to introduce wireless infrastructure throughout the Howard campus, beginning with dorms, libraries, student lounges, and classrooms. So, during the second stage of its extensive IT overhaul, Howard not only installed an integrated voice, data, and video network, it also completed the installation of 802.11b wireless LANs in all its dorms, so students with notebook computers equipped with 802.11b wireless-access cards can stay connected to the university's network while moving around in their dorms.

In addition to offering wireless LAN access in campus dorms and at the university's computer labs and libraries, Howard envisions building a "wireless canopy" to blanket the entire campus with high-speed wireless access by early summer of 2003, Moore says.

The Siemens Enterprise Networks division of Siemens AG won the bid to supply the equipment for the wired network, installing the network cabling and PBX equipment through its HiPath Professional Services business, which provided project management, integration, and installation, plus ongoing support and maintenance services.

In addition to the cabling, Siemens provided Howard with five 5,200-line Hicom 300H PBXs, which included three Hicom 300H Model 80s and two Hicom 300H Model 30s, as well as "flex shelves" that extend calling features from the PBXs to some of Howard's smaller dorms.

Under Swygert's leadership, "the university has really adapted these technologies, and now there's an expectation that we'll have it and that it'll be operational," Moore says.

With all of the technological resources at faculty and students' disposal, "they have access to any kind of information that there is," says Clint Walker, Howard's interim director of user support services and director of iLab. "When you put all these things together, it gives students a lot of resources that let them complete any sort of assignments that they have."

Simon Fraser University

Located in Burnaby, British Columbia, just outside of Vancouver, Canada, Simon Fraser University serves close to 25,000 students, faculty, and staff. The university also has campuses in downtown Vancouver and in nearby Surrey.

The IT team at Simon Fraser University has been interested in wireless networking since 1994. That was the year that one of the university's IT directors presented a research paper outlining how the school soon would need a new building just to house the computers and computer carrels the students would need in the next decade. Over the years, according to Worth Johnson, the school's co-director of operations and technical support, the school came up with a solution—to enable students to bring their own devices on campus. He added, "We immediately began asking ourselves, 'What are the technologies that are likely to work?'"

The university set about to familiarize its end-users and its technical staff with wireless computing through pilot projects. The first was a proprietary point-to-point wireless system, which was tested in 1991; and throughout the 1990s the IT department experimented with different proprietary systems.

Then in 2001, the university, with the help of a Vernier Networks System, deployed a new pilot WLAN based on the 802.11b standard. Pleased with the overall results, the university extended the network over the summer of 2002, to provide more coverage including access points in the Faculty of Education building, the Applied Sciences building, in study areas, and in other interior and exterior areas on campus. The university's IT staff and Vernier specialists also installed ten access points at the university's downtown campus and another 30 access points at its Surrey campus.

The wireless network, which consists of a combination of Vernier Networks 5000 series and 6000 series Integrated Systems and Access Managers, enables students to bring their own computers on campus, where they can connect easily to the network without worrying about cables or plugs. The university even discovered that it could use the Vernier Networks System for AAA (Authentication, Authorization, and Accounting) on both its wireless and wired networks.

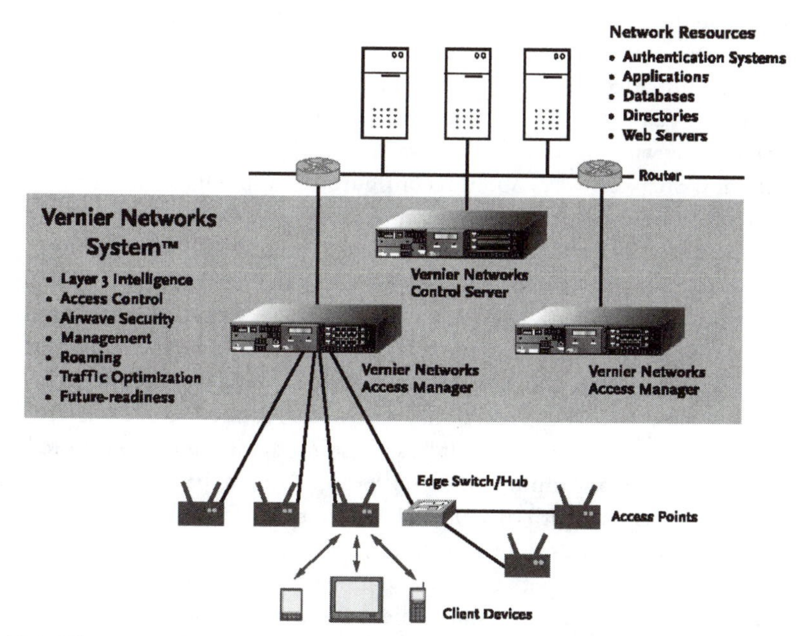

Figure 15.3 The Vernier Networks System includes a Control Server, which manages access rights for the entire network, and Access Managers that monitor, secure, and control traffic flowing through access points. *Graphic Courtesy of Vernier Networks.*

The WLAN meets the school administrator's primary requirement—whether a computing device is wired or wireless, an end-user can access the network through a simple sign-on procedure. The Vernier system allows faculty, staff, and students to use the same account and password combination (which each end-user is allowed to individualize) whether they are dialing in, or using the wireless network on campus. When an end-user moves from one location to another, they don't need to change their laptop configuration.

But the university had other requirements as well. Johnson explains, "We also wanted to support different platforms, including various PDAs. We wanted to have a single secure sign-on to authenticate users. We wanted to make sure that whatever human interfaces we provided would work across all devices, so a browser-based interface like Vernier Networks' made the most sense."

The new wireless network, with its secure login and support for roaming, has been well received. "The Faculty of Education loves the wireless network," says Johnson. The department has 35 laptops loaned out for workshops and for use by instructors.

Another user community that has responded enthusiastically to the network has been that of students working in the library with their laptops. "There's an area in the library with little rooms where people can meet to collaborate," Johnson reveals. "Those rooms are booked solid. Some of the best feedback we've received has been from students accessing the wireless network from those rooms. Now they can collaborate online. They're pleased with how well everything works."

Also pleased are the faculty at the university's business school and school of applied sciences, whose departments are located at opposite ends of the campus. These deans provided matching funds for wireless coverage in those areas. Their goal? To provide coverage zones running the length of campus, so that someone could walk from one departmental office to the other and maintain network access. The 2001 WLAN expansion meets this goal. The major traffic corridors on campus are covered by the 802.11b network, making wireless access widely available and thereby encouraging network use.

"They don't have to worry about reconfiguring computers. Everything just works. This lays the groundwork for the use of more computers in the classroom."

An important benefit of the Vernier Networks System has been its ability to provide network access to authenticated users whose devices are either configured for another network or misconfigured entirely. Johnson says, "The Vernier Networks System's adaptive networking technology and support of NAT turns out to be very important for us."

Johnson explains: "One of the inhibitors to introducing technology in the world of teaching is mistrust. Over the years, the faculty was required to use classroom equipment that was sometimes unreliable. It was not unusual for this equipment to be moved from room to room many times, and for its configuration to become inconsistent. Frustrated by their experience with this unreliable equipment, faculty members came to mistrust computers in the classroom altogether. And while they've continued to trust their own computers for research, they are wary of using computers for classroom instruction. One of our objectives is to provide network services that enable faculty to bring their own computers to the classroom, where the network can automatically adapt itself to each computer's configuration. This enables faculty to teach using the computers that they are familiar with and that they trust."

The Vernier Networks System enables faculty to trust their computers again. Johnson gives the example of a faculty member working in his office and calling up a website that he wishes to discuss in the classroom. The laptop is plugged into the wired network. It has been assigned a fixed IP address. Because Vernier Networks System supports Network Address Translation (NAT) and translates addresses from one network for another, the professor can unplug the laptop from the office's network port, walk across campus to the classroom, plug in the laptop on a different network, and continue accessing the website in front of the class. The Vernier Networks System translates the addresses between the subnets, and enables the professor to access the material he wants, without having to learn how to reconfigure the laptop for a new subnet. The professor gets to use his or her own laptop, a system with which he or she is already familiar, rather than an unfamiliar classroom system or a low-end system built from spare parts.

The faculty has responded enthusiastically to this new capability. "They can trust the network now," says Johnson. "They don't have to worry about reconfiguring computers. Everything just works. This lays the groundwork for the use of more computers in the classroom."

Once the basic wireless network was in place, the university's IT team began to develop new software that could take advantage of the Vernier Networks System's Layer 3 intelligence to improve the accounting system used for the university's print services. Like many universities, Simon Fraser University provides a central printing service for its students. When a student submits a job to be printed, the university's printing software identifies the job by the system name of the computer the student is using. Now that students bring their own computers to campus, though, the university has no way of ensuring that every computer is uniquely named.

The Vernier Networks System's Layer 3 packet inspection engine can identify the student ID associated with every packet traveling on the network, including packets headed for the printer. "We're looking into the idea of developing software based on the Vernier Networks System that would embed the student's authentication ID in the print job, so that every job would be clearly labeled with the ID of the student who sent it," says Johnson. This solution, which would provide a universal system for tracking print jobs submitted by wired or wireless computers, would not be possible without Vernier Networks' Layer 3 technology.

This software project is just the one of many in the ongoing partnership between Vernier Networks and Simon Fraser University. "Vernier Networks has been very responsive," says Worth Johnson. "They've been a good partner."

Stephen K. Hayt School, Edgewater, Illinois

This school, located near Chicago, wanted to provide students with enhanced learning opportunities wherever they might be on campus, by combining hands-on learning projects with wireless Internet access. The school administrators contacted a local consulting firm, Technocrats Consulting, Inc., that specialized in educational systems, to develop a plan to achieve this vision.

Technocrats worked with school personnel to develop the following goals:

* Implement mobile computing, so that lessons aren't confined to the classroom.

* Design a system that lets students face the teacher, not the wall where the network connection is located.
* Continue to utilize the investment made in Mac and PC computers and software.

To accomplish these goals required a single wireless networking system that could work with Hayt School's existing PCs and Macs. The solution Technocrats devised not only met the school administrator's requirements, but it was also quite cost-effective. The plan called for Proxim, Inc.'s Harmony wireless networking solution, which allows both Macs and PCs to be recognized on one wireless LAN system. With the Harmony system onboard, all of the school's 65 laptops (50 Apple iBooks; 15 Dell Inspiron PCs) can be used simultaneously on the same LAN from anywhere on the campus. Students can access the Internet, the school website, classroom files, lesson plans, and collaborative projects from any machine, anywhere on the school grounds.

The Harmony wireless network is perfect for the Hayt School. It supports 802.11b, it is flexible, it provides investment protection, and it can provide an easy and inexpensive migration path to new wireless standards. It maintains the same configuration and management interface, without disturbing the existing installation.

Deploying a Harmony-based 802.11b WLAN is also easy. The initial call to Technocrats was made on April 27, 2001, and on May 29th, students were logging onto the new wireless LAN from both Mac and Dell PC laptops. (To make sure the learning environment was not disturbed, technicians installed the entire wireless network over the three-day Memorial Day holiday.)

Of course, Technocrats first performed a thorough site survey and mapped exactly where each access point would be placed, the coverage range of each, and its channel designation. Then, all of the access point channels were pre-programmed, and Technocrats pre-customized the solution to ensure that both the PC and Apple iBook notebook computers could run seamlessly over one wireless LAN.

During the 3-day holiday, technicians positioned and installed the 49 Harmony AP (44 in classrooms and five in the auditorium, cafeteria and main office) so that they could provide coverage throughout the school. Because of their range, the 44 classroom APs actually covered 55 rooms. The Harmony system provides seamless roaming, so when an end-user walks across campus with a "logged on" laptop, he or she is not aware of the different channel or signal changes that occur as the computer moves connectivity from one AP to another.

Teachers and students really benefit from their newfound liberty. For example:

* The 3rd grade classes takes laptops into the cafeteria to take advantage of the larger space and create life-sized drawings of the human form. Then they use a correlating student Internet learning site to identify and label the different internal organs.
* Students from the 7th grade use their laptops to participate in a Northwestern University Collaborative Project. The project works with individual teachers, school project teams, and multi-school collaborations in school districts throughout Illinois. Students are able to use the resources of participating museums, libraries, and cultural institutions for innovative, web-based learning opportunities.
* One class created a "Wonderful World" environmental slide show, and saved the file on the server. They were then able to present their work in their classrooms and to other classes in the auditorium.

School administrators are looking toward the future possibilities of their new wireless LAN. They have a formal keyboarding program in the works that will enhance the students' ability to use the laptops to their full advantage. Because any hand-held Windows CE device can be recognized on their LAN, the school is also contemplating how those devices can be used to enhance the curriculum.

According to Linda Smentek, Technology Coordinator for Hayt School, "The new system has enhanced learning tremendously. The laptops integrate technology across the curriculum, and by having such mobility, you can teach in a much more hands-on way, wherever the lesson happens to be. We're excited about the new opportunities—a lot of which I'm sure we haven't even discovered yet."

CONCLUSION

Why do so many vertical industries make significant investments in wireless networking? Many experts try to make the reason for going wireless too simple by citing hardware cost savings—even the venerable research firm, Gartner, touts the total cost of ownership for wireless connections. Gartner says that a WLAN costs run a little more than $3000 per port, versus $5000 for wired links, with most of the savings coming from doing away with the need to pull cable through walls. But that's just part of the reason for the growing interest in Wi-Fi.

The burgeoning interest in all things wireless is fed by a new generation of devices and applications, which are changing technology strategies and increasing mobility opportunities. Executives are beginning to understand the efficiencies that wireless networking can bring about—resulting in impressive total cost of ownership and return on investment figures.

The trick is to not let the costs or the challenges that wireless networking presents get in the way. Instead, when considering going wireless, ask, "How can it be used to your organization's best advantage?" Once, you have the answer, you're on the Wi-Fi track.

Section VI: Optimizing A WLAN

Chapter 16: Providing Quality of Service

A typical enterprise network topology has multiple local area networks (LANs) connected by bridges or Layer 2 switches. Many times the different LANs are all of one type, typically wired Ethernet, but that is changing. Wireless networks are entering these hallowed grounds. Furthermore, today's networks are converging into a unique network infrastructure that is expected to carry data in the form of voice, video, and mission critical applications. These applications require a certain amount of bandwidth in order to provide the end-user with a quality product, and that's where Quality of Service comes into play. The term "Quality of Service" (QoS) refers to the quality of network services. It is sort of an umbrella term for all the related technologies that are put in place with the objective of providing QoS, so as to supply better network services in order to satisfy network applications' bandwidth needs. QoS is not related to a specific technology or network topology, but transparently crosses all networks—IP, Frame Relay, ATM, Ethernet, cellular, and Wi-Fi.

With the help of "Quality of Service" or "QoS" technologies, networks (and network managers) can give priority to certain data, users, and/or applications. This is typically accomplished via bandwidth management techniques that allow specific types of application to have specified bandwidth and delivery requirements satisfied. In a non-QoS-enabled network all data packets generally receive the same "best effort" service.

When QoS mechanisms are in place, an overburdened WLAN can be optimized so it can sufficiently service mission-critical applications, and queue less important data for transmission as network conditions allow. QoS allows an unresponsive, overburdened, or oversubscribed network resource to be sufficiently responsive so as to handle any necessary prioritized network traffic.

QoS techniques operate by providing not only the necessary bandwidth resources, but also controll of latency and jitter (which is required for interactive traffic), while improving a network's data loss characteristics. QoS technologies achieve these objectives by providing a large assortment of tools and protocols for managing and avoiding network congestion, applying bandwidth reservation, and instituting traffic shaping and policing techniques. For example:

* Enhanced queuing tools can differentiate incoming traffic in multiple queues to prioritize interactive applications (e.g. video packets) to provide the end-user with a quality video stream.
* QoS tools can manage and even avoid congestion by slowing bulk traffic applications such as large file transfers.
* QoS mechanisms can shape and police data traffic in order to provide a specific level of service between the data source and its end-users.
* Bandwidth reservation techniques that cross the network can provide a guaranteed end-to-end service for constant bit rate services like voice and video.

Historically, much of the data that networks move around has related to business applications where strict QoS measures are unnecessary, since these applications aren't materially affected by packet arrival latencies. But now, whether the network landscape consists solely of wired LANs, is wireless, or a mix of both, networks deliver a variety of applications, some of which are quite sensitive to QoS. So while delivering text and other relatively simple types of data around a network (wired or wireless) doesn't necessarily require complex Quality of Service mechanisms, network managers find that more and more they must deal with QoS issues, since many of today's applications differ in their network transmission requirements. Thus network managers are seeking quality of service technologies that offer guaranteed bandwidth while maintaining specific link quality, delay, and jitter parameters.

Note, however, that when considering Quality of Service in a networking environment, you also must allow for the human element. That is because, at a fundamental level, QoS services are based on human perceptual considerations.

QoS PARAMETERS

The important parameters that define QoS are bandwidth, link quality, delay, and jitter. However, not all applications have strict requirements for all of these parameters. For example, voice communication (telephony) has relaxed requirements for bandwidth (from about 8 to 64 Kbps), but has very demanding requirements for latency (from 5ms to 150 ms or so for packet communications) and jitter (5 ms).

We will now examine the four primary QoS parameters.

Bandwidth is the measurement of the communication speed at the OSI's Physical Layer. Typically it is measured in bits per second, although sometimes frames-per-second can be used. A frame, however, is an application dependent unit and varies across video, voice, and audio applications. In general, if the wireless network is to be used for the delivery of multimedia applications, it must offer guaranteed bandwidth provisions for uninterrupted delivery of the audio/video applications. Depending on the quality of video, bit rates from 300 Kbps (e.g. streaming video) to 20 Mbps (e.g. high definition video such as HDTV) are needed. This isn't as difficult as it sounds.

Link quality refers not only to the amount of bandwidth that is needed to enable quality connectivity, but also to the ability to ensure that a significant number of bits (or frames or packets) arrive error-free. The ideal network would also provide the ability for error-riden frames or packets to be corrected through such mechanisms as automatic repeat request (ARQ), frame-error-rate, or packet-error-rate. Since wireless networks are designed to be just 90 percent error-free, this can present a real quandary. Link quality is usually measured in bit-error-rate and is expressed as a negative exponent of 10 (e.g. 10-9).

Delay, also known as *latency*, refers to the time difference between the transmission and reception of packets of information, from the source to its destination. Interactive applications, such as telephony, online gaming, and videoconferencing, are particularly sensitive to delay. For such applications, a delay of less than 100 milliseconds (ms) is needed end-to-end. Here it is important to note that the wireless network receives only a portion of this end-to-end delay budget. Values in the range of 5 ms to 10 ms typically are allocated for the wireless network portion of the link.

Jitter, also known as *delay variation*, refers to the variation in arrival (latency) of data packets. Jitter is an important consideration in isochronous applications (i.e. applications whose signals are dependant on some uniform timing—many carry their own timing information embedded as part of the signal), because these applications must generate or process regular amounts of data at fixed intervals. Examples of such applications include telephones that send and receive voice samples at regular intervals, and fixed-rate video that generates data at regular intervals and thus must receive data at regular intervals. Often, the method used to minimize jitter employs buffers to control the delay variation at the receiver end. This method, however, translates the jitter problem into a latency problem, and thus is only useful for non-interactive applications such as streaming video. A typical maximum jitter requirement for high quality video is about 10 ms—this is difficult for a WLAN to provide, but not as troublesome as provisioning link quality over a WLAN environment.

IP NETWORKS

Provisioning QoS in an IP network (Wi-Fi, Ethernet and the Internet are all IP networks) is all the more difficult because these networks are "dumb." In most instances, designers focus on provisioning QoS via more bandwidth, rather than on building more intelligence into the network. Here's why.

TCP/IP is built into the UNIX operating system and is used by the Internet, making it the de facto standard for transmitting data over networks. TCP/IP is a suite of communication protocols allowing communication between groups of dissimilar computer systems from a variety of vendors. This suite of protocols is designed to ignore the kind of traffic it is carrying, and to concentrate on getting that traffic to the right place.

IP (Internet Protocol) moves packets of data between nodes. TCP (Transmission Control Protocol) is responsible for verifying delivery from client to server, and so it provides packet retransmission in case of failure. However, UDP (User Datagram Protocol), which is also part of the TCP/IP suite, doesn't retransmit missing packets, leaving it up to the application that's using UDP to decide whether it needs the missing pieces or not.

Networks (wired and wireless) are "bursty" by nature. This is because data that traverses the network isn't constant—network users do not constantly upload and down-

Application	Telnet, FTP, IRC, RPC, SNMP, etc.	This layer is where users typically interact with the network.
Transport	TCP, UDP	Provides data flow for the Application Layer. This is the also the layer where guarantees of reliability may be introduced.
Network	IP, ICMP, IGMP	Responsible for determining how data reaches its destination but no guarantee data will arrive since it only decides where the data is to be sent
Data Link	Network interface and device driver	Responsible for communicating with the actual network hardward (e.g. NIC card)

Figure 16.1 TCP/IP stack as it relates to the Layers of the TCP/IP Network Model (note that this model has fewer layers than the ISO/OSI Reference Model, but they encompass the same things).

load data. For example: if you are viewing a web page and you hit the enter key, data is transmitted, and the image and text load. Once loaded, the transmission stops while you read the web page. At that point the bandwidth is again sitting idle, and utilization is back at zero. This cycle continuously repeats itself throughout the networking environment. Furthermore, since packet loss is often the result of congestion or rerouting in the network, packet loss commonly occurs during bursts of activity.

The degree of a network's "burstiness" impacts the network's ability to provide QoS, especially voice quality. While the packet loss-concealment algorithms used by most Voice over IP (VoIP) implementations (and even other types of bandwidth-sensitive applications) can disguise isolated lost packets, they are much less effective when a series of packets is lost. It doesn't take a lot of dropped packets in a voice call to make a conversation sound choppy (or for a video stream to delay and jitter). Of course, with the right design and attention to outside influences (interferers, blockage, etc.), a substantial majority of collisions or dropped packets can be reduced, increasing a network's "quality" throughput quotient.

With VoIP, however, the problem of QoS transcends "bandwidth." For example, in the May/June 2002 issue of *IEEE Internet Computing* is a discussion that deals with VoIP QoS in detail; however, the emphasis isn't on bandwidth, but on delay and jitter. That's because just 56 Kbps of bandwidth is needed to convey an uncompressed G.711 voice signal through a network. But when you take that voice stream, chop it up into packets, and push it through a network that is also being used for data transport, video delivery and other data, problems can and do occur between the transmitting node and receiving node. This makes it difficult to guarantee that the voice packets will arrive at their destination within an average of 10 ms of each other. If you can't ensure that those packets will arrive within 10 ms of each other, the voice quality degrades below that defined by G.114 and G.131 (the standards for voice telecommunication established back in the 1950s).

Many recent packet-switched voice systems have relaxed the end-to-end packet delay tolerances to about 150 ms; such longer delays, however, should be the exception rather than the rule. (Human hearing tolerates delays up to 150 ms.)

Coping with Burstiness

In principle, IP networks, and thus Wi-Fi networks, can support interactive, multimedia traffic, and even the long-heralded convergence of the communications infrastructure. This "convergence", however, that has been delayed—in all IP networks—due to a lack of adequate capacity to support "real-time" services.

In recognition of the growing need for support of bandwidth intensive real-time services such as voice-over-IP, network managers have begun considering an investment in substantial upgrades to the transport capacity of their network backbone. But do they really need to take that step? The jury is still out on that question.

Some experts suggest that overprovisioning is the optimal approach to addressing the QoS issue. For example, Sprint says that none of their links are more than 40 percent filled with traffic, thus eliminating any congestion on the backbone. But other experts argue that capacity will remain scarce, and hence it is worth implementing bandwidth economization methods by means of QoS techniques. Also, as you will learn, overprovisioning will not always solve a QoS issue in a Wi-Fi network.

A second concern is the optimal utilization level for the network. IP network traffic is more bursty than voice telephony traffic, which suggests that the optimal utilization rate for an IP network is likely to be lower than for a telephone network, because of the need to provision to handle peak loads (or else suffer quality of service degradation during peaks). These two issues are related because the lower the optimal utilization level, the greater is the required investment in overprovisioning in order to sustain any given level of quality of service. The greater the cost of using overprovisioning to address the QoS problem, the more attractive are the mechanisms that facilitate efficient allocation of scarce capacity.

A standard mechanism for addressing the problem of bursty traffic is to aggregate sources in the hope of smoothing peaks (i.e. peaks are uncorrelated). The potential need to support high-bandwidth, time-sensitive traffic means that even relatively small-scale economies may be quite valuable. Interestingly, if the network is used extensively for high-bandwidth, time-sensitive applications (streaming audio and video in particular), the addition of such services may actually alleviate the "bursty" problem, since by their very nature these applications reduce the burstiness of the aggregate traffic.

QoS FOR WLANS

Wi-Fi effectively handles data services, but now it must show that it can support real-time traffic. This requires QoS functionality. To deliver QoS effectively via a wireless network, the network must be designed so that QoS can be maintained across the network ecosystem—wired and wireless—in an end-to-end manner (source to destination). A bottleneck at any juncture can nullify one or more of the QoS provisions.

When you consider QoS in the Wi-Fi network, you must realize that it has always been limited to the radio frequency. Today, when a network manager provisions end-to-end QoS over a wireless network, a policy management server must be installed so that it can aggregate network traffic. That may soon change.

The engineering community has decided to provide QoS via data packet prioritization so that important packets are more likely to make it through an end-to-end QoS implementation. This could be, for instance, from a Wi-Fi-based VoIP phone through a

DSL connection to a telco termination to the PSTN, or a video signal from a cable modem over an 802.11a connection to a receiver.

***Note:** A "flow" usually refers to a combination of source and destination addresses, source and destination socket numbers, and the session identifier, but the term can also be defined as any packet from a certain application or from an incoming interface, which is how the term is used in this chapter.*

This is where the Institute of Electrical and Electronics Engineers (IEEE) enters the picture. Its 802.11 Working Group realized the need for some QoS parameters for Wi-Fi to reach the comprehensive, converged communications usage that seems to be its destiny. That Working Group formed the 802.11e Task Group to draft a specification that would ensure that time-sensitive multimedia and voice applications could be sent over a Wi-Fi connection without jitter or interruptions. Thus Task Group 802.11e (also known as TGe) is an effort on the part of the IEEE 802 standards body to define and ratify new MAC functions, including QoS specifications for wireless communication protocols in the 802.11 family of standards. But at the same time, the Task Group is also ensuring that providing priority for one or more traffic flows does not make other flows fail.

Wi-Fi's QoS Contributions

Although many complain about Wi-Fi networks' lack of Quality of Service, there are some provisions for QoS in the 802.11 specifications. As Wi-Fi technology and its wireless networks gain acceptance in many different networking environments, some may

ISOCHRONOUS APPLICATIONS

Voice and video services require what's known as "isochronous data transfer" for high-quality operation. For example, transferring an analog voice signal across a digital network involves using an analog-to-digital converter to continuously sample the input voltage waveform to produce a series of digital numbers (perhaps of 8-bit precision). This series of samples is then transmitted across the network, where a digital-to-analog converter transforms the series of digital samples back into a voltage waveform.

For such a scheme to work well, the network transporting the voice data must not introduce significant amounts of delay in packet arrival times. Minimal delay is easy to achieve when there is little other competing traffic on the network. It's when the network is heavily loaded that it's difficult to deliver low levels of delay.

Not all information that needs to be transferred across a network is equally impacted by network latency. Packets containing pure data (as opposed to voice or video) can be delayed significantly, without seriously affecting the applications they support. For example, if a file transfer took an extra second because of network delay, it's doubtful that most users would even notice the extra transfer time. With voice and video, however, milliseconds count.

The general strategy for minimizing delay in voice and video packet transfer is to institute some sort of priority scheme, under which such latency-sensitive packets get sent ahead of the less time-critical traffic and receive "preferred treatment" by network equipment.

forget that the main characteristic of all 802.11 networks (a, b, and g) is their distributed approach, which provides simplicity and robustness against failures. But, as you will learn, that distributed approach sometimes hinders QoS solutions.

For example, 802.11b provides two methods of accessing the medium. One is called the distributed coordination function (DCF), and the other is called the point coordination function (PCF).

DCF could be defined as the "classical Ethernet-style network media access," i.e. individual stations contend for access. DCF is actually a type of slotted Aloha scheme that is based on a CSMA/CA rule. (CA stands for "Collision Avoidance" verses Ethernet's Collision Detection scheme.) Collision avoidance is needed because wireless devices do not "listen" to the medium at the same time they are sending, whereas wired devices do.

PCF acts upon a single point in the network (most likely, a network access point) like a centralized "traffic cop," telling individual stations when they may place a packet on the network. In other words, PCF acts a lot like a token-based procedure. The wireless device requests the medium, the request is queued, and when the access point (AP) gets to that position in the queue, it is given a token to talk for a period of time. The scheme means that a wireless device can predict when it is to get the medium and how long. That's where QoS comes in—predictability lends itself to QoS.

However, since most Wi-Fi networks and devices use DCF, there is little or no experience with PCF usage in a Wi-Fi network for QoS. Furthermore, even though PCF could support QoS functions for time-sensitive multimedia applications, this mode interjects three other issues into the QoS paradigm—any or all of which could lead to poor QoS performance:

- Its inefficient and complex central polling scheme causes the performance of PCF to deteriorate when the traffic load increases.
- Incompatible cooperation between modes can lead to unpredictable beacon delays. At Target Beacon Transition Time (TBTT), the Point Coordinator schedules the beacon as the next frame to be transmitted, and the beacon can transmit when the medium has been found idle for longer than a PCF interframe space interval. Depending on whether the wireless medium is idle or busy around the TBTT, the beacon frame may be delayed. The time for which the beacon frame has been delayed, i.e. the duration it is sent after the TBTT, defers the transmission of time-bounded MSDUs (MAC Service Data Unit) that have to be delivered in the Contention Free Period mode. However, since wireless computing devices start their transmissions even if the MSDU transmission cannot finish before the upcoming TBTT, QoS performance can be severely reduced by introducing unpredictable time delays in each Contention Free Period. In the worst case, the maximum beacon frame delays of around 4:9 ms are possible in 802.11a.
- Transmission time of the polled computing devices is unknown. A device that has been polled by the Point Coordinator is allowed to send a frame that may be fragmented into a different number of small frames. Furthermore, in 802.11a different modulation and coding schemes are specified, so in 802.11a networks the transmission time of the MSDU can change and is not under the control of the Point Coordinator. These issues can prevent the Point Coordinator from providing QoS guarantees to other stations that are polled during the remaining Contention Free Period.

Another way to create QoS is with a network access point (a telecommunications term for the location where data enters the telecommunications network, not a wireless access point) somewhere past the wireless access point and use PPP (point-to-point protocol) and/or a VPN (virtual private network) to access the network access point, and let it grant various QoS parameters.

Wired is not Wireless

Several characteristics of wireless communication pose unique problems that do not exist in wired communication. These characteristics are:

- Mobility.
- Highly variable, location-dependent, and time varying channel capacity.
- Burstiness.
- A tendency toward high error rates.
- Limitation in channel capacity, battery, and processing power of the wireless device.

To provide the Quality of Service needed in a wireless environment, there must be mechanisms in place that can account for and adapt to such wireless specific issues. For example:

- Proportion of usable bandwidth depends on the number of hosts. So, to support QoS, a limit must be placed on the number of users per cell.
- To deal with channel quality degradation, the geographical span of the cell must be limited.
- To support QoS, a limitation must be placed on the rate of traffic sources.

The root of the latency problems in WLANs is not only that all end-users share the same link, but that a collision can effectively prevent any party from accessing a channel. Furthermore, 802.11 currently has no method to decide which client to service and for how long (e.g. no idea of the flow rate or number of backlogged packets for arbitrary traffic distributions). The binary exponential back-off scheme leads to unbounded jitter.

Moreover, 802.11e only provides the hooks to implement QoS. It is up to the packet scheduler in the 802.11 access point to not only provide throughput, delay bound, and jitter bound, but also to efficiently utilize the channel. A scheduling scheme can deliver QoS from a tighter form of service differentiation to explicit guarantees within the practical channel error rates.

***Note:** "Delay bound" refers to fixed and variable delay; "jitter bound refers to variance in packet delay; and "delay-jitter bound" refers to a bound between the maximum and minimum delays that a flow's packets experience.*

However, you can support QoS in a WLAN by placing a limitation on the rate of traffic sources, by using statistical differentiation such as DiffServ's priority queuing (PQ), weighted fair queuing (WFQ), and traffic shaping mechanisms. (See Fig. 16.2.) Let's look at what these differentiations mean.

Priority queuing (PQ): This technique gives strict priority to important traffic by ensuring that prioritized traffic gets the fastest handling at each point where PQ is used. Priority queuing can flexibly prioritize according to network protocol (for example IP, IPX, or AppleTalk), incoming interface, packet size, source/destination address, and so on. In PQ, each packet is placed in one of four queues—high, medium, normal, or low—based on an assigned priority (some systems use just three categories). Packets that are not classified by this priority list fall into the normal traffic queue. During transmission, the algorithm gives higher-priority queues absolute preferential treatment over low-priority queues. But although PQ is useful for making sure that mission-critical traffic traversing various wide are network (WAN) links gets priority treatment, most PQ techniques rely on static configurations and therefore do not automatically adapt to changing network requirements.

Weighted fair queuing (WFQ): For situations in which it is desirable to provide consistent response time to heavy and light network users alike without adding excessive bandwidth, the solution is flow-based WFQ. In other words, a flow-based queuing algorithm that creates bit-wise fairness by allowing each queue to be serviced fairly in terms of byte count. For example, if queue #1 has 100-byte packets and queue #2 has 50-byte packets, the WFQ algorithm will take two packets from queue #2 for every one packet from queue #1. This makes service fair for each queue—100 bytes each time the queue is serviced. WFQ ensures that queues do not starve for bandwidth and that traffic gets predictable service. Low-volume traffic streams, which comprise the majority of the average network's traffic, receive increased service, transmitting the same number of bytes as high-volume streams. This behavior results in what appears to be preferential treatment for low-volume traffic, when in actuality it is creating fairness. WFQ is designed to minimize configuration effort, therefore it can automatically adapt to changing network traffic conditions.

Traffic shaping: Refers to techniques that create a traffic flow that limits the full bandwidth potential of the flow(s). Traffic shaping is used in many ways, for example, when a central site that normally has a high-bandwidth link (e.g. a T-1 line), while remote sites have a low-bandwidth link, (e.g. a DSL line that tops out at 384 Kbps). In such a case, it is possible for traffic from the central site to overflow the low bandwidth link at the other end. With traffic shaping, however, you can pace traffic closer to 384 Kbps to avoid the overflow of the remote link. Traffic above the configured rate is buffered for transmission later in order to maintain the configured rate.

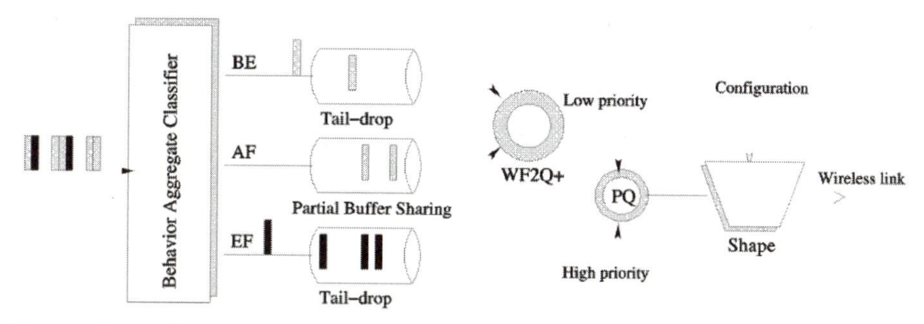

Figure 16.2 DiffServ can limit the rate of the WLAN's traffic sources in order to support the network's QoS needs. It does this by dividing traffic into three service categories: BE (Best Effort), AF (Assured Forwarding), and EF (Expedited Forwarding). BE and AF traffic is forwarded using WFQ (Weighted Fair Queuing), and EF traffic is forwarded using PQ (Priority Queuing).

The optimal solution would be to provide service differentiation that adapts itself to channel and traffic conditions while providing fair access to all wireless computing devices. This could be done by (1) QoS differentiation, (2) statistical QoS guarantees, and (3) absolute QoS guarantees.

To implement such a solution requires that there be a method to incorporate a *distributed* fair queuing technique into the QoS enabling mechanism. In distributed environments, like wireless networks, each wireless computing device operates independently of each other without a central control point where the complete knowledge of demand, available resources, and current channel characteristics could be known. Because of this distributed nature, the ideas of admission control and resource allocation need to evolve into a distributed methodology where each computing device operates independently and makes decisions cooperatively for admission, resource reservation, channel access, and data transmission.

Note: *A distributed environment distributes some of the MAC functionality onto a main processor system. This requires sufficient host processor and associated resources, such as memory, to be available to partition MAC functionality. A typical application would be Wi-Fi NICs, where the computing device's processor is available to run a driver. Thus in a distributed environment the MAC driver takes on more of the functionality of the MAC, including 802.11 MAC fragmentation/defragmentation, power management, encryption/decryption, and queue management. Also, the MAC driver requires more memory to support the power save queues, QoS queues, etc., and more MIPS (millions of instructions per second) from the system processor to perform tasks such as fragmentation/defragmentation and encryption/decryption.*

Next, adaptive protocols must be in place to assess, account, and adjust for the changes and fluctuation in wireless characteristics, mobility, error rate, and capacity. Also, energy efficiency in the MAC QoS protocol needs to be investigated. Finally, cross-layer QoS mapping should be investigated in order to accurately translate the user requirements into a suitable QoS parameter in the lower layers. And that is just for the wireless sector.

END-TO-END QoS

Voice and multimedia applications require end-to-end QoS. As the reader should now understand, in the communications industry, "Quality of Service" describes the treatment data receives over a network. QoS may be applied at various points that can serve as "bottlenecks" or "points of aggregation," to give precedence to specific data to ensure quality across those points. However, appropriate levels of QoS must be applied throughout the network to ensure consistent treatment. This is known as "end-to-end QoS."

From the overall network service perspective, QoS should provide end-to-end traffic control, so that users' applications can be properly served according to the allowable quality requirements, such as latency, jitter, and packet loss rate. To comply with the service quality requirements, user level traffic of the applications should coordinate QoS traffic control with transport level QoS at the network interfaces. 802.11e working alone can't provide this. However, an end-to-end QoS architecture across wired wide area networks, wired local area networks, and wireless LANs, which are based on DiffServ, 802.1D/Q, and 802.11e, respectively, could provide the level of service necessary to support quality real-time traffic.

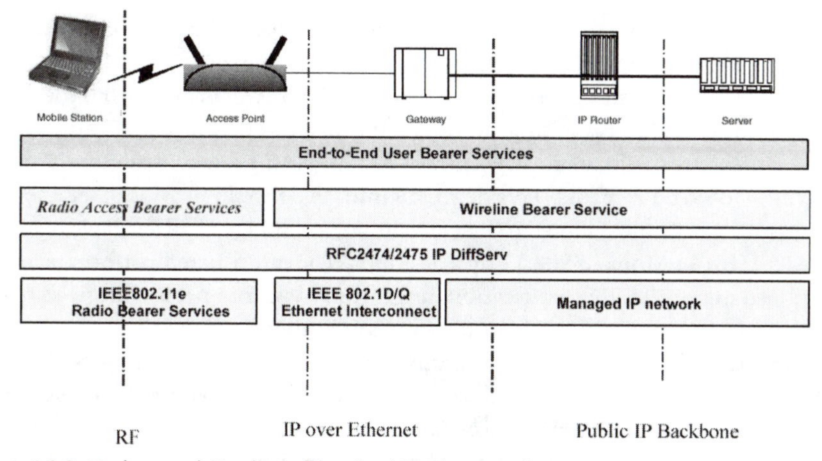

Figure 16.3 End-to-end Quality of Service (QoS) network structures.

Consider Fig. 16.3, where the user traffic QoS specifies end-to-end network traffic delay, jitter, and policing. Since most data services accessed by remote servers carry user traffic through multiple heterogeneous networking environments, the user level QoS should be decomposed into each network interface segment as follows:

Radio access network allows air interface between a mobile station (wireless computing device) and an access point (AP) as defined by 802.11e. Under the prioritized QoS paradigm of 802.11e, there is differentiated channel access to traffic with eight different priority levels.

Wired LAN is the Ethernet interconnection between AP and gateway terminating subnet traffic, including traffic through other MAC bridges. 802.1D/Q specifications define the QoS mechanisms that can be used in a wired LAN. Similar to a WAN's DiffServ, 802.1D/Q also classifies the traffic type and prioritizes on the traffic class at the bridge.

Managed IP WAN is a wireline interface managed by network service providers such as a telco. The Managed IP WAN is normally managed by a Service Level Agreement (SLA). A user's traffic DiffServ parameters can be directly reflected into the IP routers, so that traffic can be prioritized over other traffic, if necessary.

802.11 WLAN will utilize mechanisms set forth in the 802.11e specification, when it is ratified. The specification is virtually complete and should be ratified before 2004, thus there have been some reports about the utility of 802.11e. These reports, however, have been focused on QoS issues at the Layer 2 single hop, i.e., the wireless link between a computing device and an access point. Yet, most user traffic in the Application Layer traverses multiple networks, which means that end-to-end QoS is crucial.

802.11e

As the multimedia applications (e.g. voice over IP) and streaming media (e.g. audio/visual streaming via the Internet) emerge, and as portable devices such as laptops and PDAs become more and more popular, interest in having a level of service similar to those available from the conventional wired networks grows. But the wireless network's architecture must extend so it can handle such applications. The emerging 802.11e standard just may be the ticket for the air interface for such multimedia applications.

In fact, the main push for QoS features over a Wi-Fi network comes from the 802.11e specification. The 802.11 Working Group established the 802.11e Task Group and gave them the mandate of enhancing the current 802.11 MAC protocol to support applications with QoS requirements. This new QoS initiative should provide several levels of performance specifically tailored to take maximum advantage of all available data rates, from 802.11b's 11 Mbps to 802.11a/g's 54 Mbps. The multi-tiered performance levels provisioned in the 802.11e specification range from basic Internet applications to high quality video streaming and support for voice applications.

The basic Wi-Fi media access mechanism is to listen before sending, with slotted random backoff (i.e. CSMA/CA) and packet-by-packet acknowledgment (i.e. automatic repeat request or ARQ). This mechanism is used because today's radios can't transmit and receive at the same time, so that detecting collisions is difficult. Instead of collision detection (as used in 802.3 wired networks, ergo their use of CSMA/CD rather than CSMA/CA), a positive acknowledgment is sent for each correctly received packet. The sending device knows to retransmit the packet if an acknowledgment is not received. Furthermore, since WLANs often operate with packet loss rates of around 10 percent because of noise and interference, quick retransmission at the physical level is necessary. As shown in Fig. 16.4, 802.11e enhances this basic media access mechanism to provide QoS. Different traffic streams can be given different priority access to the wireless channel by adjusting their initial waiting period (i.e. Arbitration InterFrame Space or AIFS), and their random backoff contention window (i.e. CW min and CW max).

In order to support both IntServ and DiffServ QoS approaches (see Text Box) in Wi-Fi networks, 802.11e defines a new mechanism called hybrid coordination function (HCF). This mechanism is backwardly compatible with basic DCF/PCF. It has both polling-based and contention-based channel access mechanisms in a single channel access protocol. Thus the HCF combines functions from the DCF and PCF with some enhanced QoS-specific mechanisms (EDCF and EPCF) and QoS data frames in order to allow a uniform set of frame exchange sequences to be used for QoS data transfers.

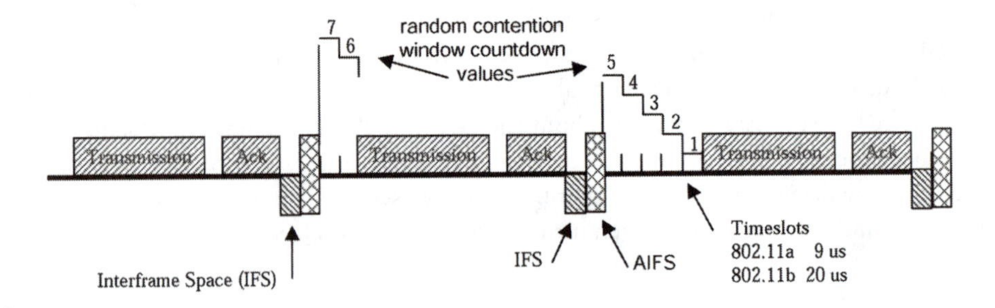

Figure 16.4 How the 802.11 MAC's basic mechanism is being enhanced in 802.11e to provide QoS.

The HCF uses Enhanced DCF (EDCF) as a contention-based channel access, which operates concurrently with a controlled channel access based on a poll-and-response protocol. EDCF is designed to provide differentiated, distributed channel accesses for frames with eight different priorities (from 0 to 7), by enhancing the legacy DCF. Each frame from the higher layer arrives at the MAC along with a specific priority value. Each QoS data frame also carries its priority value in the MAC frame header. In the context of the 802.11e standard, the priority value is called Traffic Category Identification (TCID). An 802.11e computing device implements four access categories, where an access category is an enhanced variant of the DCF with a single FIFO queue. Each frame arriving at the MAC with a priority is mapped into an access category. This relative priority is rooted from the IEEE 802.1D bridge specification.

More specifically, EDCF uses priority-based service differentiation, but is still statistical in nature (which means that no absolute quality guarantees can be made), while Enhanced PCF (EPCF) allows parameterized support of QoS. Both PCF and EPCF provide the means for guaranteed bandwidth management, including support for various types of traffic, but these are optional features formulated by the 802.11e Task Group. Such option features still require CSMA/CA legacy functions.

The 1394 Trade Association Gets Involved

The 1394 Trade Association is an industry forum that develops specifications for implementation of IEEE 1394 high performance serial bus technology based on the IEEE 1394 series of standards. The IEEE 1394 series of standards provides the necessary bandwidth for audio and video content to be sent from one 1394 device to another, essentially using bridging products. Since IEEE 1394 was originally designed as a serial bus or pathway between computer peripherals, the 1394 series defines a digital serial network for interconnect by cable for a wide range of devices that employ diverse protocols and data types, including IP, MPEG, and digital video data.

Note: IEEE 1394 provides for a single plug-and-socket connection on which up to 63 devices can be attached with data transfer speeds up to 400 Mbps. While the standard describes a serial bus or pathway between one or more peripheral devices and your computer's microprocessor, because IEEE 1394 is a peer-to-peer interface, one camcorder, for example, can dub to another without being plugged into a computer. Many peripheral devices now come equipped to meet IEEE 1394.

Wired IEEE 1394 proponents seeking the path to wireless are taking a serious look at 802.11e developments. In fact, the 1394 Trade Association's Wireless Working Group (WWG), which is chartered to deliver their industry's standard into the wireless domain, has jumped into 802.11e group activities. According to Peter Johansson, a project leader of the 1394 Working Group, this was because the WWG felt that there was a need for "a lot of explaining and education on what's needed for QoS tailored for asymmetric data traffic like audio and video." He went on to explain that the 1394 Trade Association is looking to add so-called "express traffic," which Johansson defines as "a mechanism for a prior allocation of time for access to a channel." While this isn't a guaranteed access to the radio in the strict telco sense, "we are asking that when a radio is available we get a shot at it," he said.

***Note:** Peter Johansson is also the IEEE P1394.1 chairman and author of the proposal for a method to improve IEEE 802.11 QoS for audio/visual streams by applying fundamental principals and mechanisms based on the IEEE 1394 architecture.)*

WHAT IS 1394?

The 1394 series of standards, which consists of (but is not limited to) 1394, 1394a, 1394b, and P1394.1, was originally designed to as an I/O InterConnect interface mainly for external digital devices. The IEEE 1394 interfacing standard's high data rates allow for real-time video transfer, and is extensively used in interconnectivity solutions for products such as PCs and PC peripherals, video cameras, video recorders, DVD players, and set-top boxes (STBs).

To understand the importance of the 1394/802.11 partnership you need to understand that 1394 is:

- A hardware and software standard for transporting data at 100, 200, or 400 megabits per second (Mbps).
- A digital interface—there is no need to convert digital data into analog and tolerate a loss of data integrity.
- Physically small—thin serial cable can replace larger and more expensive interfaces.
- Easy to use—there is no need for terminators, device IDs, or elaborate setup.
- Hot pluggable—users can add or remove 1394 devices with the bus active.
- Scaleable—may mix 100, 200, and 400 Mbps devices on a bus, flexible topology—support of daisy chaining and branching for true peer-to-peer communication.
- Inexpensive—guaranteed delivery of time critical data reduces costly buffer requirements.
- Non-proprietary—there is no licensing problem.

The 1394a standard is a supplement to the orginal 1394 specification. It supports data rates up to 400 Mbps over shielded twisted pair cable (not Category 5 cable) up to 15 feet, after which data rates fall off considerably. Nonetheless, a single 1394a cable could deliver data at speeds up to 100 Mbps up to 70 feet and through the use of $40 off-the-shelf 1394a repeaters, the distance could be extended and data rates accelerated to 400 Mbps.

1394b standard, which the IEEE ratified in April 2002, complements the IEEE's 1394a standard, and is intended mainly to replace multiple audio, video, and control cables with a single multiple-conductor cable to connect components in a cluster of components. The new standard also can be used to connect the clusters.

P1394.1 is a proposed high performance serial bus bridge that specifies the connection between multiple 1394 buses. It is being designed to allow high bandwidth subnets to be clustered together without affecting the performance of the overall interconnect.

Since 1394 is defined by the high level application interfaces that use it, not a single physical implementation, part of that backbone could be wireless. Magis, a supplier of wireless 802.11a chipsets, has demonstrated 802.11a equipment that simultaneously supports the 1394a and 802.11a protocols.

The lobbying effort appears to have worked. The 802.11e Task Group has embraced the initial QoS concepts for audio/video (A/V) streaming proposed by the 1394 Wireless Working Group. Acceptance enables the 1394 Wireless Working Group to develop a 1394 protocol adaptation layer for devices using the 802.11e QoS provisions.

Johansson also asserts that the IEEE 802.11e Task Group "looks at things from an Ethernet perspective." And that the 802.11e Task Group has "focused mainly on Voice over IP (VoIP) over 802.11 networks." But, as Johansson explains, "voice characteristics are very different from A/V [audio/visual]. . . and the data arrives differently." Voice doesn't require the same type of QoS guarantees as A/V because data loss is less of an issue.

The same cannot be said for multimedia, which requires a high quality, smoothly displayed multimedia stream, meaning that all of the data must arrive at the same time. But it is believed that convenient and cost effective distribution of multimedia content can be achieved by bridging 1394 content between clusters of devices using 1394/802.11 bridge devices.

The goal of especially the 1394 Trade Association, but also the 802.11e Task Group, is to apply 1394 QoS behaviors to 802.11 by addressing the problems encountered with intensive multimedia delivery over what is essentially an Ethernet standard evolved for the wireless world. Johansson says that both the 1394 and 802.11e groups "concur on the fundamental QoS concepts necessary for high-quality audio and video streams, such as scheduling and channel access. Wireless 1394 to some extent is an oxymoron." He further explains that it's more of the paradigms and behaviors of data transfer that are applicable to wireless LANs than the actual 1394 technology as it is implemented by PC manufacturers for wired devices.

Wireless 1394

Wireless 1394 over 802.11a/e could serve as the wireless backbone of a multimedia network. But for this could occur, a number of details must be worked out. For instance, the 1394 Groups must:

* Design a protocol adaptation layer (PAL) specific to IEEE 802.11.
* Specify bridges to interconnect the wired and wireless domains. This would allow firmware developed for (wired) IEEE 1394 products to migrate to wireless domain. It would also minimize reengineering between wired and wireless domains.
* Complete draft standard IEEE P1394.1 and profile the details of P1394.1 specifically applicable to bridges for IEEE 802.11.

1394 PAL would support IEEE 1394 Transport Layer functions; support isochrony and streaming data; coexist with other users of the underlying IEEE 802.11 transport; behave "like" IEEE 1394, and conceal differences between IEEE 1394 and IEEE 802.11 Physical and Data Link Layers. In turn, 1394 PAL for IEEE 802.11 will be able to permit wireless devices to talk to each other via wired to wireless bridges with the bridges isolating Physical and Data Link Layer differences in each domain from the other. It also will enable wired across wireless via bridges, enabling a wireless domain to serve as the backbone to connect wired clusters in different areas.

Toward that end, the 1394 Trade Association developed a document that specifies methods to (1) mimic IEEE 1394 infrastructure (transactions, isochrony, stream data, configuration ROM and CSR architecture) using the facilities of IEEE 802.11, and (2) implement IEEE P1394.1 bridge behaviors in the same domain. The document also states that the methods are to be compatible with the simultaneous use of IEEE 802.11 by other protocols, e.g., IP.

Steve Bard, chairman of the 1394 WWG explains, "PAL is a piece of software or firmware or an application that accepts traffic from 1394, [and] maps it to a domain such as 802.11. The ability is to bridge between previously interoperable domains." He goes on to say that, of course, to do so involves cooperative effort between the 1394 WWG and the 802.11e Task Group.

The Wireless 1394 specification is intended to spell out support for "two different logical networks over one medium," says Bard. He also points out that, "several companies in consumer electronics industry are developing a 1394/802.11 bridge," which he describes as "a small box with a 1394 socket. Plug it into your device's 1394 connection. It gets power off the 1394 connection, but also has an 802.11 radio in it. The PAL is the firmware translation layer that bridges the two domains. One box doesn't do much, but two of them solve interesting problems."

The Goal of the 802.11/1394 Partnership

It is hoped that IEEE 1394 will enhance QoS for IEEE 802.11, by establishing QoS based on scheduled management of channel access guarantees. The behaviors of 1394 that the WWG seeks to emulate for 802.11 involve things like accurate clock distribution, connection management protocols, and command sets. According to Johansson, the 1394 Working Group's goal is to "complete a spec showing how you reproduce 1394 behaviors." Johansson goes on to say that ultimately, 1394 will provide a PAL for devices using the 802.11e QoS provisions. And if you look at the 1394.1 draft specification, it provides a logical description on how to bridge 1394 into an 802.11a network. It also defines how the 1394 Trade Association's work on PAL for 802.11 (which is still a work in progress) will further smooth data construction and timing integration between wired and wireless networks.

Audio and video experts agree that 802.11 must address MAC services encompassing scheduled access to the radio channel. James Snider, the 1394 Trade Association's executive director, comments that the experience gained with wired 1394 devices can be applied, and the firmware adapted, for use in the wired or wireless environment. The final outcome of using a 1394 protocol adaptation layer in devices using 802.11e will be data transfer methods that are suitable for multimedia over WLANs. Thus, any collaboration between these two groups would prove quite beneficial.

802.11e Applications

802.11e focuses on three kinds of multimedia applications: IP-based applications, IEEE 1394-based consumer device applications, and converged IEEE 1394 and wireless IP multimedia applications. Let's look at how 802.11e does it.

IP-based multimedia applications are supported by not only the H.323 series of standards including the G.723.1 audio codec to enable interoperability between IP-based multimedia applications, but also the Session Initiation Protocol (SIP), which is used for initiating an interactive user session that involves multimedia elements such as video, voice, chat, gaming, and virtual reality. IP-based multimedia applications (e.g. web-based call centers, multicast enabled applications, and H.323-enabled and/or SIP-enabled interactive audio and video applications) are commonly found in not only the Internet, but also corporate intranets. Implementation of VoIP over Wi-Fi networks is an obvious extension of IP-based multimedia applications.

For VoIP applications, wireless handsets, softphones, or PDAs typically support either H.323 or SIP for call control. Originally developed by the International Telecommunications Union (ITU) for desktop videoconferencing, H.323 is commonly used for IP telephony phones and IP PBXs. But, because H.323 was originally a videoconferencing standard, its functionality is better suited for multimedia applications.

SIP, which was developed by the Internet Engineering Task Force (IETF) as an Application Layer signaling protocol, allows end-users to use IP networks (including Wi-Fi networks) to establish sessions, instead of just phone calls, which can include voice, video, or instant messaging. SIP can be used in "presence" applications where users can list themselves as available, similar to a buddy list. With a SIP client on a PDA, you can turn the PDA into an 802.11 phone.

Furthermore, unlike H.323, which is based on binary-encoding, SIP is a text-based protocol so applications are easier to implement, thus SIP will win out over H.323, especially for voice over WLAN applications. Another item in SIP's favor is that it can support IP mobility and thus should be able to work with 802.11e to provide quality voice delivery over a Wi-Fi network.

IEEE 1394-based multimedia applications include both consumer device applications and converged 1394 and wireless IP multimedia applications. IEEE 1394 is an evolutionary standard when compared to current I/O (input/output) interfaces, and provides a good networking foundation for consumer electronic devices, with hot plug and play, high data rate, and QoS support benefits. The 802.11e specification is being designed to support at least three simultaneous 1394-based DVD rate MPEG-2 channels, or one HDTV rate MPEG-2 channel over an 802.11a network. The IEEE 1394 applications can run directly over the 802.11e MAC, or use an IP encapsulation to run over the 802.11e MAC. Since the IEEE 1394 bus is a Data Link Layer network with isochronous transfer mode capability, it is quite natural that there must be the capability to transmit specific IP flow through a certain isochronous channel of 1394 bus, and A/V flow (such as MPEG2-TS) through a certain isochronous channel of the 1394 bus. So it is necessary to take note of the relationship between channel ID and IP flow, the bandwidth of the isochronous channel, the direction of the IP flow transmitted through the channel, and the attribute of the flow.

Wireless IP multimedia applications are supported by 802.11e when the transmission is between computers, gateways, PDAs, STB (Set-in-Box) TVs, and so on. Higher-layer multimedia applications can use the RTP/RTCP protocol and IP DiffServ to map the user priority to 802.1D-based MAC priority and then pass the prioritized data to different access categories of 802.11e MAC. They communicate to the 802.11e MAC through the 802 Data Service Access Point (DSAP) and Management SAP (MSAP).

802.1D/Q

The IEEE 802.1D MAC bridge specification (for wired networks) allows different MAC sublayers in the IEEE 802 family to interwork. (An 802.11 access point typically implements an 802.1D bridge connecting the 802.11 MAC and 802.3 MAC.) The 802.1D bridge supports up to eight user priorities by implementing multiple FIFO (first in first out) transmission queues between two MAC entities. By default, priority queuing can be used for these multiple queues. That is, a frame can be forwarded to the egress MAC only if there is no frame in the higher priority queues.

The 802.1Q Virtual LAN (VLAN) tag extends the existing 802.3 frame format for wired LANs, and it specifies the user priority of the frame. (The 802.3 MAC itself does not support any differentiated channel access to different priority traffic, but via the 802.1Q VLAN tag, the 802.3 MAC frames can carry the corresponding priority value, which in turn can be used by the 802.1D MAC bridge for a prioritized forwarding.) Since the 802.11e EDCF QoS scheme roots are in the 802.1D specification, priority parameters of 802.11e and 802.1D are interoperable.

When the 802.11e MAC frame is received at the ingress of a VLAN bridge supporting 802.1D/Q, it is classified by a VLAN ID, filtered via a filtering ID, and forwarded based on the traffic class that is mapped by the user priority. When the traffic class is mapped by user priority, 802.1D/Q frames are allocated into specific priority queues according to the traffic classes. When the traffic frames are dequeued from the forwarding process, it is transmitted to the next bridge via egress. Also note that 802.1Q has a recommended mapping table between user priorities that are defined in 3 bits of TCI (Tag Control Information which is part of the VLAN Tag Header) in 802.1Q, and traffic classes that specify the priority queue in traffic forwarding process in VLAN bridges.

End-to-end QoS Coordination via DiffServ

DiffServ currently is considered the dominant QoS protocol in the Network Layer. It does this by dividing network traffic into classes:

* EF (Expedited Forwarding)—interactive application packets (e.g. online gaming, VoIP, or videoconferencing) will reach their destination with short delay and little jitter.
* AF (Assured Forwarding)—packets will be treated with minimal sustained throughput. While this might work in some instances for interactive applications, in most instances it would not assure delivery without delay or jitter.
* BE (Best Effort)—packets will be sent via best effort, so there may be some packet loss and the packets may not arrive together. Thus there might be delay and considerable jitter if the packets carry interactive application data.

However, even though DiffServ provides support across different network interfaces—in wireless and wired LAN environments—DiffServ *alone* cannot function to control traffic for QoS over these different topologies. That is because it is difficult to map between different service domains or subnetworks such as WLANs. Instead, 802.11e and 802.1D/Q must be used to provide QoS in the MAC sublayer. But since 802.11e and 801.1D/Q are non-DiffServ domains, end-to-end QoS environment cannot be provided properly unless these different QoS techniques are coordinated under common QoS specifications.

Let's take a second to recap. The 802.11 specifications provide for two coordination functions: (1) The mandatory distributed coordination function (DCF) built on Carrier Sense Multiple Access with Collision Avoidance (CSMA/CA), and (2) an optional

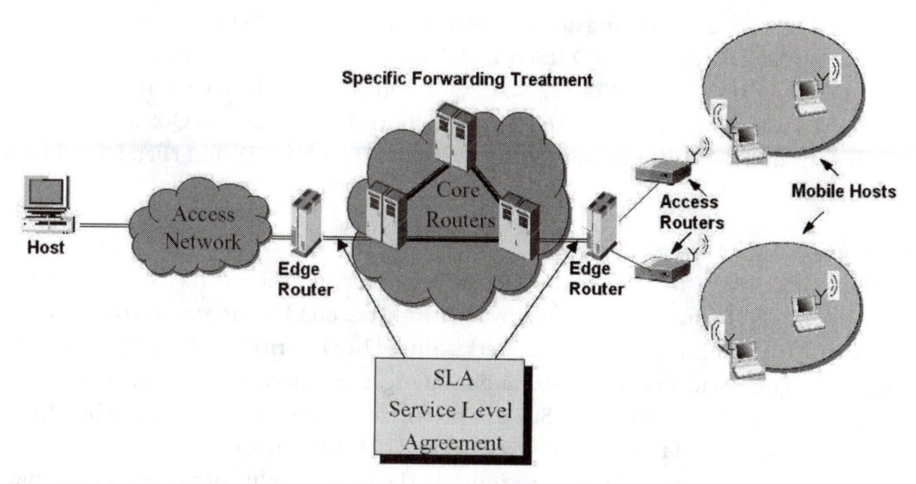

Figure 16.5 A typical end-to-end DiffServ architecture. A DiffServ edge router handles the classification, metering, and marking of the packets, while the core routers take care of queue management, scheduling, and traffic shaping.

point coordination functions (PCF) built on a poll-and-response protocol. However, most Wi-Fi devices implement the mandatory DCF mode only.

The DCF functions control traffic based on non-preemptive service (i.e. first in first out or FIFO). When the MAC frame arrives at the queue, it waits until the channel becomes idle, and defers another fixed time interval, called DCF InterFrame Space (DIFS), to avoid the potential collision with other network nodes. When the channel stays idle for the DIFS interval, it starts the random backoff counter. When the backoff counter expires, the frame is transmitted over the air. If a frame arrives at an empty queue and the medium has been idle longer than the DIFS time interval, the frame is transmitted immediately. Furthermore, each computing device maintains a contention window that selects the random backoff count. The backoff count is determined as a pseudo-random integer drawn from a uniform distribution over the interval. If the channel becomes busy during a backoff process, the backoff is suspended. When the channel becomes idle again, and stays idle for an extra DIFS time interval, the backoff process resumes with the suspended backoff counter value.

It is typical that a single computing device simultaneously services multiple sessions under different applications such as VoIP, streaming video, email, or FTP. According to the service type, traffic should be treated differently at the network node. As depicted in Fig. 16.3, three network interfaces can be defined in an end-to-end network. Each network interface has independent QoS coordination functions. However, the DiffServe Code Point (DSCP) is a single point of traffic control across multiple network interfaces, so that the end-to-end QoS can be provided transparently over all networks.

WLAN

In the wireless network, the wireless computing device performs the packet classification and conditioning in the Network Layer and forwards the packet to the AP. As illustrated in Fig. 16.3, a computing device should map QoS in the IP layer to the 802.11e. In computing devices supporting DiffServ and 802.11e, the DSCP value should be mapped to the TCID placed in the 802.11e MAC QoS control field. DSCP values are recommended by standards. According to the traffic control structure, two QoS architectures can be considered—direct mapped QoS between DSCP and TCID, and Hierarchical QoS architecture.

Incorporating the Wired LAN

As illustrated in Fig. 16.3, when the AP receives either an Ethernet (i.e. 802.1D/Q) or a WLAN (i.e. 802.11e) frame in the local area network, the 802.11e AP converts the Ethernet frame into the 802.11e frame, and vice versa. Since User Priority in 802.1D/Q and TCID in 802.11e have the identical field size and meaning, they can coordinate the QoS parameters seamlessly. That is, when the 802.11e-prioritized QoS service is used, the first three bits of the 802.1Q TCI field are conveyed in the TCID field of the 802.11e QoS Control field. Note that both sets of three bits indicate the priority value of the frame. Further, the TCID of 802.11e in AP should be coordinated with the User Priorities specified in the 802.1D MAC bridge standard.

WAN

When the 802.3 MAC frames are terminated at a gateway, as illustrated in Fig. 16.3, IP packets are reassembled and forwarded to the destination. When the IP packets arrive at the intermediate IP router supporting DiffServ, specific forwarding treatments are enforced based on the DSCP values.

Delay-sensitive IP packets of the EF class are entered in the high priority queue and forwarded with preemption, as specified in RFC 2598. When the IP packets in the AF class enter in the IP router supporting QoS, the DiffServ engine performs AF specific forwarding treatment. Each AF class (e.g. AF1x, AF2x, AF3x, and AF4x) allocates different forwarding resources, which are typically priority buffer size and bandwidth. Once the IP router experiences traffic congestion, it is determined whether or not IP packets with AF class will be dropped in accordance with the drop precedence values that represents "x." When the network congestion occurs, IP packets in the EF class are always

INTSERV AND DIFFSERV

There are two main approaches to adding QoS support in an IP network: Integrated Services (IntServ) and Differentiated Services (DiffServ). IntServ provides fine-grained service guarantees to individual traffic flows. It requires a module in every hop IP router along the path that reserves resources for each session. However, IntServ is not widely deployed since its requirements of setting states in all routers along a path are not easily implemented or scalable. Whereas DiffServ only provides a framework offering coarse-grained controls to aggregates of traffic flows. DiffServ attempts to address the scaling issues associated with IntServ by requiring state awareness only at the edge of DiffServ domains. At the edge, packets are classified into flows, and the flows are conditioned (marked, policed and possibly shaped) to a Traffic Conditioning Specification. In this way, simple and effective QoS can be built from the components during early deployments, and network-wide QoS can evolve into a more sophisticated structure.

protected, so as to maintain the low-jitter, low-loss and low-latency parameters that are defined in the SLA. When the SLA defines the QoS traffic control requirements, network administration should configure every IP router under the DiffServ Domain where a single QoS framework is managed. In case of multiple DiffServ Domains, DSCP values can be modified at the Ingress node where the IP packets arrive across network service area.

WANs and IntServ

As the 802.11e draft now stands, the QoS improvements can be implemented through software or firmware enhancements of the MAC in all 802.11 systems. Manufacturers producing 802.11a/b/g systems will be able to add these enhancements into their devices, and improve multimedia services without additional hardware or infrastructure costs. But the QoS is only guaranteed in the LAN, not across a series of unrelated networks. This leads us to the activities of the Integrated Services over Specific Link Layers (ISSLL) Working Group at the Internet Engineering Task Force (IETF). This Working Group is designing a method that can provide IntServ (see Text Box) QoS over specific link technologies by using DiffServ network segments. This solution maintains the IntServ signaling, delay-based admission, and the IntServ service definitions. The edge of the network consists of pure IntServ regions. However, the core of the network is a DiffServ region, and all flows are mapped into one of the few DiffServ classes at the boundary. Furthermore, in order to support both kinds of IP QoS approaches in WLAN links, different kinds of QoS enhancement schemes for both infrastructure and ad-hoc modes have been proposed.

APPLYING QoS

QoS can be applied at different layers of communication and aggregation points, such as networking devices where all traffic from one interface passes through to another, and on a shared medium such as Ethernet or Wi-Fi.

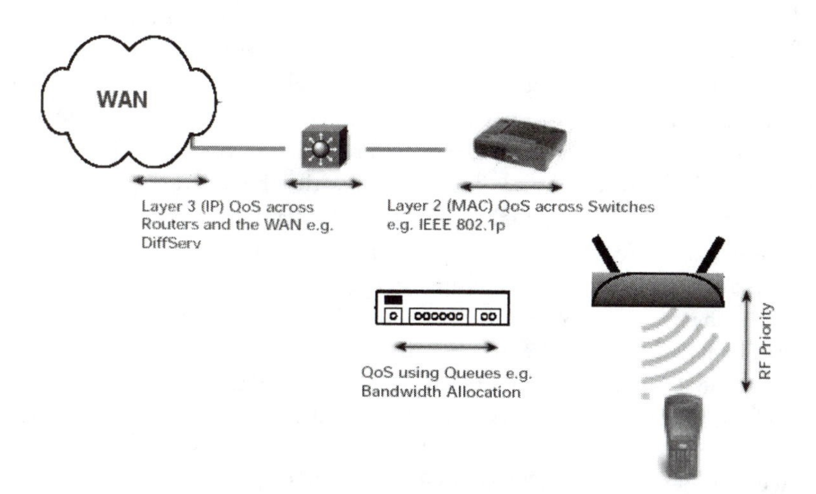

Figure 16.6 This diagram illustrates the different QoS methods that can be applied at different layers of communication and aggregation points.

In most instances, a QoS function gives an end-user or an application precedence or priority over other traffic, and applies in the case of congestion that may occur due to excessive traffic or different data rates and link speeds (e.g. 10 Mbps or 100 Mbps Ethernet, 11 Mbps or 54 Mbps Wi-Fi) that exist in the same network. If there is enough bandwidth for all users and applications, then applying QoS has very little value. But that is unlikely as excessive bandwidth comes at a very high cost. The more likely scenario is that the network's total bandwidth is shared by different users and applications, and thus QoS is required to provide policy enforcement for mission critical applications and/or users that have critical bandwidth requirements.

Different methods of QoS can be applied. The two main categories of QoS are:

QoS via queuing: A network shared by different users (e.g. a revenue-based network, network that is shared by a mixed-use office building, or a HotSpot) is typically implemented with Service Level Agreements (SLA). The SLAs are based on many different factors, such as how much each user group pays for bandwidth. Thus, one or all points of aggregation, such as a switch and some high end routers or policy managers, can allocate different percentages of the total bandwidth to different groups of users through the use of queues. Bandwidth allocation can be further divided and applied to different applications, again using queues.

Application QoS via packet marking. A network or a portion of the network's bandwidth can be shared by different applications. But one or more application may be more latency-sensitive (e.g. voice communication) or more mission-critical (e.g. videoconferencing) than others. In such a case, a priority is assigned to the traffic type by adding the appropriate QoS marking or tags to the network traffic. This method provides higher precedence while the data is passed through points of aggregation (routers, switches, and gateways) and the medium of transfer (Ethernet and/or Wi-Fi).

PROPRIETARY QoS TECHNOLOGY

A few vendors have attempted to provide proprietary QoS enhancements to Wi-Fi's technology. Sharewave's (acquired by Cirrus Logic) attempt can be found in its Whitecap technology. Whitecap has been incorporated into products from Panasonic, Netgear, and others, but since the products were not interoperable with other 802.11-based products, they didn't do too well. Nonetheless, Cirrus Logic continues to offer Whitecap technology; but this time the vendor sought Wi-Fi Alliance certification for its Whitecap2 product. Thus the latest version can interoperate with other 802.11b products.

The challenges of delivering high quality multimedia traffic over a wireless network include the distance a remote client is from the access point, impediments such as doors and walls that are typically found in most wireless networking environments and, most importantly, the number of client devices in use. Just one client device receiving video over a Wi-Fi network can use the full capacity of the available bandwidth.

Many of the large video-oriented electronics companies hope to support wireless video as adoption of wireless networks takes hold. And after many years of false starts and exaggerated promises, it appears that high-quality digital video will play a significant role in wireless networkings. The reason? At the 2003 International Consumer Electronics Show, upstart Magis Networks showed off its new "Air5 chipset." Using the same unlicensed 5 GHz spectrum that is the foundation for the 802.11a wireless networks standard, Magis Networks has its own trump card in the form of a proprietary MAC sublayer that is optimized for wireless delivery of high-quality video, TCP/IP data, and audio.

Even though the 802.11e QoS standard is just around the corner, many CES attendees felt that Magis could be holding a winning hand with its 802.11a radio and proprietary time domain multiplexed (TDM) Media Access Controller, and novel seven-antenna radio that can extend the useful range of a wireless network device. With claims that Air5-based networks can support several simultaneous streams of high-quality video, audio, and TCP/IP data throughput of up to 40 Mbps at ranges of up to 250 feet, the company has created a groundswell of interest.

Others are tackling the same issues, using different kinds of technology. ViXS Systems' Matrix 802.11a wireless communications processor optimizes multiple streams of wireless video for delivery over an 802.11a network. The Matrix processor is designed for "gateway" appliances. It uses two 802.11a channels to enable the distribution of quality video to multiple wireless devices. When combined with ViXS' Xcode MPEG/video network processor, Matrix can provide robust video distribution with guaranteed broadcast-quality video streams at 30 frames per second. According to ViXS, this combination will allow the construction of network media gateways and access points to stream video simultaneously to a variety of digital and analog devices such as laptops, PDAs, TVs, and next-generation HDTVs. All the receiving device needs is the basic off-the-shelf 802.11a network interface card (NIC).

XCode's MPEG chip is a multistream transcoder that maintains QoS for up to eight real-time digital video streams. ViXS' Matrix/XCode chipset can support two 802.11a channels in a wireless video network, which is key to supporting multiple broadcast-quality video streams with guaranteed QoS. Also included is network management software for QoS. The management software monitors and manages the video stream in real-time. In addition, it provides DES, Triple DES and AES encryption technologies for content security.

A WLAN with a Smart Edge?

Cisco believes that there should be a smart "edge" to the wireless network. Toward that end this networking giant is giving away its Cisco Client Extensions (CCX) software.

In doing so, Cisco is fundamentally opposing the trend to centralize the intelligence of wireless LANs, while leaving the edge "dumb." According to Cisco Consultant Martin Cook, such an architecture makes a WLAN's deployment and management more expensive.

"When you're rolling out a WLAN you want to take some of the pain away from the difficult job of network deployment," says Cook. "If for example you take the business of managing the radio frequency problems—how are you going to do that, if you have a dumb edge? You have to have intelligence distributed around the network. We think you can't know anything about the RF environment if you have a dumb edge. We want intelligence in the AP."

Cook listed several silicon makers who produce WLAN chips, who have signed up for the CCX software. On that list are Intel, Atheros, Intersil, Texas Instruments, Atmel, Marvell and Agere.

CCX also addresses the QoS issue. According to Cook, Cisco believes that CCX addresses the need for a more intelligent infrastructure that will be needed for mobile voice devices to use voice over Internet (VoIP) technology.

Cook also cites another example of the advantage of having intelligence distributed around the wireless LAN—fast roaming. He explains, "If we have intelligence in the PC, in the access point, in the PDA, the handshakes that make up authentication don't have to take so much time; we can leverage the intelligence out there at the edge and improve performance. We can give our WLAN smarts to other people out there, so that the device can communicate as if it were a Cisco client, to the infrastructure."

Cisco also believes it already has the key collaborator signed up in its push to claim dominance on WLAN technology. And, according to Cook, that key collaborator is Intel.

"Intel will be out largest partner with respect to delivering; they are currently buying our access points," asserts Cook. Cisco believes that all Centrino based PCs will be CCX comformant.

There probably will be other technological approaches to achieve similar solutions to the ones mentioned in this section. So it may not be long before multimedia over wireless is actually available to everyone. That, of course, will only cause the market to clamor for more.

MULTIMEDIA IN A WI-FI NETWORK

The convergence of wireless networking and multimedia seems a natural. There are a growing number of products employing 802.11 technologies to move music and video, and to present slideshows and other types of presentations hitting the market. For example, at the 2003 Consumer Electronics Show, wireless multimedia was the killer application—in fact there were six flavors of wireless video alone. But Wi-Fi enabled devices for both music and video were bountiful. Even the heavy hitters like HP, Motorola, Sony, and Toshiba offer Wi-Fi-enabled products that link televisions, DVD players, and other similar gadgets with Wi-Fi networks.

- HP's Digital Media Receiver uses Wi-Fi to bring music and pictures stored on a networked PC to your television or stereo.
- Motorola demonstrated a device it calls "Simplefi" that can link a network PC with a stereo system wirelessly.
- Sony's 802.11a-based RoomLink network streams video or music, or sends photos to a television from a Vaio PC over a standard 802.11a network. However, it uses the GigaPocket proprietary digital rights management software. That's not Sony's only foray into the Wi-Fi world—in late 2002, BroadQ announced the release of software that can convert a Sony PlayStation sporting a Wi-Fi PC card into a media center adapter for television.
- Toshiba drew quite a crowd at the 2003 CES with its 802.11b-based wireless large-screen TV.

Powering many of the consumer gadgets is software from SimpleDevices, a company with high hopes for Wi-Fi and home entertainment. "Wi-Fi is critical to the adoption of connected devices in the home," says Hanford Choy, vice president of engineering at

SimpleDevices. As more people adopt Wi-Fi, Choy says, wireless multimedia moves within reach of the average consumer.

Then there is the Wi-Fi product from ReQuest Multimedia. In April 2003, this industry leader in digital music servers launched its ARQPocket Music Browser software for PocketPC. According to Nick Carter, executive VP at ReQuest, "ARQPocket sets a new standard for user interface simplicity. Wireless networks and off the shelf PDA's offer a low-cost alternative for advanced 2-way control of AudioReQuest." ARQPocket is written in Virtual Basic for Windows CE devices such as PocketPC PDAs from companies such as DELL, Toshiba, Viewsonic, and many other manufacturers. Simply add an 802.11 NIC and access point to any Ethernet network for integration with any AudioReQuest Nitro, Fusion, Triton, or Tera Music Servers.

"Network technology is transforming the way people live, and enjoy music," says Jeff Hoover, owner of Audio Advisors and CEDIA President. "ReQuest has once again raised the bar for the industry. AudioReQuest is a great example of how Electronic Lifestyle systems remove the barriers between people and their media," adds Hoover.

VOICE OVER WI-FI

Mobile computers and wireless networking increasingly entice a mobile workforce to carry more computing power and maintain a real-time Internet connection to their company resources and customers. The next logical move is to provide voice over Wi-Fi.

Voice over Wi-Fi is already making its way into the vertical markets. According to the high-tech research firm, In-Stat/MDR, there has been additional demand from verticals such as education, healthcare, retail, and logistics for VoIP using Wi-Fi networks. While an estimated 80,000 wireless IP handsets shipped in 2002, shipments of voice over Wi-Fi handsets are expected to surpass half a million units by 2006.

Even though the Wi-Fi WLAN installed base is increasingly adding voice to existing wireless networks, and the market is projected to grow significantly, there is little demand for wireless voice beyond the previously mentioned vertical markets. Nonetheless, VoWLAN vendors bank on the fact that if the solution is easy and cheap enough to implement, it will eventually find its way into areas outside those markets. As a result, two vendors, Symbol and Spectralink, have partnered with PBX, wireless LAN, and LAN Telephony vendors to sell their Voice over Wi-Fi products. Industry experts predict that as the demand for voice over WLAN increases, more vendors will enter the market, bringing handset prices down, and pushing VoWLAN handsets out to the more mainstream business environments.

Although Spectralink has its own proprietary QoS method, even that company understands that a standard QoS specification would allow for more competitors to enter the market, driving prices down, which will result in faster market growth. So while existing VoIP technology can provide voice services over a wireless network, to make Voice over Wi-Fi a reality, engineers must tackle QoS, security, and roaming issues.

While we touched on some the issues engineers must address when provisioning QoS in a wireless environment earlier, let's now look at the same issues as they apply to Voice over Wi-Fi. Then we will consider some solutions being prepped by the industry and standards bodies to attack those issues.

***Note:** I want to thank CMP Media LLP and Ravi Kodavarti, the author of "Overcoming QoS, Security Issues in VoWLAN Designs," published on the CommDesign.com website on April 3, 2003, as the following relies on that article.*

Stacking on UDP

The 802.11 specification supports two modes of operation: infrastructure and ad hoc. In infrastructure mode, all end stations communicate to the wired network and to each other through an access point (AP). The AP must provide bridging functions to facilitate traffic either between the end stations or between end stations and the wired networks. An ad hoc network allows connectivity between two end stations, without the need for an AP, using a peer-to-peer type protocol.

The 802.11 standard controls the interface mechanisms at the OSI's Data Link MAC sublayer and the Physical (PHY) Layer, with higher layer protocol support left to the user. In the case of voice communications, an implementation using RTP/UDP/IP would reside on top of the 802.11 MAC and PHY. Fig. 16.7 shows the different layers of the UDP protocol stack that could reside on top of the MAC and PHY.

In theory, the architecture described above would provide an effective way to deliver voice capabilities over 802.11 links. But, in reality, designers still will face some QoS issues when working with an 802.11 link. Let's look at this issue in more detail.

Achieving QoS

There are significant differences between wireless and wireline networks with respect to QoS issues. QoS in wired networks range between guaranteed service and best-effort service. Guaranteed service works when bandwidth of the network is typically larger than the bandwidth of the service that is guaranteed. In best-effort service, the individual bandwidth allocated changes over time, and the user adjusts the bandwidth requested based on the congestion of the network. In effect, each of these types of network implementations enables QoS by decreasing packet loss, latency, and jitter.

The UDP protocol can be used in networks that can provide guaranteed service. UDP dumps packets on the network and hopes that it goes through to the other side. It relies on higher layers to deal with the issues of a packet that does not make it through. The TCP protocol can also be implemented in best-effort networks (e.g. IP networks). As part of TCP, there is an acknowledgement sent from the destination. If an acknowl-

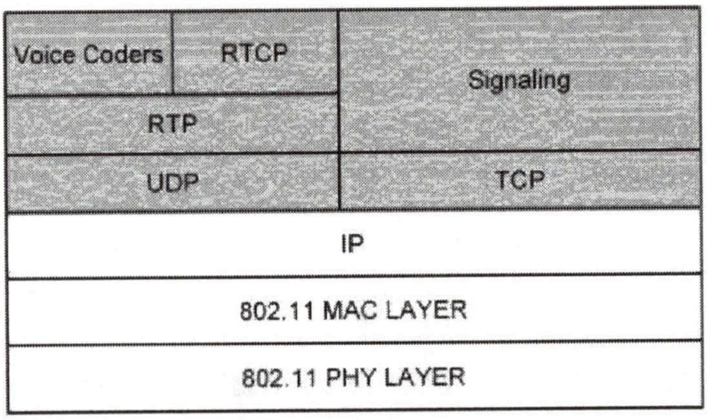

Voice Coders
RTCP
Signaling
RTP
UDP
TCP
IP
802.11 MAC LAYER
802.11 PHY LAYER

Figure 16.7 In VoWLAN systems, the UDP protocol sits on both the MAC and PHY layers.

edgement is not received, the transmission will be re-sent at a slower rate—the assumption here is that the network is congested.

Typically, VoIP implementations use UDP even for best-effort networks, and they account for the lost packets using various higher layer techniques. These implementations assume that the underlying network will be designed to account for the latency and jitter requirements of the higher-layer application.

In a wired network, accounting for latency and jitter are fairly straightforward—that's not the case in a wireless network. Unlike the wired network, WLANs must deal with tough propagation issues in order to determine channel performance. Thus, during the design of a WLAN system, engineers must combat issues such as multipath and Rayleigh fading,

Note: *Rayleigh fading occurs when an end-user device moves about while transmitting and receiving data. Rayleigh fading is caused by multipath reception. When a mobile antenna in a Wi-Fi network receives a large number of reflected and scattered waves, wave cancellation effects cause the instantaneous received power, as seen by that moving antenna, to become a random variable, dependent on the antenna's location.*

To account for the uncertainty in the wireless medium, the 802.11 MAC includes an acknowledgement (ACK) protocol. When a packet is transmitted, the sender firsts listens for any activity on the air, and if there is none, waits a random amount of time before doing a transmission. This methodology is called carrier sense multiple access/collision avoidance (CSMA/CA), which can be viewed as a "listen first, talk later" methodology. If an ACK is not received, either due to interference or collision, then the entire process is repeated. The MAC layer ACK protocol is independent of the higher layer protocol, whether it is UDP or TCP.

The ACK protocol builds a layer of reliability on the WLAN transmission, making it very useful in data transmissions. However, in voice applications, this protocol adds

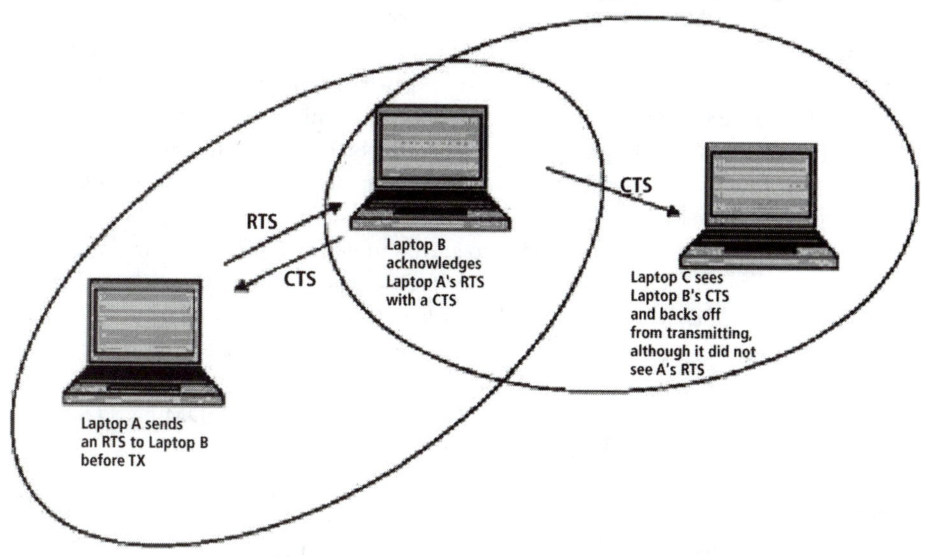

Figure 16.8 This diagram illustrates Wi-Fi's RTS and CTS mechanisms.

jitter and latency to voice traffic. In order to account for jitter, buffers need to be used, which in turn add more latency.

The ACK function is not the only QoS headache for designers looking to deliver voice services over WLAN systems. The WLAN MAC also includes a request to send/clear to send (RTS/CTS) mechanism. When used together, RTS and CTS decrease the chance of collision on a system by making sure that end stations in the vicinity of the source and destination hear the RTS and CTS respectively. RTS and CTS add robustness to the system, at the cost of adding latency to the packets that are transmitted using this protocol. Fig 16.8 shows the cone of influence of an RTS and CTS frame exchange.

Avoiding the QoS Problem

To avoid the problems caused by the ACK protocol, designers can implement other techniques to reduce retransmissions. One way to accomplish this task is by fragmenting a packet into smaller packets.

While an ACK function is still required during transmission of fragmented packets, it is expected that overall latency and jitter will decrease as the likelihood of a smaller transmission getting corrupted is reduced. This would benefit wireless VoIP implementations, especially if low-bit-rate vocoders were used to compress the voice traffic. For example, designers can use a G.729 or G.723 codec to decrease overall latency and jitter on a wireless system, even though these vocoders add some fixed latency to the voice path. Since digital signal processors (DSPs) work very well in vocoding algorithms, it would vastly improve the voice quality of a wireless VoIP implementation if a DSP were present as part of the mobile computing device.

The 802.11e draft specification provides another alternative for dealing with QoS problems. This draft specification defines an enhanced distributed control function (EDFC) that allows a WLAN access point to provide up to eight virtual channels to every computing device. In order to ensure the highest priority channel is transmitted first, each of these eight channels has associated QoS parameters.

Additionally, under 802.11e, an AP could also support a hybrid control function (HCF). Through this function, the AP can take control of the channel before any of the stations do, thus reducing collision overhead and the number of retransmissions. Since the 802.11e draft standard is supplementary to the 802.11 MAC sublayer, however, it would reduce overall latencies for wireless VoIP if implemented as part of a MAC hardware implementation.

The 802.11g specification also helps to solve some of the QoS problems caused by interference on wireless channels. This spec defines the use of either orthogonal frequency division multiplexing (OFDM), or packet binary convolutional coding (PBCC) coding schemes. To provide enhanced error protection, these modulation schemes employ convolutional coding, thus allowing them to deliver better packet error rate, latency, and jitter, due to the superior coding nature. (Note: The OFDM-based 802.11a spec also supports convolutional coding.)

Security Issues

QoS is not the only issue designers must tackle when pitching VoIP services over 802.11 links. Unlike VoIP over wired systems, VoIP over a WLAN system entails comprehensive security for all aspects of a call. The reason network administrators implementing VoIP over a wireline network are not be overly concerned about an attack on their secure network is that typically all Ethernet drops are well protected, and it is virtually impossible for a hacker to get access to the network without breaking into the facility. Whereas, the main aspects of security in a WLAN environment are the privacy of a voice call and protection from denial-of-service attacks. Thus it is imperative that authentication and packet traffic are secure in order to ensure security in these cases.

The 802.11i standard is a MAC sublayer enhancement allowing support of both packet security and authentication security. The authentication security stems from the 802.1X protocol. 802.1X does not provide any cipher support. Instead, it provides a framework for authentication and key management functions using the extensible authentication protocol. The 802.1X protocol allows for a mechanism where a server on a network can provide dynamic keys to each WLAN client. The draft 802.11i proposal also supports 802.11X enhancements with respect to the pre-authentication of clients. This work is primarily driven to support roaming on WLAN networks.

Current mechanisms of 802.11 cipher-based security methods revolve around using the Wired Equivalent Privacy (WEP) protocol. However, WEP is not considered adequate for enterprise applications, since hackers can easily decode the underlying key that is used for data traffic. Additionally, since WEP is a static implementation, it is a chore for network administrators to change the key on an AP, because this would entail changing it on every station as well. Some implementations use access control lists that authenticate based on the MAC addresses of computing devices. However, MAC addresses can be easily duplicated to spoof the AP.

To address this security issue, the Wi-Fi Alliance has adopted a subset of 802.11i for immediate certification. This program is referred to as Wi-Fi protected access (WPA).

While the security features of cipher support and authentication support in the 802.11i standard afford a layer of protection for WLAN networks, they also add complexity for voice traffic. Authentication for server-based methods adds latency to the setup of a call, and ciphering using WEP, WPA, or AES adds latency to each packet (if these were to be implemented in software). The 802.11 standard treats 802.11i as a MAC layer enhancement. Therefore, in order to minimize delay, it is imperative that silicon vendors add support for the authentication and cipher security as part of their hardware.

***Note:** Chapter 17 has more detailed information on all of the security methods mentioned in this section.*

Roaming and Interoperability

Roaming and interoperability also play a critical role in the development of WLAN systems that support voice. On the roaming front, WLAN system calls must support fast handoff and authentication between access points when handling voice calls. If fast handoff is not supported, designers will encounter delay during the probe, authentication, and re-association stages.

The Inter Access Point Protocol (IAAP) and other proprietary methods support roaming between different APs. Studies show that handoff delays between different APs can be as high as 400 ms. The 802.1X additions to pre-authentication, as part of the 802.11i draft, and the AP roaming protocols, as part of the 802.11f draft, address methods to decrease handoff delays. Silicon vendors must provide support for these features as part of their MAC implementations in order to support VoIP mobility.

In addition to dealing with roaming between WLAN devices, 802.11 designers must be concerned with roaming voice calls between cellular and WLAN systems. Right now, standards on this front are pretty crude. The cellular sector has defined some packet-based specs, but most of the efforts to date have focused on data services. Thus, to provide true VoIP roaming across cellular and WLAN networks, standards will need to be established to promote wide scale deployment and adoption. Currently, vendors are working on proprietary ways of solving this issue.

As engineers have observed, a lot of pitfalls lay ahead in the delivery of voice services over wireless links. Fortunately, the 802.11 committee, through its e, i, and f drafts, is addressing this issue head on. In the interim, however, designers looking to add voice capabilities to their WLAN designs must deal with packet loss and jitter issues carefully.

Available VoWLAN Systems

Some vendors haven't waited for the 802.11e standard to be ratified. Many have struck out on their own in order to fulfill the growing market for voice over Wi-Fi. The beneficial features Wi-Fi phones bring to the end-user include the ability to converge the telephony function directly into an already existing data network infrastructure, and the ability to operate with any access point. Customers also turn to IP telephony because it simplifies their network infrastructure and can lower an organization's overall communications costs since a mobile IP phone allows them to add mobility without paying for cell phone airtime.

Calypso Wireless recently launched its new C1250i videophone. The phone uses Wi-Fi networks for both videoconferencing and data access. This mobile videophone also is capable of seamlessly switching back and forth between the cellular networks and Wi-Fi networks.

Cisco Systems announced a "new" VoIP 802.11 handset in April 2003, which it dubbed the "7920 IP Phone." Cisco plans to ship this Wi-Fi phone to U.S. channel partners sometime in June 2003, with availability in other countries soon after.

Note, however, that the 7920 IP Phone communicates only with 802.11b technology and "is designed for use within enterprises rather than totally replacing a cell phone," at least according to Charlie Giancarlo, Cisco's senior vice-president of switching, voice, and carrier systems.

Of course, Cisco isn't new to the voice over Wi-Fi marketplace. Its 7960 IP phone, which sits on a desk and plugs into a wired Ethernet network, has an installed user-base. However, the 7920 IP phone offers the added benefit of mobility around a building or campus that has a wireless LAN.

According to Giancarlo, the 7920 could be ideal for environments such as retail stores, where employees need to move around a site during the workday. "The phone should deliver two hours of talk time and 24 hours of standby time before the need to recharge," he adds.

Motorola is also in the process of producing Wi-Fi equipped phones. Those phones supposedly will be able to switch automatically and seamlessly from one Wi-Fi location to another.

Nokia has demonstrated VoIP over WLAN. The Nokia system uses Mobile IPv6, fast handovers, and context transfer for security, QoS, buffers, and header compression state. The company's measurements indicate that the Mobile IP parts add very little to the delay overhead. What the company found problematic, though, is the device driver for the 802.11 NIC. According to the company, if it can find a solution for that little dilemma, the WLAN phone's handover time would be well under 5 milliseconds.

Intermec Technologies added VoIP capabilities to its wireless-enabled 700 series of mobile computers in a bid to keep retail clerks accessible to answer questions, make decisions, and take customer requests—whether on the floor or in the back room—improving customer service and employee productivity.

The 700's VoIP technology enables unit-to-unit voice communications over existing 802.11b wireless LAN infrastructures. Unlike current in-store voice communications technology, the 700 with VoIP provides enterprise-level communications, allowing workers to talk to anyone else on a corporate network using a 700 device—whether they are in the same store, a store across town, or the distribution center across the country.

"Now workers in every store and warehouse can have instant voice communication to locate merchandise and solve problems for customers," said Scott Medford, director of retail business development for Intermec. "And it's done with the existing wireless and wired LAN infrastructure simply by adding software to a retailer's in-store data collection device."

The 700 mobile computer, which weighs less than 16 ounces with battery, is designed to fit comfortably in the hand. It uses Microsoft's Pocket PC 2002 and runs virtually all 32-bit Windows CE applications. To support the voice application, the device comes standard with full duplex audio capability and a standard 2.5mm headset jack designed to accept the widest possible variety of headset options available. The user-replaceable, rechargeable lithium ion battery delivers 8-to-10 hours of continuous use on a single charge; more than adequate for a full shift's work. The 700's rugged case can withstand multiple four-foot drops onto concrete and its high-contrast VGA touch screen delivers excellent readability indoors and out.

The 700 with a VoIP solution package includes application software, installation guide, installation CD, and a standard cell phone-style hands-free earphone with a microphone. The package delivers consistently high, conversational quality voice throughout a WLAN's coverage area. All that's required to get VoIP up and running over an existing WLAN are: some 700 handheld devices with integrated 802.11b, the VoIP application, and, of course, the 802.11b network. No servers or gateways are needed. Furthermore, both the voice application and unit directory services can be centrally managed with all other aspects of the 700 via the Intermec 6920 enterprise device manager.

SpectraLink's NetLink IP Wireless Telephone is being tested at North Carolina State University. This phone is compatible with 802.11b wireless LAN, and uses the Skinny Client Control Protocol that is supported on Cisco's Call Manager (which provides caller ID, call forward, conference calling, multiple line capabilities, and many other common functions). When used in conjunction with the SpectraLink Voice Priority (SVP) server, the system allows QoS for voice packets. The SVP server helps to recognize voice packets being transmitted over the network, and gives these packets a higher priority to lessen the number of packets lost on the network. However, once 802.11e is ratified, the SVP server will be obsolete; but, since 802.11e will be a new standard, it could be a year or two before the standard migrates into usable service.

To incorporate the Wi-Fi phones onto the network involves three simple steps: (1) the phones need to be setup on the Cisco Call Manager in the same way that any other VoIP compatible client would be configured; (2) the MAC address of the phone and the phone's DNS information need to be configured; (3) the final step is to activate each phone when it is within the WLAN's coverage area, so it can receive its network information from the server.

Once a phone is activated, an individual could, for example, roam throughout the WLAN's coverage area with the phone, switching from access point to access point without loss of any voice packets. Since the access points are monitoring activity, they can allow the user of the VoIP wireless phones to use the phone anyplace on the campus as long as there is an access point available.

Symbol Technologies continues its wireless product and technology leadership. In 2001, Symbol offered the first wireless VoIP handset that supports worldwide both IEEE 802.11b 11 Mbps Direct Sequence (DS) wireless local area networks and International Telecommunications Union (ITU) H.323 standard-based telephony systems. The Spectrum24 NetVision Phone is offered directly to Symbol's core customer base and indirectly to its community of wireless VoIP telephony and channel partners through OEM and reseller agreements. The NetVision handset allows enterprise customers to add wireless voice connections to their in-building wireless LANs, and to achieve mobility with the same level of functionality as their existing Ethernet desktop phone systems. Symbol expects the demand for in-building VoIP wireless handsets, like NetVision, to grow proportionally with the adoption of 802.11b wireless LANs, as enterprise customers extend their wired networks to support wireless PDAs, laptop PCs, and yet-to-be developed converged voice and data devices.

The Symbol wireless VoIP product set features the NetVision family products resold with the Ericsson WebSwitch 100 G4, a four port desk top gateway for the small business. The wireless VoIP handset is also designed to integrate directly into the Nortel Networks ITG Product line, Mitel IPERA 2000, and Cisco AVVID/Call Manager, while firmware has been developed to support a host of other telephony systems from Alcatel, Innovaphone, Motorola, Vegastream, and more.

When integrated with gateway products, the NetVision wireless VoIP handset supports PBX (Private Branch Exchange) supplementary services, such as call waiting, call transfer, conferencing, call park, paging, as well as Symbol's added mobility features, including intercom mode (walkie-talkie), text messaging, and pre-emptive roaming. Furthermore, the NetVision phone supports Symbol's Quality-of-Service (QoS) voice prioritization to guarantee high voice quality.

✪ AN ALTERNATIVE TO QoS

TeleSym brings the advantages of Internet phone calling to mobile computers. The company's SymPhone System software enables clear, cost-effective voice communications on wireless mobile computers. The SymPhone codec uses TeleSym's patent-pending Edge Quality of Service technology, providing high quality without requiring QoS in the network. It can maintain a call, even when subjected to a 50 percent loss in data packets.

The SymPhone System extends the functions of the office phone to any location with Wi-Fi connectivity. Users can make and receive calls over the Internet from a mobile computer (laptop, tablet-PC or PDA), throughout a building or campus, at home and at a public Wi-Fi HotSpot. Wireless factory and warehouse users can also use SymPhone as an intercom or cordless phone, using the same computer they use to access their company email.

"In the course of the next year, wireless connectivity for computers will become the rule, rather than the exception," explains Raju Gulabani, TeleSym's chief executive. "At that point, it makes perfect sense to include voice calling as part of every networked device, particularly in the enterprise environment."

For mobile workers in wireless enterprises, the SymPhone System extends the functions of the office phone to a mobile computer, such as a laptop, a tablet PC or a PDA, allowing users to roam throughout the building or campus, into HotSpots or in the home, while making and receiving calls over the Internet. Wireless factory and warehouse users can also use SymPhone as an intercom or cordless phone, using the same device they use to access their company data and emails.

Breakthrough sound quality distinguishes SymPhone from other telephone systems and "soft phone" Internet calling software. In calls between SymPhone users across the open Internet, the audio fidelity is said to be "near-CD-quality," with no perceived latency (delay). Moreover, according to TeleSym, this quality is maintained even in the rugged wireless network environment.

CONCLUSION

QoS is on the forefront of networking technology. The future brings us the notion of user-based QoS in which QoS policies are based on the user as well as on the application. The upcoming 802.11e standard is a good start. It can be an efficient means for QoS support in WLANs for a wide variety of applications, although problems, as discussed in this chapter, still remain to be solved. In particular, there must be a continued evolution toward end-to-end services. But, thankfully, companies like Cisco are expanding QoS interworking to operate seamlessly across heterogeneous Data Link Layer technologies, and are working closely with host platform partners to ensure interoperation between networks and end systems.

Multiple powerful trends in the technical community are driving the need for distribution of high quality digital media in a wireless environment. A large number of service providers and equipment manufacturers stand to reap the economic benefits of these trends, if the necessary equipment and networking technologies can be delivered to meet both the quality requirements of these applications and the market's price points.

So while it may be some time before the corporate market will truly take advantage of multimedia data delivery over a wireless infrastructure, home use of integrated voice/video/data networks will happen much faster. But there is no reason to wait for these more exotic features. Today's products offer more than sufficient capabilities for many applications. And as long as you put some hard questions to your vendors about interoperability and upgrade paths, you can deploy a network that you can enhance as needed, over time, safely.

Chapter 17: Dealing with Security Issues

Some wireless local area networks (WLANs) are deployed just for the sake of convenience; others are implemented to transform business processes by giving employees mobile access to information. Most of us work in organizations that fall somewhere in between these two extremes. But most Wi-Fi advocates are grappling with the question of whether the convenience, mobility, and ability to process real-time information on the spot justifies the diminished security WLANs bring to an organization's overall networking environment.

When reports began circulating in mid-2001 that researchers had found the IEEE 802.11 WEP (Wired Equivalent Privacy) security system was vulnerable to attack, the news dampened the extremely hot wireless LAN market. While Wi-Fi's performance, interoperability, and manageability continue to improve, the image of security vulnerability hangs like a dark cloud over the Wi-Fi industry. In 2002, *Network Computing* magazine conducted a WLAN security reader poll wherein fewer than one-third of the respondents said they would be willing to accept a little less security in exchange for the benefits of wireless network access.

However, some organizations don't give security a second thought. In fact, in a TNS Intersearch study commissioned by Microsoft, only 42 percent of WLAN sites surveyed had implemented authentication systems. (Hopefully, these organizations implemented their internal WLANs "outside the firewall," to provide limited access to internal systems.)

Others feel that the benefits of Wi-Fi versus security hazards are acceptable tradeoffs. This same group rationalizes that when network users need more sensitive information, they can be provided with VPN connections, just like dial-up, DSL, and cable-modem users. That's all well and good, but those networks still are vulnerable to wardriving or other external attacks, in which users outside the organization gain access to the organization's Internet connection, or to insecure internal systems where they can mount further attacks.

What most people don't realize, and the press hasn't emphasized, is that you *can* design a secure wireless system. There are numerous products on the market that can help you to deploy a secure wireless network—most are based on existing 802.11X standards. In addition, a number of vendors have jumped on the obvious market opportunity and released WLAN security overlays that provide a range of enhanced services that adequately address Wi-Fi's security issues. Those vendors include Agere Systems, Cisco Systems, Proxim Corp., and Symbol Technologies, among others. However, those solutions often forsake multi-vendor interoperability. To deploy a WLAN based on this type of gear, however, will require a larger budget. In addition, the network manager must accept the burden of increased network complexity. But not every wireless network needs a "security overlay" system.

This chapter will help you to determine what is best for your wireless networking environment. As such it is written so as to help the reader to understand the key elements of a comprehensive WLAN security system. Hopefully, that will enable the reader to assess a specific WLAN's security needs, and also to assess the organization's level of risk aversion, along with the price the organization is willing to pay to achieve security. With that knowledge in hand, the reader can craft a security plan that best suits his or her organization's needs.

SECURITY BASICS

One measure of a standard's success is the degree to which it encourages competition and makes technology more cost-effective for users. By this measure, Wi-Fi has been an unbelievable triumph. But another measure of success is the degree to which a standard anticipates and addresses future implementation issues. TCP/IP, for example, which was crafted more than 30 years ago, has withstood the test of time. By that yardstick, the IEEE 802.11 series' ongoing changes at both the Physical and the Data Link Layers, together with minimal security capabilities, make it easy for experts to second-guess the designers. Of course, 802.11 isn't even ten years old. Moreover, critics must also remember that it took nearly seven years to develop the initial 802.11 standard; if the IEEE had waited until it was sure a WLAN could also be made secure, the publication of the first 802.11 standard would have taken even longer. Whatever the reason for 802.11's paltry security provisions, everyone agrees that this series of wireless standards fails to provide for security measures that can pass muster with enterprise administrators.

In Wi-Fi's early days, people considered 802.11's ESSID (Extended Service Set Identifier), a string that was defined for each access point, as a wireless password. This offered an illusion of security. But it wasn't long before implementers realized that since access points routinely broadcast these "wireless passwords," anyone could intercept them. Even when broadcasting was disabled, the strings could be extracted in clear text from the management frames passed by the wireless computing devices and the access points. Today, ESSIDs often are detected automatically by a WLAN client to allow end-users to connect to wireless networks transparently; that is as long as no other security points exist.

Since the standard doesn't provide an authentication framework, MAC (Media Access Control) address restrictions are sometimes used to control access to a WLAN. However, this approach is an administrative burden, is vulnerable to address spoofing, and ties access to the computing device (which can be stolen) rather than to the end-user.

Finally, there's Wired Equivalent Privacy—or WEP. But WEP's static shared-key architecture has little appeal for enterprise IT professionals. That's because noted security experts like Scott Fluhrer, Itsik Mantin, Adi Shamir, and others have exposed the weaknesses in WEP's underlying encryption system. Clearly, there's a need for privacy based on dynamic session keys that are distributed after a robust authentication.

THREE KEYS TO REAL SECURITY

Authentication, privacy, and access control (or authorization) are the three key services necessary for comprehensive wireless LAN security. Although each of these security services can be delivered, the challenge is to ensure they are reliable, interoperable, scalable, and cost-effective. And if you want to deliver these solutions sooner rather than later, the systems chosen must be flexible enough to integrate with existing mobile devices and network infrastructure.

Authentication: In many environments, the principal need is for a WLAN security system that authenticates users via an existing user ID and password. In some WLAN systems, authentication is transparent, with the standard Windows login information passed to a wireless authentication system. In other cases, end-users are given enough initial network access to pass credentials to a web-based authentication server, and if the process is successful, they are given extended network access. In more sophisticated

WEP'S WEAKNESSES

WEP (Wired Equivalent Privacy) is an optional IEEE 802.11 feature used to provide data security that is equivalent to that of a wired LAN without privacy-enhancing encryption techniques. WEP allows a network administrator to define a set of respective "keys" for each wireless network user based on a "key string" that is passed through the WEP encryption algorithm. Network access is denied by anyone who does not have the required key. 802.11a and b specify that WEP use the RC4 algorithm with a 40-bit or 128-bit key (152 and 512-bit key versions also now exist). When WEP is enabled, each station (computing devices and access points) has a key. The key is used to encrypt the data before it is transmitted through the airwaves. If a station receives a packet that is not encrypted with the appropriate key, the packet will be discarded. This was supposed to prevent unauthorized network access and eavesdropping.

The IEEE 802.11 Working Group designed WEP with the initial goal of providing a level of security that conformed to the difficulty of tapping Ethernet network traffic. In the case of wired Ethernet, you would need physical access to a network to sniff packets and intercept data. So while WEP's minimal security meets at least that level of protection, it fails because of flaws in the conception and implementation of the protocol.

But to be fair, some of those flaws are a result of computational limitations that were in existence when the specification was being developed. The number crunching expected to be available on the Wi-Fi cards was orders of magnitude lower than what was available even in 1999. Other flaws had to do with the then current export restrictions on strong encryption, which placed an initial limit on WEP to just 40 bits.

But WEP also has other weaknesses, including bad packet integrity checking (i.e. an interloper could insert or modify data in transit without being caught), and the requirement that all users on a network use the same keys, which must be manually entered (unless a network authentication system is in place).

implementations, a server authenticates the users, and they, in turn, authenticate the wireless network to ensure they are not being seduced by rogue access points.

The IEEE 802.1X protocol, used in conjunction with Extensible Authentication Protocol (EAP), is the key component for future standards-based WLAN authentication. EAP is an authentication protocol that supports multiple authentication mechanisms. It typically runs directly over the OSI's Data Link Layer, without requiring IP, and therefore includes its own support for in-order delivery and retransmission. While most of the enterprise-oriented WLAN vendors have built 802.1X support into their newest access points, the availability and interoperability of 802.1X clients are somewhat limited.

Privacy: Privacy (encryption) services commonly are linked to authentication such that unique per-session keys are distributed at the time of authentication. Most network managers believe encryption is mandatory, or at least desirable. Unfortunately, WEP, which is the most widely implemented WLAN encryption standard, requires frequent rekeying to be effective. Many products available today use, or offer firmware upgrades to, TKIP (Temporal Key Integrity Protocol), which is an interim fix to WEP. TKIP overcomes some of WEP's known vulnerabilities without requiring hardware replacement. But most industry experts agree that TKIP is more of a tactical bandage than a strategic cure.

In the long run, the industry will implement AES (Advanced Encryption Standard), which offers more robust encryption methods; but that transition will require new hardware. The author notes that some WLAN chipsets now ship with integrated AES encryption. Check to see if the hardware you use supports AES.

Access Control: Controlling user and group access to specific servers and applications based on credentials is an important element of many enterprise networks. But although access control is arguably one of the most critical security services, it is not effectively addressed in emerging WLAN standards.

In fairness to the IEEE 802.11 committee, access control is often seen as a component of policy-based network management, which should be applied to all wired and wireless LAN technologies at higher protocol layers. Likewise, accounting, which is important for some enterprise environments and critical to the emerging WLAN HotSpot market, is an element that will be managed up the stack, not at the Physical or Data Link Layers.

Now let's see how the reader can secure a WLAN. While the three aforementioned keys play a large part in deploying a secure WLAN, there is more that can and should be done to ensure your wireless data is secure.

DEPLOYING A SECURE WLAN

Concerns about network security can place so many roadblocks in a project's path that management can feel that the WLAN project is spiraling out of control. This can result in shelving the project altogether, as might be the case when there are concerns over compromised data. Some of the issues raised will be a one-in-a-million chance of a security violation. But the solutions to such unlikely events can also:

- Dramatically raise the cost of the WLAN project.
- Create a network that can't deliver a satisfactory data throughput speed.
- Lessen end-users' ease of access.

So it must be decided what *informed* risks the organization is willing to take.

Also, bear in mind that when any multi-level security program is implemented, it typically restricts user access to certain network assets. While this may be necessary to keep intruders at bay, it may also restrict the organization's normal work processes.

Therefore, when developing a security plan, take the time and effort to determine which security procedures are being implemented to address risk management, and which are due to risk-aversion.

That's where a sound security plan comes into play. A good, well thought out security plan can help your organization to avoid excessive risk aversion behavior, create a credible network security environment, and even to create a wider range of sound strategic security options for the future. Furthermore, as you dig into all aspects of network security, you will find that many of the numerous security measures that can and should be implemented take place on the human level, not the hardware or software level.

For example, it's the Wi-Fi-enabled computing devices that are brought in through the back door, without the IT unit's knowledge, that provide the most serious security threat. That's because these wireless devices are often plugged directly into an organization's network and transmit sensitive data that can be easily picked up by a snoop using freeware hacking tools and an inexpensive wireless NIC. Malicious or not, unauthorized access to an organization's network, wired or wireless, is never welcome. Hacking and eavesdropping are easy forms of industrial espionage. A rogue wireless set-up can do great harm. Sophisticated eavesdroppers aren't even required to be near the premises—long-range antennae can pick up wireless signals from hundreds or even thousands of feet/meters away.

A Wireless Policy

Industry experts estimate that most organizations with more than 50 employees probably have one or more rogue access points on their premise. Uncontrolled wireless access means hackers can read corporate email, sniff for super-user accounts and passwords, gain root or administrative access, drop in Trojan horses (hidden executable programs) for remote monitoring, and open back doors into an unprotected network. A good security policy could help ease this situation.

In fact, when drafting a security plan, crafting a wireless policy should be at the top of your "to do" list. To protect an organization's networks from unauthorized access, an organization should adopt policies that, on one hand, welcome wireless technology, and on the other hand, protect the organization's networks. This requires (1) putting in place a written, well-publicized corporate wireless policy, and (2) training IT personnel in how to hunt down unauthorized installations.

A wireless policy should include, but not be limited to, the following elements:

- An agreement that is executed by each and every employee which states that they will not provide (intentionally or unintentionally) WLAN configuration information to outsiders, or construct an ad hoc network.
- A requirement that all WLAN users must be trained on the ins and outs of wireless networking. For example, most employees with wireless laptops don't realize their wireless PC card can remain active when they're not transmitting or receiving data. This means an attacker could use that active link to perform a number of nefarious

deeds. User training must take into account everything from serious security issues to the more mundane idea of teaching employees that moving their wireless computing device a few inches can improve throughput drastically.

* A requirement that the organization's IT department must approve, in writing, the use of any Wi-Fi-enabled device, and the department must keep a record of all wireless devices with information concerning their NICs and the attendant APs.
* An acknowledgement that the IT department is the only department authorized to select, standardize, and approve wireless security configurations.
* An acknowledgement that the employees understand what can and can't be downloaded to a mobile computing device. (Of course, this requires that the organization set out a clear policy on what can and cannot be downloaded via a wireless connection.)
* A requirement that all wireless devices must pass a security audit before they can be approved for use. The audit should include determining if the promiscuous broadcasting of MAC addresses is turned off, and Wired Equivalent Privacy (WEP) is enabled.
* A requirement that SSIDs and passwords are to be regularly exchanged for new designations (using randomly generated alphanumeric codes).
* A requirement that the IT department perform scans, at least once every two months, for rogue wireless devices.

A wireless policy, while important, is only one in a series of steps that should be taken to help secure the data that traverses the WLAN.

Minimum Security Procedures

The next step in provisioning a good wireless security plan is implementing some basic network security procedures. Let's look at some important minimal security steps that can ensure that the organization's communications are secure as they travel the airwaves:

* Evaluate the organization's wired LAN security policies to determine how to extend them to the WLAN.
* Know the WLAN's vulnerabilities and do what you can to mitigate them. (Some who deploy a WLAN will put little effort into securing their WLAN, because in their view the data that is transmitted through the WLAN has little or no value to others apart from the network's end-users.)
* Take a complete inventory of all access points and Wi-Fi devices.
* Perform a risk assessment to understand the value of the assets in the organization that need protection.
* Consider standardizing on a single vendor's products.
* Prior to purchasing any wireless gear, ensure that the vendor provides firmware upgrades and that the gear can support such upgrades, so that security patches can be deployed as they become available.
* Fully understand the impact of deploying any security feature or product prior to deployment.
* Enable all security features on all WLAN gear, including the cryptographic authentication and WEP privacy features.

- Ensure that encryption key sizes are at least 128-bits (or as large as possible considering the capabilities of the wireless components).
- Secure all access points (APs) by:
  - Changing the default WEP encryption key that comes with the access point provided by the vendor.
  - Changing the default vendor-set SSID for the APs.
  - Disabling the "broadcast SSID" feature on all APs, so that the client's SSID must match that of the AP.
  - Validating that the SSID character string does not reflect the organization's name (division, department, street, etc.) or products.
  - Understanding all default parameters and ensuring that all have been changed.
  - Disabling all nonessential management protocols on the APs.
  - Changing the default AP administrative password and ensuring that all APs have strong administrative passwords.
  - Enabling user authentication mechanisms for the management interfaces of the APs.
  - Ensuring adequately robust community strings are used for SNMP management traffic on the APs.
  - Configuring SNMP settings on all APs for the least privilege (i.e. read only). Disable SNMP if it is not used.
  - Forbidding the installation of APs without the consent and oversight of the network manager.
  - Positioning AP on the interior of buildings, if possible, versus locations near exterior walls and windows.
  - Empirically testing AP range boundaries to determine the precise extent of the wireless coverage and then positioning all APs so that their signals are less likely to leak outside the intended coverage area.
  - Placing APs in secured areas to prevent unauthorized physical access and user manipulation.
  - Ensuring the APs are turned off when not in use, e.g. weekends, holidays and even overnight if the workday approximates a nine to five shift.
  - Ensuring AP channels are at least five channels different from any other nearby wireless networks to prevent interference.
  - Ensuring that the APs' reset function is used only when needed and invoked by authorized personnel.
  - Ensuring that APs are restored to the latest security settings when the reset function is used.
- Secure all wireless computing devices by:
  - Changing the default vendor-set SSID on all devices.
  - Installing antivirus software on all devices.
  - Installing personal firewall software on all devices.
  - Ensuring that the "ad hoc mode" is disabled unless the environment is such that the risk is tolerable.
- Choose NICs that support password-protection of attribute changes to prevent the setting of the NICs from being changed by end-users.

- Ensure that default shared keys are periodically replaced by more secure unique keys.
- Deploy MAC access control lists.
- Use static IP addressing on the network.
- Disable DHCP (Dynamic Host Configuration Protocol).
- Ensure that management traffic destined for APs is on a dedicated wired subnet.
- Use a local serial port interface for AP configuration, to minimize the exposure of sensitive management information.
- Since WLANs can be exposed to viruses and worms when end users download data from a website, implement real-time anti-virus scanning at the network gateway and ensure that it is applied at all WLAN access points to prevent infection and rapid spread of content-based attacks.
- Ensure end-users are fully trained in computer security awareness and risks associated with wireless technology.
- Stay current.
  - Evaluate and adopt the most powerful wireless standards, when pertinent, as soon as they become available.
  - Fully test and deploy software patches and upgrades on a regular basis.
  - Designate an individual to track the progress of 802.11 security products and standards (IETF, IEEE, Wi-Fi Alliance, etc.), and the threats and vulnerabilities relating to the technology.
- Perform comprehensive security assessments at regular intervals, including validation that rogue APs are not operating within the organization's facilities.
- Determine corporate-wide procedures for network authentication. Include user-based authentication, rather than device-based (so that an intruder can't gain network access by stealing or simulating a wireless device), as well as centralized management of authentication (i.e. authentication credentials are stored in a central repository and don't have to be distributed to every access point).
- Install a properly configured firewall between the wired infrastructure and the wireless network (AP or hub to APs).

Now let's take a closer look at some of the aforementioned security measures. There is no excuse not to implement the measures set out below, as they are easy to implement. When used in tandem they can help to ensure a WLAN's protection.

WEP (Wired Equivalent Privacy) refers to scrambling communications between a computing device and an AP, using a symmetrical encryption technique called RC4 on the Data Link Layer. In an attempt to provide a more secure transmission mode, Wi-Fi equipment manufacturers cooperated in the development and adoption of the encryption standard called Wired Equivalent Privacy (WEP). Almost every vendor supports WEP in some way or the other. When used correctly, WEP can prevent casual eavesdropping and unauthorized access. While WEP offers 40, 128, 152 and 512-bit encryption strengths, it also suffers from a number of drawbacks. All users have the same encryption key, so when one key is compromised the entire network is jeopardized. Many early APs and NICs only support 40-bit encryption, but newer gear supports 128-bit and greater WEP encryption. But the original 40-bit encoding key used by WEP can be broken in just a few minutes by a hacker using freeware such as AirSnort (see http://airsnort.shmoo.com), and the more powerful 128-bit WEP keys can be broken in a few days.

***Note:** The most recent form of WEP supports 512-bit encryption. Proxim, for example, has developed an 802.11a/b/g PC card, an 802.11b/g PC card, and an 802.11a/b/g PCI card for client devices. The 802.11a/b/g PC card is supplied in both a silver and a gold version. The silver card supports 64- and 128-bit WEP protocol encryption, while the gold version adds 512-bit WEP support.*

Here is another note of caution about WEP: its shared key system is often considered a management nightmare because the network administrator is responsible for distributing passphrases, hex keys, or ASCII strings that represent the encryption key. If the key is leaked, things can get messy very quickly. Not only is the data compromised, but the procedure for changing keys varies from vendor to vendor. Disseminating a new key, while not complicated, can be difficult.

Service Set Identifier (SSID) can be vulnerable. (A SSID is an ASCII string configured by network administrators into all access points and wireless stations that share a common WLAN.) Since a SSID is a relatively simple password that is common to all devices on the WLAN, it is easy to compromise. Furthermore, since the default setting of SSID is more times than not left unchanged for long periods of time, and since APs are typically configured to broadcast their SSID in the open, an intruder can easily obtain the network's SSID by using readily available tools.

PCs and laptops only respond to a network tied to their particular SSID, but SSIDs are not only transmitted, they're sent unencrypted. This issue stems from the 802.11b specification. For example, according to the SecureNet Service (SNS) security advisory at www.lac.co.jp/security/english/snsadv_e/60_e.html, Windows XP on a PC maintains a list of all the access points to which it has ever connected. If, say, a laptop boots up and it's out of range of any access point, it searches for available access points by continuously broadcasting inquiries, each of which contains the SSID of every AP it has ever encountered. Therefore, it is possible to "sniff out" these SSID values assigned to registered access points by using a packet monitoring tool for wireless LANs.

The advisory further states: "sending out packets encrypted with WEP is not a recommended security practice in an environment where the original access points are not available."

Later we'll see how to handle Wi-Fi security issues through the use not only of VPNs, but also of WPA (Wi-Fi Protected Access), which is a new interim solution for link-layer security that is based on the work-in-progress at the IEEE 802.11i Task Group.

MAC address filtering is a useful security measure for small WLANs, but it's not for large installations. Every wireless NIC has a unique MAC address, a 12 digit hexadecimal number that is unique to each and every NIC in the world. And since each NIC has its own MAC address, if you limit access to WLAN APs just to authorized MAC addresses, you can control who should and should not be accessing the network.

But while MAC address filtering is a good idea for small networks, it presents problems for larger WLAN installations. The first is the management aspect—to implement MAC address filtering, the WLAN manager must keep a database of every MAC address allowed to access the network. This database must be kept either on each AP individually, or on a special RADIUS server that each AP can access. Any time a device is added, lost, stolen, or changed in any way, the network manager must update the list. If the WLAN serves a small user-base, MAC address filtering is a viable security measure. In an enterprise network, however, it's not a practical solution because it would necessitate a full time employee just to keep up with database changes. Second, MAC address filtering complicates support for roaming between different APs.

Before implementing this security measure, know that (1) MAC addresses can be spoofed, and (2) filtering merely verifies the identity of the NIC, not the identity of the computing device or the person using the computer.

While we have laid out the outline of a comprehensive network security plan, so far we have only discussed in detail minimal security measures. Let's now up the ante.

OPTIONAL SECURITY MEASURES

Most if not all of the preceding measures should be mandatory for all WLANs. But for comprehensive security within a wireless environment, you should also consider other security measures, such as:

- Implementation of additional authentication procedures, e.g. dynamic session-based encryption keys that would be changed automatically at fixed intervals and on re-authentication.
- Deployment of other forms of authentication for the wireless network, e.g. RADIUS and Kerberos.
- Deployment of other authentication methods, e.g. biometrics, smart cards, two-factor authentication, or PKI.
- Installation of Layer 2 switches in lieu of hubs for AP connectivity.
- Deployment of IPsec-based VPN technology for wireless communications. Enhance AP management traffic security through the use of SNMPv3 or an equivalent cryptographically protected protocol.
- Upgrading existing WLAN equipment so they can use WPA.
- Deployment of intrusion detection sensors on the wireless part of the network to detect suspicious behavior or unauthorized access and activity.

Now let's look more closely at these optional network security measures.

WPA

As the reader should now understand, WEP is ineffective in securing confidential business communications. The IEEE 802.11i standard, a longer term security solution for wireless networking, has had several delays and is not expected to be ratified until toward the end of 2003. These delays caused several members of the Wi-Fi Alliance to team up with members of the IEEE 802.11i task group to develop a significant near-term enhancement to Wi-Fi's security. Together this team developed the Wi-Fi Protected Access (WPA) standard.

In November 2002, the Wi-Fi Alliance released WPA as an interim replacement for WEP and other aspects of Wi-Fi security. As an interim version of the IEEE's 802.11i security specification, WPA adopts TKIP-based fixes to WEP, adds the packet integrity upgrade, and introduces standardized support for 802.1X/EAP network authentication.

WPA is designed to run on existing hardware as a software upgrade. Plus, since it is derived from the IEEE 802.11i standard, WPA is designed to be forward compatible with 802.11i, once that standard is ratified. When properly installed, WPA can provide a high level of assurance that as network data traverse the airwaves, it will remain protected and that only authorized network users can access the network.

But what exactly does WPA offer? While no security solution can ever claim to be "absolutely secure," the protection that WPA provides is significant. In fact, many cryptographers are confident that WPA addresses all of WEP's known deficiencies.

WPA requires that clients and access points use a shared network password (technically called the "preshared secret"). But unlike the WEP's "passphrase," which merely converts ASCII into hexadecimal, the WPA password actually creates a cryptographic outcome that's sufficiently random so that breaking a key is an unlikely event. And because the implementer only needs to enter a password, this improves the likelihood it will be used. Still, it should be noted that WPA will fall back to WEP if *even a single device* on a network cannot use WPA.

Next, WPA uses Temporal Key Integrity Protocol (TKIP), which is part of the 802.11i specification. The IEEE 802.11i Task Group designed TKIP to distance the encryption key from the actual data by (1) performing several algorithms to the key before generating the encrypted data; (2) performing dynamic key management (i.e. changing the temporal keys frequently); and (3) performing message integrity checks (MICs) to prevent forgery and replay.

WPA, therefore, constructs encryption keys differently.

WPA also employs the IEEE 802.1X access control protocol (which is usable on wired networks as well). Although 802.1X requires special server software, that software is commonly found in most corporate networks. However, since WPA uses TKIP in addition to the 802.1X access control protocol, it still can be used to improve the security in home and small-office networks that don't use such servers.

Besides products from the aforementioned vendors, Cisco's AIR AP1230B access point, Intel's Pro/Wireless 2100 LAN 3B Mini-PCI Adapter (used in Centrino notebooks), and Symbol's Wireless Networker CompactFlash Wireless LAN Adapter Model LA-4137 have received the Wi-Fi Alliance's WPA certification. In the near future, WPA certification will be a standard aspect of Wi-Fi certification.

Furthermore, the Wi-Fi Alliance supports WPA with a certification program. The Alliance announced in late April 2003 that nine products from six vendors had been certified as meeting the specifications for WPA. Out of those six companies—Atheros, Broadcom, Cisco, Intel, Intersil, and Symbol—three are Wi-Fi chip vendors (Atheros, Broadcom, and Intersil) and together they supply the chips for most Wi-Fi network cards and access points. That means that the vendors whose Wi-Fi components use those chipsets can produce WPA compliant products very quickly.

Moreover, most vendors are expected to provide WPA security for their existing installed product base, through downloadable software updates. If you have an existing WLAN, check your vendors' websites to see if updates are available. Note that if you do upgrade your WLAN or build an additional WLAN around WPA security, and your organization is also using older Wi-Fi gear that isn't upgradable, that gear will not be able to access the WPA-secured network. And if the legacy gear is allowed to access a WPA network, the WPA-enabled gear will revert to its old WEP ways.

The WPA effort also has been driven by the need for enhanced Wi-Fi security that would be software-upgradeable to the more than 490 Wi-Fi certified products in existence today. Since after 802.11i is ratified, obtaining all of the advantages of the full IEEE 802.11i standard likely will require a hardware change, WPA will have a place in the Wi-Fi security picture for the near future.

The concern with WPA, though, is how such an upgrade will affect organizations that use, or want to deploy, network authentication. EAP, which is the most prevalent authentication method in use today, is not a secure protocol—it sends its messages as cleartext. While a method of tunneling EAP using TLS (Transport Layer Security) is available, it requires installing client certificates on every computer that wants to connect. And even TLS leaves some information in the clear, which potential intruders might find useful. Nonetheless, an EAP/TLS combo offers mutual authentication in which the client and authentication server can verify each other's identity before the transaction starts.

A fix to this problem is in the works. It is referred to as "EAP-TTLS," "Tunneled TLS," or just "TTLS." This solution allows the computing device first to connect using a generic TLS connection to the server, and only after the connection is obtained does it send its user information to start an EAP/TLS session, i.e. encryption is tunneled within encryption.

Another aspect of TTLS is that in addition to EAP, it supports several older messaging standards. As of this writing, TTLS is in draft form at the Internet Engineering Task Force (IETF). However, client software is available for most major platforms, including OS X, all Windows flavors, Linux, and several commercial versions of Unix.

TTLS isn't the only method available to secure EAP transactions. Microsoft and Cisco are pushing PEAP (Protected EAP). This solution uses an approach that is almost identical to the TTLS method, but it only supports EAP and CHAP, an older Microsoft authentication protocol. PEAP is also in front of the IETF. Microsoft has shipped PEAP updates to Windows XP and 2000, and is expected to follow with Windows 98 (both versions), NT 4.0, and ME.

In the ideal world, the two standards would merge into a single one. Whether that occurs is anyone's guess. Microsoft's lack of support for non-EAP protocols could be a way for the software giant to push users away from other authentication systems.

Public Key Infrastructure (PKI)

Consider PKI for networks or end-users requiring high levels of security. This comprehensive system provides public key encryption and digital signature services through a framework of services for the generation, production, distribution, control, and accounting of public key certificates. The purpose of PKI is to manage keys and certificates (also known as digital IDs) that are used for identification, entitlements, verification, and privacy. By managing keys and certificates through PKI, an organization establishes and maintains a secure and trustworthy networking environment. In the past few years, PKI technology has become the preferred means for providing stronger levels of identification, privacy, verification, and security management capabilities.

Since PKI provides strong authentication through user certificates, users can use those same certificates with application-level security, such as signing and encrypting (i.e., using encryption certificates) messages. Thus PKI provides applications with secure encryption, and authenticates network transactions as well as data integrity and non-repudiation.

It wouldn't be too difficult for WLANs to integrate PKI into their ecosystem for authentication and secure network transactions. Third-party manufacturers, for instance, provide wireless PKI, handsets, and smart cards that integrate with WLANs. Smart cards could provide even greater utility (e.g., portability, mobility) since the certificates are integrated in the card. Users requiring lower levels of security, on the other hand, need to consider carefully the complexity and cost of implementing and administering a PKI before adopting this solution.

RADIUS

RADIUS (Remote Authentication Dial-In User Service) is an IETF security management protocol that lets you easily control which remote and/or wireless LAN users connect to your network, and what resources they can access. While RADIUS isn't an official standard, the RADIUS specification is maintained by a working group of the IETF.

A RADIUS server authenticates dial-in users on a network. For example, when you log into your favorite ISP, the remote access server you talk to sends your login info to a RADIUS server which answers with your rights and configuration details. If your password is wrong, you won't be able to log in.

RADIUS is a widely deployed protocol for network access authentication, authorization and accounting (AAA). And even though RADIUS has problems relating to issues with security and transport, RADIUS is a part of the 802.11i specification. This is primarily due to the fact that RADIUS is simple, efficient, and easy to implement—making it possible for RADIUS to fit into the most inexpensive embedded devices. Furthermore, the issues with transport are most relevant for accounting in situations where services are billed according to usage. This is because RADIUS runs on UDP, with no defined retransmission or accounting record retention policy, and does not support Application Layer acknowledgments or error messages. As a result, usage-based billing is often done with SNMP, which also runs over UDP, although a TCP transport mapping has been developed.

In terms of security, RADIUS offers some Application Layer security such as:

- Trust established between RADIUS clients and servers via a shared secret, which is an encryption key used by RADIUS to send authentication information over a network.
- Support for per-packet integrity and authentication.
- Support for hiding of specific attributes such as User-Password, Tunnel-Password and Microsoft Vendor Specific Attributes.

But RADIUS security is particularly poor when dealing with the cleartext Password Authentication Protocol (PAP). IETF's RFC 2865 document requires that the RADIUS Response Authenticator be globally and temporally unique, since the stream cipher that RADIUS uses to encrypt User-Password is based on the Response Authenticator, and thus if it were to repeat, the keystream would be compromised. And, unfortunately, not all implementations of RADIUS ensure that the Response Authenticator is not repeated.

Furthermore, the RADIUS Response Authenticator and Message-Authenticator attributes are both vulnerable to dictionary attack. And simple shared secrets are commonly shared with many or even all network access servers.

***Note:** A network access server, commonly referred to as a "NAS," is a computer or a special device designed to provide access to the network. For example, a NAS can be a computer connected to the network and equipped with several modems to allow outside end-users to access the network. Also services, such as PPP, SLIP, telnet, etc., use a NAS.*

Here are a few steps you can take to lessen RADIUS's security vulnerabilities:

* To make an attack against the User-Password impossible, don't allow PAP.
* Use a credible generator for Request Authenticator as described in IETF RFC 1760.
* Include a Message-Authenticator attribute in all packets (RFC 2869 already requires this for EAP authentication).
* If possible, use authentication methods that offer protection against an offline dictionary attack. These include the aforementioned EAP TLS or two methods that are still under development—EAP SRP (Secure Remote Password) and PEAP (Protected Extensible Authentication Protocol).
* Use high-entropy shared secrets (i.e. don't limit shared secrets to 16 characters and utilize randomly generated shared secrets).
* Use a different shared secret for each RADIUS client-server pair.
* If the network access servers can support IPsec, then the best thing to do is to forsake RADIUS Application Layer security entirely and just run RADIUS over IPsec ESP (Encapsulating Security Payload) with a non-null transform, as described in IETF RFC 3162. This method supports per-packet integrity, authentication, confidentiality, and replay protection for both authentication and accounting packets. Unfortunately, many embedded systems do not have sufficient resources to run IPsec, so RADIUS/IPsec is not widely used.

Kerberos

Developed by MIT, the Kerberos security system helps to prevent individuals from stealing information that gets sent across the wires from one computer to another.

The name "Kerberos" comes from the mythological three-headed dog whose duty it was to guard the entrance to Hell. Kerberos guards the "entrance" to a network by encrypting the data so that only the computer that's supposed to receive the information can unscramble it.

Kerberos authenticates users by way of exchanging electronic tickets between clients and services. It cleverly encrypts and de-encrypts these tickets before and after transmitting them. The "heart" of a Kerberos system is the Key Distribution Center (KDC). All the computers associated with a KDC make up what's called a "strengthened realm"

With Kerberos, by exchanging time-sensitive tickets, you can make transactions secure without sending passwords in plaintext over the network. For a program to take advantage of Kerberos, it must be Kerberized, which basically means that it can obtain tickets from the Kerberos server and negotiate with a Kerberos-aware service. Just about any application can be Kerberized, and just about any computer service can be made Kerberos-aware.

Although Kerberos is a fairly complex protocol, here are a few basic characteristics:

Every user and every service has a password. Only the owner of the password and the Kerberos server know that password. (Passwords, however, must remain confidential, as Kerberos provides no inherent protection against those that are stolen.)

Kerberos works by providing a "principal" (which represents the end-user or a computerized service), "tickets" that can be used to identify that end-user or service to other principals, and "secret cryptographic keys" (like passwords) for secure communication with other principals. These tickets are similar to a personalized business card that only the principal can have, because the principal must provide its Kerberos password to get the tickets from the Kerberos server (also known as the Key Distribution Center, KDC).

A computer service that uses Kerberos receives its own ticket when the service is set up. A person requesting use of a "kerberized" service must obtain his or her own tickets by using a Kerberos initialization program called "kinit", and by providing kinit with his or her Kerberos password. When a person requests to use a kerberized service (usually a program on a personal computer which communicates with a server somewhere on the network), the program sends the ticket obtained from kinit to the computer service. The computer service inspects the user's ticket in conjunction with its own to make sure it is valid. If the ticket passes inspection, the service is confident that the person who sent the ticket is truly who and where (which computer) he or she claims to be. Conversely, the service, providing its own ticket for inspection, inherently proves its own identity. Both sides of the communication are now authenticated to each other and can securely exchange information.

Biometrics

If security is an absolute top priority for a networking environment, consider biometric solutions. Biometric devices include fingerprint/palm-print scanners, optical scanners (including retina and iris scanners), facial recognition scanners, and voice recognition scanners. Biometrics provides an added layer of protection when used either alone or along with other security solutions. For example, for organizations needing higher levels of security, biometrics can be integrated with wireless smart cards or wireless laptops or other wireless devices, and used in lieu of username and password to access a wired or wireless network. Additionally, biometrics can combine with VPN solutions to provide authentication and data confidentiality.

VPNs

You may think that a simple solution to WLAN security exists in the form of VPN technology, and some organizations do implement generic VPNs to provide wireless LAN security. But there are a number of new products that can enhance VPN capabilities to meet wireless networking's unique needs. In some cases, this may involve more sophisticated policy-based access controls. In other instances, it may include supporting VPN access while users roam between access points, not only on the same subnet but also between subnets, with session-persistence capabilities to ensure that applications are not interrupted.

Even though VPNs may solve key problems associated with WLAN security, they aren't a panacea—at least not yet.

VPN gateways are implemented within many organizations at the boundary of the enterprise LAN and the Internet, to provide secure data transmission across public network infrastructures. In recent years VPNs have allowed organizations to harness the power of the Internet for remote user access (dial-up, DSL, cable-modem). Today, VPNs typically are used in three different scenarios: remote user access, LAN-to-LAN (site-to-site) connectivity, and extranets. These VPNs are designed to provide authentication and privacy, some also can be integrated with firewall software to control access, and other VPN products come with traffic-shaping software to limit bandwidth consumption by application, user, or group.

The VPN / WLAN Connection

When VPNs are used in a WLAN environment, they provide data privacy via strong encryption to prevent eavesdropping, and facilitate authentication of wireless stations and their users. VPNs can employ simple user names and passwords in a Remote Authentication Dial-In User Server (RADIUS) directory to more sophisticated digital certificates and public key techniques; and cryptographic techniques to protect IP information as it passes from one network to the next, or from one location to the next. By the act of encapsulation and encryption of one protocol packet inside another, isolating those packets from other network traffic, you create a "VPN tunnel," which is how a VPN protects data as it traverses various network byways, including the Internet.

VPNs are commonly implemented to up a WLAN's security quotient, although VPNs can present challenges to the overall management of the network. To use a single VPN gateway to secure all WLAN traffic, all traffic must funnel through a connection to the VPN. In many organizations, a distinct wireless network is created and connected to other internal networks by a VPN gateway. But to support such a separation, you need an appropriate Ethernet network infrastructure, including plenty of bandwidth and virtual LAN (VLAN) capabilities. In addition, as with any VPN, you'll need to ensure that all users have appropriately configured VPN clients, which may require a software installation on every mobile device.

Most VPNs in use today make use of the IPsec protocol suite. IPsec, developed by the IETF, is a framework of open standards ensuring private communications over IP networks. It provides the following types of robust protection:

* Confidentiality ensures that others cannot read the information in the message.
* Connectionless integrity guarantees that a received message is unchanged from the original.
* Data origin authentication guarantees that the received message was sent by the originator and not by a person masquerading as the originator.
* Replay protection provides assurance that the same message is not delivered multiple times, and that messages are not out of order when delivered.
* Traffic analysis protection provides assurance that an eavesdropper cannot determine who is communicating, or the frequency or volume of communications.

IPsec routes the messages via an encrypted tunnel by two special IPsec headers, which are inserted immediately after the IP header in each message. The Encapsulating Secu-

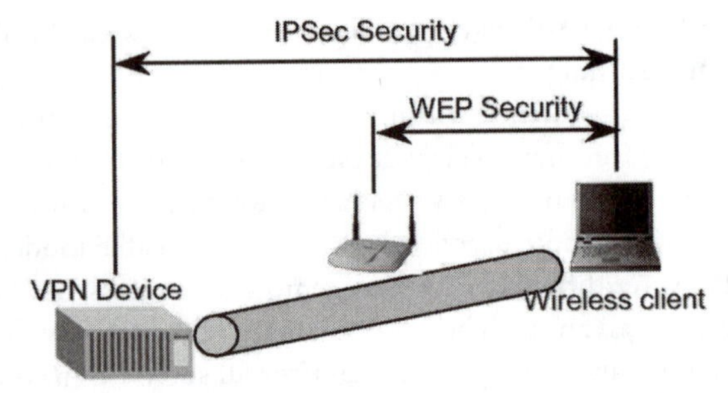

Figure 17.1 A WLAN with a VPN that uses IPsec in addition to WEP.

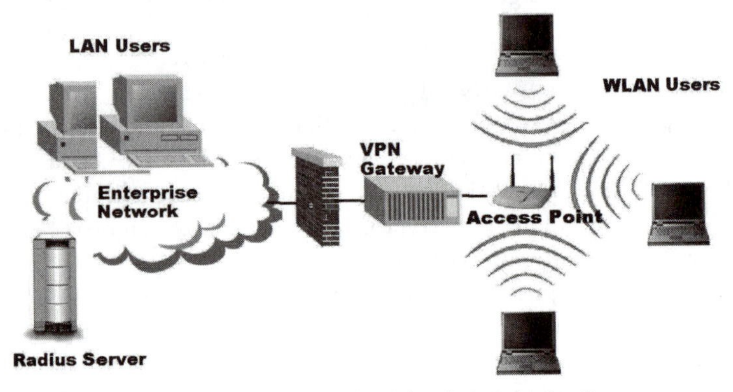

Figure 17.2 An example of a wireless network with a "VPN overlay."

rity Protocol (ESP) header provides privacy and protects against malicious modification, and the Authentication header (AH) protects against modification without providing privacy. The Internet Key Exchange (IKE) Protocol is a mechanism that allows for secret keys and other protection-related parameters to be exchanged prior to a communication without the intervention of a user.

As shown in Fig. 17.1, the IPsec tunnel is provided from the wireless client through the AP to the VPN device on the enterprise network edge. With IPsec, security services are provided at the network layer of the protocol stack. This means all applications and protocols operating above that layer (i.e., above layer 3) are IPsec protected. The IPsec security services are independent of the security that is occurring at layer 2, the WEP security. As a defense-in-depth strategy, if a VPN is in place, an organization can consider applying IPsec and WEP. With a configuration as shown in Fig. 17.2, the VPN encrypts (and otherwise protects) the transmitted data to and from the wired network.

As shown in Fig. 17.2, a VPN overlay allows wireless devices to connect securely to the enterprise network through a VPN gateway on the enterprise edge. Wireless clients (computing devices) establish IPsec connections to the wireless VPN gateway—in addition to or as a substitute for WEP. (The computing device doesn't need special hardware, although it must be provided with IPsec/VPN client software.) The VPN gateway can use preshared cryptographic keys or digital certificates (public-key based) for authentication. Additionally, user authentication to the VPN gateway can occur using RADIUS or one-time-passwords (OTP) generated with SecureID, for example.

The VPN gateway may or may not have an integral firewall to restrict traffic to certain locations within the enterprise network. Additionally, the VPN gateway may or may not have the ability to create an audit journal of all activities. An audit trail is a chronological record of system activities that is sufficient to enable the reconstruction and examination of the sequence of environments and activities. A security manager may be able to use an audit trail on the VPN gateway to monitor compliance with security policy and to verify whether only authorized persons have gained access to the wireless network.

In short, consider using VPN technology in conjunction with other security methods to up a WLAN's security quotient. But keep in mind that while the VPN approach enhances the air-interface security significantly, this approach does not completely address security on the enterprise network. For instance, authentication and authorization to a particular enterprise application are not addressed with this security solution.

Where do security-minded organizations turn next? Some organizations may seek to develope a comprehensive enterprise security strategy utilizing specialized security products that have recently appeared on the market.

SECURITY PRODUCTS

The IEEE 802.11 committee has credibly addressed workable wireless LAN standards for the Physical and Data Link Layers, but the absence of a standards-based security architecture is a big headache for organizations contemplating a large-scale rollout of WLAN services. Even before WEP's weaknesses were exposed, vendors and enterprise users recognized that something more was required for WLAN security. Numerous products specifically address a WLAN's vulnerabilities and several vendors offer combined security solutions in a single product. The four vendors whose products are described in this section are representative of such enterprise security solutions.

Each of the four solutions discussed herein provides good management capabilities that significantly enhance the security and manageability of wireless devices. But, all require careful installation planning to ensure they do not constrain the WLAN's performance. Finally, be advised that all of the following security solutions are expensive—not only to acquire, but also to implement and manage.

Three of the products discussed in this section—Bluesocket's WG-1000 Wireless Gateway, ReefEdge's Connect System, and SMC Network's EliteConnect—are hardware-based systems; the NetMotion's Mobility solution runs as services on a Windows server, but don't read too much into this distinction. The hardware offerings typically run on Pentium-based appliances—they may hide the software under a sleek cover, but it's still there.

The reason we will examine a software-based system is that if your WLAN is geographically large but the number of users is relatively small, you may find better value in a software solution that can be licensed on a per-user basis. Although perhaps the bigger distinction has to do with client software, which NetMotion's Mobility requires on the mobile devices. This client software provides some advanced functionality, but you'll need to deal with distribution and update issues, and you may be limited in terms of the supported platforms.

For the NetMotion platform, wireless traffic is routed through a central Windows server, which handles authentication, security policy-compliance, and in some cases, roaming, whereas the hardware vendors take a slightly different approach. Access controllers are installed between the WLAN's access points and the organizaton's network infrastructure—usually in communication closets at the edge of the network—and act as highly configurable firewall-VPN devices. Depending on the services, client software might be unnecessary, or you might need a standard IPsec client.

The ReefEdge and SMC products also provide roaming across IP subnets, a key feature if your users need to move between physical locations (and networks) without disrupting their network sessions. (Bluesocket's device will support roaming when version 2.0 becomes available.) However, if you have lots of users, installing hardware at your network's edge provides significant flexibility and distributes the processing load, but it adds considerably to WLAN deployment costs. Some vendors have suggested that since their products provide important services, you can cover the added cost by deploying lower-cost commodity-class access points. But low-cost access points may lack the range and reliability of enterprise-class alternatives. (There is more discussion on dumb versus smart access points in Chapter 18.)

Like that of firewalls and traditional VPN gateways, the performance of these comprehensive security products is a key concern, especially with encryption enabled. Evaluate system performance in the context of your own network. Many issues—including the number of hardware boxes deployed, the speed of servers and the application mix (which will affect the packet-size distribution)—can affect performance. And with WLANs moving to 54 Mbps and beyond, you'll need to factor such speed increases into your plans as well.

Getting Down to Details

Now to see what you can expect in the way of enterprise level security products, we will examine four products in detail.

Bluesocket WG-1000 Wireless Gateway 1.0. This product can deliver maximum control over your WLAN. It is loaded with features and it offers very good performance. Although the first-generation product lacks roaming support and offers limited centralized management capabilities, the latest version, 2.0, now provides roaming support.

The WG-1000 supports a variety of external authentication systems, including NTLM, RADIUS, and LDAP. Its web-based configuration and management is easy to navigate. Not only does it offer QoS via bandwidth throttling, it lets you see how much traffic each user generates through a particular gateway. IPsec and PPTP encryption mechanisms are supported too.

As a single-box hardware solution, the WG-1000 lacks a central server, which can complicate management in large environments where multiple systems are deployed. You can set up two wireless gateways in a master-slave configuration to provide fail-over support for added redundancy, but if you want to support a new wireless subnet, you have to configure a new gateway from scratch. Even Version 2.0 lacks a satisfactory central management system. Bluesocket says that this deliberate omission is because it wants to avoid the single point of failure that exists in the two-tier hardware solutions. But some network managers may feel that, in the long run, adequate redundancy and central management are both necessary.

The WG-1000 lets you monitor end-users from a web page, with the status automatically updating periodically. This feature can provide near real-time information on who logs into the system and how much traffic that user is generating. It also informs you of the CPU usage on the gateway. In addition, the WG-1000 can store two separate system configurations, a handy safeguard if the new configuration fails.

Since the earlier version of WG-1000 lacks roaming support, it will work best in an organization with a flat wireless network requiring only one gateway, or in environments where management of subnets and network security may be decentralized (many university setups fit this description). However, with encrypted throughput of around 30 Mbps, a single gateway might introduce performance problems on high-traffic networks. Nonetheless, if your organization can live with these performance limitations, Bluesocket's WG-1000 is an appealing and cost-effective solution.

NetMotion Mobility Server 3.50. Mobility's strongest points are its ease of installation and use, and its effective session-persistence and roaming. This software solution should scale to meet the needs of most midsize and large WLAN installations—as long as you are willing to invest in high-performance server hardware. NetMotion is a pioneer in this space, and its website includes excellent white papers on wireless security, addressing the needs of organizations planning to deploy wireless, and techniques to optimize performance.

Server installation should be a breeze. If you have a Windows network, this product will fit in easily with your existing NT Domain or Active Directory. The wizard will guide you through the input of minimal configuration information. To authenticate clients that connect to the server, the domain or AD is checked to see if the user belongs to a NetMotion user group. The Mobility server operates on UDP Port 5008, which can be allowed through your firewall or forwarded in a NAT environment. Note, though, that Mobility requires proprietary client software (available only for Windows platforms).

NetMotion implements several algorithms for encryption, including AES-Rijndael, DES, 3DES, and Twofish, which you choose on a per-user basis. The software also supports IPsec, L2TP/IPsec, and PPTP, implemented with the assistance of standard Windows 2000 and XP services. If your users' mobile devices run Windows 2000 or XP clients, they communicate with the Mobility server using Windows' integrated encryption services as an alternative to Mobility's, but you'll still need to run the Mobility client.

In a performance test, Mobility delivered 23.75 Mbps encrypted and 33.6 Mbps unencrypted performance on our 600-MHz server. Although this level of performance may be of concern if your network is very large, NetMotion says it has conducted extensive internal testing, demonstrating that performance scales in relation to CPU speed. In addition, the multithreaded architecture takes advantage of multiple CPU configurations. Performance is limited by bidirectional network traffic over a single network interface; adding multiple interfaces can dramatically increase the performance.

While the NetMotion Mobility product did not provide the author with an organization that supports thousands of concurrent users, the vendor did point the author to a number of customers with hundreds of concurrent users that are operating with a single server. Although a single-server implementation is the easiest to manage and can support a significant number of users, NetMotion lets you set up distributed Mobility servers to segment traffic. You also can enable failover by setting up a redundant server that will compensate for a failed server if there is an outage. However, you cannot manage all servers from a single console.

Mobility also handles roaming, although not as well as ReefEdge and SMC, probably because those systems let clients roam without releasing and renewing IP addresses.

Mobility works great if operating within a Windows environment with automated client software distribution. If, however, you are in an environment that doesn't allow for easy software rollouts, or requires the use of non-Microsoft client operating systems, consider a hardware-based solution that can operate with standard VPN clients.

ReefEdge Connect System 2.06. ReefEdge Connect is a two-tiered hardware solution, with ConnectBridge devices attached to access points at the network's edge and a ControlServer at the center. (The term bridge is confusing, since it suggests a Layer 2 relationship between devices, when in reality this product operates at higher layers.)

ReefEdge Connect also offers QoS (Quality of Service) capabilities that let the administrator throttle bandwidth. This could be handy in a bandwidth-constrained environment, such as a high-density WLAN, or an environment in which you want to limit the impact of bandwidth-hungry applications like streaming media.

ReefEdge's support for back-end authentication database integration is a bit limited, although it offers support for NTLM and RADIUS in addition to its own standalone authentication database. ReefEdge supports IPsec for secure sessions, but lacks support for PPTP and L2TP/IPsec.

There are a few items that network managers might find a bit irritating. Every time you make even minor changes on the ConnectServer, including defining static NAT address mappings, you are required to reboot the affected ConnectBridge or ControlServer. Also, during the time that the ConnectServer is rebooting, wireless users have no access to network resources, because the ConnectServer acts as their DNS server. And when the ConnectServer is back up, users are forced to reauthenticate. In addition, the ReefEdge product requires that you use NAT on all clients. However, you can map NAT addresses to external IP addresses statically using the management interface, which can be helpful to support the access points, which, without IP addresses, would be unreachable from the outside network.

ReefEdge offers two versions of its ConnectBridge, recommending the smaller ConnectBridge 25 for support of a single access point only. Also it is noted that the ConnectBridge 100 supports higher speed traffic, where the ConnectBridge 25 is limited in capacity—a *Network Computing* magazine's test managed only 8 Mbps of throughput unencrypted and 4 Mbps encrypted. Though in a strange twist of events, the testers discovered that they could disable a ConnectBridge 25 by sending it 6400-byte ping packets, so don't use the ConnectBridge 25 except in very-low-traffic environments.

ConnectBridge 100, however is another story. It passed a *Network Computing* magazine's tests with ease, providing wire-speed throughput in the unencrypted datastream tests. The product's roaming features are extremely efficient. When roaming between subnets, ReefEdge Connect tunnels data from a previous session started on one bridge to the next bridge. However, if you roam to a third subnet/bridge, the first bridge will tunnel to the third bridge and not the second. According to the vendor, this is to ensure that tunneling is handled efficiently and lowers the overhead on your backbone.

ReefEdge Connect has an attractive web interface and the latest version, 2.5, supports L2TP, PPTP, and LDAP. Also, ReefEdge supports a crypto-accelerated version, which is said to offer encrypted throughput in excess of 60 Mbps.

SMC Networks EliteConnect WLAN Security System. SMC EliteConnect is an OEM version of Vernier's two-tiered 6000-series system. The Secure Server authenticates users, maintains a consistent configuration across your network and manages roaming operations, while the Access Manager connects to access points at the edge of each subnet. The Access Manager enforces network policies, creates and manages encrypted tunnels, and provides secure access to the network resources. This architecture facilitates scalability to the extent that you can control how many access points connect to each Access Manager. SMC even includes an integrated four-port Ethernet switch, though most sites can define VLANs on existing closet switches to accomplish the same goal.

EliteConnect installation and configuration should take just a few minutes. To get the system running, just plug the access points into the Access Managers, connect the Access Managers and Secure Server to their appropriate switches, and point your browser at the Secure Server's default IP address. Then use the web interface to complete the configuration.

EliteConnect delivers a highly appealing feature set together with excellent configuration and management capabilities. In addition to supporting authentication, access control, and subnet roaming, EliteConnect offers other convenience features for managing WLAN resources. Network managers will appreciate the product's ability to drill down through the management interface and see what connections the users make in near real-time. And EliteConnect presents its network monitoring information in an easily readable format.

Another useful management feature is that EliteConnect lets you customize the end-user authentication web page. Its integrated NTP time server support makes it easy to keep the log's time stamps in sync, and helps to avoid problems that can occur with time-sensitive authentication protocols, such as Kerberos.

It is noted that although EliteConnect's throughput is a bit slower than that of its competitors, its distributed design makes it easy to install enough boxes so that the product does not introduce performance bottlenecks, at least when using current-generation WLAN systems. Of course, adding more boxes also adds to the cost and increases management overhead.

Some network managers, however, may find the EliteConnect's NAT (Network Address Translation) addressing scheme, which limits the number of clients per subnet, rather awkward. They can overcome this, though, by altering the subnet mask and assigning static IP addresses. EliteConnect also prevents you from pinging the Access Manager's internal address, something that may be irritating for managers that use ping tests for basic monitoring and troubleshooting. But, EliteConnect's session persistence and subnet roaming times are impressive. During a *Network Computing* magazine product test, it never dropped a packet while roaming.

The "Three Security Keys"

The reader was introduced to the three keys to real security earlier in this chapter. Now let's see how these security products deal with the three "keys."

Authentication: Most network managers want to authenticate WLAN users to ensure only legitimate users gain access to their systems. In fact, in organizations where privacy isn't so critical or where it is implemented at higher layers in the stack, authentication may be the only requirement. If this describes your situation, then note that Bluesocket, ReefEdge, and SMC Networks provide security systems that deliver web-based authentication. Before users can gain access to network resources, they fire up their browsers, which are redirected to a login page provided by the access controller. Once authentication is complete, user and group access-control policies take effect.

Access Control: In most cases, managers will want to tie access control to an existing accounts database–a Windows Domain, an Active Directory, or a LDAP service. Most products make this task simple, though the software systems that run under Windows have an easier time integrating into that environment. However, although the 802.1X protocol has garnered considerable attention in the WLAN industry as an authentication solution, none of the products in this section support that standard. The reason, at least according to the vendors, is that although 802.1X may represent the future for WLAN authentication, limited client availability and interoperability issues make it difficult to implement.

Privacy: Privacy via encryption adds another layer of complexity. The three products we examine in detail, Bluesocket's WG-1000, ReefEdge's Connect System and SMC's EliteConnect, all rely on VPN client software, which is included with many operating systems. It is noted, though that ReefEdge Connect also supports its own client, which adds functionality and simplifies installation. Some network managers may prefer to use VPN clients other than their operating system's standard versions. We found that enabling encryption hampered every solution's performance, so some managers may want to configure encryption on an application-by-application basis. Products like Columbitech's Wireless VPN, Ecutel's Viatores and NetMotion's Mobility handle encryption with special software that must be installed and configured on each client. However, you should know that this approach does add to the administrative burden.

Also, it is interesting to note that most security products enhance wireless security by offering access control through user- and group-based rules. These rules can be used to enforce use of encryption and restrict access to certain applications in much the same way a firewall does.

Supporting Subnet Roaming

One of the most valuable assets of wireless networking is the ability to support subnet roaming and session persistence. Some organizations implement WLANs using a flat address space and enforce policy where wireless and wired worlds meet. However, most enterprises want the flexibility to install wireless access points on multiple subnets. But when devices roam between subnets, problems, such as session persistence, can occur. For organizations that use WLANs primarily for email and web access, such problems may represent only a minor inconvenience, e.g. requiring users to renew their DHCP leases and reconnect to their mail servers. However, in environments that use stateful TCP-based applications (i.e. the applications require that information be "remembered" from one transmission to the next), a subnet roam will kill those programs unless the system can ensure session persistence. Thankfully, most enterprise level security products support subnet roaming and session persistence, although the specific techniques used to support subnet roaming vary from product to product, as does the speed at which the roaming takes place.

Performance and Scalability

Most security products have an adverse affect on a network's performance and scalability. To what extent, varies from product to product. To provide the reader with some intelligence on how the current crop of security products might affect a WLAN's performance and scalability, the author used some June 2002 tests that *Network Computing* magazine ran on a selected number of WLAN security solutions.

To evaluate performance, the magazine established a baseline using Ethernet-based client devices to pump as much traffic through each product as possible. Those Ethernet end points, equipped with Fast Ethernet interfaces, could transfer more than 94 Mbps, aggregate, using large frame sizes. Then the tests added, in turn, Bluesocket's WG-1000, ReefEdge's Connect System, NetMotion's Mobility, and SMC's EliteConnect products to the network and performed the same performance measurements again. The testers quickly learned that the network's wire-speed performance deteriorated dramatically with default factory encryption (usually 3DES or AES) enabled. However, both the Bluesocket and ReefEdge systems managed wire-speed performance when encryption was disabled, whereas the SMC's EliteConnect throttled throughput back about 10 percent. The NetMotion's Mobility solution took the biggest performance hit with throughput of around 33 Mbps with *encryption disabled.* The ReefEdge solution provided the *best encrypted* throughput at around 32.4 Mbps.

What are the implications of the above for scalability? The answer depends on how the system is designed and implemented. Where systems are engineered using a dis-

BEST PRACTICES FOR THE ROAD WARRIOR/HOTSPOT USER

The process of untangling wireless components deployed outside your IT department's control can be a headache, but it must be done to ensure the organization's data stays secure. Here are some tips on how best to ensure that end-users who access outside networks do so securely:

- Educate the end-user on the importance of disabling file sharing capabilities any time they access a network outside their "home" network.
- Be sure the end-user has both a personal firewall and antivirus software installed on their mobile computing device.
- Teach the end-user how important it is to use a secure transport system, such as the corporate VPN.
- Impress upon the end-user the importance of disabling the mobile computing device's wireless NIC when it is not in use. Then show them how.
- Ensure that the IT department stays abreast of updates, new security software, and OS patches and that the same are deployed as soon as advisable.
- Do not re-use passwords to sensitive systems.
- Clean up after security events.

tributed architecture, as with ReefEdge's and SMC Networks' products, you can control the number of access points whose traffic is funneled through an access controller. If you want more performance, you install more boxes at the network's edge. Some of the vendors' products will let you install multiple servers to distribute load, but the process may not be seamless. And while some products support fail-over for high availability, none support dynamic load-balancing.

It is also noted that all the vendors the author contacted gave long lists of customers, but none would (or perhaps could) offer names of customers that have implemented enterprise-scale environments with thousands of concurrent users. Of course, these products are relatively new to the market so perhaps that's not surprising. However, it does indicate that you must adopt a "prove it" mentality when evaluating options for a large-scale deployment.

These four products give the readers an idea of what to expect when they begin to investigate the numerous products available to secure a WLAN environment. Before investigating any potential solution, however, carefully consider the security features offered by various products and make sure that the residual risk, after the countermeasures are applied, is acceptable.

CONCLUSION

Today's WLAN security measures primarily comprise third-party infrastructure overlays. In the future, we hope to see more wireless security capabilities built into the client operating system and the infrastructure equipment. Microsoft, for example, has implemented 802.1X authentication in Windows XP, though interoperability challenges still loom. And the major enterprise-oriented access point vendors are including 802.1X support in their products.

Various industry polls have found that the vast majority of respondents feel that they shouldn't need to turn to third-party products for security. Are those respondents naive, or are vendors and standards bodies simply trailing demand? The answer probably falls somewhere in the middle.

Although most IT managers understand the unique security challenges WLANs pose, implementing robust wireless security doesn't make much sense without taking adequate steps to secure more traditional systems. Thus, you should look to establish comprehensive security policies backed by integrated systems that address all your needs, not just the protection of mobile users. Even if you've been lax in implementing security on legacy systems, perhaps because it would inconvenience users, you now have a chance to take a stand on wireless. In this sense, wireless may provide an opportunity for security officers seeking to correct past ills.

Meantime, your choices are somewhat more tactical in nature, and they aren't cheap. However, if the benefits of mobility are visible on the bottom line, you can indeed engineer a system that won't keep you awake at night.

Section VII: The Hardware

Chapter 18: The Access Point

The deployment of a wireless local area network (WLAN) requires very little hardware. An ad hoc network only needs two computing devices, each equipped with a wireless network interface card (NIC). Add an access point (AP), and a bit of cabling, and you have a simple wireless infrastructure setup. Even a corporate WLAN, which may need sophisticated antenna equipment or an additional "box" here and there, still has the access point as its centerpiece.

However, as with many things in life, when the needs are few, sometimes making the right choice can be difficult. Only the WLAN deployment team knows which of the following components (including their various permutations) are necessary in order for the users of a specific WLAN to obtain an optimal wireless networking experience.

Selecting access points for deployment within a corporate WLAN system is complex. These devices, which are about the size of the typical office telephone, provide a wireless interface to the client devices and a wireline to the LAN. Up until the year 2002 or so, the technical differences between various vendors' access point products were minor. That's all changed. Today, access points run the gamut, including the following:

- Simple RF receivers.
- Sophisticated devices that provide RF excellence.
- Intelligent instruments that not only offer great RF transmission and reception, but also can connect two network segments and keep traffic local by filtering traffic based on MAC addresses—allowing access when necessary to other parts of the network (akin to a Layer 2 bridge).
- Smart tools that can perform, in addition to bridging between wired and wireless networks, higher functions that are normally reserved for routers or switches in wired networks. For example, in 802.11 (a, b or g) networks, some of these intelligent devices can function as network access servers.

Note that the latter two types of access points may lock you into a specific vendor solution, because these devices offer functions that fall outside the province of the 802.11

series. For instance, products that offer functionalities, such as special security or management features, are provided via proprietary solutions.

Access points that function only as RF receivers are sometimes referred to as "thin" or "dumb" APs, especially if they are offered along with some kind of "intelligent box" that feeds these APs with any "smarts" they might need. Access points that perform more exhaustive tasks, e.g. user authentication, are referred to as "fit," "fat," "smart," or "intelligent" APs. These devices may or may not work in tandem with some kind of intelligent box.

SIMPLE APS

Simple APs aren't crammed with user and network configuration data, instead they are designed to concentrate on the radio functions. These devices rely on all of the necessary controlling intelligence to be provided elsewhere on the network, e.g. by an access controller, a wired switch, or even a router. These APs are usually described as being "thin," "dumb," or "lite." Whatever they are called, they range from devices that merely satisfy the 802.11 (a, b, or g) specification to sophisticated gear that offers RF excellence. Either way these simple devices cost less and need less management than their more complex counterparts—their basic radio functions are unlikely to need attention, and the security and other features that need regular attention are in the traditional location—some type of central control device, i.e. an intelligent box.

Furthermore, since these APs aren't loaded with network and user configuration data, they are of little or no interest to hackers. And if one of these simple devices is stolen, it's rendered inoperable as soon as it's unplugged.

SOFT ACCESS POINTS

Some vendors offer software that, when loaded onto a computing device that is connected to a wired LAN, enables the computing device to act as a hardware access point. These software products are reminiscent of the old Winmodem software, in that they use the computing device's CPU to process the data flow. For example, a laptop can be easily configured to function as an AP with commonly available software, such as the freeware tool Host AP for Linux systems, or PCTEL, Inc.'s Segue SAM, a soft access point that permits Wi-Fi-enabled computing devices to function as APs.

The large vendors are also getting in on the act. Intel plans to integrate "soft access points" for wireless access into PCs. This extra software will let home laptop users connect to the Internet by using a wired home PC as a bridge. And Microsoft is supposedly hard at work on what it calls a "Soft Wi-Fi," which, according to Microsoft's press relations services, is a new driver model. The Soft Wi-Fi will allow processing currently done by the 25 MHz chip used by hardware-based access points to be performed within Windows. Thus, according to Microsoft, any Windows-based PC can function as an access point.

Beware, though, that when soft APs are used within a corporate networking system, it is difficult for the IT department to keep track of them because the soft AP appears as an authorized station to all wire-side network scans.

Vendors such as Aruba, Proxim, and Symbol advocate the simple AP/intelligent box approach. There is some logic to using one of their solutions:

- In many enterprise-sized networks, APs are deployed in volume, so per-access point savings add up as the wireless network size increases.
- Bundling software intelligence into an access controller or switch device (instead of distributing it out to the access points) enables the IT department to make upgrades and changes to only one device—the smart one—rather than on a per-access point basis.
- It is perfect for campus environments, which require a lot of access points, because it can inexpensively increase wireless coverage.

But the downside to designing a WLAN around the AP/intelligent box approach is that the APs must be able to contact the intelligent box. Oft-times that box is located somewhere within the wired network environment, thus packets of data are required to go through twice as many plug devices before they get to the end-user. As illustration, data is transferred from the LAN switch, to the WLAN intelligent box, to the LAN switch, and then to the simple access point. Whereas, in a WLAN designed around smart APs, the data is transferred directly from LAN switch to AP.

Richard Bravman, CEO of Symbol Technologies, believes that the use of catchall switching in combination with simple access points will dominate the next generation of WLAN technology. "We have put all the essential capabilities into one system," says Bravman, "so that organizations no longer need to buy separate pieces from multiple vendors, or costly pieces from one vendor, to achieve secure application-specific wireless networking."

Symbol's Mobius Axon Wireless switch offers Layer 2 and 3 WLAN functions (e.g. IP inspection and load balancing), and certain Layer 4 features (e.g. HTTP, instant messaging, and security solutions such as Kerberos authentication). The layers are managed by an XML or command line interface. The switch is paired with Mobius' "thin" APs. The MAC software layer has been taken out of these APs and put into the wireless switch, leaving only the Physical Layer for the AP. This arrangement allows the Mobius APs to act simply like an Ethernet port—a wireless socket through which data packets are passed.

Let's now look at what a smart AP has to offer. (There is more on intelligent boxes later in this chapter.)

SMART APS

A smart access point (also known as a "fat" or "thick" AP) provides radio functionality and has most of its network intelligence in the same box, thus these devices can handle most of the protocols for roaming, encryption, management, user authentication, and so forth. Such industry giants as Cisco and Enterasys back the smart AP approach. A smart AP presents the end-users it serves to the wired network switch as if they were physically connected. Furthermore, smart APs reduce the load on central switches within the wired LAN, albeit at the cost of needing to be managed.

One of the downside with these APs is that the smarter the AP, the higher the cost. Another is that these smart devices present very tempting targets to thieves in that they require horsepower in the form of memory and processing power. Also, if the WLAN is large with many smart APs, upgrading these devices with new firmwire or security features means that a technician must manually connect to each device to perform the upgrade. This can result in IT personnel lugging around ladders, screwdrivers, flashlights, etc. just to get to the various APs.

However, integrating network services directly into the AP enables important services to be pushed out to the first point of contact with the wireless user. The thought is that by provisioning access control lists and policies directly from the radio, end-users can move, for example, onto another subnet in another corporate location, and still retain all their access rights.

THE IN-BETWEEN "FIT" AP

There are also APs that offer an amalgam of simple and smart. These devices are typically referred to as "fit" APs. They handle the radio side and encryption, and the network switch or an access controller takes care of the authentication, roaming, access limitations, and other by-user functions. Airflow, Extreme Networks and Trapeze are vendors offering "fit" AP products.

Let's examine Extreme Networks' intelligent box/fit AP solution, which Extreme refers to as the "Unified Access Architecture." By treating wireless as just another switch port, Extreme Networks' Unified Access Architecture establishes a single enterprise infrastructure that serves both wireless and wired users. It also features end-to-end management and mobility, along with the security and scalability demanded in today's organizations. Further, the Unified Access Architecture speeds and simplifies planning, configuration and troubleshooting. It also centralizes network management on a single console and can implement advanced policies.

The Unified Access Architecture is based on two building blocks: the Summit 300-48, a switch that can simultaneously handle wireless and wired applications, and the Altitude 300 wireless port, which is basically a simple access point with a bit of added functionality. Thus the Altitude 300 wireless port has the criteria of a "fit" AP—it is unburdened by expensive CPU, memory, system software, or power supply, and it implements the Advanced Encryption Standard (AES) and the Wi-Fi Protected Access (WPA) standard.

The Altitude 300 contains virtually no configuration data until it's connected to the Summit 300-48, which allows it to immediately download all pertinent information. That makes it easy to set up the Altitude 300s.

The Summit 300-48 and Altitude 300 are a powerful team. The Summit 300-48 can even detect and shut down rogue WLANs that can compromise enterprise security. The Unified Access Architecture relies on ExtremeWare to manage the fully integrated enterprise. Advanced policies, like Quality of Service (QoS), intrusion detection, and authentication, are defined and maintained using EPICenter software. Since these policies are heuristic, they keep tabs on end-users even after they've been granted network access.

***Note:** The issue of whether network intelligence should be centralized or pushed to the edge of the network isn't new. In the early 90s 3Com Corp. tried to introduce what it termed "distributed management intelligence" by providing much of the network's intelligent in an adapter card that sit inside your everyday PC. The idea never caught on.*

✪ COMPARING THE DIFFERENT OPTIONS

The typical smart AP found in most corporate WLANs acts as a gatekeeper and a switch—both snatching and spewing forth packets in and out of the air, checking for authorization, applying various services to them, forwarding them to the wired network and, then, of course, relaying packets back in the other direction. In functionality these devices perform much in the same way as network switches on the edge of a wired network. Examples of this type of AP model include Cisco's Aironet, Proxim's Orinoco, and Symbol's Spectrum24.

In many instances, however, it is better to let an access point do what it does best—radio frequency control. That's why some enterprise-class access point vendors are backing away from the smart AP model. These innovative vendors, while taking different approaches, are building simple or fit APs and pairing them with some type of intelligent box. The box provides all or most of the "smarts" needed for deploying and managing large-scale enterprise wireless networks. With this model, to access any necessary additional information, features, or functions, the simple APs tap into the intelligent box, which is sitting somewhere inside the wired LAN (e.g. a wiring closet).

There are advantages to using stripped down APs, and pairing them with a centralized intelligent box that can handle the majority of the WLAN's system functions (with the help of the wired network's switches). One such benefit of using this model in a corporate WLAN is that by centralizing the configuration and management of the APs, the management of those APs is greatly simplified.

As the 802.11 series of standards matures, network managers will need to upgrade or even replace access points to keep pace with the evolving standards (which, in all likelihood will affect such things as performance and security). A wireless LAN architecture that is designed around the simple AP and intelligent box model, whether the vendor dubs its "box" an access controller (Proxim, etc.), a switch (Symbol, etc.), a router (Chantry Networks) or a gateway (Bluesocket), just may be the wave of the future. Such an architecture can *always* reduce management costs and complexity, and many times can even lower deployment costs (depending on the size of the WLAN). These intelligent devices

PORTAL

To integrate the IEEE 802.11 architecture with a traditional wired LAN, the IEEE introduced a final logical architectural component. This component, which the IEEE calls a "portal," functions as a "translation-bridge" between the WLAN and an existing wired LAN. In other words, a portal can provide logical integration between existing 802-type wired LANs (e.g. Ethernet and Token Ring) and 802.11 LANs.

The 802.11 series of standards does not constrain the composition of the network's distribution system; therefore, it may be 802 compliant or non-standard. If data frames need transmission to and from a non-IEEE 802.11 LAN, then those frames, as defined by the 802.11 standard, enter and exit through a logical point, the portal. Where the "intelligent box" concept falls under the IEEE's definition of a portal is up for debate, especially since the portal and access point can be on a single physical entity or provided by separate devices.

typically use a virtual LAN architecture and policy-based networking to deliver bandwidth, security, and networking services by device, by user, by application, and by location, all from a single access port.

When considering whether to go with a network design that utilizes APs paired with an intelligent box, versus using smart APs alone, keep the following in mind:

- Smart APs cost more per unit, but since they are connected directly to the wired network, no additional hardware is needed (except for the NIC).
- Simple APs are inexpensive, but must be paired with some type of intelligent device, e.g. an access controller or a switch. But if the WLAN requires a lot of APs, costing for this method is less, even when you factor in the expense of the extra box to manage them.
- APs can be "smart" and still be centrally managed—centralized management is a must for large WLANs, regardless of whether the APs are simple, smart, or fit. Thus, the issue may be only whether the "smart" AP's per-unit cost is low enough to make it cost-efficient to buy them plus a centralized management system. For example, Cisco's IOS (a router operating system) is run in Cisco's smart APs, like the Aironet 1000 series, to distribute network services out as far as possible to the network edge. But the Cisco Wireless LAN Solutions Engine (a special management appliance that resides in a data center, plugged into a backbone switch) still can be used to manage the smart APs.
- In terms of cost alone, it is the size of the network that may determine whether to look to a smart or simple WLAN infrastructure. For a small network, investing in simple APs and an intelligent box may be too costly, but for a large network, such a solution may actually be very cost-effective.
- If the decision is to go with smart APs, ask the vendor what technical benefit there is to applying network services in the AP, rather than in a LAN switch, be it wired or wireless. Those benefits are not yet entirely clear.

While the smart AP will continue to thrive in some WLAN environments, the simple AP paired with an intelligent box will scale better in others. Before making your final decision, have potential vendors put in writing *exactly* what they and their product(s) can and will deliver. Have them delineate what their solution offers that you can't get elsewhere, e.g. technical advantages (e.g. Quality of Service, heightened security, performance) and business benefits (e.g. scalability, lower cost of management). Finally, ask them to prove how they can deliver all they promise.

INTELLIGENT BOXES

As you should now understand, if there is an intelligent box at the helm, upgrading a WLAN can be greatly simplified. As most WLANs stand today, when it is time to upgrade or replace an access point, it is a hands-on operation—access point by access point. But when a WLAN is designed around a simple APs/intelligent box combo, most if not all management activities can take place in the box instead of at the access points. For example, to upgrade the WLAN's access points settings to, say, enhance security, all that might be needed is a one-time upgrade to the intelligent box, and then that box will automatically upgrade all of the APs, with no need for a personal visit from a technician. And when it is time to replace some of the WLAN's APs, say, to upgrade the network from 802.11b to 11a or 11g, all an IT staffer may be required to do is to go to the old access point's location, remove it and plug in the new one—everything from that point on is simplified since the management settings for the old access point are stored in the intelligent box—the new access point automatically inherits the old access point's settings.

Ease of management is only the start of lowering your total cost of ownership (TCO) when you design a WLAN using an "intelligent box" architecture—at least so claim vendors such as Aruba, Extreme, and Symbol. Because the cost of producing a simple access point is lower than the cost of producing a smart AP, as access points need replacement, or as the network grows, lower cost access points will contribute to an overall lower TCO. However, note that the intelligent box is more costly than the average wired hub or switch to which the typical access point connects.

The intelligent box vendors have done their homework. All have some sort of analysis that they can trot out, that will show the number of access points at which an intelligent box architecture will start saving money over traditional WLAN designs. Ask to see that information. Then use it in your decision-making process.

Wireless Switches

How and where vendors implement the intelligence to run the APs vary. For example, the aforementioned Aruba, Extreme, Proxim, and Symbol are adding centralized intelligence via a wireless LAN switch that centralizes control of access points and wireless switching—it's similar to what intelligent switching did for wired LANs. The basic idea of wireless switching is to take the intelligence normally embedded in the access point (e.g. security) and move it upstream to a wireless switch to which the access point connects. Some benefits can be derived from designing a WLAN using this type of architecture.

***Note:** Regardless of whether you use a "wireless switch" approach or not, for the wireless network to connect to a wired network, you must still use a wired Ethernet LAN switch. You need the Ethernet switch whether the APs are smart, simple, or in-between. So when performing cost-of-ownership comparisons, you must presume the use of some number of wired Ethernet switch ports in your equation. Of course, the number of ports required will depend on the size and configuration of your deployment.*

These vendors may be on the right track—promoting cooperation between the network's switching system (a more robust control point) and the access point (a price-driven product). Many experts think that it might be a good idea to make access points more of a dumb receiving and transmitting device and the switch more of the brain. Their conventional wisdom is that everything a WLAN needs can be done better with a switch/access point duo, rather than in one very narrowly focused wireless appliance.

Put another way, the beauty of wireless switch technology is that it provides a structured blueprint, along with the centralized troubleshooting tools that are normally needed to scale and secure WLANs beyond departments and across all of an organization's facilities.

Without an architecture built around some kind of centralized intelligence, access points must act as isolated systems that provide 802.11(a, b, or g) functions such as encryption and authentication. But if you move these functions into a switch, the access points connected to the wireless switch can do what they do best—perform RF radio duties, which require virtually no management.

Wireless switch technology provides the network's "intelligence" by (1) controlling each access point's power and channel settings, and (2) storing configuration data. For instance, if an access point failure occurs, the wireless switch can automatically detect the failure and instruct nearby access points to adjust power and channel settings to compensate. And, if a new access point is installed, the switch can automatically discover it and upload the appropriate power and channel settings.

This intelligent technology can also protect against the security threat of rogue access points, since the wireless switch is charged with the duty of validating any and all access points as they access the network. When a rogue access point tries to access a WLAN built upon a wireless switch architecture, the switch checks a trusted list of allowed devices, users, and user policies, and when the switch determines the device is "illegal," it proactively shuts down the rogue access point and alerts the network manager.

A wireless switch architecture also addresses some of the challenges a network manager faces when trying to combine security with mobility. Wireless switching technology can integrate mobile IP (a standard that solves roaming issues across IP subnets), while maintaining user authentication state, since the switch can transparently reauthenticate users as they move from access point to access point. This is possible because, for example, stateful policy engines can enforce predefined rules on a per-user basis. As users move, their policies follow. The other advantage of such functionality is that network managers can provide some users, such as guests, just with HTTP access, while others, e.g. employees, can receive access to a wider range of ports and services.

A wireless computing device accesses a WLAN by making an association with the access point that has the strongest signal. With a wireless switch at the helm, the access point is connected to a wireless switch, which is located in a wiring closet or in a data center. Acting as a repeater, the access point forwards the 802.11 association request to

Wireless Switching - How It Works

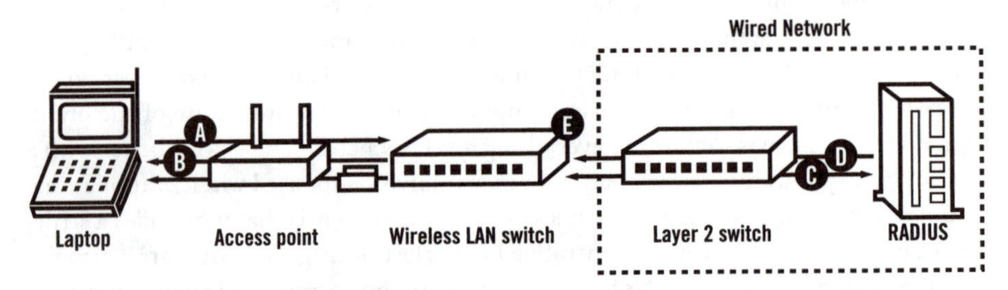

Figure 18.1 In most WLANs, access points act as isolated systems providing 802.11 functions, but when you add a wireless switch to the mix, many of the APs functions are taken over by the switch, allowing the AP to do what it does best, receiving and transmitting radio signals.

the wireless switch. The switch acknowledges the request and authenticates the wireless user via the 802.1X protocol, e.g. validating user credentials through Remote Access Dial In User Service (RADIUS). The RADIUS server then passes encryption keys to the wireless switch. The wireless computing device independently derives the keys on its own and begins sending encrypted data. Thus the wireless switching technology maintains the end user's identity across the wireless infrastructure, so that services and security can be delivered seamlessly as the end-user moves from access point to access point.

The advantage of wireless switching is that these intelligent devices can serve as the brains of a WLAN system. They can constantly monitor air space, network growth and user density, and dynamically adjust bandwidth, access control, quality of service, and other parameters as mobile users roam through the corporate facilities. A wireless switching architecture also gives network managers the flexibility to mix and match security capabilities ranging from Layer 3 VPNs to Layer 2 authentication and encryption schemes such as 802.1X, Wireless Equivalent Privacy (WEP), Temporal Key Integrity Protocol (TKIP), and Advanced Encryption Standard (AES), without having to upgrade or reconfigure access points.

Although the design of the wireless switch varies from vendor to vendor, all are designed to perform traditional Layer 2 tasks. But some vendors, like Symbol, which offers the Axon wireless switch that plugs into a wired LAN switch technology, offer

MULTI-LAYER SWITCHING

To understand how multi-layer switching works, you have to understand the OSI (Open Systems Interconnection) Reference Model. (See Appendix I.) For many years, the OSI Model has been the reference layering paradigm for data networking. It provides a powerful architecture that includes not only well-defined Layer n/Layer n+1 protocols and rich peer-to-peer protocols, but also network addressing via the Data Link Layer's MAC sublayer. As such, the OSI Model provides a layered network design framework that enables devices (including bridges, routers, switches, and access points) from different vendors to work together.

In early local area networks, there was no need for switching devices, because (1) networks were simple affairs, and (2) they were relatively slow. But as networking technology evolved, there eventually came the need for high-speed switching. That's because in today's networks, switches perform some of the most important functions on a network: moving data efficiently and quickly from one place to another. As data passes through a switch, it examines addressing information attached to each data packet, which allows the switch to determine the packet's destination on the network. It then creates a virtual link to the destination and sends the packet there.

The efficiency and speed of a switch depends on its algorithms, its switching fabric, and its processor. The layer at which the switch operates determines the switch's complexity, and the layer at which the switch operates is determined by how much addressing detail the switch reads as data passes through. Thus, the designation of a switch as a Layer 2, 3, or 4 switch simply refers to the functions the switch performs as they pertain to the OSI's seven-layer model of networking.

devices that can perform Layers 2, 3 and 4 tasks. The Layer 3 and Layer 4 switching functions are managed through either an XML-based or command-line user interface.

Wireless Router

Another device that vendors find attractive for use as an intelligent box is the router. Chantry Networks has introduced its BeaconWorks, a Layer 3 router that is paired with simple access points. The router uses a proprietary protocol to tunnel over the wired IP network to connect to the WLAN's access points.

The beauty of putting the intelligence in a router is that such devices are adept at bypassing failed nodes and finding new pathways to keep a network up and running. According to Robert Myers, chief technical officer for Chantry, "When mobile users connect, they are assigned an IP address on a subnet managed by the BeaconMaster, which can then provide the routing. We support an array of routing protocols. A BeaconMaster can support hundreds of BeaconPoints and several thousand mobile users. We can handle the whole address space for the customer."

Unlike the wireless switches, which are typically designed to focus on Layer 2 switching functions and Layer 3 quality of service features, and thus are normally placed in a wiring closet, the Chantry wireless router is built for the data center or network operations center.

Chantry engineers designed a protocol that lets the APs communicate with the router over an IP network. "Mobile users connect to the access point and can roam to any other BeaconPoint in the network and maintain the same IP address," Myers says.

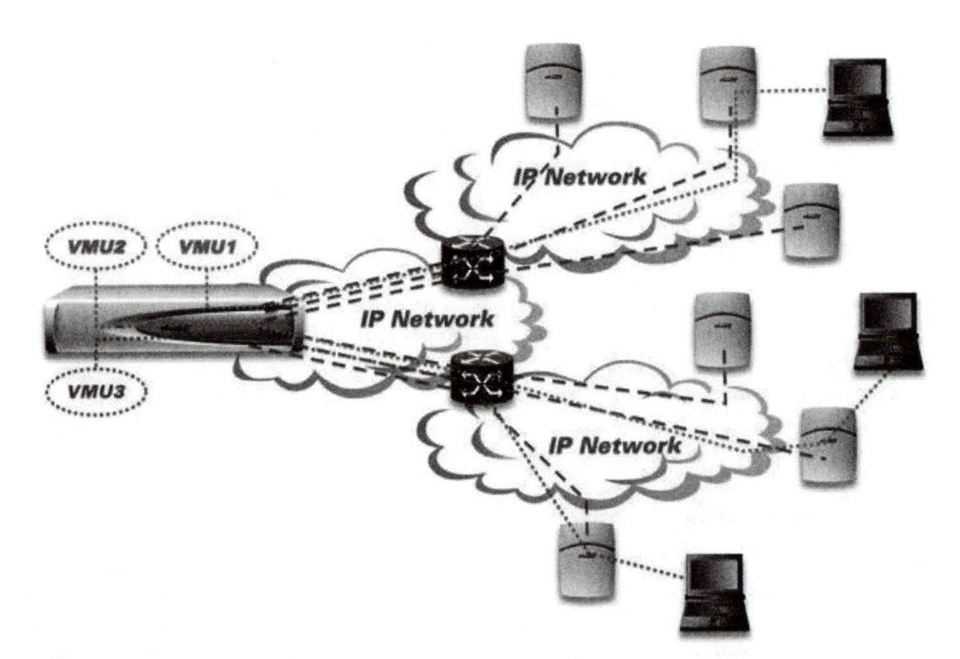

Figure 18.2 The Chantry BeaconWorks solution combined with its VNSWorks product allows a WLAN to create multiple virtual networks over a single WLAN infrastructure. *Graphic courtesy of Chantry Networks.*

Chantry also offers Virtual Network Services (VNSWorks), which allows an organization to create virtual WLANs easily. As discussed in Chapter 8, separate, protected virtual networks within a single physical WLAN infrastructure are ideal for many large organizations that have network traffic that needs to be segmented or prioritized differently to enhance security or enable different classes of service.

Aruba also has a new product in development that does something similar, only it uses the Generic Routing Encapsulation protocol, a technique that also lets the Aruba device be installed in a data center. Users pass through an access point to the switch, which handles authentication, access policies, and encryption, and creates a personal firewall for each user.

Access Controller

While many HotSpot installations use access controllers to authenticate and authorize end-users based on a subscription plan, an access controller also has a place in the corporate network ecosystem. An access controller (whether hardware or software), installed on the wired portion of the network between the access points and the protected side of the network, delivers value-added functionality. Among the forms this functionality can take are those of the first line of defense against unauthorized access, mobility (since most access controllers also offer subnet roaming capabilities), and a method for managing and administering the WLAN.

The use of an access controller should be considered whether the organization's network architecture involves a small, medium, or large WLAN installation. Companies such as Cranite Systems, Proxim, ReefEdge, and Vernier Networks offer access control solu-

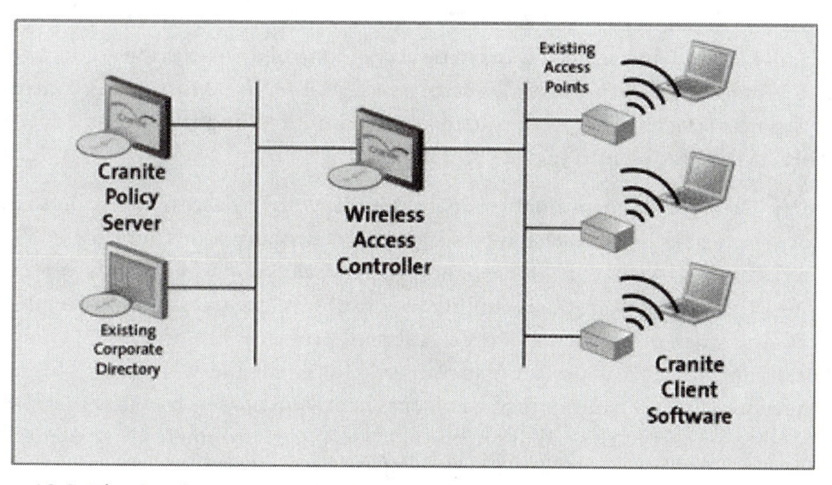

Figure 18.3 The Cranite Systems' WirelessWall Software Suite consists of the WirelessWall Policy Server to support the creation of policies that control the characteristics of each wireless connection; WirelessWall Access Controller, to enforce policies for each wireless connection, to encrypt and decrypt authorized traffic, and to provide mobility services to users as they move across subnets throughout the network; and Cranite Client Software which operates on each mobile device accessing the network to terminate one end of a secure tunnel (the other end terminates at the Access Controller), and encrypt and decrypt data for that device's connection. *Graphic courtesy of Cranite Systems.*

tions that can strengthen any network architecture that hosts a resident wireless LAN. Access controllers provide centralized intelligence behind the access points to regulate traffic between the relatively open wireless LAN and important wired network resources.

Some of the reasons for designing a WLAN's architecture around an access controller include:

Security. The most compelling reason for adding an access controller to a corporate network is that it segregates the enterprise's wireless access infrastructure from the protected corporate network. If implemented correctly, such a controller offers security that can block most if not all intruders.

Network security begins with network access, usually from the edge of the organization's network. By acting as a gateway between the WLAN components and the wired network, an access controller, for example, can block a hacker lurking outside corporate headquarters from getting entry to sensitive data and applications. It can also offer Mobile IP capabilities, such as allocating IP addresses, maintaining a list of authenticated IP addresses, and acting as a traffic filter.

Most network managers want to authenticate WLAN users in order to ensure that only legitimate users gain access to their systems. In fact, in organizations where privacy isn't so critical or where it is implemented at higher layers in the stack, authentication may be the only requirement. If this describes your situation, the products from Bluesocket, ReefEdge, and SMC make life easy by delivering Web-based authentication. With an access controller at the helm, before users can gain access to network resources they must access a login page, which is automatically provided by the access controller when end-users fire up their browser. Once authentication is complete, user and group access-control policies take effect.

In most cases, managers will want to tie access control to an existing accounts database—a Windows Domain, Active Directory, or LDAP service. Most access controllers make this task simple, though the systems that run under Windows have an easier time than others integrating into such an environment.

Mobility. One of the most valuable capabilities provided by access controllers is support for subnet roaming and session persistence. Some organizations implement WLANs using a flat address space and enforce policy where wireless and wired worlds meet. However, most enterprises want the flexibility to install wireless access points on multiple subnets. And when devices roam between subnets, problems can occur.

For organizations that use WLANs primarily for email and Web access, this shortfall may represent only a minor inconvenience that requires users to renew their DHCP leases and reconnect to their mail servers. However, in environments that use stateful TCP-based applications, a subnet roam will kill those programs unless a system can deal with this issue. Many products support subnet roaming and session persistence; however, the specific techniques used to support subnet roaming vary from product to product, as does the speed at which the roaming takes place. So make the appropriate inquiries before making a purchase.

Simple Access Points. An added benefit of using an access controller in a WLAN that needs more than a dozen or so APs is that it reduces the need for "smart" or "enterprise-grade" access points, many of which are relatively expensive, owing to the fact that they offer numerous non-802.11 features. But with an access controller at the helm, the WLAN deployment team can focus on finding a vendor that offers high-quality simple access points, i.e. RF excellence and low cost, since the centralized access control functions in an access controller can serve all access points. These simple APs primarily implement the 802.11 standard, and not much more.

An Open System. Another incentive for spending the bucks for an access controller is that these devices can provide the means for building an open system. One of the problems with smart APs is that they offer enhancements (i.e. security, performance) over and above the basic wireless connectivity required by 802.11 specifications. But these enhancements can only be realized if the system is designed around one vendor's products—all network interface cards and all access points must be from the same vendor. While this might not be a problem during the initial design and deployment stage, such a closed system can result in difficulties down the road. On the other hand, simple access points can easily communicate using the basic 802.11 protocol with radio NICs made by multiple vendors while the access controller transparently provides such enhancements as better security, quality of service, and roaming. Thus, even if the current WLAN is designed using only a limited number of APs, you may still want to consider an access controller-based architecture if you envision expanding the WLAN in the future.

Costs. For large WLAN systems, using an access controller-based architecture is a "no brainer"—the costs are lower than going without an access controller and using smart APs. This is especially true for networks requiring a large number of access points, such as an enterprise system, since thin or dumb access points cost less. The cost savings are generally in the range of $400 per access point. Of course, a best-of-breed access controller carries a hefty price—on average about $5000 (although the prices seem to be coming down). Do the math—if the WLAN design plans call for more than a dozen or so access points, the less expensive way to go is an access controller and simple APs. For a smaller WLAN, the costs savings aren't so clear-cut. However, if you factor in the annual expenses incurred by the IT department in managing, maintaining, and troubleshooting a WLAN built upon a smart AP architecture, it's also easy to see that an access controller can save costs even for smaller WLAN installations.

Note: *The reader should remember that most smart APs connect directly to a network's wired Ethernet switch (e.g. a Cisco Catalyst switch), whereas most simple or dumb APs connect to some type of intelligent box that contains some of the network-service intelligence that the smart APs contain. Then that box is connected to the wired network's Ethernet switch. So you need an extra box in the middle in the simple AP/intelligent box scenario.*

Centralized Management. The more management intelligence that is moved out of the AP and into a central device, the easier it is to upgrade, monitor, troubleshoot, and manage the WLAN, because you don't have to visit each individual device to perform any of these tasks. By placing the WLAN's intelligence in a central device such as an access controller (or other type of intelligent box) rather than the access points, the overall network is easier to manage. Centralized management is a must for enterprise-size WLANs, regardless of whether the APs are smart or dumb. (Note that an AP can be "smart" and still can be centrally managed.)

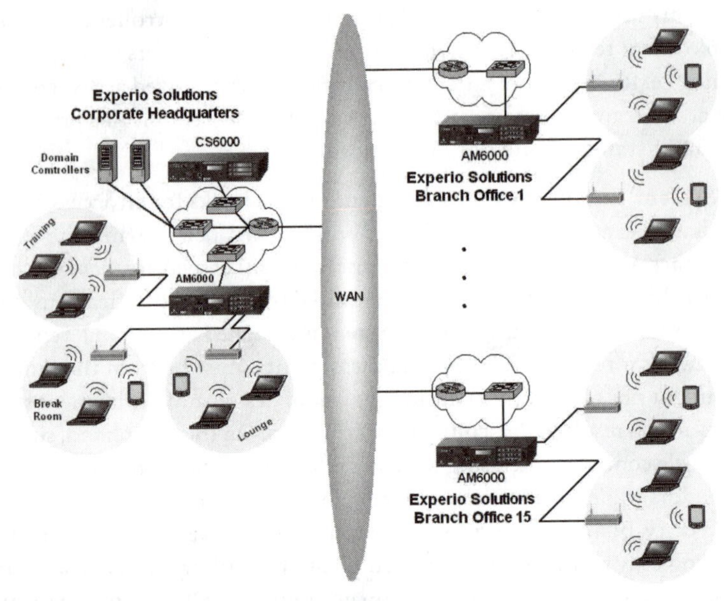

Figure 18.4 This diagram depicts the Vernier Networks System, which consists of the CS 6000 Control Server as a centralized security configuration and management system, and the AM 6000 Access Manager as used in a network designed for Experio, a consulting firm with nearly 1000 employees and 16 offices across the U.S. The WLAN's architecture, as depicted in this graphic, provides centralized control over multi-site networks, and uses a two-tier architecture that can scale to support even the most distributed enterprise networks. *Graphic courtesy of Vernier Networks.*

***Note:** In the near future, expect to find a single switch that can apply network services to both the wireless and wired user populations.*

DUAL MODE OR SINGLE MODE?

While 802.11b computing devices and their networks are pervasive, especially in the HotSpot market, support for 802.11a is coming on strong as well. Organizations that can't decide whether to go with 802.11a or 802.11b/g can have both, via products that integrate both 802.11a and 11b/g functionality on a single access point. These products are designed so that their performance is at least as good as their single-mode counterparts. Furthermore, many of these products, rather than using dual-mode chipsets, have integrated multiple radios that can share a single Ethernet backbone connection in a single box.

While dual-mode access points have their advantages, and performance variations among brands of access points are generally minimal, dual-mode APs should not be used in every WLAN installation. A decision first must be made as to whether a dual-mode architecture is a workable solution for the organization's proposed WLAN. To help in the decision-making process, consider the following:

* If the WLAN supports a conference room that will be used by multiple departments then installing at least one dual-mode access point will provide the WLAN the flex-

ibility to accommodate both 11a and 11b standards. And because the physical coverage area is confined to a single room, you don't really need to worry about radio-propagation differences between the 11b/g (2.4 GHz) and 11a (5 GHz) radios.

- If wireless access is needed in areas where there is simply too much interference for 11b/g devices to work optimally, dual-mode access points may be the way to go.
- Dual-mode APs can protect investments in legacy APs while offering the advantages of newer technology; that is, if the long-range plan is to upgrade the WLAN to 802.11a. However, you may want to consider the possibility of bypassing 11a and going straight to 11g, in which case, building a new WLAN around dual-mode technology is unnecessary.
- When designing a network around 802.11a technology: a dual-mode system will allow the organization's employees to take advantage of HotSpots during their travels. (The vast majority of HotSpot technology is based on 11b technology, not 11a.)

For more ambitious projects, such as establishing coverage throughout an office building, the design problems may outweigh the flexibility that dual-mode products offer.

Numerous vendors offer quality, enterprise-class, single-mode access points, but when looking for dual-mode access points, the choice is more limited. Consider products such as Cisco Systems Aironet 1200 Series AP, Proxim Orinoco AP-2000 Dual-Mode Access Point, Intermec MobileLAN WA22 AP, Intel 5000 LAN Dual AP, Linksys WAP51AB Dual-Band Access Point, and D-Link AirPro DI-764 AP.

AP SELECTION CRITERIA

So far the AP discussion has centered on the various types of AP design, but there are other, "specific" selection criteria that might also need to be considered when determining which access points to use. Features that may seem insignificant when deploying a pilot WLAN take on more importance during a full-scale WLAN deployment.

NICs. For example, when designing an 802.11b WLAN for common usage, such as office tools or web browsing, the rule of thumb is that an AP equipped with one 11 Mbps card serves 20-50 users. But in instances where, say, the WLAN is to support videoconferencing, it is advisable to select APs that can accommodate two NICs. Then, when a videoconference via the WLAN is in session, it will be possible to group the conferencing wireless computing devices so that they use one NIC, while the non-conferencing users access the WLAN via the other card (this is possible because each NIC has a different network name).

Power-over-Ethernet. Another selection criterial may revolve around Power-over-Ethernet (PoE). Does the WLAN project need APs that offer the PoE feature? While WLANs were supposed to lower infrastructure costs and reduce IT workload, electrical cabling has proven to be a weakness in the WLAN model. The hassles of delivering electrical power to wireless access points create an ongoing headache for WLAN deployment teams—increasing costs, slowing ROI, and tying up valuable IT resources.

PoE technology, based on the IEEE 802.3af draft standard, delivers power to network devices over two pairs of wires in Ethernet cabling. A PoE power-sourcing device

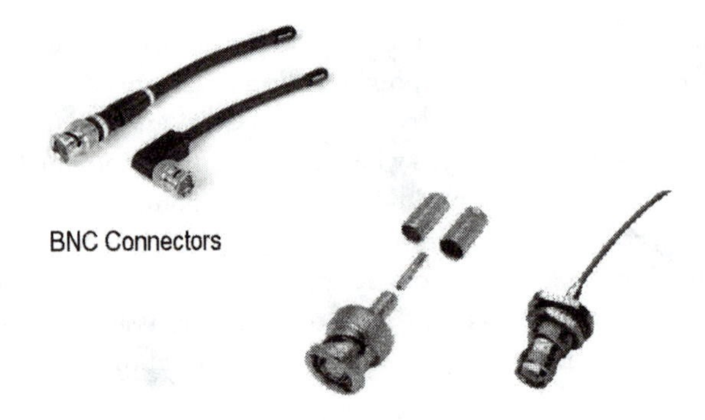

Figure 18.5 The left graphic depicts whip antennae with a BNC connector. They provide quick-connect half turn, same as on the old 10base2 Ethernet cables. (In case you are wondering, BNC stands for "Bayonet Neill-Concelmann".) The right grapic depicts TNC connectors. These connectors are basicially just a threaded version of a BNC connector (ergo the "T" rather than the "B").

sends both network data and power down an Ethernet cable to a PoE-ready AP that has an Ethernet port designed to accept both data and power. Because the PoE-ready AP receives power through its network cable, it does not need additional power cabling or even a local power source. As a result, the device can be installed wherever the Ethernet cable can reach: in a ceiling, on a pole, in a corridor, etc.

Many vendors offer PoE-ready APs.

Antennae. Another purchasing consideration is the antenna. An antenna or even two antennae may be provided with each AP deployed. However, these antennae may be inadequate. In such cases, purchase separate antennae, either from the AP vendor or from a third-party vendor. The site coverage requirements will dictate the exact antenna system(s) required for the best connectivity.

For instance, if you are installing 802.11b APs and know that the antennae will be, or might be, supplemented with another antenna, many times it is advisable to purchase APs that offer an antenna connector option, typically a BNC connector. (See Fig. 18.5.) Popular opinion is that BNC connectors work okay for 2.4 GHz networks as long as they're new and of good quality, otherwise they can be subject to fairly high signal loss. (Note: if you notice that the connector is in the least bit loose, replace it.) Since the BNC connector is the most common connector used by AP antennae, you can easily mix and match antenna types to suit the WLAN's environment, experimenting until you achieve optimal signal strength.

It is noted, though, that Cisco uses an RTNC (reverse TNC) connector, which is basically just a threaded version of a BNC connector, on its APs rather than a "genuine" BNC connector. Furthermore, some vendors use the "N" connector, which is a threaded, larger connector. The "N" connector performs quite well, and is perfect if you use a thicker cable, like LMR-400.

While access point vendors generally sell antennae as separate components, don't overlook third party vendors that specialize in antennae. The next chapter provides more detailed information on antennae usage in the WLAN environment.

CONCLUSION

Don't stint on hardware. If you deploy a WLAN that requires more than 20 access points, seriously consider one of the new AP/intelligent box solutions. In many instances such solutions ease the complexity of maintaining and managing the network.

But also realize that the jury is out on which WLAN architecture is better—the one built upon smart APs, or one that is based on the dumb AP/intelligent box technology. One of the reasons for the hesitancy is that not all switched-based WLAN products are equal. For example, products such as those offered by Extreme Networks combine the functionality of a 48-port 10/100 Ethernet wired switch with a system that can manage dumb APs that operate under the 802.11a, 802.11b and 802.11g standards. Whereas switch-based products from Symbol and Trapeze Networks lack the tight integration between wired and wireless networks offered by the Extreme Networks system. However, unlike Symbol's or Extreme's products, the Trapeze solution offers the ability to detect unauthorized users and rogue APs on the network.

Many times the final decision regarding the AP selection will be based on the size of the WLAN. If deploying a small WLAN that needs only one or two APs, the decision will most likely be to go with one of the smart APs. But for other WLAN teams the access point selection process will be frought with controversy—smart, fat APs; or dumb, thin APs. Do you use an access controller, wireless switch, or wireless router device? For many, while the simple AP/intelligent box vendors take different approaches, the concern is that the additional intelligence built into many of today's enterprise APs puts too much burden on a single device. Just keep in mind that regardless of the type of AP chosen, centralized configuration and management is possible whether the underlying APs are smart, dumb or somewhere in-between.

Chapter 19: The Antenna

Antennae are so important to a WLAN's success that the reader should understand antenna technology. This chapter will provide a quick tutorial on antennae in a WLAN environment.

An antenna has two purposes: (1) transmitting data via radio waves, and (2) receiving data via radio waves. Metal is generally used in the construction of antennae, in particular those metals that are electrical conductors, as they will absorb radio signals that occur in the space around it. Conversely, given enough feed power, a metal object will emanate radio signals into the atmosphere. The shape and structure of the "metal" affects the behavior of the antenna.

Antennae increase the range of a WLAN. However, choosing an antenna involves many factors such as coverage area, maximum distance, and location. A detailed site survey is essential.

Providing adequate RF coverage is a primary concern when deploying a WLAN. This is where the antenna plays a vital role. When a WLAN is designed around an effective antenna solution, it's possible not only to increase the WLAN's range and corresponding coverage, but also to decrease its overall costs due to the need for fewer access points.

Ironically, the development and adoption of wireless networking for PCs was initially hampered by the fact that major PC companies were data and digital experts, but not experts in high-speed analog and antenna design. Now, of course, the technology has become popular (first in Europe and now in the U.S.), and many PCs and laptops have small PCMCIA ("PC Card") network interface cards equipped with small antennae or even hidden, built-in antennae.

ANTENNA USAGE

Antennae are generally classified by the directivity of the antenna. "Directivity" refers to the ability of an antenna to focus energy in a particular direction when transmitting. Conversely, it also has the ability to better absorb energy from a particular direction when receiving. The radio coverage of the wireless LAN can be maximized with the selection of an appropriate antenna. Antennae are available with omni-directional or directional radiation patterns, and with various gain ratings.

Restrictions

To understand how antennae work within a WLAN environment, the WLAN designer must understand the limitations due to indoor propagation. Any radio transmission in an indoor office environment is subject to interference, including multipath interference, and thus excess signal loss.

Multipath interference occurs when an RF signal has more than one path between a receiver and a transmitter. This occurs in sites that have a large amount of metallic or other RF reflective surfaces. Just as light and sound bounce off of objects, so does RF.

Multipath interference can cause the RF energy of an antenna to be very high, but the data would be unrecoverable. Changing the antenna type and location can eliminate multipath interference.

***Note:** Perhaps the best way to explain multipath is to relate it to something that commonly occurs when listening to an FM or AM radio in a car. As the driver pulls up to a stop, he or she may notice static on the radio. But as the car moves forward a few inches or feet, the station begins to come in more clearly. This occurs because by rolling the car forward a bit, the car antenna's position changes only slightly, but just enough so that it is out of the zone where the multiple signals converge.*

A radio that could link up over a range of several hundred feet in "free space," can only achieve 50-150 feet or so indoors, because the propagation path is rarely direct when the signal must travel via waves that are contained within a building. This is because most of the time the signal must pass through or bounce off walls and other objects. Additionally, when using 802.11b (or g) technology that operates in the 2.4 GHz frequency band, you run into signal absorption by water that exists as moisture in the air, and in water-bearing foods (that's why microwaves use this frequency). Human bodies are the equivalent to "water bearing food," meaning that there will be a difference in range as people move about an area during their work day.

However, while changes in humidity can affect signal integrity, and in an outdoor environment rain can produce a measurable effect on signal strength, snow and hail are usually not a problem. Here's a bit of food for thought if deploying an antenna out-of-doors: rain falling at a rate of 6 inches per hour can cause additional signal losses of about 0.02 dB per .6 miles (1 kilometer) between links. Thus, unless you try to connect

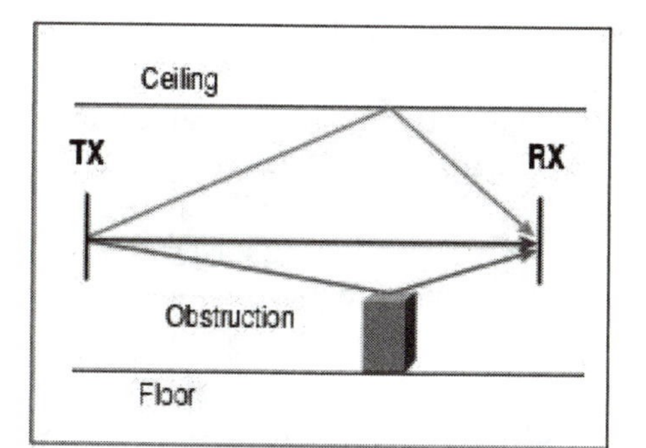

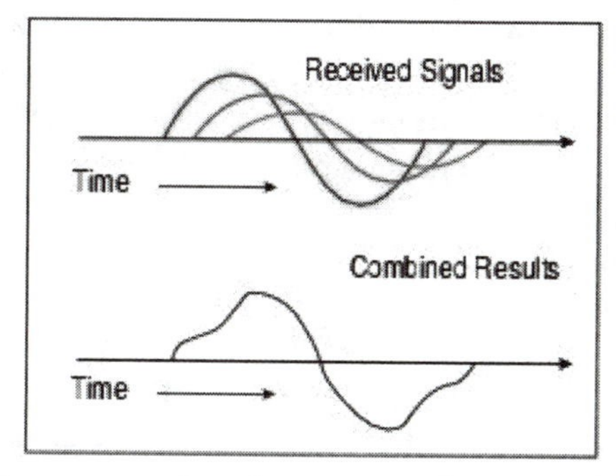

Figure 19.1 Multiple signals combine in the RX antenna and receiver to cause distortion of the signal. There can be more than one path that RF takes when going from a TX (transmit power) to a RX (receiver sensitivity) antenna. *Graphic courtesy of Cisco Systems.*

very long links, rain is not really much of an issue—it is just another factor that needs to be accounted for when it comes to potential interference.

For the most part, many of the restrictions in indoor environments do not exist for outdoor sites. In outdoor or free space, the propagation is limited by the standard free space path loss. Normally, the loss in free space is equal to range squared, but indoors, the loss exponent is more like the 4th power—when a sheetrock wall is penetrated, there is approximately 6 decibles (dB) of loss, whereas a concrete wall causes around a 20 dB loss. Once the direct path experiences too great a loss, the alternate multipath rays will carry the signal. Thus, it is hard to know with certainty what path the signal will follow at any given time.

Note: *An increase in coverage within an antenna's beam width is called "antenna gain," whereas a decrease in coverage is referred to as a "loss." Both are measured in decibels (dB).*

There are two types of signal fade caused by multipath propagation. One is caused by signals bouncing off of walls to get to the RF receiver. The other is due to the summation of local signal paths, which can lead to cancellation of the signal.

Whatever the cause, a situation often occurs where there is a great discrepancy between predicted range and actual range. This can result in difficulty determining a WLAN's cell size. If you have propagation in a long hallway, the loss can be small, but a multipath signal can still arrive with almost the same level and out of phase, effectively canceling a portion of the signal. Even small movements can substantially affect reception.

Note: *A radiated signal is comprised of waves that oscillate simultaneously during transmission. Waves that move within the same oscillation frequency and thus reinforce each other are said to be "in phase." When waves are in phase and encounter no obstructions, the received signal is strong and clear. However, when waves encounter an obstruction (metal, wall, trees, etc.), they can be reflected, bent or absorbed causing the waves to reach the target with different path lengths and at different times. This phenomenon is referred to as "multipathing," and results in the fading or nullification of the signal at the receiving end. The longer the path of the transmitted waves, the more likely it is that they will suffer multipath interference.*

ANTENNA TYPES

There are many different types of antennae, but for most WLAN designs, you only need to consider two:

One is the *omni-directional* antenna, which has a 360 degree coverage pattern on a horizontal plane. This type of antenna is ideal for square or square-like coverage areas (such as found in most office buildings), because its coverage pattern is configured in both horizontal plane and torus-shaped (i.e. doughnut-shaped) forms. Included within the omni-directional antenna group are the plane antenna and the dipole antenna.

The other is the *directional* antenna, which concentrates the coverage pattern in one direction, producing an almost conical-shaped coverage pattern (similar to that of a flashlight). With a directional antenna, the directionality is specified by the angle of the beam width. Typical beam width angles are from 90 degrees to as little as 20 degrees. A directed beam is ideal for elongated areas (such as warehouse aisles), corners and outdoor point-to-point applications since they allow for a longer but narrower coverage

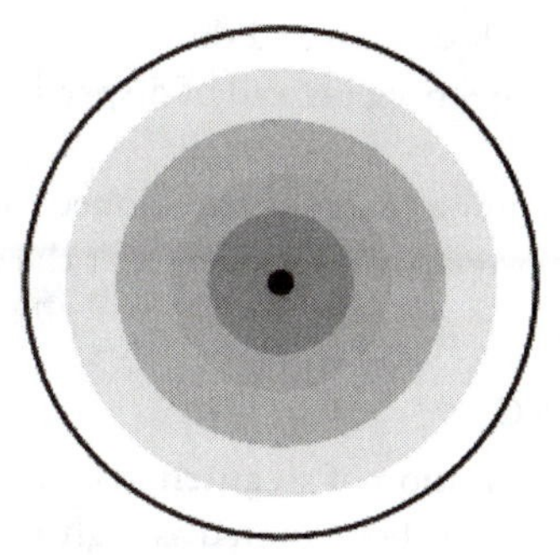

Figure 19.2 Omni-directional antenna coverage pattern.

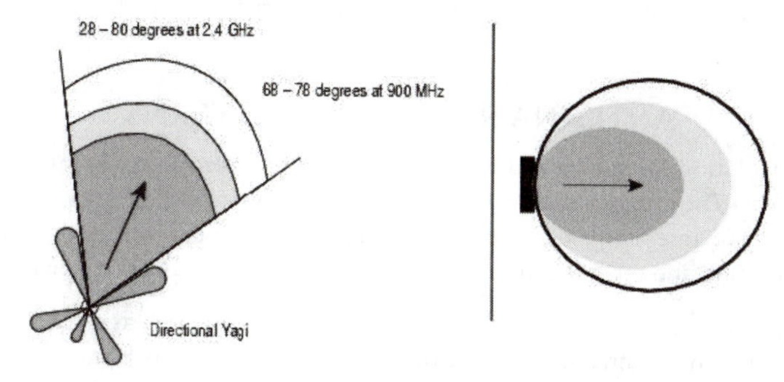

Figure 19.3 Examples of the beam coverage of a yagi antenna (left) and a patch antenna (right).

pattern. Included in the directional antenna group are patch antennae for wider directed beam coverage, and panel antennae and yagi antennae for narrow beam coverage.

In the early days of mobile radio and other wireless systems, the base station antenna radiation pattern was determined primarily by the required gain to ensure reliable communication over the required coverage area, and usually had a simple omni-directional pattern. The evolution of modern wireless communications very much involves the story of how electrical engineers were able to figure out ways to precisely control the antenna radiation pattern, in terms of envelope shape and electrical tilt.

The relatively high frequencies of the WLAN bands allow for a more directional or "beam" form of transmission, which is reflected in some antenna designs. An omni-directional antenna's horizontal radiation pattern can have some directivity added in the vertical direction to increase the power, sensitivity, or "gain" (such as changing the "down-tilt" of the signal). Simple changes in an antenna can shape the radiation such that the "doughnut is flattened" and the outer-most energy can be focused, resulting in a decibel gain in the radiated energy along the horizontal plane.

***Note:** In a sense, "gain" can also be considered as "loss," since increasing the gain of an antenna results in a smaller effective angle of use.*

This ability to provide a more directional or beam form of transmission means that WLAN users need not consciously "point" their mobile computing device in a particular direction. It is also why antenna placement is so important.

Yagi antennae have tremendous gain and directivity, since they can direct almost their entire signal in just one direction, instead of dispersing all of the energy around a 360-degree partial sphere.

Panel or sector antennae allow a cell to be partitioned into sectors, each with different frequencies (sectors that overlap cannot operate at the same time on the same frequency and polarization).

ANTENNA POSITIONING

The proper positioning or orientation of an antenna helps to ensure maximum coverage area. Antennae should generally be mounted as high and as clear of obstructions as is practically possible. Best performance is attained when both the transmit and receive antennae are located at the same height and within direct line of sight of each other.

ANTENNA RADIATION PATTERNS

Antenna radiation patterns are affected by polarization, free space loss, and propagation in solids. The transmission loss between a transmitting and receiving antenna is a function of the antenna gain, the distance, and the frequency. For best performance, the transmitting and receiving antennae must have similar polarization alignments.

Additional "free space loss" occurs because of signal spreading: as a signal radiates outward from an antenna, its radiated power is spread across an expanding spherical surface, with the power level inversely proportional to the distance from the source antenna. This antenna radiation must pass by and through solid objects, so it will be subject to losses from reflection and absorption. For example, oxygen atoms in the atmosphere cause a prominent peak in the attenuation effect of the atmosphere. Clandestine inter-satellite communications are performed on this frequency, so that the signals will not reach earth. The atmosphere's attenuation almost vanishes at 94 GHz (the W-band), which is why many radar systems operate around this frequency.

Water, however, is a very good signal absorber above 2 GHz. Fog, rain, the leaves of trees, people, etc. can rob energy from a microwave signal. This is why microwave ovens produce electromagnetic waves in the 2.4 GHz range; 2.4 GHz penetrates food very well, but the water molecules cannot vibrate as fast as the microwaves are pushing them, so the molecules absorb the microwave energy and release it as heat.

Reflections from objects (metal objects in particular) give rise to multipath distortion (fading). Some paths can converge and become constructive (adding to signal strength) or destructive (fading). Elevation pattern shapes are controlled, to keep the maximum response at or slightly below the horizon for best far and near field coverage. The dimensions and height of a communication sector will determine an antenna's azimuth and elevation beam width requirements. These can be derived from the established beam area formula for the approximate gain needed:

G(dBi) = 10 log 10 / 29,000 / antenna azimuth * the antenna elevation

Spacing in excess of .75 wavelengths from a large conducting surface leads to deep nulls. Antennae placed in the middle of a roof are prone to diffraction loss if they are not high enough to clear the roof edges.

In most instances, access point antennae should be positioned on or close to the ceiling. Also, omni-directional antennas should be placed in the center of the coverage area whenever possible. If, for some reason, it is necessary to mount an omni-directional antenna below the coverage area (e.g. on a desk, table, window seal, etc.), point the top of the antenna up. But if the antenna is mounted above the coverage area (e.g. in the ceiling), the top of the antenna must be pointing down. On the other hand, directional antennas always should be pointed in the direction of the coverage area.

✪ ANTENNA DIVERSITY

Antenna diversity has long been promoted as a means of improving overall signal reception. Diversity is a technique that takes multiple observations of a signal coming off a transmit station, in order to recover that signal with greater accuracy.

Not all forms of diversity are the same. For example, dynamic diversity is optimized for low power consumption and high performance; this technique requires close attention to link-quality on the receive side for an effective implementation.

A diversity antenna system can be compared to a switch that selects one antenna or another, never both at the same time. A radio in "receive" mode will continually switch between antennae listening for a valid radio packet. After the beginning sync of a valid packet is heard, the radio evaluates the sync signal of the packet, on one antenna, then switches to the other antenna and evaluates. The radio then selects the best antenna for the remaining portion of that packet.

On "transmit," the radio selects the same antenna it used the last time it communicated to that given radio. If a packet fails, it switches to the other antenna and retries the packet. Thus a diversity antenna system can improve signal strength, by dissipating the negative effects of multipath loss

One caution concerning diversity–it is not designed to cover two different cells. If antenna #1 communicates to device #1, while device #2 (which is in the antenna #2 cell) tries to communicate, antenna #2 is not receptive (because of the position of the switch), and the communication fails. Diversity antennae should cover the same area, but from slightly different locations.

However, through antenna diversity, a receiving station obtains multiple observations of the same signal sent off a transmitting station. The redundancy built into the multiple observations can be used to recover the transmitted signal with a higher degree of accuracy at the receiver. It should also be noted that antenna diversity offers substantial benefits to a WLAN implementation, providing the luxury of more than one antenna and the ability to select the best antenna for usage.

While antenna diversity has many advantages, it also has its drawbacks. Antenna diversity requires more extensive signal processing, which in turn leads to increased power dissipation. Hardware overhead also increases, although a fair amount of circuits and building blocks can be shared among multiple signal paths.

Fortunately, advanced techniques such as dynamic diversity can optimize the power/performance trade-off and permit the technology to further penetrate into WLAN applications. But to properly implement dynamic diversity and achieve its full benefits, designers must perform a full and accurate characterization of the receive-side channel.

Dynamic diversity delivers several advantages: it provides a high throughput rate without excessive overall power consumption, it does not interfere with traditional transmitter-side link-enhancing techniques, such as transmitter power control and packet retransmission; and it maintains backward compatibility and interoperability with currently deployed wireless communication equipment, since it is a receiver-side-only enhancement. Moreover, the receive-side link-quality assessment techniques play a key role in realizing the dynamic-diversity concept. The dynamic-mode selection can be driven by various Physical Layer as well as Data Link Layer's MAC sublayer parameters that can indicate a change in the quality of the communication link.

Receive Modes

To understand dynamic diversity, you must understand that there are a number of receive modes.

* *No diversity (single-antenna mode):* This is an obvious case where there is only one receive antenna. It allows the simplest implementation and results in the lowest power consumption of all cases.
* *Switched diversity*: Only one receive antenna is chosen at any given time during reception, based on some prescribed selection criterion. The antenna connection is switched when the perceived link quality falls below a certain prescribed threshold.
* *Selection diversity:* One antenna is chosen, whose receive path yields the larger signal-to-noise ratio (SNR) or signal power. The SNR or signal-strength measurement can take place during a preamble period at the beginning of the received packet. So, a single antenna connection is maintained most times, but during the measurement of the SNR/signal strength, both antennae connections need to be established. The actual selection/switching process can also take place in between packet receptions, and can be done on a packet-by-packet basis or can take place once in a number of receptions or prescribed time period.
* *Full diversity*: Both antennae are connected at all times. Since both received paths must be powered up, this mode consumes the largest amount of power, but it also offers the largest performance gain compared with other configurations, especially in severe fading environments with large delay spread. The digital front-end techniques—signal detection, frame synchronization, and carrier frequency offset estimation/correction, for instance—can also benefit from the availability of multiple receive paths.

Link Assessment

The next item to consider when contemplating a diversity antenna set-up is the "link assessment." In a conventional indoor WLAN, the data rate of a given communications link is typically adjusted at the transmitter side, based on some measure of the successful packet transmission rate. As the channel condition worsens (for example, as the receiving station moves away from the transmitting station, or the antenna orientation changes in a mobile station), the link data rate is adjusted downward, since reliable communication at the initial rate is no longer feasible.

Dynamic diversity enables a higher-link rate in more adverse channel conditions than is possible in conventional systems, while keeping the transceiver/ modem com-

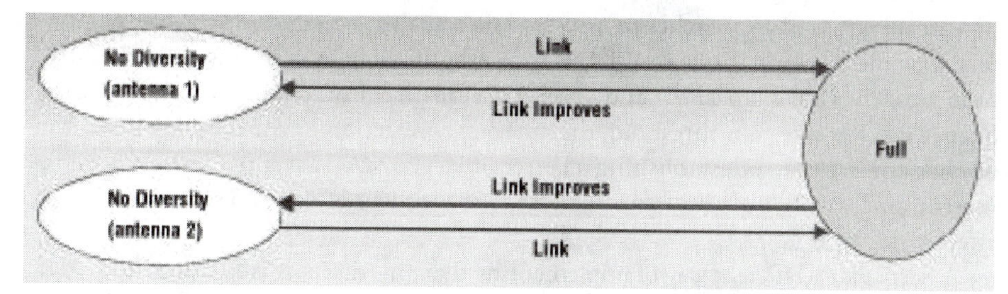

Figure 19.4 Dynamic diversity enables a higher link rate in more adverse channel conditions than is possible in conventional systems, while avoiding excessive overall power consumption by the transceiver/modem components.

ponents from consuming excessive overall power. In contrast to the conventional WLAN system based on transmitter-side link-quality assessment, dynamic diversity requires a receiver-side link-quality measure.

The state transition diagram in Fig. 19.4 illustrates one particular strategy that allows dynamic selection between full diversity and no diversity. As the link quality deteriorates, the receiver transitions from a single-antenna connection to the full-diversity mode. The transition back to a single-antenna connection is triggered by an indication of significantly improved link quality. When going back to a single-antenna configuration, a comparison of the received-signal strengths in the two antenna paths can easily lead to a preferable antenna connection. In this sense, the scheme of Fig. 19.4 also incorporates a "slow" form of selection diversity.

A transition between diversity modes and antenna connections is signaled by a change in the perceived quality level of the link. The transmit-side link-quality assessment is typically based on the estimated dropped-packet rate, via the observation of the acknowledgement packet and the number of retries attempted. However, the link-quality assessment at the receiver side, as required for dynamic diversity, is based on any combination of three factors: SNR, modem-detection quality measure, and MAC-layer link-quality measure.

Another useful PHY Layer parameter that can lead to a good link-quality measure is a detection-quality measure (DQM). For example, the detection quality is reflected in an averaged magnitude of the soft decisions captured at the Viterbi detector input. Since the functional relationship between such a DQM and the bit error rate or the packet error rate can be obtained empirically, the DQM can drive the mode selection, the antenna selection, or both. No matter which method is adopted, the MAC Layer functions must always verify the Receiver Address field in the MAC header, to ensure the packets are intended for the receiving station under consideration.

***Note:** The requirements for a Viterbi detector (also known as a Viterbi decoder) is a processor that implements the Viterbi algorithm (which is commonly used in a wide range of communications and data applications) depend on the applications where the detector is used. For instance, in addition to antennae, Viterbi detectors are used in cellular phones with low data rates (e.g. below 1 Mbps) and very low energy dissipation requirement. They also are used for trellis code demodulation in telephone line modems, where the throughput is in the range of tens of Kbps, with restrictive limits in power dissipation and the area/cost of the chip.*

Another class of approaches to assessing link quality is that of MAC Layer parameters. One method applicable to WLANs is to examine the Retry Subfields in the MAC headers of the received packets, and observe the frequency of retried packets. As the frequency reaches a certain threshold, the antenna connection or the diversity mode can be changed in hopes of establishing a better link. The MAC Layer parameters are convenient, since the link quality can be assessed independent of the data rate or multipath effects.

A particularly efficient way of implementing dynamic diversity is to utilize both PHY and MAC Layer parameters. For example, the receiver can rely on the inspection of the Retry Subfields to sense degradation in the link quality, and signal a transition to the full-diversity mode. On the other hand, the reverse transition from the full-diversity mode to the simpler antenna setting can be triggered when the Received Signal Strength Indication (RSSI) level increases by some prescribed amount (which could be a data-rate-dependent value).

Diversity antenna systems are standard in many APs, because a diversity antenna system is the best way to reduce multipath loss. But although an antenna or even two antennae may be provided with an AP, as the reader now understands, different antennae provide different coverage patterns and detailed implementations vary widely. Thus the AP(s), antennae, and implementation techniques should be selected according to site coverage requirements. Furthermore, it is vital that the decision to use antenna diversity be made upfront, since that decision has an impact on how a site survey is performed.

For more information on antenna diversity read "Antenna Diversity Strengthens Wireless LANs," published in the January 2003 edition of CMP Media LLP's publication, *Communications System Design*. The article can also be found on the CommsDesign website at www.commsdesign.com/design_library/OEG20030103S0053.

For a good tutorial on antenna basics check out this website:
www.ictp.trieste.it/~radionet/1997_workshop/wireless/notes/sld027.htm.

ANTENNA SELECTION CRITERIA

We will now look at some of the features you should consider during antenna selection. They include:

Gain. Antennae transmit and receive radio waves; the focused strength of this radiated energy is measured in terms of gain in decibels (dB). It is important to take gain into consideration when selecting network antennae. The gain in the antenna focuses the transmitted signal towards the targeted area of coverage. It also focuses incoming energy on the receive side. Gain is expressed relative to the performance of a theoretical isotropic antenna that radiates equally in all directions. By definition an isotropic reference antenna has a gain of 1 (0 dB). An antenna that has a gain of 2 (3 dB), compared to an isotropic antenna, would be written as 3 dBi. (Note that while power is expressed in dB, gain is expressed in dBi.) For every 3 dB increase, you double the emitted power out of the antenna. You can use higher gain antennae with lower powered radio cards to increase signal strength, link quality and distance, but be sure you don't go over the regulatory EIRP (Equivalent Isotropically Radiated Power) limit for your locale. The following table lists the power levels permitted in the three primary geographical regions. Operation in countries within Europe, and regions outside Japan or the United States, may be subject to additional or alternative national regulations. For example, in Australia the ACA or Australia Communications Authority defines an antenna's EIRP limit. Check your local regulatory agency for their specific wireless networking restrictions, which include antennae EIRP limitations—any deviation from the permissible settings is an infringement of national law and may be punished as such.

Transmit Power Levels for Different Regions

Maximum Output Power	Geographic Location	Compliance Document
1000 mW	United States	FCC 15.247
100 mW (EIRP)	Europe	ETS 300-328
10 mW/MHz	Japan	MPT ordinance 79

But also see Appendix III: Regulatory Specifics re Wi-Fi.

Note: ***An antenna's power is equal to the total amount of power actually transmitted through the system's antenna, including the transmitter's power output, the cable's power loss, and the antenna's gain capability. This power is measured in Watts.***

You'll want an antenna that provides enough gain on both the transmitting and receiving side to establish stable links while staying within regulatory limits. Antenna manufacturers typically include various gain models in their product offering, to accommodate differing access point gain requirements.

VSWR. Short for "Voltage Standing Wave Ratio," this is the ratio of the maximum/minimum values of the standing wave pattern along a transmission line to which a load is connected. VSWR value ranges from 1 (matched load) to infinity for a short or an open load. For most WLAN antennae, the maximum acceptable value of VSWR is 2.0, but try to find an antenna with a VSWR of 1.5 or less, which offers approximately a Return Loss of 14.5 dB. This means that most of the signal from the transmitter to the antenna is radiated (96% radiated and 4% reflected). A VSWR of 2.0 (return loss of 9.5 dB) means that 90% of the signal is radiated and 10% reflected.

Radiation pattern. An antenna's radiation pattern indicates how the transmitted radio wave energy is distributed in space. The vertical cut of a radiation pattern, also known as the elevation cut, is measured across the antenna's elevation plane. The horizontal cut of a radiation pattern, known as azimuth, is measured across the horizontal plane. Narrowing either the elevation or the azimuth pattern will increase the antenna's gain. Radiation patterns can also be altered according to the antenna polarization. This refers to the orientation of the antenna's radiated signal, which can be circular, horizontal, vertical, etc.

Note: ***If you deploy a diversity antenna solution, keep in mind that since a pair of antennae must be within each other's radiation patterns in order to communicate; uniform radiation patterns across the frequency band are crucial to the network's performance and reliability.***

Beamwidth. The beamwidth of an antenna describes how a signal spreads out from the antenna, as well as the range of the reception area. Beamwidth is measured between the points on the beam pattern at which the power density is half of the maximum power. A high gain antenna has a very narrow beamwidth and may be difficult to align.

Polarization. Radio waves are a combination of both an electric field and a magnetic field, which are perpendicular to each other. Together these fields are known as the electromagnetic field. The position and direction of the electric field with reference to the earth's surface determines the wave polarization. A horizontal polarized antenna is positioned to have the electric field parallel to the ground, whereas in the case of a vertical polarized antenna, the electric field is perpendicular to the ground. For a link to work, antennae at both ends must have the same polarization. And if the site has other APs in the area that are causing interference, you need to polarize your antennae differently to those other antennae so as to decrease the interference.

The gain, VSWR, and radiation pattern characteristics of an antenna design are closely related; therefore, if you alter one, the other two will be affected, which, in turn, will directly impact the antenna's performance.

Some other factors to consider when selecting an antenna include the following:

- The range of the antenna.
- The coverage requirements. For example, an antenna that serves a distribution center with numerous vertical aisles would have different coverage requirements than an antenna serving a corporate office.
- The environment. When antennae are used indoors, the building construction, ceiling height, and internal obstructions must be considered. In outdoor environments, obstructions such as trees, vehicles, buildings, and hills must be considered, along with salt air, moisture, ice, high heat, etc.
- The antenna's lifetime expectation. Antennae that are placed out-of-doors or in harsh environments will have a different life expectancy than an antenna in a library or corporate office.
- The mounting options. Does the antenna need to be hidden, or does it need to be mounted at a height that allows its signals to avoid interference from, for example, trucks loading and unloading on a docking bay?
- Whether the antenna's appearance should be a factor in the selection process. If the antenna can't be hidden under ceiling tiles, does it need to blend with its surroundings? Many times corporate offices, HotSpot venues, etc., require that the AP and its antenna be unobtrusive.

Check out the Webcats Wireless website, http://webcatswireless.com/toolbox/. It offers a number of very helpful antenna placement tools.

Note: *Cisco Systems provides some great charts on antenna gain and power limits for 802.11a and 802.11b. Their web site also contains a handy band plan. (Cisco also seems to update the site quite regularly.) Those documents can be found at http://www.cisco.com/univercd/cc/td/doc/product/wireless/airo1200/accsspts/ap120scg/bkscgaxa.htm#1013020. Or, if the URL is just too troublesome to type in, go to Google and search for "Channels, Power Levels, and Antenna Gains"—be sure to use the quote marks.*

CONCLUSION

Don't ignore the importance of the antennae. Choosing the right antenna and matching its characteristics to the best propagation path are the two most important factors in setting up a communications circuit. The weakest link in the communications circuit is the wrong propagation path. The best transmitter, antenna, and receiver are of little use if the the propagation path is improper.

The role of a wireless antenna is to direct radio frequency (RF) power from a radio into the coverage area. Different antennas produce different coverage patterns, however, and need to be selected and placed according to site coverage requirements.

Finally, remember that the choices made during the antenna selection process can make or break a WLAN system, just like the choice of speakers can make or break a stereo system. For example, a good antenna properly deployed can reduce stray RF radiation, by making the signal up to 100 times lower outside of the work area, and thus much harder to surreptitiously intercept.

Chapter 20: Client Devices

These are the most visible components of a wireless network—the physical platform for wireless applications that provide services such as data capture and display, information processing, location detection, and even voice communications. These devices may be carried by end-users, mounted within shipping containers, or installed inside a vehicle.

The purpose of any network, whether wired or wireless, is to exchange information and share resources beyond one device. In general, the types of devices found *using* (not comprising) a wireless network may include some or all of the following:

- Personal computers (PCs).
- Laptop/notebook computers.
- Personal digital assistants (PDAs).
- Tablet computers.

Figure 20.1 Some of the client computing devices that most commonly access the typical corporate WLAN.

- Information storage devices—CD-ROMs, DVDs, Zip Drives, portable hard drives, etc.

Figure 20.2 Storage devices and more. This graphic depicts from left to right the Apple iPod MP3 player, Canon digital camera, Zip drive, HP CD Player, Labtec Web cam, Pocketec portable hard drive, Phillips Tivo device, and Toshiba's portable DVD player.

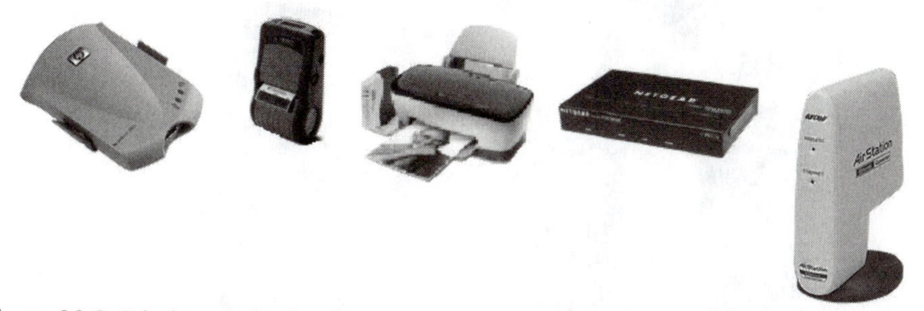

Figure 20.3 Printing wirelessly. This graphic depicts from left to right the HP Jetdirect 380x 802.11b wireless external print server, Zebra QL 320 direct thermal mobile printer, Epson Stylus C8OWN printer, Netgear PS111W 802.11b Wireless Ready Print Server, and the Buffalo AirStationTM Ethernet Converter (WLI-T1-S11G), which allows you to connect a network-ready Brother Printer/MFC to a Wi-Fi network in ad-hoc or infrastructure mode. Print servers help small workgroups to easily share a broad range of network-capable printers and multi-function peripherals across wireless networks.

- Printers—any device that makes an impression on media, usually paper, and is connected to a computer (wired or wirelessly). These devices range from a simple printer in a cash register to advanced printers that are capable of magazine quality photo images. The printers most often found in a corporate environment include inkjet, laser, and the new LED (Light Emitting Diodes) printers.
- Scanners—these devices come in many different forms, including hand-held, feed-in, and flatbed types; some can scan black-and-white only, others can also scan color. Very high-resolution scanners are used for scanning photos and graphics for high-resolution printing. Lower resolution scanners are just fine for the average corporate office's scanning needs.
- Devices for the vertical market. These include ruggedized computers (used in field, warehouse and service operations, barcode readers (commonly used in the trans-

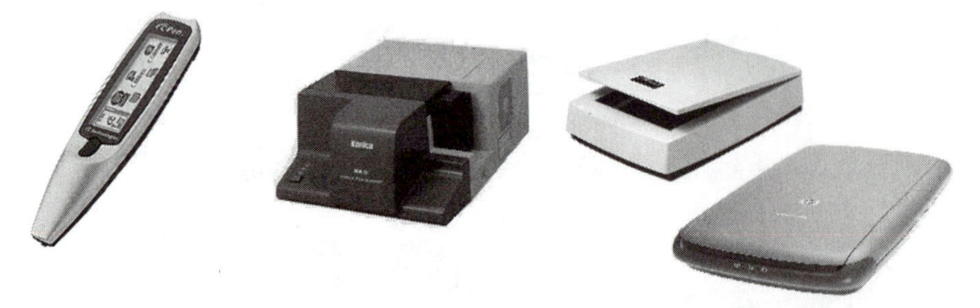

Figure 20.4 This graphic depicts from left to right the C-Pen 800C Handheld Scanner (can scan directly into a cursor position on a PC), Konica Film Scanner for Professional RX-3, Trust SCSI Connect 19200 high resolution scanner (good for scanning graphics), and HP 3500c Scanjet color scanner.

Figure 20.5 This graphic depicts from left to right Intermec ScanPlus 1800 (barcode reader), passive RF tags that require no battery and are inexpensive and long lasting, Symbol Technologies SPT1700 handheld computer with barcode scanner, Panasonic Toughbook 01 ruggedized handheld computer, and Miltope TSC-750 ruggedized computer.

portation, warehousing and distribution industries), RF tags (used to identify and track goods and equipment–most commonly used in the transportation, warehousing and distribution industries, but large organizations may use RF tags to keep track of its hardware and equipment assets), and handheld PCs (devices that are normally a little larger than the palm-sized PDAs, use the Windows CE operating system, have a form factor that is more rugged than the typical PDA, need a larger battery than the typical PDA, are used with work-related applications like inventory management systems, are not pre-loaded with personal organization software, and offer specialized keys or touch-screens for easy input of regularized form-information).

- Smartphones–a combination PDA and mobile phone offering wireless Internet access as well as voice communications.
- Wi-Fi phones.

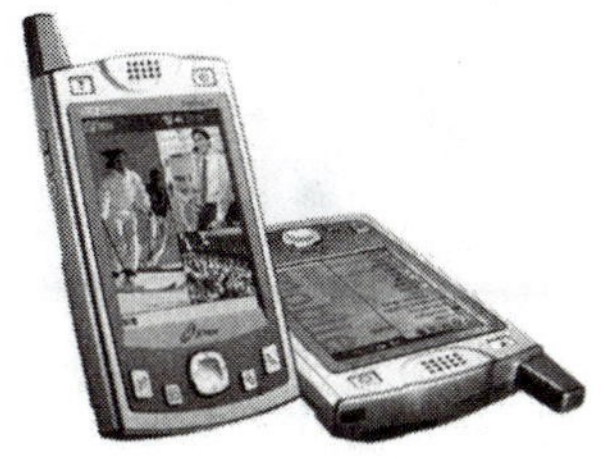

Figure 20.6 This graphic shows an Eten Smartphone P600. These devices can serve dual duty as both wireless phone and PDA.

✪ THE SELECTION CRITERIA

Each WLAN user should be equipped with at least one of the following devices: a laptop, tablet computer, PDA, handheld computer, a phone based on one of the 802.11 flavors, a wireless printer, etc.

***Note:** Processing power counts for a lot, so if you plan to add wireless capabilities to computing devices that have low processing power, consider upgrading first. When you access a WLAN—using any vendor's NIC—the processor utilization will become quite high.)*

Don't try to support any and every device that crosses the organization's threshold. Instead, create a "cafeteria plan" (a term coined by Gartner Inc.) that offers employees a limited choice of mobile computing devices, and a variety of support options. Such a plan allows the IT department to provide the employees with device options that more closely suit their personal preferences, yet contains costs and development complexity.

The best way to implement a cafeteria plan is to categorize all available mobile computing devices into three levels of support.

* *Level One:* The organization's standard mobile computing devices. If the employee uses one of these devices, the IT department will support it in the same way that PCs are supported.
* *Level Two*: Employees will be permitted to access the WLAN via level two devices, but the IT department will not support them in any way whatsoever.
* *Level Three*: Banned devices. If the organization catches an employee using any of these devices within the IT department's jurisdiction, that person will be asked to remove the banned device from the organization's facilities.

Allocate a budget. Give the employees a list of supported devices, and let them choose from that list up to the amount of the budget. This way you can satisfy the employees' desire for the latest technology, but keep costs within budget.

Keep abreast of new devices, manufacturers, and form factors. Not only will it help to anticipate new opportunities for applying technologies within the organization, but it will also help keep everyone on top of what's occurring in the mobile and wireless technology marketplace.

✪ THE WIRELESS NETWORK CARD

Also known as a Network Interface Card (NIC), PC card, or PCMCIA card, these devices contain the radio, which allows mobile computing to access a wireless LAN. For an

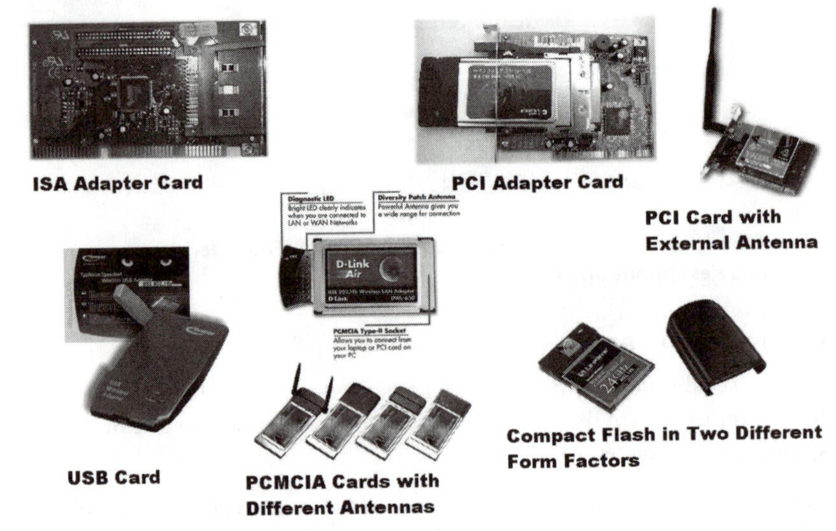

Figure 20.1 Wireless network interface cards come in many different forms.

employee to use a WLAN, he or she must have a computing device that is equipped with wireless 802.11a, b, g, or dual-mode (depending on the specification the WLAN uses) capability. This capability can be provided via a wireless network interface card. Wireless NICs come in a variety of forms–ISA, PCI, USB, PCMCIA Card, and Compact Flash.

To add wireless networking capabilities to a stationary PC, the PC will need a wireless NIC. These wireless NICs can come in a variety of form factors, e.g. ISA, PCI, USB or PC adapter (if a PCMCIA card, more commonly referred to as a PC card, is to be used with the stationary PC), since such a card allows you to use a PCMCIA card with the average desktop workstation.

To add wireless networking to a mobile computing device such as a tablet PC, laptop, PDA, or handheld computer, you will need a PCMCIA card (or Compactflash card in the case of some PDAs) that can plug into one of the device's free PCMCIA slots (or in the case of some PDAs and handheld computers, a module that can be slipped into or connected to the device). Note that since 2002, many mobile computing devices ship from the factory with embedded Wi-Fi capability.

Selection Criteria

There are many different computing devices and diverse computing environments and all require consideration when equipping a computing device with a wireless network interface. This list should help the deployment team with their selection process.

- If the WLAN is based on 802.11a technology, consider purchasing dual-mode cards for all mobile computing devices, since 802.11b is the technology used in most HotSpot locations and in most home wireless networks.
- When equipping a desktop computer with a wireless NIC, opt for a card with an external antenna. If you use a NIC with an internal antenna, the path from the antenna to the access point must go through the machine, impairing performance.

An external antenna will normally provide about 15% more signal strength, if the external antenna is placed in a position so that signals don't encounter obstacles.

* Be sure that the NIC's drivers support the device's operating system. While virtually all cards support Windows OS, and some support Apple, it may be more difficult to find cards that support Linux, Windows CE, Palm, etc.
* Be cognizant of the fact that you can have choices in antennae; for example, sometimes an extension antenna is available as depicted in Fig. 18.15's PCI card with external antenna. When adding wireless capability to a stationary computer, such an antenna is desirable because it allows the signal to avoid some of the clutter that surrounds these types of computing devices.
* Cards from various vendors can provide very different range limits. Access points may provide coverage to one card and deny it to another in the same location. This is the result of the basic radio frequency (RF) performance of the radio's transmitter/receiver. Some vendor radios can boost their power for greater coverage performance.
* Features over and above the IEEE 802.11a/b/g standards, such as EAP security are not interoperable among different vendors; features such as load balancing will not work with a mix of station radios. This means that you must take the effort to differentiate between what a card offers in the way of 802.11 standard features and vendor proprietary features.
* The functionality that all wireless NICs should have in common includes: modulation (translate baseband signal to a suitable analog form); amplification (raise signal strength); synchronization (carrier sense); error checking.
* Because most WLANs are based on the features and functionality of the access points, it's important to ensure that whatever vendor product you settle on, the NIC's radio can be made available in a wide range of clients—laptops, PDAs, handhelds computers (if applicable), inventory tracking terminals, and perhaps even telephones
* What flavor—a, b, g—dual mode? Dual mode, while having some drawbacks, may be the answer if the mobile computing device is to be used in different network settings, i.e. home, HotSpot, and corporate.
* If you are equipping a desktop computer with a wireless NIC, the card should have an external antenna. If you use a NIC with an internal antenna, the path from the antenna to the access point must go through the machine, impairing performance. An external antenna will normally provide about 15% more signal strength if the external antenna is placed in a position so that signals don't encounter obstacles.
* Consider using an external antenna, even with mobile computing devices. Users that use mobile computing devices (e.g. laptops, tablet PCs, and PDAs) with either a NIC with an internal antenna or one with a built-in NIC, may find that moving the device only a short distance can cause a noticeable difference in signal strength.
* When considering NICs with external antennae, note that a quarter-wave external antenna for a PC card is about 1.25 inches long and a full-wave antenna is about 5 inches long.
* Choose NICs that support password-protection of attribute changes, to prevent the settings of the network cards from being illegally or accidentally changed by users.
* Performance among various types and brands of NICs differ because of the individual NIC's drivers and setup routine.

* If you need antenna diversity, look for cards that offer that choice; you may also want to consider cards that have a plug that can be removed, so that you can plug in an external antenna to improve range.
* If buying PCI adapters for stationary PCs, determine in what environment the adapter will and will not work—some PCI adapters will only work in an all-PCI system, so if the computer has ISA and PCI slots, the adapter won't work.
* When purchasing PC card adapters, check the card ejector button on the adapter to determine if it works smoothly; you don't want an adapter that makes it hard to put the card into the PC.
* If you are purchasing wireless NICs for handheld computers and PDAs, be aware that most (but not all, e.g. HP's iPaq is equipped with a PCMCIA slot) require the purchase of not only the NIC but also additional equipment, such as the Xircom Wireless LAN module for Palms, the Springport Wireless Ethernet Module for Handsprings, or a PC Card sleeve.

CONCLUSION

Since the purpose of the WLAN is to exchange information and share resources, there are usually a wide variety of devices that need to access the network. However, since new client devices are hitting the market almost daily, the selection process can sometimes be confusing.

To find your way through the maze of new devices, read hardware reviews, check out what people have to say on user forums such as the ZDNet Networking/Connectivity Forum, which can be found on the ZDNet website (www.zdnet.com) or the practicallynetworked forms, which can be found at (www.practicallynetworked.com). Also check out the CNET message board which can be found on the CNET website (www.cnet.com) and news groups. Another good place for good information is friends and colleagues; ask them about their experiences with wireless equipment.

But, don't consider just the computing device and its connectivity, also use your end-users' needs as a guide in your decision-making process.

Finally, take the time to determine how you want to handle your road warriors' and teleworkers' networking needs.

Chapter 21: Cabling, Connectors and Wiring

Don't forget connectivity. Consider how the deployment team will connect the APs to the wired network, and how the antennae will connect to the APs and PC cards.

Some may wonder why there is a chapter on cabling since Wi-Fi is wireless. But cables and connectors are still needed even for Wi-Fi installations. For instance, you need a Category 5 or 5e cable to connect the access point to the wired portion of the network. You also may need to connect an antenna to an AP, or a device's wireless NIC to a removable antenna.

CATEGORY 5 AND 5e CABLE

While Wi-Fi allows you to build a data network without the cost and congestion of cabling, you will still need Category 5 or 5e cabling. Category 5 cabling connects your access point to the wired data network.

Category 5 and 5e cable is the most common type of cable used to connect a network of computers or to connect a computer system to the Internet.

Category 5, which is made up of eight color-coded, twisted copper phone wires encased in a sheath, has been the network cabling standard since 1994. In 1998, however, Category 5 was superseded by Category 5e, which is short for "Enhanced Category 5." Similar to Category 5 cable, Category 5e is enhanced to support speeds of up to 1000 Mbps, more commonly referred to as 1 Gbps.

Testing is more stringent for Category 5e cables than for Category 5 products. Its testing criteria include several additional measurements, some of which help to better quantify the UTP cable's noise characteristics. Furthermore, despite its 100 MHz specified bandwidth, category 5e is usually tested to a bandwidth of 350 MHz.

Category 5 and Category 5e cable are the preferred cable types for structured cabling systems. For instance, Category 5e is designed to support Gigabit Ethernet Networks, but either Category 5 or 5e cabling is typically used in Fast Ethernet Networks (10Base-T, 100Base-T4 and 100Base-TX), FDDI networks, and even some ATM networks.

RJ45 connectors and Jack Modules are used on both Cat 5 and 5e cabling.

Access Points

Cabling for access points is easy, since the only answer is to use Category 5 or 5e cabling to connect the AP to its wired component. But note that both CAT 5 and 5e cabling have length limitation of 328 ft. per cable run. Therefore, if the coverage area is more than 328 feet away from the pipe to the wired network, you will need one or more intermediate devices (hub, router, switch) along with the necessary power source, to connect the backhaul point with the coverage area.

Be sure to label the cabling (at each end of the run) as it is installed, describing each drop point location. Draw up and maintain a cabling diagram, to indicate exactly where the cabling is located in reference to the facility's floorplan/layout.

Many locations, such as the executive office suite, will not want the cabling (or access points for that matter) to be visible. That means that the installer must run the individual drop points in such a manner that the cabling is hidden from view, or aesthetically acceptable.

ANTENNA CABLE

Attached to most antennae (except the standard dipoles), is an interconnect cable. This cable provides a 50 Ohm impedance to the radio and antenna, with a flexible connection between the two items. (As a point of reference, antenna equipment for televisions is 75 Ohm.)

RF cabling has a high loss factor and should not be used except for very short connections. The cable length should have been determined during the site survey. The typical length is three feet, more or less, although it is possible to run an antenna cable as far as 40 feet and still obtain adequate signal strength. However, the reader should understand that the longer the cable, the more signal loss there is, so select the most direct route from the antenna to the radio.

Low loss/ultra low loss cable provides a much lower loss factor than an interconnect cable, and should be used when the antenna must be placed at any distance from the radio device. Also, although the cable is low loss, it should still be kept to a minimum length. Thankfully, many times low loss cabling is the only cable type supplied by vendors (e.g. Cisco) for mounting the antenna away from the radio unit.

Low loss/ultra low loss cable is usually offered in four different lengths with one RTNC plug and one RTNC jack connector attached. This allows for connection to the radio unit and to the interconnect cable supplied on the antennae. While they are similar to the normal TNC connectors, they cannot be mated to the standard connectors.

***Note:** If you are not careful you may be tied to one vendor insofar as AP / antennae are concerned. For example, you are installing Cisco Aironet products, to ensure compatibility with the Cisco products, you are required to use antennae and cabling from Cisco.*

AP Antennae

The deployment team also must consider how the antennae are to connect to the access point. It is assumed that the access points will be placed in out-of-the-way locations, such as being hidden by ceiling tiles, up high on a beam, over the top of a window, etc. Many times the positioning of the access point isn't conducive to its antenna placement. But when positioning an antenna apart from the access point, keep in mind that antenna cabling can cause signal loss. This is true at both the transmitter and the receiver. As the length of cable increases, so does the amount of loss introduced. To operate at optimum efficiency, keep cable runs as short as possible.

Wireless NIC Antennae

Correctly connecting the radio to a removable antenna will make the system more flexible, and it can increase both the device's range and throughput speed. But many wireless NICs come with the antenna built-in; thus they cannot be removed. Such a card may prevent you from connecting, if your location requires more "gain" to cover the distance. So if any of your end-users need a wireless set-up that provides more gain, look for a PCMCIA card with external antenna connectors. They are identified by a small round connector on the end or side of the card—the connectors are sometimes covered with small plugs, so look carefully.

Antenna Cabling Selection Process

To reiterate, any time you run cable to connect an antenna to its associated device, there will be signal loss. So make your runs as short as possible. To help you in your cabling decision-making, use an attenuation calculator. Times Microwave Systems provides attenuation calculators at its website, http://www.timesmicrowave.com/telecom/cable_calcultors.

Don't be confused if your equipment has several different types of connectors. There are adapters and pigtail cables available to connect to just about anything.

Once you've tested all the cabling and connectors, you can fasten the cable to form a secure installation. The best way to do this is to use cable clips that fit over the cable and allow you to mount the cable against, for example, a wall for a professional-looking job. These clips are very inexpensive, usually running less than 20 cents each. The clips should be spaced evenly to provide sufficient support. Trusty cable ties (zip ties) can be used to secure the cable to the antenna mast. Right angle "N" connectors also can be used to alleviate stress on your coax cable if the mounting orientation is awkward.

Lightening arrestors, signal splitters / combiners and perhaps even amplifiers (although amplifiers are strictly regulated) may also form part of your RF connection hardware rig.

When selecting antenna cable, the author's best advice:

* The shorter, the fatter—the better.
* Plan the location of your radio to minimize the length of antenna cable needed.
* As the length of your cable increases, compensate by using lower loss (thicker) cable.
* Connect the antenna directly to the radio to eliminate cables whenever possible.
* Check with the manufacturer of the WLAN system hardware before adding new cables and connectors!

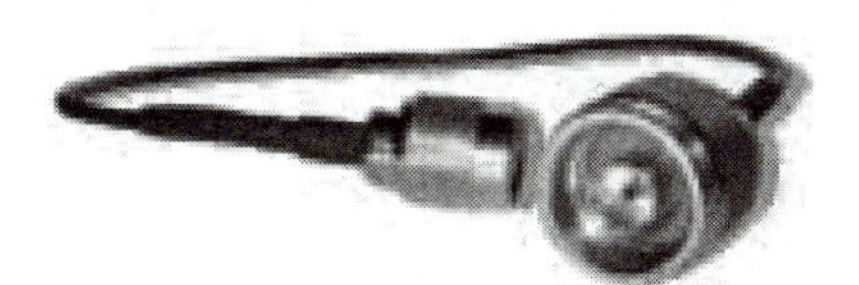

Figure 21.1 A pigtail cable, which is simply a small length of cable with adapting connectors to join a proprietary socket on a device to an external antenna cable.

THOSE PESKY CONNECTORS

Antennas typically have N connectors although Cisco antennae will most likely have a ReverseTNC to avoid Pigtails. Most APs and PCI-based NICs with external antenna jacks will have either RP-TNC or RP-BNC connectors, while most PC-card-based NICs don't have any connectors. The few that do (e.g. Lucent and Cisco) normally will have Lucent or RP-MMCX connectors (older devices may have MMCX connectors).

Use connectors sparingly. Each connector in a system introduces some loss, so try to avoid adapters and unnecessary connectors whenever possible. Try to buy cable with the proper connectors already attached, and, of course, use the shortest length possible, Wi-Fi equipment doesn't put out much power, and every little bit helps extend the AP's range and reliability.

Also bear in mind that when matching cables, you may encounter connectors of reverse polarity (male + female swapped, with same threads), reverse threading (left-hand instead of right-hand thread), or even reverse polarity reverse threading (both). Make sure you know what you're getting before you buy.

Note: *Ever wondered why WLAN cards have these annoying, special connectors? The FCC (and other regulatory agencies) is the culprit. Various regulatory agencies have requirements that unlicensed equipment be shipped with connectors "not available to the general public." The actual FCC rules—Part 15: Unlicensed RF Devices, Title 47 of the Code of Federal Regulations specifically 15.203—say that class licensed RF equipment must be designed to ensure that no antenna other than that furnished by the responsible party shall be used with the device. Once hard-to-find connectors, like MMCX, become too popular, at least in the regulatory agency's opinion, they are banned as being too available to the general public.*

Always check the connectors even when purchasing products from the same vendor. While some device families use the same type of connectors through their whole range (particularly those from same OEM - for example, Lucent is the same as Compaq), others change their connectors across their product families (for example Cisco offer two types of external connectors).

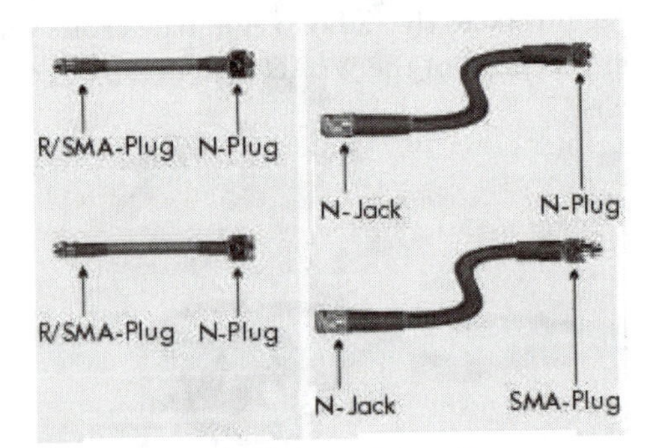

Figure 21.2 Examples of low loss, weatherproof, and phase-stability coaxial antenna cables. All of the depicted cabling feature 50 Ohm impedance for connecting between access point and 2.4 GHz antenna.

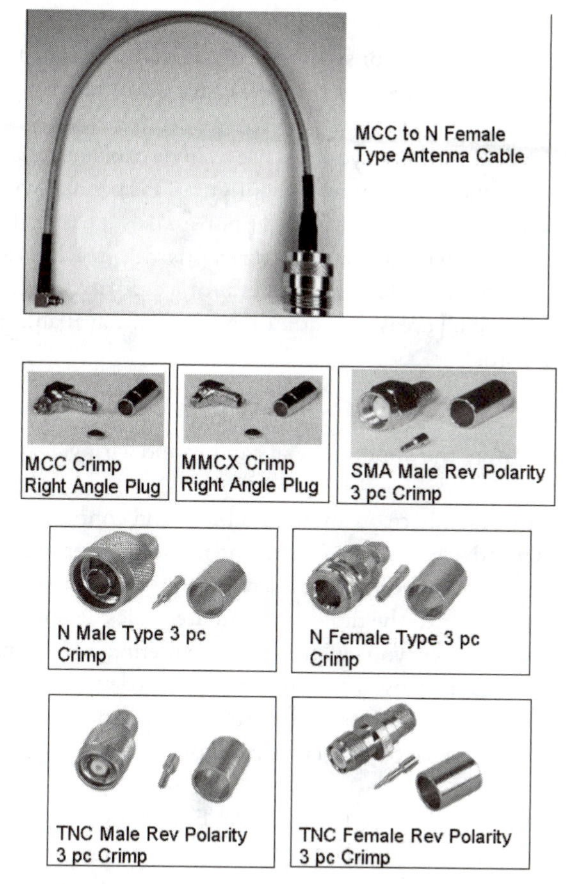

Figure 21.3 An example of the various types of connectors you might need when deploying a WLAN.

✪ LIGHTNING ARRESTORS

When using outdoor antenna installations, it is always possible that an antenna will suffer damage from potential charges developing on the antenna and cable, or surges induced from nearby lightning strikes. Lightning arrestors are designed to protect radio equipment from static electricity and lightning induced surges that travel on coaxial transmission lines. So either purchase an antenna with a lightening arrestor that protects the equipment up to 5000 Amperes, or install one separately. Note, however, that a lightning arrestor will not prevent damage in the event of a direct lightning hit.

You can install the lightning arrestor indoors or outdoors. Follow the regulations and/or best practices applicable to lightning arrestors in your region. When installing a lightening arrestor outdoors, ground the arrestor by using a ground lug attached to the arrestor and a heavy wire (#6 solid copper) and connect the lug to a good earth ground. When installing an arrestor indoors, place the wireless LAN device near a good source of ground, such as structural steel or the ground on an electrical panel, and ground the arrestor using one of those grounds.

ELECTRICAL

The deployment team also must consider the AP's electrical needs. For large deployments, consider using APs with Power over Ethernet (PoE) capability. APs with PoE obtain their power via the standard Category 5 or 5e cabling.

When you go with APs with PoE, you can use a single cable installation, which dramatically improves your choice of mounting configurations because you no longer need to consider AC power outlet locations. PoE support makes it easier than ever to overcome installation problems with difficult-to-wire or hard-to-reach locations.

If PoE is not an option, then lay out the cash for a certified electrician to run the proper electrical wiring to all areas that don't have a readily available electrical outlet. Don't use extension cords!

CONCLUSION

Most people tend to ignore the importance of cabling and wiring. Don't be among that group.

Here's the author's best advice on antenna cabling and connectors.

The most *important* advice is to check with the manufacturer of the WLAN system hardware before adding new cables and connectors!

Connect antenna directly to the radio to eliminate cables whenever possible. There will be instances, though, where you will need to use antenna cabling, plan the location of your radio to minimize the length of antenna cable needed.

To help make your WLAN a success, remember that when using antenna cable—go shorter and fatter. As the length of your cable increases compensate by using lower loss (thicker) cable.

Next, if possible purchase APs with Power over Ethernet capabilities.

Last but not least, use lightning arrestors.

Section VIII: Final Thoughts

Chapter 22: Wi-Fi's Future

Wi-Fi and its 802.11 series of specifications are traveling a similar evolutionary path to their wired equivalent, Ethernet and the 802.3 standards. Over the next two or three years we will see increases in transmitted bit-rates, and additional functionality in the areas of QoS, security, and power control. But the question that begs to be asked is, "Can Wi-Fi, like its wired counterpart, achieve the goal of the 'ever present' wireless network?" The answer is, "maybe." It certainly seems that the industry is moving in the right direction.

At this very moment battles are raging between IT departments, with their concern about security, and employees who demand networking freedom. Slowly, mobility is winning. "Once people have wireless inside their offices," says Frank Keeney, co-founder of the Southern California Wireless Users Group, "they never want to go back. It's a tremendous productivity tool." That goes double for road warriors, the real source of potential wealth in the wider Wi-Fi world. These corporate nomads, who number 11 million or more in the U.S. alone, encompass field sales representatives, insurance adjusters, real estate agents, delivery managers, and more.

Even the most technophobic CEO can imagine the benefit of a salesperson tapping into a database minutes before he or she sits down with a client; or a field service technician dialing up and instantly accessing a copy of a technical manual.

A wide variety of new devices are now available, such as Voice-over-Wi-Fi phones, Wi-Fi-enabled printers, projectors that offer the convenience of wireless connectivity, laptops with Wi-Fi capability built-in, and much more. Products such as these continually add to the usefulness of wireless networking. This trend enlarges people's notions of what is possible and helps to create a well-founded desire for the increased mobility that wireless networks can bring to their lives.

Wi-Fi also is spreading within academia as more and more schools make technology investment a top priority. Even when a school can't put a PC in every classroom, school administrators have discovered that Wi-Fi enables them to use computer carts. These allow teachers to schedule a period during the day or week when their students can have access to the computers, the Internet, and all that that entails. Thus even in these cash-strapped times, the maximum possible number of students can benefit from a school's technology investment. Wi-Fi enables the computer and cart to roam seamlessly from access point to access point throughout the school, to portable classrooms and even the assembly room, eliminating the time normally spent shutting the devices down between sessions.

Educational institutions can even lease laptops to students, so they can take them from classroom to classroom and then home at night. In these institutions, students can connect to the Internet, school printers, and other network resources during the school day, since access points are positioned throughout the school campus, eliminating the need for each student to connect to an Ethernet port.

Another Wi-Fi trend emerging today is voice/data convergence. Traditionally, voice and data networks have been separate, because no single technology meets the reliability and performance requirements of both. That, too, is rapidly changing. The focus of most converged network strategies is Voice over IP (VoIP), a technology that can transmit telephone conversations over any packet-switched IP network, whether it's a wired local area network, a Wi-Fi network, or the Internet. Standards, such as H.323 and the Session Initiation Protocol (SIP), provide the necessary functionality. Of course, Quality of Service is another key requirement for successful VoIP communications, but with wireless connectivity now a reality, the stage is set for seamless, end-to-end voice/data solutions.

✪ A DISORDERLY INFRASTRUCTURE

How does a Wi-Fi network become ubiquitous? Where will the ever-pervasive wireless infrastructure come from? To understand how Wi-Fi could blanket the earth, consider an interesting quote that comes from Alessandro Ovi, the technology adviser to European Commission president Romano Prodi. He recently uttered a perfect, profoundly beautiful, description of Wi-Fi's capabilities. "These are water lilies; little ovals of connectivity that are not centrally deployed, that sometimes overlap a little, whose stems lead to the Internet usually through a broadband connection. If you happen to be near one of these water lilies, depending on how the access point is configured, you might get an Internet connection."

However, there are diverse ideas about who should pay for these pretty "lilies" and how they should work. The companies that tried initially to build proprietary wireless systems across the U.S. have bitten the dust, but other companies, such as Cometa, Boingo, and T-Mobile, are stepping into the breach. These companies are building a network of HotSpots—small local venues where anyone who can ante up a small fee can use their Wi-Fi-equipped mobile computing device to access the HotSpots' Wi-Fi network for a bit of high-speed Internet surfing.

Some businesses (and individuals) install Wi-Fi networks so they can offer free wireless networking to their customers and/or the public-at-large. Still others install what they believe is a private network, but inadvertently let their network spill over their property lines. Thereby making access "available" to those within their proximity. There are even generous individuals who have started building their own "lilies" (wireless networks) so that other people can use them. These networks are oft-times referred to as "FreeSpots" or "community" networks. And in order to link all of these "free" networks together, individuals have begun to map these "lilies of connectivity" so that folks can locate and access them.

History teaches us many different ways that infrastructures can begin, and grassroots movements have a long history in advancing technological innovations (e.g. the telephone, the PC, and the Internet)—which often are "disorderly." So in Mr. Ovi's words, "Let these water lilies grow for a while longer... It could be beautiful."

HOW FAST CAN IT GO?

There will most likely be a new Wi-Fi standard to feed these many trends sprouting around the Wi-Fi phenomenon. Engineers are playing with Wi-Fi technologies in an effort to boost its speed. Some have even succeeded, testing products that can run at 72 and even 108 Mbps, although the maximum "official" Wi-Fi speed is 54 Mbps. But that sanctioned limit may soon be a thing of the past. Stuart Kerry, chairman of the IEEE 802.11 Working Group, has indicated that a collection of members has formed a "High Throughput Study Group" to work on a potential high-performance standard that would boost both 802.11g and 802.11a standards.

Although there is as yet no official Task Group, as discussed in Chapter 6, this standard for increased throughput might be called 802.11n. Proposals say the new technology could achieve a bit-rate of at least 108 Mbps, and perhaps as high as 320 Mbps. Look for this new standard to be completed sometime in 2005 or 2006.

In addition, in early 2003, the IEEE tentatively announced the creation of a new wireless ultrabroadband standard, which it has dubbed 802.16a, or more memorably, "Wi-Max." It covers a square mile comfortably, meaning it would take just 49 transmitters to blanket San Francisco. As Larry Brilliant of Cometa fame says, "Now it gets interesting." If you can cover an entire city with wireless Internet access, you suddenly have a very cheap alternative to cellular networks.

Tomorrow there will be new products, many with multiple capabilities. Of course, interoperability is the key. As the technology improves and the engineering community continues to find new ways to exploit this exciting technology, interoperability will provide investment protection for Wi-Fi's users.

TWENTY PREDICTIONS

It's interesting to hear what other Wi-Fi advocates predict for the technology's future. Here are 20 predications on Wi-Fi's future from some of the industry's most ardent supporters.

Prediction No. 1: The low cost of wireless communications will drive a growing market for embedded machine-to-machine communications in industrial equipment and major appliances for monitoring and problem notification. As illustration, British Gas is placing wireless devices on their bottled gas so the bottles can "phone home" when they are empty.—Sumit Deshpande and Don LeClair, technology strategists, Computer Associates International Inc.

Prediction No. 2: The automobile will become a platform for mobile applications. The vehicle has a big battery, with the ability to generate electricity and the space for all kinds of gadgetry. Furthermore, people are spending more time than ever in their vehicles, so it's not unreasonable to look for in-car telematics to include GPS, data storage, docking for multiple types of handheld devices, hard-copy output, and so on. This already exists in law enforcement—and the new bus-based, 48-volt auto system standards will accelerate the vehicle telematics explosion.—John Parkinson, chief technologist, Cap Gemini Ernst & Young U.S. LLC.

Prediction No. 3: 2003 will see the introduction of the first cellular handset with Wi-Fi capability. This will be the start of a major acceleration in access device penetration. The first devices will be Microsoft Smartphone-based handsets, using SD card-based 802.11 NIC, such as the one soon to be launched by SyChip and SanDisk. Further support will come from Microsoft itself. Within six months integrated Wi-Fi and cellular chipsets will be available for sampling from Texas Instruments, Qualcomm, and Philips.—Martin North of Sydney, Australia-based digitalplays.com.

Prediction No. 4: Wi-Fi will be invisible to most consumers, but key in the enterprise. Retailers, warehouses, deliverypeople, medical practitioners, fleet managers, venue operators (such as sports arenas and conference centers), hospitality employees, campus-based field forces such as security personnel and facilities staff, and even public safety responders will use Wi-Fi as a primary conduit for data exchange (to/from laptops, PDAs, desktops, handheld scanners, point-of-sales terminals, and kiosks) using mostly private networks but some public networks as well. This will let the service industry's personnel serve staff, customers, and each other with the same benefits of as-needed data exchange and deep data retrieval that their white-collar colleagues have enjoyed for a decade on networked desktop PCs.—Galen Gruman, Editorial Director, IT Wireless, a website that specializes in leveraging wireless technologies for the enterprise.

Prediction No. 5: Wi-Fi will become the #2 prepaid telecom application. A proliferation of Wi-Fi uses will arise quickly, leading to many service providers offering "Prepaid Wi-Fi" services. Travelers, students, and an increasingly mobile population will be able to access these Wi-Fi applications wherever they are, at home, in the office, or on the road—Gene Retskey, a telecommunications author, motivational speaker, professor, and Editor-In-Chief of *The Prepaid Press*, the leading trade publication on prepaid telecommunications.

Prediction No. 6: Wi-Fi's potential as a last mile solution has generally being overlooked. WLAN technologies will become very widespread in the distribution of broadband access from DSL/cable nodes to local communities. It is relatively cheap to purchase/install, global standards are in place, equipment is widely available, there are no licenses to purchase, and there is a nice fat pipe to the Internet to offer.—Tony Crabtree of Hampshire, England-based Juniper Research,

Prediction No. 7: Within the next five years, more nonvocal transactions will traverse the airwaves, causing a trend for mobile devices such as smartphones and PDAs to become the holder of our identities within the next five years. Much like the role of a driver's license, our mobile communications device will serve to authenticate us and securely contain credentials and certificates. Biometrics, embedded appropriately, would thwart identity theft.—Peter Athanas, associate professor, Department of Electrical and Computer Engineering, Virginia Polytechnic Institute and State University.

Prediction No. 8: By as early as 2004, more than one million remote and mobile devices will be integrated with enterprise applications. Early adopters will include the industrial, oil and gas, manufacturing, and utilities industries. Typical applications will include homeland defense sensors, the monitoring of flow and pressure of petroleum production, meter reading, and field communications.—Bob Ross, WebSphere integration program director, IBM Software Group.

Prediction No. 9: By 2004 there will be more than 50,000 publicly accessible HotSpots around the world. Established wireline carriers will create the vast majority of the HotSpots, although packet wireless carriers and cheap "Wi-Fi-in-a-box" products also will be responsible for creating some of these public access points. Aggregators will be the glue that binds together all these Wi-Fi "islands." Also by 2004, different network variants will begin to merge into a seamless, "wireless broadband" global network for roaming purposes. The public doesn't care what acronym or standard is used; they'll just want "wireless broadband."—John Rasmus, vice president, GRIC Communications Inc.

Prediction No. 10: By 2004, CRM vendors will need to help companies manage customer interactions via wireless devices. CRM vendors that don't offer that functionality—such as the ability for retailers to push information like coupons to their customers' wireless devices—will find themselves left on the dock as the ship sails off without them.—Bud Michael, executive vice president, Kana Inc.

Prediction No. 11: If quality of service, roaming, and security issues can be worked out—and all of these straightforward technically—there is a killer app for companies to run "wireless PBXs." In other words, route their voice traffic within the company over the private Wi-Fi network. This would save mucho costs, and give the companies features and management capabilities they don't have today. We're probably talking two to four years for even early adopters to go this way, but the numbers could get very big very fast.—Kevin Werbach, an independent technology analyst, author, and founder of werbach.com.

Prediction No. 12: Wi-Fi will be everywhere by 2007; it will be delivered either by new installations or via the conversion of old pay telephone sites into Wi-Fi access points. However, although sending voice and data over Wi-Fi will become popular, making money off of such installations will face similar if not greater challenges as that other grassroots phenomenon, the Internet. Popularity alone does not guarantee the success of a business model. Even so, we can be certain that some sort of personal communications device that allows one to roam from cellular to Wi-Fi networks will appear. Indeed, much of the communications technology of the future will be wireless in nature.—Richard "Zippy" Grigonis, editor-in-chief of VON (Voice on the Net) magazine, www.vonmag.com.

Prediction No. 13: By 2008 there will be 100 million Wi-Fi devices seeking 10 million wireless broadband customers. The result will be a wide variety of businesses, from incumbent wireline and cellular service providers to truck stops and delis, profiting from wireless Internet traffic.—Dr. Larry Brilliant, vice chairman of Cometa Networks

Prediction No. 14: Within the next five years, all front-end user computing interfaces will be wireless.—Sumit Deshpande and Don LeClair, technology strategists, Computer Associates International Inc.

Prediction No. 15: In the next five years, wireless access will revitalize a lot of back-end applications that have had limited success in past years, e.g. sales force automation.—Dale Gonzalez, vice president of wireless engineering, Air2Web Inc.

Prediction No. 16: Within five years, wireless PDAs will become the most popular handheld device sold, combining cell phone voice plus Internet access and email in a single device.—J. Gerry Purdy, principal analyst, MobileTrax LLC.

Prediction No. 17: By 2007, PDAs and cell phones will have merged into single devices. These diminutive devices will have 802.11 (whatever flavor), Bluetooth, 3G, and, possibly, direct satellite capability. They'll be voice-controlled and use a heads-up holographic display. Laptops will become unnecessary for most folks.—Doug Jackson, director of technology customer services, University of Texas at Dallas.

Prediction No. 18: By 2013, your kids will point to an Ethernet cable and say, "Mommy, what's that."—Paul Gillin, editor, SearchDatabase.com.

Prediction No. 19: By the year 2020, it will be common for healthcaregivers to use mobile computing. Implanted wireless devices will monitor people's health continuously, allowing the medical profession to recognize and treat most diseases in their infancy. Mobile computing also will be used to monitor the population's diet, control unhealthy habits such as smoking and alcohol consumption, and enable people to maximize the effects of exercise. For instance, diseases such as diabetes will be virtually controlled through wireless monitoring and corrective-action devices that will automatically adjust insulin levels, without the patient even knowing.—Phil Asmundson, deputy managing director of the Technology, Media & Telecommunications Group, Deloitte & Touche LLP.

Prediction No. 20: Today's web services are about machine-to-machine communications, integrating different computing systems together using open standards. But we shouldn't forget the thousands of other devices such as cell phones, PDAs, and pagers that need to talk and coordinate with each other. By 2006, web services on small devices will become increasingly important for giving the sales force and other workers access to corporate information behind the firewall. These devices will also begin to see usage in automotive and healthcare applications.—Bob Sutor, director of Web services strategy, IBM Software Group.

CONCLUSION

At this stage of the game, Wi-Fi's future is wide open. While technical hurdles remain, the market for "all things Wi-Fi" is strong. As Wi-Fi technologies continue to evolve, they will spawn a wide range of new capabilities for enterprises and institutions across a number of industries. These capabilities can offer many organizations a key competitive advantage.

But as the widespread deployment of WLAN technology moves forward, it is vital that there be a solution that can provide end-to-end connectivity. This means that wireless networks must easily integrate with wired and cellular networks to form a seamless entity. A requirement for this end-to-end solution is a common set of advanced features such as security, manageability, VoIP, and QoS. With such features, IT managers will be able to deploy a universal bandwidth management strategy across their organizations.

The one thing that is crystal clear is that Wi-Fi's vulnerabilities are far outweighed by its great revolutionary potential for the future.

Chapter 23: Some Final Musings

Although skeptics point to the many challenges that Wi-Fi faces including security, congestion, interference, and a lack of a credible billing or roaming infrastructure to support the HotSpot market, don't let it detour you from adopting a wireless infrastructure. Wi-Fi's flaws are nothing more than a speed bump, given the billions of dollars the industry is pouring into research and development.

Furthermore, the history of technology has proven time and again that when a certain open architecture gains escape velocity, there is no turning back.

So despite its shortcomings, Wi-Fi is becoming pervasive. That's because it offers a strong value proposition, provides multiple and expanding uses, and its supporters have taken the necessary steps to ensure industry and global standardization.

Wi-Fi networks also are the core to building out a wirelessly connected nation. Far flung Wi-Fi networks just need to be brought together to build a bottom-up community network that can provide low-cost, mass market connectivity and applications.

In fact, Wi-Fi makes an excellent candidate for building out mainstream data networks not only in the developed world, but also in emerging markets, such as Africa, India, and China, where cost is a very important factor. With a Wi-Fi-enabled wireless infrastructure serving as the cornerstone, it is now possible for many areas of the world to leapfrog over more developed nations into a high-speed environment.

WI-FI'S TIME HAS COME

Howard Strauss of Princeton University, in preparation for a discussion on ubiquitous computing, searched the archives at Princeton and discovered the following quotes from various faculty meetings concerning the introduction of another technological marvel of the early 20th Century—electricity—onto the Princeton campus:

- Electricity is just a fad, just as silly as the Wright brothers flying, it won't survive.
- Do we really need an electric point in every classroom? Maybe just connecting a few of them will be enough.
- There are too many standards at the moment (DC, AC, 110v, 220v, etc.). We should wait till there is just one standard.
- We could buy a years' worth of candles for the cost of cabling the campus.
- It isn't reliable, so we will have to have candles anyhow.

* Some students will want more than one light, we should ration it and only allow them one light each.
* We should limit each student to 25W, they don't need any more than that.
* They will use it for non-academic purposes like reading trashy novels, and we shouldn't be providing for their non-academic life.
* It will allow night classes, so we will have to work more hours, and it will turn universities into learning factories.

In short, in spite of the myopic views of those comfortable with the status quo, you can't stop progress. Wi-Fi is here to stay.

CONCLUSION

Treat Wi-Fi-enabled communication within your organization's domain in the same way as you would treat an Internet connection. Subject it to the same security considerations. Consider adding (if already a part of the network ecosystem) VPN, Remote Authentication Dial-in User Server (RADIUS), or other remote user authentication protocols, as well as a strong firewall between the wireless network and the internal company LAN. (VPN and RADIUS each provide authentication and encryption that's much stronger than WEP.)

There is plenty of running room as the technology moves from the home to the enterprise and even (potentially) to a citywide, and even nationwide, level (if companies like Cometa Network have their way). Wi-Fi is truly the next big thing.

Finally, get used to thinking of wired remote access networks and the WLAN as a single network.

Appendix I: The Open Systems Interconnection (OSI) Model

The OSI is a standard description or reference model for how information from one point in a network is transmitted from one endpoint, through a network, to another endpoint. The OSI reference model is purely a conceptual model, it does not "communicate" itself. The OSI model is composed of an architecture or framework of seven layers, each specifying particular network functions. Everything from a cable to a web browser fits into this layered framework.

OSI was the first worldwide effort to standardize the entire field of computer communications, or data networking, in the form of a networking framework for implementing hardware and protocols. The OSI model was developed by the International Organization for Standardization (ISO). OSI was originally a detailed specification of computer internetworking interfaces formulated by representatives of major computer and telecommunication companies during committee meetings held beginning in 1983. But the committee established a common reference model for which others could develop interfaces, which in turn could become standards.

***Note:** Because "International Organization for Standardization" would have different abbreviations in different languages ("IOS" in English, "OIN" in French for Organisation internationale de normalisation), it was decided at the outset to use a word derived from the Greek isos, meaning "equal." Therefore, whatever the country, whatever the language, the short form of the organization's name is always ISO, not to be confused with the Open Systems Interconnection reference model's acronym, OSI.*

The OSI model, completed in 1984, is still considered the chief architectural model for intercomputer communications. OSI continues to be administered by the ISO, so any new standard that seeks validation as an ISO standard for computer communications must be compatible with the OSI reference model. The model also can be used to guide product developers so that their products will consistently interoperate with other communications products. Finally, since the OSI reference model is a common point of reference for categorizing and describing network devices, protocols, and issues, it has value as a recognized, single view of communications that gives everyone a common reference point for education and discussion about communications.

The tasks that move information between networked computers or communicating devices are divided into seven task groups, with each task or group of tasks then assigned to the appropriate OSI layer. The layers are in two categories: The upper layers, sometimes called the *Application Layers*, and the lower layers, or *Data Transport Layers.* The upper three or four layers are used whenever a message passes from or to an end-user and are generally implemented only in software. The lower three layers (up to the Network Layer) handle data transport issues and are used when messages travel through the host computer or device. The bottom two layers, the *Physical Layer* and *Data Link Layer*, are implemented in hardware and software, though with IP networks only the bottom layer (the Physical Layer) need actually be hardware, since it is closest to the physical network medium and is responsible for putting information on it.

Each layer is basically self-contained and has its own function and, so that the tasks assigned to each layer can be implemented independently. Data going to and from the network is passed layer to layer. Each layer is able to communicate with the layer immediately above it and the layer immediately below it. This way, each layer is written as an efficient, streamlined software component. When two computers or other devices communicate on a network, the software at each layer on one device assumes it is communicating with the same layer on the other device. This can occur because, when a layer receives a packet of information, it checks the destination address, and if its own address is not there, it passes the packet to the next layer. One layer's functionality can thus be updated without affecting adjacent layers.

Although manufacturers and telecom / datacom product developers do not always strictly adhere to OSI in terms of keeping related functions together in a well-defined layer, practically all communications products are described in relation to the OSI model. Different network devices are designed to operate at certain protocol levels, and each network protocol can be mapped to the OSI reference model.

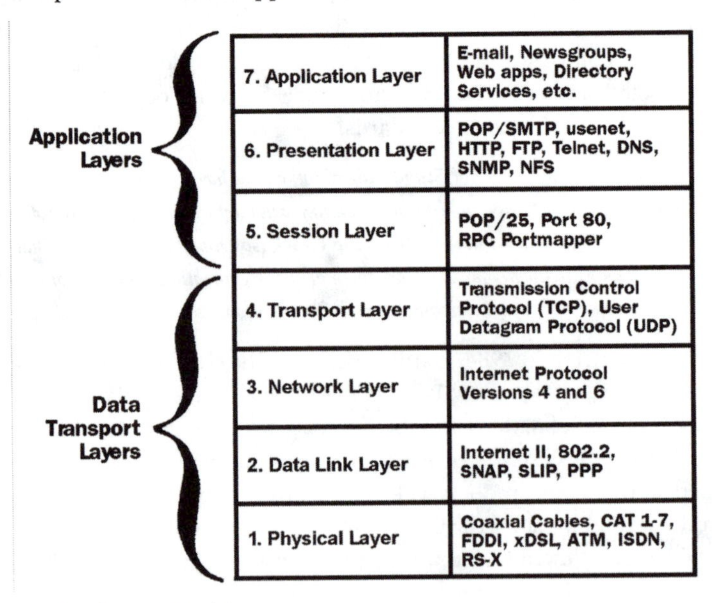

Figure A.1 The OSI Model.

The OSI model assumes that each communicating user is at a computer or device equipped with hardware and software adhering to these seven functional layers. When one person sends a message to another, the data at the sender's end will pass down through each layer in that device to the bottom layer, then over the channel, and at the other end, when the message arrives, data will flow back up through the layer hierarchy in the receiving device, and through the application to the end user.

The seven layers are (from the top, downward):

Layer 7. The Application Layer

This highest layer in the OSI architecture ultimately leads the outside world —the end-user. However, the Application Layer is not itself the user application; it only provides the system independent interface to the actual data communications application and its own user interface (though applications may indeed perform Application Layer functions).

Although the highest and seemingly the most abstract layer, the Application Layer actually consists of a complex set of standards and protocols. Application Layer Protocols are classified into Common Application Specific Elements (CASE) and Specific Application Specific Elements (SASE).

Layer 7 is where communication partners are identified, quality of service is identified, and user authentication and privacy are determined. This layer contains functions for applications services, such as file transfer, database processing, e-mail, remote file access and virtual terminals.

Layer 6. The Presentation Layer

Sometimes called the *Syntax Layer*, this layer is usually that part of an operating system that is concerned with the representation (syntax) of messages' data associated with an application during the transfer between two application processes. Applications routing data are simply routing binary streams, which has no meaning without a definition as to how it is to be formatted. A raw binary representation alone isn't good enough, since two computers communicating with each other may have totally different configurations: One could be using a 16 bit word size, the other 32; a PC could be using an ASCII character set, while an IBM mainframe could be using EBCDIC, etc.

The Presentation Layer must therefore do its part to provide transparent communications services by masking the differences of varying data formats between dissimilar systems and, in general, converting incoming and outgoing data from one presentation format to another (for example, from a text stream into a popup window).

This layer is also concerned with methods of data encryption and data security, and compression algorithms that may have also changed the data format.

Here's how the Presentation Layer works: The Presentation Layer in one computer will attempt to establish a "transfer syntax" with the Presentation Layer in another computer by negotiating a common syntax that both applications can use. Failure to do so results in a non-connection. A widely used standard for the Presentation Layer is ISO 8824 and 8825

Layer 5. The Session Layer

This functions as an interface between the hardware (the bottom four layers) and the software (the top two layers). This layer is the "dialog manager." The Transport Layer directs the data stream, but it is the Session layer that controls how data flows. It negotiates and creates a connection between two Presentation Layers when requested, then controls the data flow. More technically, it establishes, maintains and otherwise controls the use of the basic communications channels provided by the Transport Layer, by setting up, coordinating, and terminating conversations, exchanges, and dialogs between the applications at each end. It handles session and connection coordination. A session establishes the connection between communicating devices, providing synchronization, security authentication, and network naming. Protocols that function at this layer include RPC, XNS and LDAP. This layer is often combined with the Transport Layer.

Layer 4. The Transport Layer

This layer defines the rules for information exchange, manages the end-to-end flow control and delivery (for example, determining whether all packets have arrived), and error-checking / recovery within and between networks. It ensures complete data transfer.

This is the highest of the lower layer protocols in the OSI protocol stack. It provides the means to establish, maintain, and release transport connections on behalf of session entities. It provides a connection-oriented or connectionless service. In a connection-oriented session, a circuit is established through which packets flow to the destination (most protocols at this layer are connection-oriented). Error and flow control are dealt with at this layer (ACK). And most gateway functions are found at this level.

The Transport Layer is the first of what's called the *"peer-to-peer" layers,* which means that once the lower layers are implemented, a Transport Layer can transparently communicate directly with the Transport Layer of another data entity. It provides an idealized full duplex bit pipe to the upper layers in which the binary stream sent from one end, makes it, in order, to the other end. The result is that the upper layers get an idealized bit pipe and need only be concerned with what the data is, not how it arrived. Some protocols that reside at this level are TCP, UDP, SPX and TFTP.

Layer 3. The Network Layer

This layer handles how data is routed from the source computer to the destination computer, sending it in the right direction over the correct intermediate nodes to the proper destination, and receiving incoming transmissions at the packet level. The Network Layer does routing and forwarding within and between individual networks and provides a Connection Oriented Network Service (CONS) or a Connectionless Network Service (CLNS). No matter what the route of the actual connection, the data arrives at the destination as if the two data entities were directly connected. It is at this layer that network net traffic problems are managed such as Quality of Service (QoS) and Type of Service (TOS). Devices operating at this layer include routers, brouters (a combination bridge and router), and Layer-3 switches. Protocols working in this layer include ARP, DLC, ICMP, IP, IPX, NetBEUI and RARP. Routing Protocols at this level include BGP, EGP, EIGRP, IGMP, IGRP, OSPF, and RIP.

Layer 2. The Data Link Layer

This layer concerns itself with the procedures and protocols for operating the communications channels (transmission protocol knowledge and management), and provides error detection / correction and synchronization for the Physical Level. Devices at this layer include switches, bridges, and intelligent hubs. The Data Link layer consists of two sublayers: The *Logical Link Control (LLC)* and the *Media Access Control (MAC)*. The LLC sublayer is in charge of establishing and maintaining links between communicating devices. Ethernet and Token Ring protocols operate in the Data Link Layer, as do protocols such as ATM, CSMA/CD, FDDI, PPP and SLIP. The MAC sublayer is responsible for framing data. Computers or other communicating devices identify themselves via MAC addresses and Network Interface Cards (NICs), while the Data Link Layer at the same time organizes information from the higher layers into "blocks of bits," called frames (such as the "Synchronous Data Link Control" frame), for orderly transfer and error control. Sometimes this involves bit-stuffing to pad out strings of 1's.

Layer 1. The Physical Layer

This bottom layer is responsible for activating, maintaining, and deactivating the physical connection for bit transfers between "Data Link Entities." It provides the hardware means of sending and receiving a bit stream (a "bit pipe") through the network at the most basic electrical, mechanical, functional and procedural levels. An example of a physical layer ISO standard is the RS-232 interface. Devices at this level include cables, connectors, hubs, multiplexers, repeaters, receivers, switches, the 802.11 access point and other hardware.

Here is a simple mnemonic phrase to remember the correct order of the layers:

7 **A**pplication–**A**ll
6 **P**resentation–**P**eople
5 **S**ession –**S**eem
4 **T**ransport–**T**o
3 **N**etwork–**N**eed
2 **D**ata Link–**D**ata
1 **P**hysical–**P**rocessing

Appendix II: Understanding RF Power Values

Radio Frequency signals are subject to various losses and gains as they pass from the transmitter through the cable to the antenna, through the air (or solid obstruction), to the receiving antenna, cable and receiving radio. With the exception of solid obstructions, most of these figures and factors are known and can be used in the design process to determine whether an RF system such as a WLAN will work.

DECIBELS (dB)

Decibel measurements are found in the electronics, audio and telecommunications fields. They have wide applicability because they are not measurable quantities of anything; they simply represent two signal power ratios. When a decibel figure is positive, then the second signal is stronger than the first signal. When the decibel figure is negative, then the second signal is weaker than the first signal.

The Decibel (dB) scale is a logarithmic scale denoting the ratio of one power value to another—for example:

dB = 10 log10 (Power A/Power B)

An increase of 3 dB indicates a doubling (2x) of power. An increase of 6 dB indicates a quadrupling (4x) of power. Conversely, a decrease of 3 dB is a halving (1/2) of power, and a decrease of 6 dB is a quarter (1/4) the power. Some examples are shown in the table below.

Decibel Values and Corresponding Factors

Increase	Factor	Decrease	Factor
0 dB	1 x (same)	0 dB	1 x (same)
1 dB	1.25 x	-1 dB	0.8 x
3 dB	2 x	-3 dB	0.5 x
6 dB	4 x	-6 dB	0.25 x
10 dB	10 x	-10 dB	0.10 x
12 dB	16 x	-12 dB	0.06 x
20 dB	100 x	-20 dB	0.01 x
30 dB	1000 x	-30 dB	0.001 x
40 dB	10,000 x	-40 dB	0.0001 x

dBi

The performance of the system's antennas, however, are measured in decibels specified with respect to some standard reference element such as the "isotropic radiator" of a dipole. (An isotropic radiator is an imaginary, ideal transmitting antenna that radiates a signal equally well in all directions–theoretically a sphere. The sun is often given as an example of an isotropic radiator.) One of the actual measurement units is dBi, which means "decibels isotropic" or "gain over isotropic"–a measurement of signal gain often used in radio antenna design, which refers to signal gain in an isotropic radiator. Such an imaginary perfect antenna is a point suspended in space far from the Earth, with nothing nearby to interfere with the signals as they expand in perfect spheres.

EIRP

For non-isotropic, highly directional dipole antennas (e.g. point-to-point antennas used in bridging systems, satellite dishes and antennas used in cellular phone cell sites where the cell is divided up into sectors), the power radiated inside the footprint or area of coverage of these antennas is known as the equivalent (or effective) isotropic (or isotropically) Radiated Power (EIRP). EIRP should be distinguished from the simple, raw, Effective Radiated Power (ERP). The EIRP is defined in the direction of strongest antenna radiation–in line with the beam axis or "biggest lobe" of antenna coverage. EIRP in a particular direction is the power that would have to be radiated by an imaginary isotropic antenna to achieve the same field strength as the real, more directional, dipole antenna provides, after taking into account all cable losses and antenna gain. Thus the EIRP will be a higher figure than the ERP.

Given the ERP, one can derive the EIRP using the following formula:

EIRP = ERP + 2.14 dB

or EIRP (in watts) = ERP (in watts) * 1.64

The radius by which a receiving site may be separated from the transmitting station is proportional to the square root of the EIRP of the transmitter-antenna unit. Thus, if the EIRP is suddenly reduced by 50 percent, the maximum receiving distance reduces to 70 percent of its original radius. Therefore, it is important that the cable that connects the transmitter to the antenna dissipate as little of the signal power as possible, since such power dissipation (attenuation), reduces the power available to the antenna. A high quality waveguide tube (instead of a "lossy" coaxial cable) may only lose up to 0.49 dB per 100 feet, yet a 600 ft. run of such a waveguide would have an attenuation of approximately 3 dB and reduce the transmitter power to only 50% of the value leaving the transmitter. Nevertheless, coaxial cables almost always connect a transmitter and antenna, as long as the run between the two is kept very short.

dBm

Although the Radio Frequency (RF) power level at either transmitter output or receiver input is normally expressed in Watts, electrical engineers often express it in dBm, or "decibel, milliwatts," which is a value referenced to 1 milliwatt (mw) in a 600 Ohm (or in the case of some radio calculations, 50 Ohm) system. The power relationship between dBm and Watts can be expressed as follows: PdBm = 10 x Log PmW, so it is quite easy to convert Watts to dBm.

For example:
1 Watt = 1000 mW, so
10 * Log 1000 = 30 dBm
therefore, 1 Watt = 30 dBm

likewise:
.000000001 = -90 dBm
.000001 Watt = -30 dBm
.001 Watt = 0 dBm
.01 Watt = 10 dBm
.1 Watt = 20 dBm
100 Watts = 50 dBm

To convert dBm to Watts, you could us the following formula:

PdBm = 10 log (P/0.001), where P equals Watts.

Thus, 36 dBm = 10 log (P/0.001)

P = 3981 * 0.001, therefore
P = 3.981 Watts.

Wi-Fi and dBm

Since the dBm convention is found to be more convenient than the Watts convention, in the Wi-Fi world, dBm is also useful for "link budget" calculations. These are calculations of what environmental and equipment factors that will add and subtract from the signal power along the transmission path.

For example: specifications for equipment that operates in the Industrial Scientific and Medical (ISM) band generally allow 100 mW (20 dBm) of EIRP, which is the final power of the radiated signal after any gain produced by the antenna. This yields a useable link of about 1 kilometer (0.62 miles) in length, depending on atmospheric conditions. Maximum permissible transmitter output power in the licensed bands can vary throughout the world. Some units with dish antennas are adjustable to 50 watts (47 dBm) EIRP, including the gain of the antenna. The practical range at these levels can be up to 15 kilometers (9.32 miles) or more, again depending on local conditions. In other circumstances, e.g. the tropics, that range may come down to 5 km (3.1 miles). Higher power options (+57 dBm and +67 dBm EIRP) are now available (longer link paths may be achievable using repeaters).

Furthermore, Wi-Fi equipment is usually specified in decibels compared to known values. For example, Transmit Power (Tx) and Receive Sensitivity (Rx) are specified in "dBm," where m means 1 milliWatt (mW). So, 0 dBm is equal to 1 mW; 3 dBm is equal to 2 mW; 6 dBm is equal to 4 mW, and so on. For example, a Cisco Aironet 350 Series Access Point at 100 mW transmit power is equal to 20 dBm. dBw is occasionally used for the same purpose, but as a comparison against 1 Watt (1000 mW).

Common mW values to dBm values are shown below in the table below.

Common mW Values to dBm Value

dBm	mW	dBm	mW
0 dBm	1 mW	0 dBm	1 mW
1 dBm	1.25 mW	-1 dBm	0.8 mW
3 dBm	2 mW	-3 dBm	0.5 mW
6 dBm	4 mW	-6 dBm	0.25 mW
7 dBm	5 mW	-7 dBm	0.20 mW
10 dBm	10 mW	-10 dBm	0.10 mW
12 dBm	16 mW	-12 dBm	0.06 mW
13 dBm	20 mW	-13 dBm	0.05 mW
15 dBm	32 mW	-15 dBm	0.03 mW
17 dBm	50 mW	-17 dBm	0.02 mw
20 dBm	100 mW	-20 dBm	0.01 mW
30 dBm	1000 mW (1 W)	-30 dBm	0.001 mW
40 dBm	10,000 mW (10 W)	-40 dBm	0.0001 mW

Appendix III: Regulatory Specifics re Wi-Fi

The various regulatory agencies that govern Wi-Fi usage impose many different restrictions on the frequency range, channel support, and power usage of Wi-Fi devices. Most of those restrictions are set out in specific regulatory standards, for example:

Australia
2.4 GHz: AS/NNS 4771
5 GHz: AS/NZS 4771

Canada
2.4 GHz: RSS-210
5 GHz: RSS-210

European Union
2.4 GHz: EN 301.328
5 GHz: EN 301.893

Japan
2.4 GHz: Std 33A and Std 66
5 GHz: N/A

New Zealand
2.4 GHz: AS/NZS 4771
5 GHz: AS/NSZ 4771

United States
2.4 GHz: FCC part 15.247
5 GHz: FCC Part 15.401

Here are some pertinent examples of what you might find during your investigation into the restrictions individual jurisdictions may place on Wi-Fi:

Frequency Range

North America
2.412 - 2.484 GHz
5.15 - 5.35 GHz, 5.725 - 5.825 GHz

Europe
2.412 -2.484 GHz
5.15- 5.35 GHz, 5.47 - 5.725GHz

Japan
2.471 - 2.497 GHz
5.15 - 5.25GHz

Modulation Techniques

2.4 GHz
DSSS (CCK, BPSK, QPSK)

5 GHz
OFDM (BPSK, QPSK, 16-QAM, 64-QAM)

Channel Support

2.4 GHz
North America
11 (1 - 11)

European Union (EU)
13 (1 - 13) except for France, which is 4 (10 - 13)

Japan
14 (1-13 or 14th)[1]

5 GHz
North America
12 non-overlapping channels (5.15 - 5.35Ghz, 5.725 - 5.825Ghz)

European Union
19 non-overlapping channel (5.15 - 5.35Ghz, 5.47 - 5.725Ghz)

Japan
4 non-overlapping channels (5.15 - 5.25Ghz)

Optimal Power Consumption

Transmission mode <1155mW (estimated)
Receive mode <1221mW (estimated)
Standby mode <297mW (estimated)
Power saving mode <39.6mW (estimated)

Output Power

2.4 GHz

Worldwide

18 dBm (-65 mW) peak power

5 GHz

U.S.

a) 5.150 - 5.250: indoor only peak power to 50mW (17dBm) per FCC 15.407 specification (UNII band operation).
b) 5.250 - 5.350: peak power to 250mW (24 dBm) per FCC 15.407 specification (UNII band operation).
c) 5.470 - 5.725: not allowed.
d) 5.725 - 5.825: peak power to 1W (30 dBm) per FCC 15.407 specification (UNII band operation).

European Union

a) 5.150 - 5.250: indoor only authorized transmit power is 50 mW
b) 5.250 - 5.350: authorized transmit power is 250 mW EIRP (24 dBm).
b) 5.470 - 5.725: 1W EIRP (30 dBm) allowed.
c) 5.725 - 5.825: calibrated to provide 20 dBm peak power.

Japan

a) 4.900 - 5.000: To be determined.
b) 5.150 - 5.250: indoor authorized power is 200 mW EIRP (23 dBm).
c) 5.250 - 5.350: not allowed.
c) 5.470 - 5.725: not allowed.
d) 5.725 - 5.825: not allowed.

Maximum RF Power Level

North America

1W (at ERP, and maximum 6 dBi antenna gain) per FCC* Part 15, but also see Canada's RSS 210—Canada's Maximum RF Power Level differs a bit from what the U.S. allows.

*Federal Communications Commission

European Union (EU)

100 mW (at EIRP) per ETS* 300.328, but also see EU 301.328 and 301.893

*European Telecommunications Standard

Japan

Not specified in the governing ordinances—MPT* Ordinance 78 or 79 (also see Standards 33A and 66)

* Ministry of Posts and Telecommunications

Glossary

2G. The second generation of cellular systems. 2G systems use digital encoding. 2G networks support high bit rate voice, limited data communications, and different levels of encryption. 2G networks utilize either GSM, TDMA or CDMA technology. Note that 2G networks were the first cellular networks capable of supporting short message service (SMS) applications.

2.5G. New hybrid "always on" cellular networks such as GSM/GPRS and cdmaOne. 2.5G networks can transfer data at speeds of up to 144 Kbps (faster than traditional 2G digital networks, but slower than true 3G networks). A phone with 2.5G services can alternate between using the Web, sending or receiving text messages, and making phone calls without losing its connection. 2.5G networks actually extend existing 2G systems by adding features such as packet-switched connection and enhanced data rates. These networks also support the Wireless Application Protocol (WAP), multimedia messaging service (MMS), short message service (SMS), mobile games, and more.

2.75G. Typically refers to either an EDGE cellular network because it supports both more and better quality voice and faster data transfers via GPRS or a cdma2000 1X network because it can double the voice capacity of cdmaOne networks and delivers peak packet data speeds of 307 Kbps in mobile environments.

3G. The third generation of mobile systems. 3G systems provide high-speed data transmissions of 144 Kbps and higher. 3G networks should be able to support multimedia applications such as full-motion video, video conferencing and Internet access. Technologies touted as 3G include Wideband CDMA (commonly known as W-CDMA) and cdma2000 1xEV-DV (Data Voice) because both can provide integrated voice and simultaneous high-speed packet data multimedia services at speeds that have tested as high as 3.09 Mbps.

3G/UMTS (Third Generation/Universal Mobile Telecommunications System). See W-CDMA.

A

AAA (Authentication, Authorization, and Accounting). A term for a framework that enables intelligent access control to computer resources by enforcing policies, auditing usage, and providing the information necessary to bill for services. AAA is typically provided via a suite of network security services.

Access Point (AP). A transmitting device that acts as a wireless hub and typically interfaces between a wireless network and a wired network. Access points combined with a distribution system (e.g. Ethernet) support the creation of multiple radio cells (basic service sets) that enable roaming. Sometimes referred to as a "Base Station."

Access Controller. A device that provides centralized intelligence behind a wireless LAN's access points so as to regulate traffic between a wireless LAN and other network resources.

An access controller typically resides on the wired portion of a network between the wireless network's access points and the protected side of a wired network.

Ad Hoc Network. A wireless network composed of only stations (computing devices) and no access point. Also referred to as an "Independent Basic Service Set." See also Basic Service Set.

ADSL. See Asymmetric Digital Subscriber Line.

Advanced Encryption Standard (AES). A symmetric 128-bit block data encryption technique developed by Belgian cryptographers, Joan Daemen and Vincent Rijmen. The U.S government adopted the algorithm as its encryption technique in October 2000, as a replacement for the outdated DES encryption.

AES. See Advanced Encryption Standard.

AGC. See Automatic Gain Control.

Aggregator. A term that is used to describe a group of virtual network wholesalers that help to tie individual HotSpots networks into a larger, cohesive roaming network. Other terms for the aggregator group are "Virtual WISPs" as most do not really have an actual physical network; "brokers," since they broker deals between companies within the WISP Industry; and "managed service providers," since their business model is based upon managing Internet access service for others. An aggregator typically provides centralized authentication services in order to compute and validate traffic; fix the airtime prices in which they trade; operate as intermediaries between HotSpot operators through buying and selling airtime minutes; and fix the tariffs for roaming between HotSpot operator networks.

AM. See Amplitude Modulation.

American National Standards Institute (ANSI). A private non-profit membership organization founded in 1918. ANSI coordinates (within the U.S.) the voluntary standard system that brings together interests from the private and public sectors to develop voluntary standards for a wide array of U.S. industries. ANSI develops and publishes standards for transmission codes, protocols and high-level languages used in the telecommunications industry.

Amplitude Modulation (AM). This refers to a method of signal modulation in which the amplitude of the carrier voltage is carried in proportion to the changing frequency value of an applied voltage. AM impresses data onto an alternating-current (AC) carrier waveform. The highest frequency of the modulating data is normally less than 10 percent of the carrier frequency. The instantaneous amplitude (overall signal power) varies depending on the instantaneous amplitude of the modulating data.

ANSI. See American National Standards Institute.

AP. See Access Point.

AppleTalk. A proprietary set of local area network protocols developed by Apple for its Macintosh processors. An AppleTalk network supports up to 32 devices and data can be exchanged at a speed of 230 Kbps. AppleTalk's Datagram Delivery Protocol corresponds closely to the Network layer of the Open Systems Interconnection (OSI) communication model.

ASCII (American Standard Code for Information Interchange). A standard developed by the American National Standards Institute (ANSI) to define how computers write and read characters. The ASCII set of 128 characters includes letters, numbers, punctuation, and control codes (such as a character that marks the end of a line). Each letter or other character is represented by a number: an uppercase A, for example, is the number 65, and a lowercase z is the number 122. Most operating systems use the ASCII standard, except for Windows NT, which uses the suitably larger and newer Unicode standard.

Asymmetric Digital Subscriber Line (ADSL). A method for getting reasonably high data rates (around 1 Mbps) using the existing telephone twisted pair wiring by using spare capacity not needed for voice traffic. The asymmetric simply refers to the different data rates provided upstream and downstream.

Automatic Gain Control (AGC). A process or means by which gain is automatically adjusted in a specified manner as a function of a specified parameter, such as received signal level.

Automatic Identification and Data Collection (AIDC). A term that refers to the direct entry of data into a computer system, programmable logic controller (PLC) or other microprocessor-controlled device without using a keyboard. AIDC technologies provide a reliable means not only to identify but also to track items. It is possible to encode a wide range of information, from basic identification to comprehensive details about the item or person. AIDC includes a number of technologies, which provide different solutions to data collection problems. These include: barcode, radio frequency identification (RFID), magnetic stripes, voice and vision systems, optical character recognition (OCR), biometrics, as well as others.

B

Backbone. A central cabling or wireless system that attaches servers and routers on a network and handles all network traffic. Because of its configuration, the network backbone often decreases the time needed for transmission of packets and the amount of traffic on a network. At the local level, a backbone is a cable, set of cables, set of wireless connections, or a combination of wireless and cables that local area networks connect to for a wide area network connection or within a local area network to span distances efficiently.

Band. Electromagnetic waves can be classified by their wavelengths. Those wavelengths are classified into sections called *bands*. The electromagnetic spectrum is the collection of these bands.

Bandwidth. Technically, the difference, in Hertz (Hz), between the highest and lowest frequencies of a transmission channel is called bandwidth. But in a general sense, the term describes a medium's information-carrying capacity. It can apply to radio frequency signals, as well as networks, system buses, and monitors. However, as typically used, the term refers to the amount of data that can be sent through a given communications circuit. Although bandwidth is most accurately measured in cycles per second, or Hertz (Hz), it's also common to use bits or bytes per second to measure bandwidth.

Barcode. A sequence of rectangular bars and intervening spaces used to encode a string of data. A barcode symbol typically consists of five parts: a leading quiet zone, a start character, data character(s) including an optional check character, a stop character, and a trailing quiet zone; together these five parts form a complete, scannable entity to identify a carton or individual item.

Barcode Scanner. These electronic devices read barcodes and convert them into electrical signals understandable by a computer device.

Barker Code. Originally developed for radar, Barker codes, also known as "spreading codes" or "chipping codes," are short (13 bits or less) sequences that are normally used in one-shot schemes, as compared to most other spreading codes, which run continually. For example, one code might be used as a preamble to a long Pseudonoise (PN) sequence for the sole purpose of simplifying synchronization. The most notable property of Barker codes is that the minor peaks of their autocorrelation functions always consist of -1,0, and +1. Barker sequences are not the natural product of linear feedback shift registers, but rather they are hard-coded. The complete list of Barker codes are as follows: R2: 10 (or 11), R3: 110, R4: 1011 or (1001), R5: 11101, R7: 1110010, R11: 11100010010, R13: 1111100110101.

Base Station. In an 802.11 network, the term refers to the access point. In cellular networks, base station refers to a networking installation which houses the equipment needed to set up and complete calls on cellular phones—transmitter and receiver equipment, antennas, and computers. The base station works along with the subscriber's handset and the Mobile Switching Center (MSC) to complete call and/or data transmission. In a personal communications system network, the base station is comprised of a base station controller (BSC) and a base transceiver station

(BTS). A base station is also sometimes referred to as a "cell site."

Base Station Controller (BSC). This is a networking component of a cellular network's base station system. A BSC supervises the functionality and control of multiple Base Transceiver Stations and acts as a small switch by managing handoffs.

Base Transceiver Station (BTS). This networking component of a cellular network's base station system consists of all radio transmission and reception equipment. A BTS provides coverage to a geographic area, and is controlled by a Base Station Controller.

Basic Service Set (BSS). A term that refers to a set of 802.11-compliant stations that operate as a fully connected wireless network. Each set of wireless devices communicating directly with each other is called a basic service set (BSS). BSS is that includes at least one access point sometimes referred to as "Infrastructure Basic Service Set" or "Infrastructure Basic Service Set Mode." Compare with Ad Hoc Network and Extended Basic Service Set.

BER. See Bit Error Rate.

Binary Phase Shift Keying (BPSK). A technique employed to shift the phase of the carrier signal by 0 or by 180 degrees, depending upon the bits to be transferred. While BPSK is simple to implement and is robust, it's inefficient in terms of using the available bandwidth.

Bit. This term is actually an abbreviation for "**B**inary Dig**it**." A bit is the smallest unit of computerized data. It is also the basic unit in data communications and all computer communication is in bits. A bit is either a 1 or 0, reflecting the use of a binary numbering system (only two digits) and bits are used because the computer recognizes either of two states: ON or OFF. If you go to the lowest level, a bit is the presence or absence of an electrical (signal) charge. To understand how bits work, think of a bit as a switch. If the switch is in the "on" position, it is a 1, and if the switch is "off," it is a 0. The speed at which bits are transmitted or bit rate is usually expressed as bits per second or bps; for instance, bandwidth is usually measured in bits-per-second. Eight bits make up one byte.

Bit Error Rate (BER). The percentage of bits that have errors relative to the total number of bits received in a transmission, usually expressed as ten to a negative power. For example, a transmission might have a BER of 10 to the minus 6 (10e-6), which means that out of 1,000,000 bits transmitted, one bit was in error. The BER provides an indication of how often a packet or other data unit must be retransmitted because of an error. A high BER may indicate that a slower data rate would actually improve overall transmission time for a given amount of transmitted data. By reducing the data rate, you reduce the BER, which in turn means that the number of packets needing to be resent is reduced, thus the overall transmission rate is increased.

Bluetooth. A short range wireless protocol that operates at 2.4 GHz. Bluetooth allows mobile devices to share information and applications over a short distance (less than 30 feet or 9.14 meters) without the worry of cables or interface incompatibilities. Bluetooth is named for a Danish Viking King Harald Blåtand (Bluetooth in English) who united the countries, much as the wireless counterpart is designed to unite individual system.

bps (bits per second). When the "b" is lower case (bps) it means bits per second. When the "B" is upper case (Bps) it means bytes per second. Be careful, many people get these two terms and their corresponding acronyms confused. One "bit" of advice to keep you on track–in the telecommunications industry, LANs, WANs, USB, local loops, etc., the term should always refer to "bits per second" regardless of whether an upper case "B" or a lower case "b" is used.

BPSK. See Binary Phase Shift Keying

Bridge. This data communications device connects two or more network segments and forwards data packets between them.

Broadband. This term refers to a type of data transmission in which a single medium (wire or radio frequency) can carry several channels at once. While initially the term "broadband" referred to data transmission over 1 Mbps, the FCC and others have lowered the criteria. In many instances, the definition of broadband is any system capable of transmitting data in excess of 200 Kbps upstream and downstream. All communications systems that operate at a slower speed than broadband are called "narrowband."

Broadband Connection. This term refers to the ability to carry multiple signals by dividing the total capacity of the medium into multiple, independent bandwidth channels, where each channel operates only on a specific range of frequencies.

BSC. See Base Station Controller.

BSS. See Basic Service Set.

BTS. See Base Transceiver Station.

Byte. A set of bits of a specific length that represent a value in a computer coding scheme. A byte is comprised of eight bits. A byte is to a bit what a word is to a character.

C

Cable Modem. As used in this book, the term "cable modem" refers to a service that delivers high-speed Internet access to businesses and consumers over cable lines. A separate cable modem or router is usually required to use the service.

Carrier Sense Multiple Access with Collision Avoidance (CSMA/CA). The principle medium access method employed by Wi-Fi networks. It is a "listen before talk": method of minimizing (but not eliminating) collisions caused by simultaneous transmission by multiple radios. The 802.11 standards state that a collision avoidance method rather than a collision detection method must be used, because the standard employs half duplex radios—radios capable of transmission or reception, but not both simultaneously. Unlike conventional wired Ethernet nodes, a WLAN station cannot detect a collision while transmitting. If a collision occurs, the transmitting station will not receive an ACKnowledge packet from the intended receive station. For this reason, ACK packets have a higher priority than all other network traffic. After completion of a data transmission, the receiving station will begin transmission of the ACK packet before another node can begin transmitting a new data packet. All other stations must wait a longer pseudo-randomized period of time before transmitting. If an ACK packet is not received, the transmitting station will wait for an opportunity to retry transmission.

CCK. See Complimentary Code Keying.

CDMA. See Code Division Multiple Access.

CDMA Direct Spread. See W-CDMA.

cdmaOne. A brand name trademarked and reserved for the exclusive use of the CDMA Development Group (CDG) member companies. It is a digital mobile phone standard based on the CDMA principle, which is used in North America, Korea, and Japan. For migration to third generation mobile telephony, cdmaOne networks can be upgraded to the cdma2000 broadband standard. The following frequencies are supported by cdmaOne: Cellular and PCS Bands: 800MHz, 900 MHz, 1700MHz, 1800MHz, 1900 MHz and all IMT-2000 bands (cdma2000).

cdma2000. A name identifying the third generation technology that is an evolutionary outgrowth of cdmaOne. It offers operators deploying a second generation cdmaOne system a seamless migration path that economically supports upgrade to 3G features and services within existing spectrum allocations for both cellular and PCS operators. Regardless of technology, cdma2000 supports the second generation network aspect of all existing operators (cdmaOne, IS-136 TDMA, or GSM). This standard is also known by its ITU name IMT-CDMA Multi-Carrier (1X/3X). Operators in the U.S., Japan, and South Korea have adopted it. Although cdma2000 is similar to W-CDMA, the two technologies are incompatible.

cdma2000 1X. The first phase in the evolution to 3G is cdma2000 1X, (or 1 times 1.25 MHz carrier). This technology improves packet data transmission capabilities and speeds in the network, while boosting voice capacity by nearly two times over today's CDMA capacities. It provides a data transfer speed of up to 144 Kpps. It is known also as 1xRTT.

cdma2000 1xEV. A cdma2000 High Rate Packet Data Air Interface Specification that represents the second step in the evolution of cdma2000. This technology brings data rates of up to 2 Mbps to the network. The CDMA 1xEV specification was developed by the Third Generation Partnership Project 2 (3GPP2).

cdma2000 1xEV-DV. The 1xEV-DV (cdma-2000 1x Evolution for Data & Voice) technology evolution provides integrated voice with simultaneous high-speed packet data, video and video conferencing capabilities. 1xEV-DV is backward compatible with IS-95A/B and cdma2000 1X, allowing an operator evolution for their existing cdma2000 systems.

CDPD. See Cellular Digital Packet Data.

Cell. The *cell* is the basic geographic unit of a wireless system. In the Wi-Fi industry the term refers to the area of radio range or coverage in which wireless devices can communicate with an access point. The size of the cell depends upon the speed of the transmission, the type of antenna used, and the physical environment, as well as other factors. In the cellular telephone industry, the term "cell" refers to the basic geographic unit of a cellular system and is the basis for the generic industry term "cellular." Cells can vary in size depending on terrain and capacity demands.

Cell Site. A term used in the cellular communications industry. It refers to individual locations where there is a specific compilation of equipment: a network transmitter, receiver, antenna signaling, and related base station equipment. Not to be confused with "cell", which is the basic geographic unit of a Wi-Fi system. See also Base Station.

Cellular Digital Packet Data (CDPD). A term used to describe a wireless radio frequency communication service that delivers data packets over existing cellular phone networks that have been upgraded for CDPD. CDPD is capable of transfer speeds of up to 19.2 kbps. The CDPD packets are sent between pauses in the cellular phone conversations.

Channel. This term refers to the physical medium or set of properties that distinguishes one communication path from another. As used in this book, the term "channel" refers to an individual communication path that carries signals at a specific frequency. For example, Wi-Fi channels refer to particular frequencies at which radio waves are transmitted. Channel can also refer to a communication path between two computers or devices.

Chip. A tiny electronic circuit on a piece of silicon crystal is oft times referred to as a "chip." These chips contain hundreds of thousands of micro-miniature electronic circuit components. Components are packed and interconnected in multiple layers within a single chip. The surface of the chip is overlaid with a grid of metallic contacts used to wire the chip to other electronic devices. And all of this is done in an area less than 2.5 millimeters square. A chip's components can perform control, logic, and memory functions. Chips are found in the printed circuits of, for example, personal computers, televisions, automobiles, and appliances.

Chipping Code. A redundant bit pattern for each bit that is transmitted, which increases a network signal's resistance to interference. By using a chipping code, if one or more bits in the pattern are damaged during transmission, the original data can be recovered due to the redundancy of the transmission.

Client/Server. The "client" is a computer or program "served" by another networked computing device in an integrated network which provides a single system image. The "server" can be one or more computers with one or more storage devices.

Code Division Multiple Access (CDMA). A digital communications technology commonly used for 2G and 3G services. Originally developed for military use, CDMA is a multiple access technique, which uses code sequences as traffic channels within common radio channels. There are two CDMA common air interface standards: Cellular (824-894 MHz)–TIA/EIA/IS-95A, and PCS (1850-1990 MHz)–ANSI J-STD-008.

Codec (Coder/Decoder or Compression/Decompression). A hardware device or software program that converts analog information streams into digital signals, and vice versa; generally used in audio and video communications where compression and other functions may be necessary and provided by the Codec as well. The various codecs available may specify straight voice (PCM) or compressed voice (ADPCM).

CompactFlash. A popular memory card that uses flash memory to store data on a very small card format to make data easy to add to a wide variety of computing devices, including digital cameras, audio devices, computers, PDAs, and photo printers. CompactFlash cards are based on the Personal Computer Memory Card International Association (PCMCIA) PC card specifications. The typical CompactFlash card measures 43 x 36 mm and is available with storage capacities ranging up to 1 gigabyte. There are two types of CompactFlash cards: Type I cards are 3.3mm thick, Type II cards are 5.5mm thick.

Complementary Code Keying (CCK). A set of 64 eight-bit code words used to encode data for 5.5 and 11 Mbps data rates in the 2.4 GHz band of 802.11b wireless networking. The code words have unique mathematical properties that allow them to be correctly distinguished from one another by a receiver even in the presence of substantial noise and multipath interference. The 5.5 Mbps rate uses CCK to encode 4 bits per symbol, while the 11 Mbps rate encodes 8 bits per symbol (actually, to attain 11 Mbps CCK modulation, 6 bits of the 8 are used to select one of 64 symbols of 8-chip length for the symbol and the other 2 bits are used by QPSK to modulate the entire symbol. This results in modulating 8 bits onto each symbol.

Convolutional Code. A code generated by inputting a bit of data and then giving the commutator (an apparatus for reversing electrical currents) a complete revolution. A convolutional code is produced by convolutional coders—coders that have memory.

Convolutionally Encoded. By taking a convolutional code and repeating the process for successive input bits you produce convolutionally encoded output, such as found in a PLCP preamble and signal field. A convolutional coder (a coder that has memory) accepts k binary symbols at its input and produces n binary symbols at its output, where the n output are affected by v+k input symbols. Memory is incorporated because v>0. Code rate R=k/n. Typical values: k, n: 1 - 8; v: 2—60; R: 0.25—0.75

CSMA/CA. See Carrier Sense Multiple Access with Collision Avoidance.

D

Datagram. See Packet.

dB. See Decibel.

dBi (dB isotropic). The gain a given antenna has over a theoretical isotropic (point source) antenna. (An isotropic antenna is an imaginary, ideal transmitting antenna that radiates a signal equally well in all directions—theoretically a sphere.) Although isotropic antennas are only theory, they provide a useful tool for calculating theoretical fade and System Operating Margins. For instance, a dipole antenna has 2.14 dB gain over a 0 dBi isotropic antenna.

dBm (dB milliWatt). A signal strength or power level; 0 dBm is defined as 1 mW (milliWatt) of power into a terminating load such as an antenna or power meter. Small signals are negative numbers (e.g. -83 dBm). For example, typical 802.11b WLAN cards have +15 dBm (32mW) of output power. They also spec a -83 dBm Rx sensitivity (minimum Rx signal level required for 11Mbps reception). Additionally, 125 mW is 21 dBm, and 250 mW is 24 dBm.

Decibel (dB). The difference (or ratio) between two signal levels; used to describe the effect of system devices on signal strength, which varies logarithmically, not linearly. Since the dB scale is a logarithmic measure, it produces simple numbers for large-scale variations in signals, allowing the use of whole numbers to calculate system gains and losses. Every time you double (or halve) the power level, you add (or subtract) 3 dB to the power level. This corresponds to a 50 percent gain or reduction. A 10 dB gain/loss corresponds to a tenfold increase/decrease in signal level. A 20 dB gain/loss corresponds to a hundredfold increase/decrease in signal level. Thus the decibel scale allows big variations in signal levels to be handled easily with simple digits.

DHCP. See Dynamic Host Configuration Protocol.

DiffServ. Short for "differentiated services," this IETF standard protocol is a small, well-defined set of per-packet building blocks from which a variety of services may be built, thereby providing a framework for delivering Quality of Service (QoS) in networks. The protocol relies on traffic conditioners sitting at the edge of the network to indicate each packet's requirements. As such DiffServ can specify and control network traffic by class, which allows designated types of traffic to get precedence in the network flow. DiffServ might be used within enterprise networks, for example, to give preferential treatment to mission-critical data. DiffServ-capable routers need only track a small number of per-hop behaviors, and they service packets based on a single byte. See also IntServ, Quality of Service.

Digital. A description of data, which is stored or transmitted as a sequence of discrete symbols from a finite set, most commonly this means binary data represented using electronic or electromagnetic signals. In this book, "digital" most commonly refers to the use of a binary code (bits) to represent information. The main benefit of transmitting information digitally is that the signal can be produced precisely and although it will pick up interference or "garbage" along the way, the telecom industry has found ways to regenerate the signal back to crystal clarity. The signal is put through a Yes-No exercise–is this part of the signal a "one" or a "zero"? Then the signal is reconstructed to what it was at the beginning of the transmission, amplified and sent on its merry way. This all means that digital-based transmission is "cleaner" than analog transmission.

Digital Signal Processor (DSP). This term refers to a specialized type of processor that is optimized for performing detailed algorithmic operations on analog signals after they have been digitized. For examples DSPs handle line signaling in modems.

Digital Subscriber Line (DSL). Point-to-point public network access technologies that allow multiple forms of data, voice, and video to be carried over twisted-pair copper wire on the telco local loop between a network service provider's central office and the customer site.

Direct Sequence Spread Spectrum (DSSS). A wide bandwidth signal with low amplitude, which allows it to appear to be "noise" when received on a non-spread spectrum receiver. DSSS uses a radio transmitter to continuously spread its signal containing data packets over a fixed range of a wide frequency band. DSSS is also a transmission technology used in wireless network transmissions where a data signal at the sending station is combined with a higher data rate bit sequence, or chipping code, that divides the user data according to a spreading ratio. Compare with Frequency Hopping Spread Spectrum.

Distributed Coordination Function (DCF). A form of carrier sense multiple access with collision avoidance (CSMA/CA). A class of coordination functions where the same coordination function logic is active in every station in the basic service set whenever the network is in operation.

DNS. See Domain Name System/Service.

Domain Name System/Service (DNS). Since IP addresses, which are numeric, are difficult for people to remember, host names or domain names such as cmp.com are generally used to identify the address of computers that are connected to the Internet. Because computers on the Internet only understand numeric IP addresses, not domain names, every webserver requires a DNS server to translate domain names into IP addresses. Note, however, that a domain name may identify one or more IP addresses.

DSP. See Digital Signal Processor.

DSSS. See Direct Sequence Spread Spectrum.

Due Diligence: A comprehensive investigation and assessment of all attributes. Issues. and variables inherent in a target entity/person/product/service, which will impact upon the target's ability to achieve its strategic objectives.

Dynamic Channel Selection. See Dynamic Frequency Selection.

Dynamic Frequency Selection (DFS). This

technique, which is also known as dynamic channel selection (DCS), allows client devices to detect the clearest channels within the radio waveband.

Dynamic Host Configuration Protocol (DHCP). A protocol in the TCP/IP suite that allocates IP addresses automatically to any DHCP client (any device attached to your network, such as your computer) so that addresses can be reused when the client no longer needs them.

Dynamic Rate Shifting. Refers to the ability to adjust connection speeds for more reliable connections. The IEEE 802.11 series of standards defines dynamic rate shifting, allowing data rates to be automatically adjusted for noisy conditions. This means Wi-Fi devices will transmit at lower speeds (e.g. 5.5 Mbps, 2 Mbps, and 1 Mps in the case of 802.11b networks) under noisy conditions. When the devices move back within the range of a higher-speed transmission, the connection will automatically speed up again.

E

EDGE. See Enhanced Data rates for Global Evolution.

Equivalent (or effective) Isotropic (or isotropically) Radiated Power (EIRP). Refers to the technical value that evaluates the strength of receive signals; in other words, the equivalent power of a transmitted signal in terms of an isotropic (omni-directional) radiator. Normally the EIRP equals the product of the transmitter power and the antenna gain (reduced by any coupling losses between the transmitter and antenna).

EIRP. See Effective Isotropic Radiated Power.

Electromagnetic Spectrum. See Spectrum, but also see Electromagnetic Waves.

Electromagnetic Waves. Visible light is made up of electromagnetic waves, vibrations of electric and magnetic fields that propagate through space. In contrast to slow-moving ocean waves to which they are analogous, electromagnetic waves travel at the speed of light: 300 million meters per second, or 669.6 million miles per hour. Furthermore, every electromagnetic wave exhibits a unique frequency and wavelength associated with that frequency. All electromagnetic waves are classified according to their characteristic frequencies, into what is known as the Electromagnetic Spectrum.

Electronic Serial Number (ESN). Every wireless phone has a unique numbered assigned to it by the manufacturer. That unique number is known as the electronic serial number. According to the Federal Communications Commission, the ESN is to be fixed and unchangeable– a sort of unique fingerprint for each wireless phone.

Embedded. In this book the term "embedded" generally refers to a combination of computer hardware and software, and perhaps additional mechanical parts, designed to perform a dedicated function–Wi-Fi connectivity. In some cases, embedded systems are part of a larger system or product, as in the case of PDAs, laptop/notebook computers, home entertainment centers, etc.

Encoding. In this book, "encoding" refers to a process for conveying or storing electronic data in accordance with a standard format. So-called "line encoding" is the waveform pattern of voltage or current used to represent the 1s and 0s of a digital signal on a transmission link. Every periodic signal has both a time and frequency domain representation, and both of these can be exploited to encode information into a signal.

Enhanced Data rates for Global Evolution (EDGE). An intermediate technology that brings second generation GSM closer to third generation capacity for handling data speeds up to 384 Kbps. EDGE uses a new modulation scheme, enabling these higher data throughput speeds using existing GSM infrastructure. EDGE represents the final evolution of data communications within the GSM standard.

ESN. See Electronic Serial Number.

ESS. See Extended Service Set.

Ethernet. The IEEE's 802.3 local area network protocol is commonly referred to as "Ethernet." This Ethernet-like standard was published by the IEEE 802.3 Working Group in 1985 and carried the name of "IEEE 802.3 Carrier Sense

Multiple Access with Collision Detection Access Method and Physical Layer Specifications." However, the IEEE doesn't refer to 802.3 as "Ethernet," because Ethernet is a specific product trademarked by Xerox whereas 802.3 is a set of standards. The vast majority of wired networks that are in operation today use 802.3's "Ethernet" technology. Standard Ethernet networks use CSMA/CD (Carrier Sense Multiple Access/Collision Detection), which enables the two devices to detect a collision and institute a system for collision avoidance.

ETSI. See European Telecommunications Standards Institute.

European Telecommunications Standards Institute (ETSI). A European non-profit organization whose mission is to produce telecommunication standards for today and for the future, and whose standards are recognized throughout the world.

Extended Service Set (ESS). Several basic service sets (BSSs) can be joined together to form one logical WLAN segment, which is referred to as an "extended service set." See also Basic Service Set. Compare with Infrastructure Basic Service Set.

Extensible Authentication Protocol (EAP). This 802.1x standard is a Point-to-Point Protocol extension that provides support for additional authentication methods within PPP. This general protocol for authentication also supports multiple authentication methods, such as token cards, Kerberos, one-time passwords, certificates, public key authentication, and smart cards. IEEE 802.1x specifies how EAP should be encapsulated in LAN frames. As such, EAP allows developers to pass security authentication data between RADIUS and the access point and wireless client. EAP has a number of variants, including: EAP MD5, EAP-Tunneled TLS (EAP-TTLS), Lightweight EAP (LEAP), and Protected EAP (PEAP).

F

FCC. See Federal Communications Commission.

FEC. See Forward Error Correction.

FDMA. See Frequency Division Multiple Access.

Federal Communications Commission (FCC). The government agency responsible for regulating telecommunications in the United States.

FHSS. See Frequency Hopping Spread Spectrum.

File Transport Protocol (FTP). This Internet protocol is used to copy files between computers. FTP is also the standard method for downloading and uploading files over the Internet. FTP allows users to transfer files to and from a distant or local computer, list directors, and to delete and rename files on the distant computer.

Firewall. Hardware and/or software that sit between two networks, such as an internal network and an Internet service provider. It protects the network by refusing access by unauthorized users. It can even block messages to specific recipients outside the network.

Firmware. The basic instructions that equipment needs to function properly. Firmware is actually software that is constantly called upon by a computer so it is stored in semi-permanent memory called PROM (Programmable Read Only Memory) or EPROM (Electrical PROM) where it cannot be "forgotten" when the power is shut off. It is used in conjunction with hardware and software and shares the characteristics of both. When there is a firmware upgrade it is to modify these instructions in order to adapt to new software releases, environments, or new hardware. It may also resolve potential bugs and may increase reliability or performance.

Flash card. See CompactFlash.

Flash Memory. This term refers to non-volatile memory that does not need a constant power supply to retain its data. It offers extremely fast access times, low power consumption, and relative immunity to severe shock or vibration. Flash memory is similar to EPROM (Electronic Programmable Read Only Memory) with the exception that it can be electrically erased, whereas EPROM must be exposed to ultra-violet light to erase. These qualities combined with its compact size, make it perfect for portable devices like PDAs, barcode scanners digital cameras, cellular phones, etc.

FM. See Frequency Modulation.

Forward Error Correction (FEC). This term refers to a methodology that uses error correction coding to improve the proper transmission of digital data. This is the opposite of ARQ (automatic repeat request), which uses retransmission of data to improve the accuracy of digital data.

FreeSpot. Places where free Wi-Fi networks exist. They are usually put up around public spaces by local governments, in local venues by the venue operator, or by individuals whose philosophy is to share their high-speed Net connections with neighbors.

Frequency. As used in this book, frequency refers to the rate at which an electromagnetic waveform (electrical current) alternates, i.e., the number of complete cycles of energy (electrical current) that occurs in one second. It's usually measured in Hertz. See also Hertz.

Frequency Division Multiple Access (FDMA). A technique used in cellular communications in which channels are assigned specific frequencies. This digital radio technology divides the available spectrum into separate radio channels. It is generally used in conjunction with Time Division Multiple Access (TDMA) or Code Division Multiple Access (CDMA).

Frequency Hopping Spread Spectrum (FHSS). One of two types of spread spectrum radio technology used in Wi-Fi networks. FHSS modulates the data signal with a narrowband carrier signal that "hops" in a predictable sequence from frequency to frequency as a function of time over a wide band of frequencies. Interference is reduced, because a narrowband interferer affects the spread spectrum signal only if both are transmitting at the same frequency at the same time. The transmission frequencies are determined by a spreading (hopping) code. The receiver must be set to the same hopping code and must listen to the incoming signal at the proper time and frequency to receive the signal. Compare with Direct Sequence Spread Spectrum.

Frequency Modulation (FM). This term refers to a method used to add voice or data to a radio frequency transmission by varying the carrier frequency. Both broadcast stations in the 88-108 MHz "FM" band and television station sound channels use this modulation technique. FM has grown in popularity because of its relatively insensitivity to the static sources.

Fresnel Zone. Pronounced "fre-nel," this term refers to the area around the visual line-of-sight that radio waves spread into after they leave the antenna. When designing 802.11 networks, this area must be 80% clear or else signal strength will weaken.

FTP. See File Transfer Protocol.

G

Gain. All antennas possess a figure of merit called "gain." Gain is measured by comparing the performance of a given antenna to a theoretical ideal antenna–the isotropic antenna which has a gain of 0 dB. Thus, gain is measured in dBi–the "dB" portion stands for decibels and the "i" stands for isotropic. Gain is important to antenna selection because, all other things being equal, the antenna with the higher gain figure will have the better performance.

Gateway. In a network the term generally refers to an electronic repeater device that intercepts and steers electrical signals from one network to another. Thus a gateway device provides the entrance and/or exit paths to and from a communications network. A typical gateway is a combination of hardware and software; the hardware device (typically a "bridge") and the software work in tandem to perform the necessary translations to allow traffic to flow smoothly between networks.

General Packet Radio Service (GPRS). A radio technology for GSM networks that adds packet-switching protocols. As a 2.5G technology, GPRS enables high-speed wireless Internet and other data communications. GPRS networks can deliver Short Message Service (SMS), Multimedia Message Service (MMS), email, games, etc. See also 2.5G.

Gigahertz (GHz). The international unit for

measuring frequency is hertz (Hz), which is equivalent to cycles per second. A gigahertz is a unit of frequency that is equal to one billion hertz or cycles per second.

GPRS. See General Packet Radio Service.

GSM. Originally the acronym stood for Groupe Speciale Mobile, but it has been anglicized to "Global System for Mobile Communications," an international digital cellular standard, although it is more widely used in Europe than elsewhere.

H

H.323. An International Telecommunication Union (ITU) standard that describes how multimedia communications occur over IP networks, specifically between terminals, network equipment and services. H.323 is part of a larger group of ITU recommendations for multimedia interoperability called H.3x. Today H.323 is considered to be the standard for interoperability in audio, video and data transmissions as well as voice-over-IP because it addresses call control and management for both point-to-point and multipoint conferences as well as gateway administration of media traffic, bandwidth, and user participation.

Hacker. A person with deep knowledge of and great interest in computer and network systems and who delves into the inner workings of a system to find out how it works. Originally, hacker was a term of respect among computer designers, programmers, and engineers for those among them who created truly original and ingenious programs, devices, or sometimes very clever practical jokes. However, the current popular meaning of the term is to describe those who break into systems, destroy data, steal copyrighted software, and perform other destructive or illegal acts with computers and networks.

Handheld Computer. These devices are normally a little larger than the palm-sized personal digital assistant (PDA). They typically use the Windows CE operating system, have a form factor that is more rugged than the typical PDA, need a larger battery than the typical PDA, are used with work-related applications like inventory management systems, are not pre-loaded with personal organization software, and offer specialized keys or touch-screens for easy input of regularized form-information. Compare with Personal Digital Assistant.

Hertz (Hz). The international unit for measuring frequency is hertz (Hz), which is a measure of cycles per second.

HiperLAN. A set of wireless local area network (WLAN) communication standards that provide features and capabilities similar to those of the IEEE 802.11 wireless local area network (LAN) standards. The HiperLAN's two main standards, HiperLAN/1 and HiperLAN/2, are primarily used in European countries and both have been adopted by the European Telecommunications Standards Institute (ETSI).

HomeRF. Short for "Home Radio Frequency," this is a home networking standard developed by Proxim Inc. It combines the 802.11b and Digital Enhanced Cordless Telecommunication (DECT) portable phone standards into a single system. HomeRF uses a frequency hopping technique to deliver speeds of up to 1.6 Mbps over distances of up to 150 feet—too short a range for most business applications, but suitable for the home market that HomeRF was developed to serve.

HotSpot. In the Wi-Fi industry, this term refers to a specific geographic location in which an access point provides public wireless broadband network services to mobile visitors through a wireless LAN. HotSpots typically require that the end-user pay a fee before they can access the network and are usually situated in places that are heavily populated with mobile computer users, e.g. coffee houses, airports, restaurants, convention centers, hotels, etc.

HTTP. The acronym for **H**yperText **T**ransfer **P**rotocol. See Hypertext Transfer Protocol.

Hub. In data communications, a hub is a hardware device that is used to network computers together. These devices serve as a common wiring point so that information can flow through one central location to any other computer on the network.

HyperText Transfer Protocol (HTTP). This application protocol defines how messages are formatted and transmitted over the World Wide Web, and what actions web servers and browsers should take in response to various commands.

I

IBSS. See Independent Basic Service Set.

iDEN. Short for "Integrated Digital Enhanced Network," this wireless technology merges the capabilities of a digital cellular telephone, two-way radio, alphanumeric pager, and data/fax modem into a single network. iDEN operates in the 800 MHz, 900MHz, and 1.5 GHz frequency bands and is based on time division multiple access (TDMA) and GSM architecture.

IETF. See Internet Engineering Task Force.

IM. See Instant Message.

IMAP. See Internet Message Access Protocol.

IMT-2000. Short for International Mobile Telecommunications-2000, the International Telecommunication Union's "vision" of a global family of 3G mobile communications systems.

IMT-2000 DS (International Mobile Telecommunications-2000 Direct Spread). Also known as "direct spread W-CDMA," "Wideband-CDMA" or "W-CDMA." See W-CDMA.

IMT-2000 MC (International Mobile Telecommunications-2000 MultiCarrier). Also known as cdma2000. See cdma2000, cdma2000 1X, cdma2000 1X-EV and cdma2000 1X-EV-DV.

Independent Basic Service Set (IBSS). See Ad Hoc Network.

Infrastructure Basic Service Set. A term used to describe a Basic Service Set with an access point at the helm. See Basic Service Set.

Instant Message (IM). An application that provides the ability for end-uses to see whether a chosen friend or co-worker is connected to the Internet; and, if so, to exchange messages with them. Instant messaging differs from ordinary email in the immediacy of the message exchange. IM also makes a continued exchange simpler than sending emails back and forth.

Institute of Electrical and Electronic Engineers (IEEE). An international professional organization for electrical and electronics engineers, with formal links with the International Organization for Standardization (more commonly known as the "ISO"). This nonprofit organization develops, defines, and reviews standards within the electronics and computer science industries and is the standards body responsible for the 802.11 series of specifications.

Integrated Services Digital Network (ISDN). Because our circuit-switched telephone system has difficulties handling large quantities of data, in 1984 the ISDN specification was released to allow for wide-bandwidth digital transmission using the telephone system's existing copper telephone wiring. Under ISDN, a phone call can transfer 64 kilobits of digital data per second.

International Mobile Telecommunications-2000. See IMT 2000 DS and IMT 2000 MC.

International Organization for Standardization. See ISO.

International Telecommunications Union (ITU). The formal name given to an international organization within which governments and the private sector coordinate global telecom networks and services. Although the ITU doesn't have the power to set standards, if its members agree upon a standard, that standard effectively becomes a world standard. The ITU consists of three major sectors: The Radiocommunication Sector (ITU-R); the Telecommunication Standardization Sector (ITU-T); and the Telecommunication Development Sector (ITU-D).

Internet. A term used to refer to a global decentralized network of computers that exchange data. Each host computer on the Internet is independent and can opt in and out of the various Internet services it uses and/or offers to the global Internet community.

Internet Engineering Task Force (IETF). This is the formal name given to the protocol engineering and development arm of the Internet. For instance, the IETF defines standard Internet operating protocols such as the TCP/IP suite of protocols. The Internet Society Internet Architecture Board (IAB) supervises the IETF and its members are drawn from the IAB's membership. Standards are expressed in the form of "Requests for Comments" (RFCs).

IP Address. A 32-bit binary number that uniquely identifies a host (computer) connected to the Internet or to other Internet hosts, for the purposes of communication through the transfer of data packets.

Internet Engineering Task Force (IETF). The main standards body responsible for establishing internetworking protocols. The IETF is an open international body concerned with the evolution of Internet architecture and the operation of the Internet. The IETF is open to any interested individual.

Internet Message Access Protocol (IMAP). A standard protocol that provides a means of managing email messages on a remote server. Although IMAP is similar to the Post Office Protocol (POP), it offers more options than POP, including the ability to download message headers, create multi-user mailboxes, and build server-based storage folders.

Internet Protocol (IP). The most widely used method for transporting data within and between communications networks is IP. This middle level protocol can interact with a variety of different lower layer carriers such as Ethernet, Wi-Fi, ATM and SONET. IP is as useful for the growing field of intranets (networks internal to an enterprise or organization and not connected to the outside world; e.g. a network used for classified processing) as it is for the geographically distributed, highly heterogeneous Internet. In more detail, IP provides a connectionless, unreliable, best-efforts packet delivery system. It does this by concentrating on only one task–to find a route for data packets. It doesn't care what's in the data packets, IP only wants to know the destination addresses.

Internet Service Provider (ISP). This term refers to a company that provides its customers with access to the Internet and the World Wide Web via a user-friendly front end. Most ISPs have a network of servers (mail, news, Web, etc.), routers, and modems attached to a permanent, high-speed Internet "backbone" connection.

Internetwork Packet Exchange (IPX). An IPX is a datagram or packet protocol that interconnects networks that use Novell's NetWare clients and servers. IPX works at the Network Layer and is connectionless, e.g. it doesn't require that a connection be maintained during an exchange of packets.

Intranet. This term refers to an internal TCP/IP-based network behind a firewall that allows only users within a specific enterprise to access it.

IntServ. Short for "integrated services, a set of IETF standards that cover how application services define their QoS requirements, how this information is made available to routers on a hop-by-hop basis, and ways of testing and validating that the contracted QoS is maintained. With the IntServ approach, each network element is required to identify the coordinated set of QoS control capabilities it provides in terms of the functions it performs, the information it requires, and the information it exports. IntServ-capable routers must classify packets based on a number of fields and maintain state information for each individual flow. See also DiffServ, Quality of Service.

IP. See Internet Protocol and TCP/IP Suite.

IP Address. A 32-bit numeric identifier for a computer or device on a TCP/IP network is referred to as the "IP address." The IP address is written as four numbers separated by periods (commonly referred to as "dotted decimal"), with each 4-number set being within the range of zero to 255; i.e., 192.168.0.10 could be an IP address.

IPsec (Internet Protocol Security). A framework for a set of protocols for security at the network or packet processing layer of network communication. IPsec is commonly used for implementing virtual private networks (VPNs) and for remote user access through dial-up connection to private networks.

IPX. See Internetwork Packet Exchange.

ISA (Industrial Standard Architecture). This standard bus (a computer's electrical pathways along which signals are sent) architecture allows 16 bits at a time to flow between the motherboard circuitry and an expansion slot card and its associated device(s).

ISDN. See Integrated Services Digital Network.

ISM (Industrial, Scientific, and Medical) Band. A set of radio frequencies centered around 2.4 GHz that is universally acknowledged to be available for unlicensed use by wireless technologies. The ISM band is very attractive for wireless networking because it provides a part of the spectrum upon which vendors can base their products, and end-users do not have to obtain FCC licenses to operate the products.

ISO. From the Greek word for equal, it is the commonly used term used to refer to the International Organization for Standardization, an international organization composed of national standards bodies from more than 75 countries. ISO has defined numerous computer standards, but the most significant is perhaps OSI (Open Systems Interconnection), a standardized architecture for designing networks. See also Open Systems Interconnection.

ISP. See Internet Service Provider.

ITU. See International Telecommunications Union.

J

Jitter. In network parlance this term refers to a type of communication line distortion caused by abrupt, spurious signal variation from a reference timing position, and capable of causing data transmission errors, particularly at high speeds. The variation can be in amplitude, time, frequency, or phase.

K

Key. In a communications network, the term "key" usually refers to a sequence of random or pseudo-random bits used to direct cryptographic operations and/or for producing other keys. The same plaintext encrypted with different keys yields different cipher texts, each of which requires a different key for decryption

Key Hashing. Hashing is the transformation of a string of characters into a usually shorter fixed-length value or key that represents the original string. Thus key hashing refers to the method in which a long key is converted to a native key for use in the encryption/decryption process. Each number or letter of the long key helps to create each digital bit of the native key.

Killer App. A term that is commonly used to refer to a specific application that practically compels the general populace to embrace its related technology.

Kbps (KiloBits Per Second). A term that refers to one thousand bits per second; as such Kbps is a standard measure of data rate and transmission capacity. Also referred to as kbits/s. Note that KBps refers to kilo*bytes* by second—one thousand *bytes* per second—and is a measurement for physical data storage on some form of storage device: hard disk, RAM, etc.

Kilohertz (kHz). A term that refers to one thousand cycles per second; as such kHz is a common unit used in measurement of signal bandwidth—digital and analog. See also Hertz.

L

LAN. See Local Area Network.

Laptop. A term that the consumer and corporate marketing departments use to describe a battery- or AC-powered personal computer that is generally smaller than a briefcase; thus such devices are easily transported and convenient to use in temporary spaces such as on airplanes, in libraries, temporary offices, and at meetings. A laptop typically weighs less than 5 pounds and is 3 inches or less in thickness.

Layer 2 Bridge. A device which connects two or more networks at the

Data Link Layer, which is Layer 2 of the OSI Model. See Bridge, OSI Model.

LCD. See Lead Crystal Display.

LDAP. See Lightweight Directory Access Protocol.

Lead Crystal Display (LCD). An alphanumeric display using liquid crystal sealed between the pieces of glass.

Lightweight Directory Access Protocol (LDAP). A standard designed to query and update a directory. The IETF introduced LDAP in order to encourage adoption of X.500 directories (a set of ITU-T standards covering electronic directory services) and to address some of the deficiencies of X.500's Directory Access Protocol.

Line-of-Site (LOS). When speaking of RF, LOS means more than just being able to see the receiving antenna from the transmitting antenna's position. In order to have true line-of-site, no objects (including trees, houses or the ground) can be in an area known as "the Fresnel zone." The Fresnel zone is the area around the visual line-of-sight that radio waves spread out into after they leave the antenna. This area must be clear or else signal strength will weaken. See Fresnel Zone.

Local Area Network (LAN). A short distance data communications network consisting of both hardware and software and typically residing inside one building or between buildings adjacent each other—thus allowing all networked devices to share each other's resources.

LOS. See Line-of-Site.

M

MAC (Medial Access Control) Address. This is a network interface card's hardware address; every NIC comes from the factory with a MAC address burned into it. A MAC address consists of a unique 48-bit address which is expressed in units of 4 bits (called nibbles) as a series of numerals in the range of zero through fifteen.

MAC Address Filtering. This term refers to the ability of a network manager to block all incoming packets from a specific source or sources. The MAC address is a unique series of numbers and letters assigned to every networking device. With MAC address filtering enabled, wireless network access is provided solely for wireless devices with specific MAC addresses. This makes it harder for a hacker to access a wireless network using a random MAC address.

MAC Protocol Data Unit (MPDU). The unit of data in an IEEE 802 network that two peer MAC entities exchange across a physical layer.

MAC (Medium Access Control) Sublayer. One of two sublayers that make up the Data Link Layer of the OSI model. The MAC sublayer is responsible for moving data packets to and from one Network Interface Card to another across a shared channel. See OSI Model.

MAN. See Metropolitan Area Network.

Mbps (Megabits per Second). A term that refers to one million bits per second; as such Mbps is a standard measure of data rate and transmission capacity. Also referred to as mbits/s.

Megahertz (MHz). One million cycles per second. A common unit used in measurement of signal bandwidth—digital and analog. See also Hertz.

Message Integrity Check (MIC). A security method used to prevent bit-flip attacks on encrypted packets. During a bit-flip attack, an intruder intercepts an encrypted message, alters it slightly, retransmits it, and the receiver accepts the retransmitted message as legitimate. The client adapter's driver and firmware must support MIC functionality, and MIC must be enabled on the access point.

Metropolitan Area Network (MAN). Two or more LANs linked together so resources between the LANs can be shared. See also Local Area Network.

MIC. See Message Integrity Check.

Microprocessor. This term refers to a chipset that handles the logic operations (e.g. addition, subtraction and copying) in a computer. A set of instructions in the chip design tells the microprocessor what to do, but different applications can give instructions to the microprocessor as well. Chip speeds are measured in megahertz (MHz), so a 120 MHz chip is twice as fast as a 60 MHz chip. However, that doesn't mean a 120 MHz computer will run all tasks twice as fast as a 60 MHz computer, as speed is also influenced by other factors, such as the design of the software running on the machine, the operating system being used, and so forth. See also Chip, Megahertz.

Middleware. A term used to describe software products designed to serve as the "glue" between two otherwise separate applications so that data can pass between the two applications.

Mobile IP. An IETF communications protocol designed to allow users with mobile computing devices to move from one network to another while maintaining their permanent IP address. Mobile IP is an enhancement of the Internet Protocol (IP) that adds mechanisms for forwarding Internet traffic to mobile devices when they are connecting through other than their home network.

Mobile Identification Number (MIN). This term refers to a wireless carrier's identifier (i.e. phone number) for a phone in its network. The MIN is meant to be changeable, since the phone could change hands or a customer could move to another city.

Mobile Switching Centre. See Mobile Switching Office.

Mobile Switching Office (MSO). A switch that provides services and coordination between mobile users in a network and external networks. Also known as Mobile Telephone Switching Office (MTSO), and Mobile Switching Centre (Europe).

Mobile Telephone Switching Office (MTSO). See Mobile Switching Office.

Modem. Short for MOdulator/DEModulator, a modem is a system that converts computer data into signals that can be transmitted over telephone. If the data is to be sent over a POTS (Plain Old Telephone Service) line (versus DSL or Cable), the sending modem converts digital data to analog for transmission and on the receiving side another modem converts the analog transmission back to digital data.

MPDU. See MAC Protocol Data Unit.

MPEG (Moving Pictures Experts Group). A standard for compressing sound and video files into an attractive format for downloading or streaming across the Internet. The MPEG-1 standard streams video and sound data at 150 Kbps, while MPEG-2 provides compressed video that can be shown at near-laserdisc clarity with a CD-quality stereo soundtrack. Digital satellite services and DVD use MPEG-2.

MP3. The term refers to a specific type of codec (an algorithm that encodes and decodes or compresses and decompresses various types of data) that compresses standard audio tracks into much smaller sizes without significantly compromising sound quality.

MSO. See Mobile Switching Office.

MTSO. See Mobile Telephone Switching Office.

N

Network Address Translation (NAT). A methodology that translates an Internet Protocol address (IP address) used within one network to a different IP address known within another network. One network is designated the inside network and the other the outside. Typically, an organization maps its local inside network addresses to one or more global outside IP addresses and unmaps the global IP addresses on incoming packets back into local IP addresses. This helps ensure security since each outgoing or incoming request must go through a translation process that offers the opportunity to qualify or authenticate the request or match it to a previous request. NAT also allows an organization to use a single IP address in its communication with the world.

NetBEUI (Network BIOS Enhanced User Interface). This standard transport protocol provides a set of rules that an operating system can use to control how computers on small and medium networks talk to each other. Although NetBEUI is efficient for small workgroups, but is not routable, and is a more broadcast-based protocol than other transports protocols, e.g. IPX or TCP/IP. Thus networks running NetBEUI can become clogged with packets if there are more than 20 networked computers.

Network Interface Card (NIC). An add-in

board that enables a computer to connect to some form of computer network.

NextGen. This term, used in the cellular industry, refers to networks that provide capabilities that are beyond the means of a 2G network. 2.5G and 2.75G are examples of a NextGen networks. See 2G, 2.5G, and 2.75G.

Node. An individual computer / machine or address on a network.

Notebook. The manufacturing sector usually refers to a lightweight laptop computer, as a "notebook computer." See Laptop.

Null-modem Serial Cable. A cable that connects two computers together via their serial ports; such a connection allows the two machines to communicate. Some network hardware manufacturers, such as hubs and routers, include serial ports for programming their devices. A null-modem cable is used as the communication pathway. Physically, a null-modem cable is just a modem cable with a couple of connections reversed.

O

Open Systems Interconnect Reference Model. See OSI Model.

Orthogonal Frequency Division Multiplexing (OFDM). A special method of multicarrier modulation for encoding data onto a radio frequency signal. A single high-frequency carrier is replaced by multiple subcarriers, each operating at a significantly lower frequency. OFDM works by transmitting multiple high data rate signals concurrently on different frequencies. The channel spectrum is passed into a number of independent non-selective frequency sub-channels and these subchannels are used for one transmission link between the access point and mobile terminals. OFDM is the modulation method of choice for both 802.11a and 802.11g.

OSI (Open System Interconnection) Model. Introduced in 1984 to be an abstract model for internetworking, the OSI model defines internetworking in terms of a vertical stack of layers. The upper layers of OSI represent software that implements network services like encryption and connection management. The lower layers of OSI implement more primitive functions like routing, addressing, and flow control. Although the OSI model was designed to be an abstract model, the model remains a practical framework that many of today's key network technologies fit into, as such the OSI model provides a conceptual understanding of LAN/WAN internetworking and is an essential aspect of computer networking theory and practice.

P

Packet. A discrete packet of data which contains the addresses, and which is the basic transmission unit on an IP network. It can also be called a "datagram."

Packet Binary Convolutional Coding (PBCC). A modulation/coding technique first developed by Texas Instrument for use in 802.11b networks as a means of doubling the signaling rate of the 11 Mbps standard to 22 Mbps while maintaining backward compatibility with legacy 802.11b 11 Mbps wireless equipment. In its simplest form, PBCC reduces packet overhead through the removal of extraneous information while optimizing transmission through the use of smaller data packets, thus cutting the response time in processing those packets, which in turn allows for a greater amount of data to be transmitted between networked devices. PBCC is an optional modulation technique for 802.11g.

PC Card. A term commonly used to refer to special card-like interface devices that are approximately the size of a credit card and can be plugged into a computing device's PCMCIA slot to add functionality to the computing device. For instance, PC Cards can add another hard disk, increase RAM, add modem capability or network interface card functionality, etc. The original Type I PC Card is 3.3mm thick, which is used mainly to add RAM. Type II cards are thicker (5.0mm) and often are used for modems and NICs (although they're also used to increase a computing device's RAM). Type III cards are much thicker (10.5mm) and often are used for hard disks and radio devices. See also PCMCIA and PCMCIA Slot.

PCI (Peripheral Component Interconnect).

A mezzanine bus (a computer's electrical pathways along which signals are sent) standard developed by Intel Corporation. Most modern PCs include a PCI bus in addition to a more general ISA expansion bus.

PCMCIA (Personal Computer Memory Card International Association). A trade association that establishes standards for expansion cards for portable computers. The PCMCIA's specifications for the PC card enable the computer industry to manufacture small cards (that resemble a credit card) for the purpose of adding RAM, modems, network interface cards, hard disks, and even radio devices (e.g. pagers and global positioning systems) to computing devices. Many people call PC cards by the longer name PCMCIA cards. The PCMCIA association, however, has trademarked the term "PC card," so that's the preferred usage. See PC Card.

PCMCIA Slot. Mostly found on the side of a laptop/notebook computer, the PCMCIA slot uses PC cards that fit into it for various Input/Output (I/O) functions such as networking.

PCF. See Point Coordination Function.

PCS. See Personal Communications Services.

PDA. See Personal Digital Assistant.

Personal Communications Services (PCS). A mobile communications system interconnected with the PSTN. In Canada and the United States PCS spectrum has been allocated for use by public systems at the 2.0 GHz frequency range.

Personal Digital Assistant (PDA). A handheld computer that combines personal organizer features with computing, telephone/fax, and networking capabilities. A PDA can also have cell phone and fax sender features.

Picocell. Describes a physically small communications coverage area (less than 0.5 km in diameter).

PHY. See Physical Layer.

Physical Layer. The Physical Layer is the lowest layer within the OSI Model. Commonly written as "PHY." Wi-Fi's Physical Layer is split into two parts. One is the the PLCP (Physical Layer Convergence Protocol) and the other is the PMD (Physical Medium Dependent) sublayer. The PMD takes care of the wireless encoding and the PLCP presents a common interface for higher-level drivers to write to as well as to provide carrier sense and CCA (Clear Channel Assessment), which is the signal that the MAC (Media Access Control) layer needs so it can determine whether the medium is currently in use. See OSI Model.

Physical Layer Convergence Protocol (PLCP). Prepares MAC protocol data units (MPDUs) as instructed by the MAC Layer for transmission and delivers incoming frames to the MAC Layer.

Physical Medium Dependent (PMD). Provides the actual transmission and reception of Physical Layer entities between two stations via the wireless medium.

Ping. A diagnostic test that sends a packet of data to another computer. The "pinged" computer then sends the packet back to the source computer. "Pinging" is commonly used to see if another computer on a LAN or across the Internet is up and running and how healthy are the pathways between the two.

Plain Old Telephone Service (POTS). Standard wireline telephone service. See Public Switched Telephone Network.

PLCP. See Physical Layer Convergence Protocol.

PLCP Service Data Unit (PSDU). This term refers to the contents of an 802.11 frame.

PMD. See Physical Medium Dependent.

Pocket PC. This term refers to Microsoft's mobile device platform that is based on the Windows CE operating system. The Pocket PC platform is used for standard PIM functionality, games and multimedia web browsing. The platform is capable of running custom enterprise applications built in Visual C++, Embedded Visual Basic, or .NET Compact Framework. See also Windows CE.

Point Coordination Function (PCF). An IEEE 802.11 mode that enables contention-free frame transfer based on a priority mechanism. Enables time-bounded services that support the transmission of voice and video.

Point-of-Presence (POP). A physical location that provides an access point to the Internet. A POP can be either part of the facilities of a telecommunications provider that a service provider such as an Internet Service Provider (ISP) rents, or a separate location built and maintained by the service provider that houses servers, routers, ATM switches and digital/analog call aggregators.

Post Office Protocol (POP). A protocol that retrieves email from a mail server. There are two versions of POP. The first, called POP2, became a standard in the 1980s and requires Simple Mail Transfer Protocol (SMTP) to send messages. The newer version, POP3, can be used with or without SMTP.

POP. This is a commonly used acronym for both Point-of-Presence and Post Office Protocol. See Point-of-Presence, Post Office Protocol.

POTS. See Plain Old Telephone Service.

Protocol. A set of rules or standards designed to enable computers to connect with one another and to exchange information.

Proxy. A mechanism whereby one system "fronts for" another system in responding to protocol requests. Proxy systems are used in network management to avoid having to implement full protocol stacks in simple devices, such as modems.

PSDU. See PLCP Service Data Unit.

PSTN. See Public Switched Telephone Network.

Public Switched Telephone Network (PSTN). One of two common terms used to refer to the world's collection of interconnected voice-oriented public telephone networks, both commercial and government-owned. The other term that is often used to refer to this wired circuit-switched telephone network is "Plain Old Telephone Service" or "POTS."

Q

QAM. See Quadrature Amplitude Modulation.

QPSK. See Quadrature Phase Shift Keying.

Quadrature Amplitude Modulation (QAM). This is method that is used to combine two amplitude-modulated (AM) signals into a single channel, thereby doubling the effective bandwidth.

Quadrature Phase Shift Keying (QPSK). This is a modulation technique whereby a RF carrier is modulated via phase shifting, usually at 0, 90, 180, and 270 degrees. Basically, a digital data stream is taken two bits at a time to generate one of four possible phase states of the transmitted carrier. Also known as Quaternary Phase Shift Keying.

Quality of Service (QoS). This term refers to a guarantee of a network's speed and performance. A QoS measurement should reflect the network's transmission quality and service availability. QoS can come in the form of traffic policy in which the transmission rates are limited; thereby guaranteeing a certain amount of bandwidth will be available to applications. Or QoS may take the form of traffic shaping, which are techniques to reserve bandwidth for applications but not to guarantee its availability.

R

Radio. A device that either makes, or responds to, radio waves, which are part of a larger group of electromagnetic waves.

Radio Access Network (RAN). The portion of a mobile network that handles subscriber access, including radio base stations and control and concentration nodes.

Radio Frequency (RF). Alternating current having characteristics such that, if the current is input to an antenna, an electromagnetic field is generated that is suitable for wireless broadcasting and/or communications. Radio frequencies cover a significant portion of the electromagnetic spectrum, extending from nine kilohertz (9 kHz) to thousands of gigahertz (GHz).

Radio Receiver. A device for converting radio waves into perceptible signals; a radio receiver may be a component of a complete radio set consisting of a combined transmitter-receiver, i.e., a transceiver.

Radio Spectrum. Another term for "radio frequency." See Radio Frequency.

Radio Transmitter. A device that can take some kind of information (in this book it is data) and convert it into radio waves that can be transmitted through the air via an antenna attached to the radio transmitter. A radio receiver then intercepts the radio waves (via its antenna) and changes the radio waves back into usable data.

RADIUS. See Remote Authentication in Dial-in User Service.

Receive Sensitivity (Rx Sensitivity). The minimum level signal a radio can demodulate. Transmit power and receive sensitivity together constitute what is know as "link budget." The link budget is the total amount of signal attenuation you can have between the transmitter and receiver and still have communication occur. See also Transmit Power.

Remote Authentication in Dial-in User Service (RADIUS). This term refers to a client/server protocol and software that enables remote access servers to communicate with a central server to authenticate dial-in users and authorize their access to the requested system or service.

Return on Investment (ROI). A business process that measures the profit or cost savings that an organization realizes from an investment.

RF. See Radio Frequency.

RF Tag. An identification label capable of transmitting data via RF. Some tags also receive and store data.

RIPE Network Coordination Centre (RIPE NCC). This is one of three Regional Internet Registries that provide allocation and registration services supporting the operation of the global Internet. The RIPE NCC performs activities primarily for the benefit of the membership in Europe and the surrounding areas; mainly activities that its members need to organize as a cohesive group, even though they may compete in other areas.

Roaming. A service offered by both mobile communications network operators and HotSpot operators to enable subscribers to use telephone or computing device while in the service area of another operator. Roaming requires an agreement between operators of technologically compatible systems in individual markets to permit customers of either operator to access the other's systems.

Robust Header Compression (ROHC). This is an Internet Engineering Task Force standard that takes advantage of the fact that consecutive data packets often have identical headers. ROHC is a robust and efficient header compression scheme for RTP/UDP/IP (Real-Time Transport Protocol, User Datagram Protocol, Internet Protocol), UDP/IP, and ESP/IP (Encapsulating Security Payload) headers. ROHC can reduce header size by around 95 percent.

Router. A device that is responsible for making decisions about which of several paths network (or Internet) traffic will follow. To do this, the device uses a routing protocol to gain information about the network and algorithms to choose the best route based on several criteria known as "routing metrics." In OSI terminology, a router is a Network Layer intermediate system. Compare with Bridge and Gateway.

Rx. Receive/receiver.

Rx Sensitivity. See Receive Sensitivity.

S

Scanner. A device that reads printed text or illustrations and translates the information into a form that a computer can use, i.e. a scanner "digitizes" printed information. A small handheld scanner can be rolled across the printed medium, while a larger flatbed scanner works rather like a photocopying machine in that users place the item to be scanned onto the glass as would be done for a photocopier, close the lid and scan. A scanner's resolution is measured in dpi (dots per inch), which refers to the number of dots in a one-inch line–the more dots per inch, the higher the resolution and the sharper the image.

Secure Sockets Layer (SSL). A protocol designed by Netscape to enable users to transmit private documents via the Internet. The SSL protocol provides connection security via three basic properties. (1) The connection is private since encryption is used after an initial handshake to define a secret key and symmetric cryptography is used for data encryption (e.g. DES, RC4, etc.). (2) The peer's identity can be authenticated using asymmetric, or public key, cryptography (e.g. RSA, DSS, etc.). (3) The connection is reliable since message transport includes a message integrity check using keyed MAC and secure hash functions (e.g. SHA, MD5, etc.) for MAC computations. The programming for keeping a message confidential is composed of two layers. At the lowest level is the SSL Record Protocol, which is layered on top of some reliable transport protocol, typically TCP/IP. The SSL Record Protocol is used for encapsulation of various higher level protocols. One such encapsulated protocol, the SSL Handshake Protocol, allows the server and client to authenticate each other and to negotiate an encryption algorithm and cryptographic keys before the application protocol transmits or receives its first byte of data. One advantage of SSL is that it is application protocol independent. A higher level protocol can layer on top of the SSL Protocol transparently.

Service Set Identifier (SSID). An identifier attached to packets sent over the wireless LAN that functions as a "password" for joining a particular radio network such as a basic service set (BSS). All radios and access points within the same BSS must use the same SSID, or their packets will be ignored.

Session Initiation Protocol (SIP). An Internet Engineering Task Force (IETF) standard protocol for initiating an interactive user session that involves multimedia elements such as video, voice, chat, gaming, and virtual reality. Like HTTP or SMTP, SIP works in the Application Layer of the OSI Model. Since the Application Layer is the level responsible for ensuring that communication is possible, SIP can establish multimedia sessions or Internet telephony calls, and modify, or terminate them.

Short Text Messaging. Better known as Short Message Service or SMS. See Short Message Service.

Short Message Service (SMS). A messaging service that is similar to paging. SMS sends messages consisting of no more than 160 characters (224 characters if using a 5-bit mode) to mobile phones that use Global System for Mobile (GSM) communication. However, unlike paging, SMS messages do not require the receiving device to be active and within range since the SMS message can be held (for a specific time period) until the phone is active and within range.

SID. See Switch Identification Code.

Silicon. The second most abundant element on Earth is silicon (oxygen ranks higher). Silicon, a component of common sand, is the raw material that is commonly used for making computer chips.

SIM Card. Also known as a "Smart Card." See Subscriber Identification Module.

Simple Management Transfer Protocol (SMTP). A member of the TCP/IP protocol suite and as such it is used to send and receive email messages. However, since SMTP is limited in its ability to queue messages at the receiving end, it's usually used with either POP3 or Internet Message Access Protocol (IMAP), either of which allow the end-user to save messages in a server mailbox and download them periodically from the server. See also Internet Message Access Protocol and Post Office Protocol.

Simple Network Management Protocol (SNMP). A protocol that reads and sets both standard and manufacturer variables in network equipment. SNMP enables a management station to configure and monitor network devices. Furthermore, an SNMP agent can generate a "TRAP" (message) to an SNMP manager to send back information related to one or more variables for which thresholds have been associated

SIP. See Session Initiation Protocol.

Smart Card. Also known as a "SIM Card." See Subscriber Identification Module.

Soft Client. As used in this book, the term "soft client" refers to software that runs on a computing device to provide voice services over an IP network, typically using the computing device's built-in microphone/speaker components and sound card.

SOHO (Small Office/Home Office). An acronym used to distinguish small businesses from mid-sized and large businesses. However, many SOHOs are home-based businesses, which is where the "home office" in Small Office Home Office comes from. Still, the term SOHO has a broad sweep—including everyone who works in a small office environment, whether as employer or employee.

SQL. Acronym of Structured Query Language, and pronounced as "sequel." The pronunciation came about because SQL was developed as a result of an IBM project called "Structured English Query Language." The SQL acronym is typically used instead of its progenitor. See Structured Query Language.

Spanning Tree Protocol. An IEEE protocol designed to prevent a condition known as a "bridge loop," which can occur when a network uses two bridges to interconnect the same two computer network segments. The Spanning Tree Protocol and the Spanning Tree Algorithm enable the bridges to exchange information so that only one of them will handle a given message sent between two computers within the network.

Spam. A term used to refer to an email message sent to a large number of people without consent. Also known as Unsolicited Commercial Email (UCE) or junk email. Spam is usually sent to promote a product or service. It is also found in newsgroups, where people post identical and irrelevant messages to many different newsgroups that have nothing to do with the content of the posting.

Spectrum. A conceptual tool used to organize and map a set of physical properties to delineate electromagnetic waves that are produced by electric and magnetic fields, and which move through space at different frequencies. See also Electromagnetic Waves and Radio Frequency.

Spectrum Allocation. This term refers to the governmental function of apportioning bands to services such as radio, TV, and WLANs. Although various governments have declared their control over all spectrum and set up laws and rules to govern the use of such spectrum, realistically speaking, no one can really "own" spectrum.

Spread Spectrum. This term refers to a type of wireless signal modulation that scatters data transmissions across the available frequency band in a pseudo-random pattern. This results in a much greater bandwidth than the signal would have if its frequency were not varied and spread. Spreading the data across a frequency spectrum makes the signal resistant to noise and to interference.

Structured Query Language (SQL). A type of programming language that is commonly used to construct database queries and perform updates and other maintenance of relational databases. Despite its name, SQL is not a full-fledged language, in that it cannot create standalone applications. However, it is strong enough to create interactive routines in other database programs.

SSID. See Service Set Identifier.

Statement of Work (SOW). A document primarily used to specify the work requirements for a project or program. A SOW typically is used in conjunction with specifications and standards as a basis for a contract and often is used to determine whether the contractor met stated performance requirements once the contract is completed.

Stream Control Transmission Protocol (SCTP). Also known as next-generation TCP or TCPng, SCTP is designed to make it easier to support a telephone connection over the Internet (and specifically to support the telephone system's Signaling System 7, commonly known as SS7, over an Internet connection). SCTP is invaluable when transmitting simultaneous multiple streams of data between two end points with an established connection in a network.

Subnet. A contiguous string of IP addresses. The first IP address in a subnet is used to identify the subnet, the last IP address in the subnet is used as the broadcast address, thus anything sent to the last address is sent to every host on the subnet. See also IP Address.

Subscriber Identification Module (SIM). Commonly referred to as a "SIM Card" or

"Smart Card"; a SIM is a small printed circuit board, which is commonly found in GSM phones and other devices. For instance, when used in GSM phones, the SIM identifies the user via information, stored on the card. A typical SIM includes subscriber details, security information and the personal directory of numbers. The module or card holds a microchip that stores information and encrypts voice and data transmissions, making it close to impossible to listen in on calls. Without the SIM, the mobile phone cannot operate.

Switch. A network device that selects a path or circuit for sending a data packet to its next destination is called a "switch." However, a switch can also function as a router (a device or program that can determine the route and specifically to what adjacent network point data should be sent). In general, a switch is a simpler and faster mechanism than a router, since a router requires knowledge about the network and how to determine the route. A switch is not always required in a network since many local area networks are organized as rings or buses in which all destinations inspect each message, but read only those intended for that destination.

System Identification Code (SID). A 5-digit number assigned to each cellular carrier by the FCC. The SID is programmed into the cell phone when a service plan is subscribed to and is part of the activation process.

T

Tablet Computer/PC. A design for a fully-equipped personal computer that allows a user to take notes using natural handwriting on a stylus- or digital pen-sensitive touch screen instead of requiring the use of a keyboard. A tablet computer/PC is similar in size and thickness to a yellow paper notepad.

TCP. See Transmission Control Protocol.

TCPng. Short for TCP next-generation, which is another term used to describe Stream Control Transmission Protocol (SCTP). See Stream Control Transmission Protocol, but also see TCP.

TCP/IP Suite. This refers to a collection of communication protocols allowing communication between groups of dissimilar computer systems from a variety of vendors. The TCP/IP suite provides the basic framework for communication over the Internet. The TCP/IP suite consists of a number of different protocols, e.g. Address Resolution Protocol (ARP), File Transfer Protocol (FTP), User Data Protocol (UDP), Simple Management Transfer Protocol (SMTP), Telnet, Domain Name System (DNS), and more.

TDMA. See Time Division Multiple Access.

Telephony. The science of translating sound into electrical signals, providing a process for transmitting the signals and then converting them back to sound. Although the term could apply to POTS, it is more commonly used to refer to computer hardware and software that performs functions traditionally performed by telephone equipment. For example, telephony hardware and/or software can to turn a PC into a telephone and/or sophisticated answering service. The most popular telephony application to date is voice mail.

Temporal Key Integrity Protocol (TKIP). A protocol that is a part of the IEEE 802.11i encryption standard for wireless LANs. TKIP is oft-times referred to as the next generation of WEP (Wired Equivalency Protocol), which is currently used to secure 802.11 wireless LANs. TKIP provides per-packet key mixing, a message integrity check and a re-keying mechanism. See Also Wired Equivalency Protocol.

Temporary Local Directory Number (TLDN). A number that is used by cellular networks to deliver wireless calls to roaming wireless customers.

Terminal Node Controller (TNC). As used in this book, TNC refers to a device which assembles and disassembles frames and usually includes some form of a user interface and command set. The TNC is used in conjunction with a radio, modem, and terminal for packet radio applications. The TNC may be implemented in hardware or software.

Time Division Multiple Access. This is a method of digital wireless communications transmission that allows a large number of users to access a single radio frequency channel without interference. With TDMA, each user is given a unique time slot within each channel.

TKIP. Pronounced "tee-kip," this acronym of Temporal Key Integrity Protocol is commonly used in place of its progenitor. See Temporal Key Integrity Protocol.

Token Ring. A type of computer network in which all the computers are arranged (schematically) in a circle and a token (a special bit pattern) travels around the circle. To send a message, a computer catches the token, attaches a message to it, and then lets the packet continue its travel around the network.

Total Cost of Ownership (TCO). A type of calculation designed to help consumers and enterprise managers to assess direct (and indirect) costs and benefits related to the purchase of an IT component. The intention of the TCO is to arrive at a final figure that will reflect the effective cost of purchase, all things considered.

Touchscreen. A display device with a clear glass panel (typically liquid crystal) overlay that acts as an input device by responding to the touch of the user.

Transmit Power (Tx Power). The amount of RF power that comes out of the antenna port of the radio. Transmit power is usually measured in Watts, milliwatts, or dBm. See also Receiver Sensitivity.

Transmission Control Protocol (TCP). The Transport Layer protocol used in IP networks, which includes the Internet, local area networks, wireless local area networks, intranets, and extranets. TCP breaks data streams such as messages or file transfers into small "packets" and hands them to IP for transmission across the network. On the receiving end, TCP collects the packets and puts them in the correct order, restoring the original data stream.

Transmission Power Control (TPC). A technique used to limit the transmitted power to the minimum needed to reach the furthest end-user.

Tuple. A term that refers to an ordered sequence of fixed length of values of arbitrary types. For example, the two end points of a session or application use the IP address and the TCP port number at each end point as a tuple to form a connection

Tx. Transmit/transmitter.

Tx Power. See Transmit Power.

U

UDP. See User Datagram Protocol.

Ultrawideband (UWB). This wireless technology enables the transmittal of large amounts of digital data over a wide spectrum of frequency bands with very low power (less than 0.5 milliwatts) for a short distance (up to 230 feet). Ultrawideband also has the ability to carry signals through doors and other obstacles that tend to reflect signals at more limited bandwidths and a higher power. Also referred to as "ultra wideband."

UMTS (Universal Mobile Telecommunications System). This refers to the European Union's version of IMT 2000. This third-generation (3G) technology is designed to provide broadband, packet-based transmission of text, digitized voice, video, and multimedia at data rates up to 2 Mbps. UMTS also is designed to offer a consistent set of services to mobile computer and phone users no matter where they are located in the world. It is based on the Global System for Mobile (GSM) communication standard and endorsed by major standards bodies and manufacturers. See also IMT 2000.

Unified Messaging. This term refers to solutions that enable systems to handle voice, fax, and regular text messages as objects in a single mailbox that an end-user can access either with a regular email client or by telephone. If using a computing device (versus a telephone), the end-user can open and play back voice messages (assuming the computing device has multimedia capabilities), and faxed images can be saved or printed.

U-NII (Unlicensed National Information Infrastructure) Bands. In the U.S., the FCC has set aside three bands of spectrum in the 5 GHz range for unlicensed use. Those bands, which are commonly referred to as the U-NII bands, consist of three 100 MHz bands between 5.15 GHz and 5.8 GHz. The U-NII bands are very attractive for wireless networking because they provide a part of the spectrum upon which vendors can base their products, and end-users do not have to obtain FCC licenses to operate the products.

Universal Datagram Protocol (UDP). Part of the TCP/IP suite, UDP is an alternative to the Transmission Control Protocol (TCP), but when compared with TCP, UDP offers only a limited amount of service when messages are exchanged between computers in a network that uses the Internet Protocol (IP). For example, UDP does not provide the service of dividing a message into packets (datagrams) and reassembling it at the other end, and it doesn't provide sequencing of the datagrams.

Universal Power Supply (UPS). Generally a device that allows a system to maintain operation when changes to the power supply would otherwise interrupt the function of that system. They can range from a 9 Volt battery backup to a generator.

Unlicensed Spectrum. Airwaves that a regulatory body, such as the FCC, hasn't allocated for exclusive use by one user.

UPS. See Universal Power Supply.

V

Value-Added Reseller (VAR). A company that takes an existing product, adds its own "value," usually in the form of a specific piece of hardware or specialized application and resells it as a new product.

VAR. See Value-Added Reseller.

Virtual Local Area Network (VLAN). A local area network that maps workstations on some basis other than geographic location (e.g. by department, user class, or primary application).

Visitor Location Register (VLR). A database that contains temporary information concerning mobile subscribers who are currently located in a given mobile switching office serving area that is different from their Home Location Register (HLR). When a mobile subscriber roams away from his home location and into a remote location, SS7 messages obtain information about the subscriber from the HLR, and create a temporary record for the subscriber in the VLR. See also Mobile Switching Office.

VLAN. See Virtual Local Area Network.

Vocoder. A voice codec used for the compression/decompression of voice signals in order to reduce and thus conserve the bandwidth or bit-rate of the voice stream. See also Codec.

Voice-over-IP (VoIP). A term used in IP telephony to describe a set of facilities for managing the delivery of voice information using the Internet Protocol (IP). In general, this means sending voice information in digital form in discrete packets rather than in the traditional circuit-committed protocols of the public switched telephone network (PSTN).

Voice-over-Wi-Fi. Voice-over-IP technology optimized for a Wi-Fi network. See Voice-over-IP.

W

WAN. See Wide Area Network.

WAP. See Wireless Application Protocol.

Warchalker. According to the Warchalking Organization (www.warchalking.org), a warchalker is a person that makes a practice of marking a series of symbols on sidewalks and walls to indicate nearby wireless access. That way, other computer users can pop open their laptops and connect to the Internet wirelessly. It was inspired by the practice of hobos during the Great Depression who used chalk marks to indicate which homes were friendly.

Wardriver. A person who travels with a specially equipped mobile computing device in search of free bandwidth, which they can tap into and access the Internet.

Warwalker. See Wardriver.

W-CDMA. See Wideband CDMA.

WDS. See Wireless Distribution System.

WEP. See Wired Equivalent Privacy.

Wideband CDMA (W-CDMA). A third generation standard offered to the International Telecommunication Union by GSM proponents. This wideband Code Division Multiple Access (CDMA) protocol is expected to be used for third-generation mobile cellular systems to support very high-speed multimedia services such as full-motion video, Internet access and video conferencing. Also sometimes referred to as CDMA Direct Spread.

Wi-Fi Alliance. An organization, formerly known as the Wireless Ethernet Compatibility Alliance (WECA), whose mission is to certify all 802.11-based products for interoperability and promoting the term Wi-Fi as the global brand name across all markets for any 802.11-based wireless LAN products.

Wi-Fi Protected Access (WPA). An industry-supported, pre-standard version of 802.11i utilizing the Temporal Key Integrity Protocol (TKIP), which fixes the problems of WEP, including using dynamic keys. WPA will serve until the 802.11i standard is ratified.

Wide Area Network (WAN). A geographically dispersed communications network. The term distinguishes a broader communication structure than a local area network; although the term also usually connotes the inclusion of public (shared user) networks such as the Internet.

Windows CE. This term refers to a Microsoft Windows operating system that is specifically designed for inclusion in mobile and other space-constrained devices. See also Pocket PC.

Wired Equivalent Privacy (WEP). An encryption method designed to offer wireless LANs some measure of security, although recent studies have shown that WEP fails to offer the necessary security required by most wireless networks. WEP scrambles data sent between a client device and an access point.

Wireless Adapter. A term used to describe any device that links a computer with a network. Most wireless adapters come in either a network interface card or PC card form factor. See Network Interface Card, PC Card.

Wireless Application Protocol (WAP). This is an open specification that supports Internet protocols on wireless cellular devices such as mobile phones, pagers, two-way radios, smart phones and communicators to easily access and interact with Internet-based services. Simply put, WAP is a special way of formatting content so that it can appear on small screens, like those on cellular phones.

Wireless Distribution System (WDS). A network design used to create access point-to-access point communications when a CAT5 cable cannot be used or is unavailable. Similar to repeating, it is primarily used to extend the reach of the WLAN.

Wireless Hub. A term that is sometimes used to refer to a wireless access point.. See Access Point.

Wireless Internet Service Provider (WISP). Services run by wireless Internet access providers. These services can resemble traditional Internet service, in that they can be designed for a range of commercial clients to get their public facility online. There are many WISP business models. Some are start-up regional (and national) service providers, and others are venue owners (hotels, airports, cafés, convention centers) who have outfitted their own facilities to accommodate their visitors/customers. See also Internet Service Provider.

Wireless Local Area Network (WLAN). This refers to a local area network in which the end-user can connect via a wireless (radio) connection. The wireless connection, in turn, allows the end-user to access the network while on the move.

Wireless Network Management System (WNMS). A term that refers to software tools that provide WLAN network managers with the network's performance data and information in a mouse driven, windowed environment. Most WNMS's provide the actual performance of wireless network components in easily understood graphs and reports.

Wireless Wide Area Network (WWAN). This term generally refers to digital cellular phone networks that serve an extensive geographic area. Unlike WLANs, which are unlicensed and typically administered privately by the customer, WWANs generally are operated by public carriers, use open standards (e.g. GSM, CDMA), have a range in miles rather than feet, and provide data transfer speeds from 5 Kbps to 60 Kbps.

WISP. See Wireless Internet Service Provider.

WLAN. See Wireless Local Area Network.

WNMS. See Wireless Network Management System.

WPA. See Wi-Fi Protected Access.

WWAN. See Wireless Wide Area Network.

X

XML (Extensible Markup Language). A system used to define specialized markup languages used to transmit formatted data. XML's name is misleading; although conceptually related to markup languages, it is actually a meta language, i.e. a language used to create other specialized languages.

XOR (eXclusive OR). A fundamental binary operation frequently used in cryptographic algorithms. It operates on binary digits, bits, which take the value 0 or 1. For example, a XOR B is equal to 0 if and only if A = B, i.e. 0 XOR 0 = 0, 0 XOR 1 = 1, 1 XOR 0 = 1, 1 XOR 1 = 0. The usefulness becomes more apparent for wireless networking when you look at XOR as: plaintext XOR keystream = ciphertext. Thus, ciphertext XOR keystream = plaintext, which therefore implements a simple and robust reversible transformation of plaintext into ciphertext using a keystream—the fundamentals used in stream ciphers.

References

"2.4 GHz and 5 GHz WLAN: Competing or Complementary?," Mobilian Corporation whitepaper.

"3G and 3G Alternatives: 3G vs. Wi-Fi vs. 4G," Visant Strategies study (2003).

"Assessing Voice Quality in Packet-based Telephony," J. Janssen, *IEEE Internet Computing* (May/June 2002)

"Broadband Migration III: New Directions in Wireless Policy," M. Powell, FCC Commission speech (October 12, 2002).

Code of Federal Regulations, Title 47, Part 15: Unlicensed RF Devices.

"Coexistence between Bluetooth and IEEE 802.11 CCK Solutions to Avoid Mutual Interference," A. Kamerman, Lucent Technologies Bell Laboratories, Jan. 1999 (also available as IEEE 802.11-00/162, July 2000).

"Diving into the 802.11i Spec: A Tutorial," D. Eaton, CMP Media LLP CommsDesign website (November 26, 2002).

FCC's Spectrum Policy Task Force Report (2002)

Newton's Telecom Dictionary, H. Newton (CMP Books, New York, 2002).

Notice of Proposed Rule Making from the FCC: ET 99-231; FNPRM & ORDER 05/11/01 (adopted 05/10/01); FCC 01-158 *Amendment of Part 15 of the Commission's Rules Regarding Spread Spectrum Devices, Wi-LAN, Inc. et al.*

"Measurements and Models for Radio Path Loss and Penetration Loss in and Around Homes and Trees at 5.85 GHz," G. Durgin, T. S. Rappaport, H. Xu, *IEEE Transactions on Communications* (Vol. 46, No. 11, November 1998, pp. 1484-1496.)

"Overcoming IEEE 802.11g's Interoperability Hurdles," M. Wentink, T. Godfrey, J. Zyren, *Communication Systems Design* (May 2003).

"Overcoming QoS, Security Issues in VoWLAN Designs," R. Kodavarti, CMP Media LLP CommDesign.com website (April 3, 2003).

"Properties of a TDMA Picocellular Office Communication System," D. Akerberg, *IEEE Globecom* (December 1988, 1343-1349).

"Radio Coverage in Buildings," J.M. Keenan, A.J. Motley, *British Telecom Technology Journal* (Vol. 8, No.1, January 1990, 19-24).

"The Corner Internet Network vs. the Cellular Giants," J. Markoff, New York Times (March 3, 2002).

"The Origins of Spread Spectrum Communications, R. Scholt, *IEEE Transactions on Communications* (Vol. Com-30, No. 5., May 1982, 822-854).

"Time Synchronization for VoIP Quality of Service," H. Melvin, *IEEE Internet Computing* (May/June 2002)

"Understanding Wireless LAN Performance Trade-Offs," J. Yee and H. Pezeshki-Esfahani, *Communication Systems Design* (November 2002).

Unleashing the Killer App, L. Downes and C. Mui (Harvard Business School Press, 1998).

"Wi-Fi—802.11: The Shape of Things to Come," Wiley Rein & Fielding LLP paper.

Wireless Communications, T. Rappaport (Prentice Hall, New Jersey, 1996).

Wireless Communications: Principles & Practice, T. Rappaport (Prentice Hall, 2000)

Wireless Information Networks, K. Pahlavan, A. Levesque (J. Wiley & Sons, Inc., New York, 1995).

"Wireless LAN Benefit Study," conducted by NOP World – Technology on behalf of Cisco Systems (Fall, 2001).

"Wireless LANs," Intel Information Technology white paper (May 2001).

Index